eering Hydrology

Engineering Hydrology

Dr. Balram Panigrahi

Ms. Kajal Panigrahi

CRC Press
Taylor & Francis Group
Boca Raton London New York

CRC Press is an imprint of the
Taylor & Francis Group, an **informa** business

NEW INDIA PUBLISHING AGENCY
New Delhi – 110 034

CRC Press
Taylor & Francis Group
6000 Broken Sound Parkway NW, Suite 300
Boca Raton, FL 33487-2742

First issued in paperback 2023

© 2019 by New India Publishing Agency
CRC Press is an imprint of the Taylor & Francis Group, an informa business

No claim to original U.S. Government works

ISBN 13: 978-1-03-265394-5 (pbk)
ISBN 13: 978-0-367-20055-8 (hbk)

Print edition not for sale in South Asia (India, Sri Lanka, Nepal, Bangladesh, Pakistan or Bhutan)

Publisher's Note
The publisher has gone to great lengths to ensure the quality of this reprint but points out that some imperfections in the original copies may be apparent.

Library of Congress Cataloging in Publication Data
A catalog record has been requested

Visit the Taylor & Francis Web site at
http://www.taylorandfrancis.com

and the CRC Press Web site at
http://www.crcpress.com

Preface

Soil and water are the two important natural resources for agricultural production and sustenance of life in this earth planet. Out of these two, water is very important. It is vital to life and development of all concerns in all parts of this world. In the developing countries, it plays a crucial role in agriculture and economic sector. With increasing population coupled with industrialisation and urbanisation, the pressure on water for agriculture and domestic uses is increasing at a fast rate. There is a strong need to optimally allocate water for various uses and at the same time creating innovations in research in increasing water resources potential. Hydrology plays a vital role in the development and management of water resources. The basic inputs in the evaluation of water resources are from hydrological parameters. It therefore, forms a part of curricula at under graduate as well as post graduate level in agriculture, agricultural and civil engineering, meteorology, forestry, environmental engineering, geology and earth sciences.

Keeping the above points in view, the authors felt a strong necessity to write a text book in simple and lucid style that can help students who do not have sufficient knowledge and exposure to the subject before. The text book contains a lot of basic knowledge in the field of hydrology. Besides, a number of sample calculations in each chapter are presented in the book which will help the students to understand the subject matter very easily. The various chapters of the book are well designed, written in systematic way and are prepared from the class notes for the students besides utilising long practical field experiences of the authors. The book is written covering syllabus of hydrology and hydrologic engineering which is taught in various state agricultural universities and civil engineering departments in various engineering colleges. Besides, the book will also help students in the streams of meteorology, forestry, environmental engineering, geology and earth sciences. Besides serving as a text book, the book is intended to be very helpful for persons dealing in the areas of agriculture, agricultural and civil engineering. It will serve as an invaluable resource for all academicians, planners, designers, practicing and field engineers in the area of water resources evaluation, development and management.

The book contains 102 sample calculations, 576 short and long type questions, 105 tables and 154 figures and more than 145 references which will be of immense help to the students and practitioners. To make the subject more interesting to read, several field experimental results are presented in the book at various chapters. The book contains 11 chapters and each chapter contains several sections and sub-sections. The title of various chapters and contents of each chapter are as follows:

Chapter 1: Introduction: This chapter deals with importance of water and hydrology in day to day life, scope, history and division of hydrology, hydrologic cycle and its importance, world water balance and water resources of the country. Water budget equation and its application in estimation of water resources in surface as well as groundwater reservoirs are also dealt in this chapter.

Chapter 2: Precipitation: Chapter two deals with various forms of precipitation, causes and occurrence of precipitation, different types of precipitation including cyclonic, convective and orographic precipitation, clouds and its various types, characteristics of south-west monsoon and post monsoon rainfall and also winter and summer rainfall. Measurement of rainfall is very important in estimation of water resources and so measurement of rainfall by different types of rain gauges including procedures to select suitable site for setting up of rain gauges are discussed in this chapter. Estimation of missing rainfall data by different methods, different methods employed to estimate the average rainfall measured by various rain gauges in a catchment, methods to test the consistency of rainfall records are especially discussed in chapter two. Presentation of rainfall data and relationships among intensity, duration, frequency of rainfall are further presented in this chapter. A special section entitled "*snow contribution*" is finally presented in this chapter in the book.

Chapter 3: Runoff: Runoff is a major part of hydrologic cycle which is discussed in chapter three. Factors affecting runoff, estimation of runoff by various methods, estimation of peak runoff rate by Rational method, Cook's method and SCS Curve Number method are also discussed in this chapter. Characteristics of different types of streams like perennial, intermittent and ephemeral streams and yield estimation of streams are briefly presented in chapter three. Finally the chapter discusses flow mass and flow duration curve, their characteristics and use in field of hydrology.

Chapter 4: Abstractions from Precipitation: Evaporation, evapotranspiration, interception, depression storage and infiltration are the losses from precipitation that decide generation of runoff of a catchment. All these losses are briefly described in this chapter. Different techniques to estimate evaporation loss from water bodies including analytical, empirical and pan evaporation methods are dealt in this chapter. Different types of evaporation pans that are used to measure evaporation of water bodies are discussed in chapter four. Evapotranspiration including soil evaporation and transpiration by plants, measurement of soil evaporation and plant transpiration, reference crop evapotranspiration, potential and actual crop evapotranspiration are further narrated in this chapter. Various methods to measure crop water requirement like lysimeter, field experimental plot, soil moisture depletion studies and water balance model have been discussed in chapter four. Estimation methods including aerodynamic, combination and empirical methods like Thornwaite, modified Blaney-Criddle and Pan evaporation methods are further discussed and placed in this chapter. Estimation of actual crop evapotranspiration from potential evapotranspiration using values of crop coefficient is briefly discussed in chapter four. Interception and depression losses constitute initial loss which is also prese*nted here. Finally infiltration loss including factors affecting it, its measurement procedures and various types of infiltration indices like φ and W-index are presented in this chapter four.*

Chapter 5: Streamflow Measurement: Measurement of streamflow or otherwise called as discharge is very important in hydrology. In this chapter different methods of streamflow measurement like velocity-area and slope-area method, discharge measurement by weirs (rectangular, trapezoidal as well as V-notch), flumes (parshall as well as cut-throat flume), orifices, mouth pieces, proportional weirs, meter gate, water meter, spillway, siphon spillway, chemical dilution method, radioisotope method, electromagnetic method and ultrasonic method are discussed. Velocity measurement by current meter and different types of current meters are presented in chapter five. Relationship between stage/gauge and discharge is also described and method to compute discharge from the gauge is explained.

Chapter 6: Hydrograph: Chapter six describes hydrograph, different components of hydrograph, factors affecting hydrograph including physiographic and climatic factors, base flow separation techniques, effective rainfall and effective rainfall hyetograph in brief. Unit hydrograph, construction and application of unit hydrograph form the main part of this chapter. Derivation of unit hydrograph of varying duration by S-Curve and superposition method, use and limitations of unit hydrograph are further described in this chapter. Synthetic unit hydrograph, dimensionless unit hydrograph, distribution graph, instantaneous unit hydrograph and geomorphological instantaneous unit hydrograph are discussed and presented in chapter six also. .

Chapter 7: Flood: Flood, its causes and effects are vividly described in Chapter seven of the book. Estimation of flood peaks for both un-gauged as well as gauged catchments, flood frequency analysis by Foster, Chow's regression and stochastic methods are described with worked out problems so that the readers can find it easy to understand. Moreover, flood forecasting and control measures are discussed in this chapter so that necessary steps may be taken to tackle the situation. Flood control measures by different techniques like storage reservoir, dykes, diversion channels, spurs/groynes, soil conservation measures as well as flood plain measurement are especially described in chapter seven of this book.

Chapter 8: Statistics and Probability in Hydrology: In chapter eight, statistical and probability analysis of hydrologic data especially the discharge data are discussed. Various statistical parameters like measures of central tendency, measures of dispersion, measures of symmetry and measures of peakedness are discussed and well presented in this chapter. Theoretical probability distributions of both discrete and continuous data series are narrated in the chapter. Various continuous probability distributions like normal, log normal, gamma, Pearson type III, log Pearson type III, extreme value distribution, Gumbel's distribution and exponential distributions are explained with worked out examples so that the students and other users can easily understand the subject matter. Regression and correlation analysis, standard error of estimate, multivariate linear regression and correlation and analysis of time series including trend analysis, jump, oscillation and periodicity are also discussed in chapter eight of this book.

Chapter 9: Flood Routing: Flood routing is the main chapter of this book. The chapter contains description of different types of flood routing like reservoir routing and channel routing. Various hydrologic storage routing methods like Modified Pul's method, Goodrich method and Pul's method; hydrologic channel routing like Muskingum and Muskingun crest method as well as hydraulic channel routing are also discussed. Different flood routing machines like mechanical flood router, electric analog routing machines and digital computers are briefly described in chapter nine. Flood routing in conceptual hydrograph development using Clark's model and Nash's conceptual model are also dealt in this section.

Chapter 10: Groundwater, Wells and Tubewells: Groundwater contributes immensely to irrigation and domestic use in all regions. Geologic formations of groundwater supply like aquifer, aquiclude, aquifuge and aquitard, different types of aquifers including confined, unconfined, semi-confined and perched aquifer are dealt in this section. Well and tubewell irrigations, their advantages and disadvantages, various types of wells, proper site selection for installation of wells, and yield tests are discussed and presented in this chapter. The chapter also includes site section for construction of tubewells and various classifications of tubewells like cavity type and strainer type tubewells. Various types of strainers used in tubewells are also discussed. Design of tubewells including design of gravel packed tubewell is further described in this section. Moreover, the chapter contains various investigations for development of groundwater, methods to construct tubewells and procedures to develop tubewells. Life of tubewells and causes of failure of tubewells are discussed in this section. At the end of the chapter, the authors have highlighted a section of artificial groundwater recharge for enhancing groundwater resources.

Chapter 11: Well Hydraulics: Chapter 11 contains description of different properties of aquifers including porosity, void ratio, degree of saturation, specific yield, specific retention, permeability, hydraulic conductivity, hydraulic resistance, leakage factor and storativity of aquifer, groundwater movement, groundwater flow direction using flow net and general flow equations under both transient and steady state saturated flow conditions. Measurement of hydraulic conductivity by different methods and determination of equivalent hydraulic conductivity for layered soils are discussed and presented in this section. Construction and use of flow net for assessment of groundwater potential is also discussed in this section. Well hydraulics including steady radial flow to a well in unconfined and confined aquifers and unconfined aquifer with uniform recharge are further dealt in chapter 11. Discharge of wells and combined drawdown under interference case are also presented in this final section. Moreover, chapter 11 of this book includes discussions on well loss, specific capacity and efficiency of wells.

Balram Panigrahi
Kajal Panigrahi

Acknowledgements

The authors are woeful to Prof. S.N. Pasupalak, Hon'ble Vice-Chancellor, Orissa University of Agriculture and Technology (OUAT), Bhubaneswar for his kind blessings and encouragements to write this book.

The moral supports and encouragements provided by the Dean, College of Agricultural Engineering and Technology (CAET), Dean of Research, Dean of Extension Education, Dean PGF-Cum-DRI, Director Pharming, Monitoring and Evaluation and Chief Librarian of Central Library, Orissa University of Agriculture and Technology, Bhubaneswar for writing the book is highly acknowledged.

The authors are grateful to Dr. K.K. Khatua, Associate Professor; Dr. K.C. Patra, Professor and Dr. Shishir Kumar Sahu, Professor and Head, department of Civil Engineering, National Institute of Technology, Rourkela for their continued supports and suggestions for completing this book.

Special thanks go to Prof. S.K. Sarangi, Director, National Institute of Technology, Rourkela for his blessings and moral support to write this book.

The authors are thankful to Prof. P.K. Patra, Principal and Prof. F. Baliarsingh, Head of the department of Civil Engineering, College of Engineering and Technology, Bhubaneswar and Dr. B.K. Mishra, Professor & Head, Departments of Civil Engineering, Trident Academy of Technology, Bhubaneswar for their inspirations and moral supports to undertake this great job of writing the book.

The authors are indebted to Prof. S.N. Panda, Prof. B.C. Mal, Prof. M.K. Jha and Prof. K.N. Tiwari, Professors of Agricultural and Food Engineering

Department, Indian Institute of Technology, Kharagpur for their helps and supports to complete this job of writing the book.

The inspirations and encouragements provided by Prof. D.P. Ray and Prof. K. Pradhan Ex-Hon'ble Vice-Chancellors, Orissa University of Agriculture and Technology, Bhubaneswar; Distinguished Prof. Megh R. Goyal, Prof Gajendra Singh, Prof. M.S. Bavel of AIT, Thailand, Prof. V.M. Mayandee, Dr. S.K. Ambast, Director, IIWM, Bhubaneswar, Dr. P.K. Mishra, Director, IISWC, Dehradun for their moral supports to accomplish this job of writing the book cannot be forgotten.

It will be injustice if the authors do not pay their heartfelt obligations to all the faculty members of the College of Agril. Engg. & Tech., OUAT especially to Dr. J.C. Paul, Dr. B.P. Behera, Dr. J.N. Mishra, Dr. N. Sahoo, Dr. A.P. Sahu without whose helps, this book would have not been published.

A word of praise and forbearance is also due to the family of the authors for their blessings to take this hard job of writing the book.

Special thanks are due to the **New India Publishing Agency, New Delhi,** for kindly accepting our proposal to publish this book and for their tireless efforts in bringing out this book in the present shape in a short time.

The authors gratefully acknowledge the helps drawn from various publishers, authors and copyright owners for their kind permission to reproduce the materials in the book and beg apology to those from whom it could not be taken. Suggestions and constructive criticisms for improvement of this book are most welcome.

Balram Panigrahi
Kajal Panigrahi

Contents

Conversion Table

Unit	Multiply by	To obtain
Acre	4.05×10^3	Square meter, m^2
Acre	0.405	Hectare, ha ($10^4 \, m^2$)
Acre	4.05×10^{-3}	Square kilometer, km^2
Angstrom	0.1	Nanometer, nm (10^{-9} m)
Atmosphere	0.101	Mega pascal, M pa (10^6 Pa)
Bar	10^{-1}	Mega pascal, M pa (10^6 Pa)
British thermal unit	1.05×10^2	Joule, J
Calorie	4.19	Joule, J
Calorie per square centimeter / minute	698	Watt per square meter, $W \, m^{-2}$
Centimeter per day (elongation rate)	0.116	Micrometer per second, $\mu \, ms^{-1}$ ($10^{-6} \, ms^{-1}$)
Cubic feet	0.028	Cubic meter, m^3
Cubic inch	1.64×10^{-5}	Cubic meter, m^3
Cubic inch	16.4	Cubic centimeter, cm^3 ($10^{-6} \, m^3$)
Curie	3.7×10^{10}	Becquerel, Bq
Degrees (angle)	1.75×10^{-2}	Radian , rad
Dyne	10^{-5}	Newton, N
Erg	10^{-7}	Joule, J
Foot	0.305	Metre,m
Foot-pound	1.36	Joule, J
Bar	10^3	Kilo pascal, K pa
Gallon (U.S)	3.78	Litre, L ($10^{-3} \, m^3$)
U.S. gallon per acre	9.35	Liere per hectare, lit ha^{-1}
Gram per cubic centimeter	1.00	Mega gram per cubic metre, $Mg \, m^{-3}$

Inch	25.4	Millimeter, mm (10^{-3} m)
Micromole (H_2O) per square centimeter second (transpiration)	180	Milligram (H_2O) per square metre per second, mg m^{-2} s^{-1} (10^{-3} g m^{-2} s^{-1})
Micron	1.00	Micro metre, µm (10^{-6} m)
Mile	1.61	Kilometer, km (10^3 m)
Miles per hour	0.477	Metre per second, m s^{-1}
Milli mhos per centimeter	1.00	Deci siemens per metre, ds m^{-1}
Ounce	28.4	Gram, g (10^{-3} kg)
Ounce (FLUID)	2.96 x 10-2	Litre, (10^{-3} m^3)
Pint (fluid)	0.473	Litre, (10^{-3} m^3)
Pound	454	Gram, g (10^{-3} kg)
Pound per acre	1.12	Kilogram per hectare, kg ha^{-1}
Pound per acre	1.12 x 10-3	Mega gram per hectare, Mg ha-1
Pound per cubic feet	16.02	Kilogram per cubic metre, kg m^{-3}
Pound per cubic inch	2.77 x 104	Kilogram per cubic metre, kg m^{-3}
Pound per square feet	47.9	Pascal, pa
Pound per square inch	6.9 x 103	Pascal, pa
Quart (liquid)	0.946	Litre, (10^{-3} m^3)
Quintal (metric)	102	Kilogram, kg
Square centimeter per gram	0.1	Square metre per kilogram, m^2 kg^{-1}
Square feet	9.29 x 10-2	Square metre, m^2
Square inch	645	Square millimeter, mm^2 (10^{-6} m^2)
Square mile	2.59	Square kilometer, km^2
Square millimeter per gram	10^{-3}	Square metre per kilogram, m^2 kg^{-1}
Temperature (^{0}F-32)	0.556	Temperature, ^{0}C
Temperature (^{0}C+273)	1	Temperature, K
Ton (metric)	10^3	Kilogram, kg
Ton (2,000 lb)	907	Kilogram, kg
Ton (2,000 lb) per acre	2.24	Mega gram per hectare, Mg ha^{-1}

Conversion Factors for Different Units of Length

1 metre (m)	3.208 feet (ft)
1 metre (m)	39.37 inches (in)
1 centimetre (cm)	0.3937 inch
1 kilometre (km)	1,000 metres (m)
1 kilometre (km)	0.621 mile (statute mile, mi) = 0.5396 nautical mile (n mi)
1 foot	0.305 metre
1 inch (in)	2.54 centimetre
1 mile (mi)	5,280 feet
1 mile (mi)	1.609 kilometre

1 micrometre (μm.)	10^{-6} m
1 angstrom (^{0}A)	10^{-10} m
1 metre	1.09361 yard
1 kilometre	1093.61 yard
1 inch	0.02778 yard
1 foot	0.33333 yard
1 mile	1760 yard

Conversion Factors for Different Units of Area

1 square metre1 square feet	10.764 square fee t0.093 square meter
1 square centimeter1 square inch	0.155 square inch 6.452 square centimeter
1 square kilometre	100 hectares
1 square kilometer	0.3861 square mile
1 hectare	10,000 square meres
1 hectare	107,640 square feet
1 hectare	2.741 acres100 are
1 square foot	0.0929 square metre
1 square inch	6.452 square centimeter
1 acre	43,560 square feet
1acre	0.4047 hectare
1square mile	640 acres
1 square mile	258.99 hectares
1 square mile	2.59 square kilometres
1 square metre	1.19599 square yard
1 acre	119.599 square yard
1 hectare	11960 square yard
1 acre	4840 square yard
1 hectare	100 acre
1 square metre	0.01 acre
1 square kilometre	10000 acre

Conversion Factors for Different Units of Volume

1 cubic metre	35.314 cubic feet
1 cubic metre	1.308 cubic yards
1 cubic metre	1,000 litres
1 cubic metre	106 cubic centimeters (cm^3)
1 litre	0.035 cubic feet
1 litre	0.2642 U.S. gallon (gal.)
1 litre	0.2201 imperial gallon
1 litre	10^{-3} cubic metre (m^3)
1 cubic centimetre	0.061 cubic inch
1 cubic foot	0.0283 cubic metre
1 cubic foot	23.32 litres
1 cubic foot	7.4 U.S. gallons
1 cubic foot	6.23 imperial gallons
1 cubic inch	16.39 cubic centimeters
1 cubic yard	0.7645 cubic metre
1 U.S. gallon	3.7854 litres
1 U.S. gallon	0.833 imperial gallon
1 imperial gallon	1.201 U.S. gallon

1 imperial gallon	4.5436 litres
1 acre foot	43,560 cubic feet
1 acre foot	1233.5 cubic metres
1 acre inch	3,630 cubic feet
1 acre inch	102.8 cubic metres

Conversion Factors for Different Units of Weight

1 gram (gm)	0.0353 ounce (oz)
1 gram (gm)	0.0022 pound (lb)
1 kg	2.205 pounds (lb)
1 quintal	100 kg
1 metric ton	2,205 pounds (lb)
1 metric ton	1,000 kg
1 metric ton	10 quintals
1 metric ton	1.1 ton
1 kilogram (kg)	1,000 gm
1 ounce	28 gm
1 pound	0.45 kg
1 ton	0.9 metric ton (m.t)

Conversion Factors for Different Units of Pressure

1 bar	10^6 dynes / cm^2
1 bar	1,023 cm of water column
1 atmosphere	1,032 cm of water column
1 atmosphere	76 cm Hg
1 atmosphere	1.013 bar
1 inch Hg	0.0334 atmosphere
1 inch H$_2$O	2.49 millibar (mb)
1 mbar (mb)	0.75mm Hg
1 lb/in2 (psi)	51.72 mm Hg.
1 atm	14.70 lb/in^2 (psi)
1 atm	29.92 in Hg
1 millibar (mb)	0.0145 lb/in^2 (psi)
1 millibar (mb)	1,000 dynes/cm^2
1 millibar (mb)	1000 newton/m^2 (N/m^2)
1 millibar (mb)	100 Pascal (Pa)

Conversion Factors for Different Units of Velocity

1 m/sec	3.60 km/hr
1 m/sec	2.24 mi/hr
1 m/sec	1.94 knots
1 knots	1 n mi/hr
1 knots	1.15 mi/hr
1 knots	0.515m/sec

Conversion Factors for Different Units of Energy

1 calorie (cal)	4.187 joules (J)
1 calorie	4.187 x 10^7 ergs
1 calorie	1.16 x 10^{-6} killowatt hour (Kwh)
1 calorie	3.97 x 10^{-3} British thermal units (Btu)

1 calorie 1 Joule (J)1 Btu1   1 langely-centimetre^2 (ly-cm^2)1 N.m = 10^7 ergs 0.293

  CHU1 kWh1 J   Wh 0.527 Wh 3.60 x 10^6 J0.1020 kg(f).m

Conversion Factors for Power

1 watt (W)	1 joule / sec
1 watt (W)	0.239 cal/sec
1 watt (W)	0.0569 Btu/min
1 watt (W)	0.00134 electrical horse power
1 watt (W)1 HP1 metric HP	2.39 x 10-5 ly-m^2 / sec550 ft-lb(f)/sec = 746 W75
1 watt (W)1 kilo watt (kW)	kg(f)-m/sec= 0.986 HP = 735.5 W1 J/sec = 1 N.m/sec = 10^7 ergs/sec860 kcal/hr

Conversion Factors for Radiation to Equivalent Depth Evaporation

1 cal/cm^2	1/59 mm
1 cal/cm^2 /min^{-1}	1 mm/hr
1 cal/cm^2/ min (24 hr)	24 mm / day
1 mw /cm^2	1/70 mm/hr
1 mw/cm^2 (24hr)	0.344 mm/day
1 joule / cm^2 / min (24 hr)	5.73 mm/day

Conversion Factor for Energy Flux Density

1 cal cm^{-2} min^{-1} 1 langely min^{-1} = 69.8 milli watt cm^{-2}

CHAPTER 1

Introduction

1.1 Water and its Importance

Water has played a crucial role in sustainace of lives of all living beings including animals, plants and human beings in this earth since time immemorial. It is there since the creation of the nature and will probably remain till the destruction of the nature. It is used in multifarious uses including domestic, industrial, agricultural, recreational and many other uses. It occurs in the nature in practically three forms i.e. solid, liquid and gaseous forms. All the three forms are important to living beings; but the liquid form of water is more beneficial. The weight of the water helps in running hydroelectric turbines, the density of water helps in floating the ships, the steam energy produced by heating the water helps to run the locomotives and thus is beneficial in many ways. It is considered as the most important resource of the society. It has become part and partial in our life so that we cannot live without water. That is why water is called as life.

With the development of urbanisation, industrialisation and mechanisation, the demand of water is increasing in different domains. It is understood by some people that it is an infinite source. That is why they use it indiscremently. But this wrong concept has to be removed from the brains. In actual sense, it is a finite resource. If we do not use it judiously and rationally then time will come, when the earth planet will suffer drastic shortage of this precious resource and we all will suffer a lot and may face a looming war. It is, therefore, very urgent to focus innovative research for development and management of the resource.

1.2 Definition, Scope and Importance of Hydrology

It is very interesting to study the science of water including its origin, movement, circulation and all form of distribution in and above the earth surface. The study of occurrence, distribution and circulation of water is called as hydrology. The term hydrology is derived from two Greek words "hydor" meaning water and "logos" meaning science. That is hydrology refers to the study of science of water. Hydrology helps to understand the behaviour and distribution including circulation of water not only on the earth's surface but also in the earth's atmosphere. It forms an important branch of earth science. It is concerned with the study of water in lakes, reservoirs and streams, various forms of precipitation including rainfall and snow fall and also water occurring below the earth's surface in the pores of the soils and rocks. It is concerned with various allied sciences like meteorology, geology, statistics, physics and chemistry etc. Thus, it forms a broad spectrum of study and is inter-disciplinary in nature.

There are several classifications of hydrology. But one of the frequently referred classifications is based on its applicability. According to its applicability, it is classified as:

(i)   **Scientific Hydrology**: The study concerning mainly with academic aspects.

(ii) **Engineering Hydrology/Applied Hydrology**: The study concerning with applications in engineering.

In general sense, engineering hydrology is the science that deals with the estimation of water resources including runoff and its transmission from one place to the other, besides dealing with the properties of water in nature. For, an engineer, surface and groundwater are the two most important components to deal with. But due to lack of data of stream flow, it is usually necessary to study precipitation, evaporation and percolation to indirectly determine the availability of surface and groundwater. For planning and management of water resources systems, an engineer has to estimate

(i)   Water resources potential of a basin,

(ii)   Dependable yield for irrigation and hydel power generation and industrial water demand,

(iii) Probability of occurrence of maximum floods with their magnitude,

(iv)   Sediment yield and transportation,

(v)   Hydrometeorology,

(vi)   Reservoir capacity determination for assured water supply for irrigation and municipal water supply,

(vii)   New sources of water for irrigation,

(viii) Stream flow forecasting by use of precipitation data,

(ix)   Water control, navigation, control of soil erosion and control of pollution of waterbodies and

(x)   Operation and maintenance of river valley projects.

All these above mentioned facts are dealt with the scope of study of engineering hydrology (Garg, 1987). Measurement and publication of hydrologic data, analysis of the data for development of relationships and application of empirical solutions for solving the problems related to hydrology are also dealt with the scope of study of engineering hydrology (Suresh, 2008).

The basic knowledge of this science is a must for all civil and agricultural engineers who deal with the design, planning and management of irrigation systems including reservoirs, navigational and flood control structures, construction of bridges, piers, culverts, weirs, barrages, spillways etc. The civil engineers and other engineers who are responsible for planning, design and construction of hydraulic structures are required to address some of the answers if some problems occur in real life situation.

1.3 History of Hydrology

Hydrology is considered as a comparably new branch of science. However, many ancient philosophers focussed their attention on study of origin and occurrence of water in various stages including its circulation from the ocean to the atmosphere and from atmosphere back again to ocean. Homer, an ancient Greek philosopher, reported that there was a large subterranean reservoir that supplied water to the rivers, seas, springs and deep wells. He thought the dependency of flow to the above mentioned aqueducts was governed by both conveyance cross section and velocity approach. This hypothesis was lost to the Romans and the proper relationship between area, velocity and discharge remained unknown until Leonardo da Vinci discovered it during the Italian Renaissance (Meinzer, 1949).

During the first century B.C. Marcus Vitruvius, in Volume 8 of his treatise *De Architectura Libri Decem* (the engineer's chief handbook during the Middle age) set forth a theory generally considered to be the predecessor of modern notions of the hydrologic cycle. His hypothesis was that the rain and snow falling in the mountainous areas infiltrated the earth's surface, moved laterally in the sub surface and again appeared in the low lying areas as streams or springs (Viessman *et al.,* 1977). Practical application of various hydrologic principles was carried out by some researchers long past. For example, about

4000 B.C. a dam was constructed across the river Nile to permit reclamation of previously barren lands for agricultural production. Several thousand years later a canal to convey fresh water from Cairo to Suez was built. Historically, civilization has followed the development of irrigation. All ancient civilizations have grown up on the banks of rivers. The Horrapan and Mesopotamian civilization also has left some foot prints on evidence of systematic research of application of hydrology in flood control works by high earthen walls. Early Chinese irrigation and flood control works were also based on application of hydrology.

In the Vedas, the earliest sacred books of Aryans mentions have been made about the wells, canals and dams. In "Rigved", four types of water sources are mentioned i.e. the water that comes from the sky or rainwater; those which flow in the rivers and streams; those which are obtained by digging and those which ooze out of springs. In "Yajurved", there are mentions about the canals and dams. "Atharvaved" gives description of digging canals from the rivers. River is mentioned as cow and canal as calf. The ancient historical work mentions an account of gigantic feat of king Bhagiratha and his engineers of diverting the course of water of the sacred Ganga from the altitude of the Himalayas towards the present Indo-Gangetic plain, the granary of India.

Writings at 300 B.C. indicate that all the main lands in India were under irrigation. Besides the various ancient books which confirm the above fact, there are ruins of various ancient irrigation works, a few of which still exist. The most famous old irrigation works which is functioning even today is the "Grand Anicut" which was built by Chola rulers in the first century A.D. on the river Cauvery (Garg, 1987).

In China, where reclamation was begun more than 4000 years ago, the success of early kings was measured by their wisdom and progress in water-control activities. King Yu of Hsia dynasty (2200 B.C.) was elected king by the people as a reward for his outstanding work in water control. The famous Tu-Kiang dam still a successful dam of today, was built by Mr. Li and his son in the China dynasty (200 B.C.) and provides irrigation to about one-half million acres of rice lands. The water ladder, a widely used pumping device in China and neighbouring countries, is believed to have been invented about the same time. The Grand canal 1125 m long was built by the Sui Empire during 589 to 618 A.D (Hansen *et al.,* 1979).

Egypt claims to have had the world's oldest dam built about 5000 years ago to store water for irrigation and domestic uses. The Spaniards on their first entrance into Mexico and Peru found elaborate provisions for storing and conveying water supplies which had been used for many generations. Extensive irrigation works also existed at that time in the south western United States (Tripathy, 2008).

Near the end of the fifteenth century, the trend towards a more scientific approach to hydrology rather than purely philosophical reasoning based on the observations of hydrologic phenomena became evident. Nevertheless, until the seventeenth century, little effort was directed towards obtaining quantitative measurement of hydrologic variables. Modern scientific approach to hydrology with measurement of rainfall, stream flow data etc. were directed in the seventeenth century by pioneers like Perrault, Mariotte and Halley (Meinzer, 1949). In the seventeenth century, Perrault was successful to measure evaporation and capillarity. Halley measured the rate of evaporation of the Mediterranean sea and concluded that the amount of water evaporated was sufficient to account for the outflow of rivers tributary to the sea. The above mentioned facts on measurements of the hydrologic variables throw reliable conclusions to be drawn regarding the hydrologic phenomena being studied. Numerous advances in hydraulic theory and instrumentations were made in the eighteenth century. During the nineteenth century, experimental hydrology flourished. Significant were made in the field of groundwater hydrology and measurement of surface water. Darcy's law of flow in porous media and the Dupuit-Thiem well formula were developed during this period. From about 1930 to 1950, rational analysis replaced the empirical formulations. During this period, Horton's infiltration theory and Theis's non-equilibrium approach to well hydraulics were developed. Since 1950, a theoretical approach to hydrologic problems has largely replaced less sophisticated methods of the past.

1.4 Division of Hydrology

Hydrology includes the following sub-divisions (Suresh, 2008).

- **Hydro-meteorology**: It is that part of hydrology which studies about hydrology and meteorology.
- **Geo-hydrology:** It deals with the study of sub-surface water.
- **Limnology:** It deals with the study of surface lakes.
- **Potamology:** It is the study about surface streams.

1.5 Hydrologic Cycle

Water which is an essential element of hydrology is dynamic in nature. It can occur in solid, liquid as well as in gaseous forms. It is a continual cycle in its round from atmosphere to the earth and return by evaporation from water bodies, snow surfaces and land surfaces and all these are examples of dynamic aspects of water. The continual cycle of distribution of water is governed by the process called as hydrologic cycle which is sometime simply called as water cycle. It is to be remembered that hydrologic cycle is basic part of study of hydrology.

A convenient starting point to describe the hydrologic cycle is the water sources or water bodies. Earth's water sources are mainly rivers, oceans, lakes and underground sources. All these sources get water by precipitation including rain. The rain depends on evaporation caused due to heat energy provided by the solar radiation. The evaporated water moves upwards and subsequently condenses forming clouds and hence rain. While much of the clouds falls back as rain on the water bodies from which water gets evaporated, some portion of it gets drifted by wind several kilometres away and falls back as rain or other forms of precipitations like snow, hail, sleet etc. either on the water bodies or on the land surfaces.

While precipitation is falling, a part of it gets evaporated back to the atmosphere. Another part may be intercepted by the canopy covers of the plants, storage structures, buildings etc. from which it may be either evaporated back to the atmosphere or move down to the land surface. The portion of the intercepted water retained by the vegetations which falls back to the land surface is called as through fall. A portion of this through fall as well as precipitation falling on the land surfaces enters into the ground as infiltration which enhances the moisture content of the soil and subsequently adds to the groundwater body. There is also some loss of water in the form of evapotranspiration by the crops/plants which moves back to the atmosphere and subsequently again falls back as precipitation to the earth's surface.

Precipitation reaching the ground surface after meeting the needs of infiltration and evapotranspiration moves down the natural slope over the land surface as surface runoff. The surface runoff joins the gullies, nallahs, streams, rivers etc. and finally meets the seas and oceans. Groundwater also moves laterally below the ground surface and joins back to the water bodies on the surface through springs and other outlets as interflow after a long period. Both the surface runoff and interflow constitutes together the runoff which is an important component of hydrological cycle. The hydrologic cycle consisting of a number of processes as discussed above is shown in Fig. 1.1 (Eagleson, 1970). The hydrologic cycle useful from engineering point of view is represented in Fig. 1.2 (Eagleson, 1970). It is to be noted that hydrological cycle is a very vast and complicated cycle in which there are a large number of paths of varying time scales. Further it is a continuous recirculation cycle in the sense that there is neither a beginning nor an end or nor a pause. It has been occurring since the formation of earth and will continue for infinite period.

Hydrologic cycle has important influences in a variety of fields including agriculture, forestry, economics, geography, physics, chemistry etc. The engineering applications of the knowledge of the hydrologic cycle are found in the design and operation of projects dealing with water supply, irrigation and drainage, flood forecasting and its control, navigation, power generation, recreational purposes etc. The essential components of hydrologic cycle as outlined above are discussed in brief as follows while vivid descriptions are presented in subsequent chapters in this book.

Precipitation: The term precipitation refers to all forms of water derived from atmospheric vapour and deposited on the earth's surface. It occurs due to condensation of water vapors that have evaporated from the water bodies like ocean, sea, rivers, lakes, ponds etc. Different forms of precipitation are rain, snow, hail, mist, dew, sleet and fog. Of all these, only the first two contribute significant amount of water. Rainfall is the predominant form of precipitation causing stream flow in India. Very often the term precipitation is used indiscriminately to mean rainfall.

Evaporation: Evaporation is the process in which there is conversion of water in the liquid form to vapour. Evaporation occurs due to rise in temperature such that increasing kinetic energy of the water molecules at liquid-air interface overcomes the cohesive force of water molecules that hold them together and as a result the water molecules gets evaporated. Evaporation occurs from both

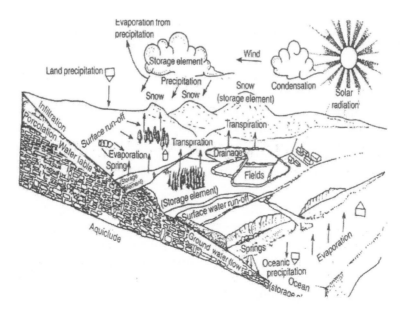

Fig. 1.1: Schematic representation of hydrologic cycle (Eagleson, 1970)

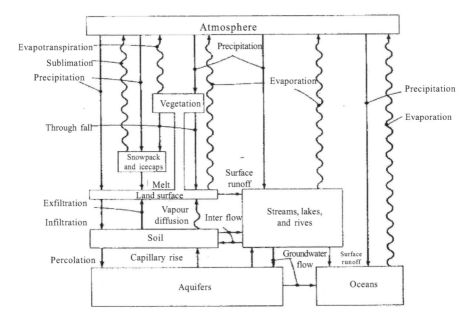

Fig. 1.2: Hydrologic cycle from engineering point of view (Eagleson, 1970)

water surfaces as well as from the land surfaces. Moreover, the part of precipitation that is intercepted by the plants' canopy and gets intercepted by other sources like storage structures, buildings etc. such that they do not reach the land surfaces gets evaporated and returned back to the atmosphere. About 80% of precipitation in India is assumed to be contributed by evaporation from seas and oceans.

Transpiration: Transpiration is the process by which water in the soil is transferred to the atmosphere by plants as water vapour. Plants require water for their growth. This water along with minerals is absorbed by the roots from soil through diffusion process. Out of the amount of water absorbed by the plants, a substantial amount is returned to the atmosphere by transpiration whereas a little amount of about 1% is retained by the plants for meeting metabolic growth.

Infiltration: When precipitation falls on land surfaces, not all the portion flows over the land. The part of the precipitation that does not flow over the land but moves down the land surface is known as infiltration. Infiltration occurs due to the porosity of the soil mass and plays a crucial role in the runoff process by affecting the timing, distribution and magnitude of the surface runoff. It is also a primary stage in the natural groundwater recharge process.

Percolation: Water that infiltrates into the soil is retained by it as soil moisture content. Some of this soil moisture is evaporated while the other is used by plants for transpiration. When the soil moisture content is more than the field capacity, then the soil moisture in excess of field capacity moves down as percolation. The percolated water adds to the ground water reservoir.

Runoff: The portion of the precipitation that makes its way towards streams, lakes, rivers or any other water bodies as surface or subsurface flow is known as runoff. Surface runoff refers to the flow of water over the land surface whereas the subsurface runoff refers to the flow below the land surface. Very often, the term runoff when used alone refers to the surface runoff only. Surface runoff is sometimes referred as overland flow. The infiltrated water that enters into the land surface flows as subsurface runoff. A part of the subsurface runoff flows laterally as interflow and subsequently joins the stream. The other part of the infiltrated water moves vertically down as groundwater and subsequently adds to groundwater reservoir (Murty, 1998).

1.6 Water Budget Equation

As discussed above, hydrologic cycle constitutes a number of events or paths. Each path of the cycle involves one or more aspects. The quantities of water going through various individual paths of the hydrologic cycle can be described by the continuity equation which is called as water budget equation or hydrologic equation. The water budget equation is based on law of conservation of mass. It is important to note that total water resources of the earth are constant and the sun is the principal source of energy for the hydrologic cycle. Human beings, if interfere in one path or process, then it consequence is reflected in one or more of the rest paths of the hydrologic cycle.

For a given location, the continuity equation of water in its various phases can be written as:

Mass inflow − Mass outflow = Change in mass storage   (1.1)

If the density of the inflow, outflow and storage volumes are same (say for water it is same as 1gm/cc), then Eqn. (1.1) can be written as:

$$V_i - V_o = \Delta S \tag{1.2}$$

where V_i = inflow volume of water into the location area during a given time period, V_o = outflow volume of water into the location area during the same time period and ΔS = change in storage volume of water over and under the given location area during the given period. In hydrologic calculations, the volumes are often expressed as average depths over the area. The average depth is computed as the ratio of the volume of water over an area to the area contributing water. The various components of the hydrologic cycle like rainfall, evaporation, runoff etc. can be conveniently expressed in units of depth over an area. The water budget equation relating to various components of the hydrologic cycle can be mathematically written as:

$$P - R - I - E - T - G = \Delta S \tag{1.3}$$

where P = precipitation, R = surface runoff, I = interception loss, E = evaporation from both land and water surfaces, T = transpiration loss by plants, G = net groundwater flow and ΔS = change in storage on or below the surface of the earth. All the terms of Eqn. (1.3) can be in any units of volume or equivalent depth. However, all the terms of the above water budget equation expressed in units of depth (for eg. in centimetres) over an area are the convenient and common unit. Infiltration component of the hydrologic cycle does not occur explicitly in the water budget equation (Eqn. 1.3)) as the infiltration, which is a loss to the runoff process is a gain to the groundwater system. Following few examples illustrate the use of water budget equation for estimation of parameters of hydrologic cycle.

Eqn. (1.3) can be conveniently expressed in two forms i.e. hydrologic budget/water budget equation above the surface and that below the surface as follows:

Hydrologic Budget Equation above Surface:

$$P + (R_1 - R_2) + R_g - I - E_s - T_s - I_f = \Delta S_s \tag{1.4}$$

Hydrologic Budget Equation below Surface:

$$I_f + (G_1 - G_2) - R_g - E_g - T_g = \Delta S_g \tag{1.5}$$

Hydrologic Budget Equation for the Earth: (Sum of Eqn. 1.4 and (1.5)

$$P - (R_2 - R_1) - (E_s + E_g) - (T_s + T_g) - (G_2 - G_1) - I = \Delta S_s + \Delta S_g \tag{1.6}$$

Sometimes interception loss which does not occur explicitly is neglected since the component contributes both to evaporation and as throw fall once strikes the ground surface contributes to different components like infiltration, runoff etc. Neglecting interception, the water budget equation for the earth surface is expressed as (Eqn. (1.7) :

$$P - (R_2 - R_1) - (E_s + E_g) - (T_s + T_g) - (G_2 - G_1) = \Delta S_s + \Delta S_g \tag{1.7}$$

In Eqns. (1.4) to (1.7), the terms P, R, E, T, I, I_f and G refer to precipitation, surface runoff, evaporation, transpiration, interception, infiltration and groundwater contribution, respectively. The sub script 1 and 2 refers to inflow and outflow to and from the area under consideration of water budget which may be a watershed, region etc. Similarly the sub script s and g refers to above the surface and below the surface, respectively. ΔS_s and ΔS_g refer to change in surface and groundwater storage, respectively. If the subscripts are dropped from Eqn. (1.7) so that letters without subscripts refer to total precipitation, and net values of surface flow, underground flow, evaporation, transpiration and storage, the hydrologic budget equation then becomes as:

$$P - R - E - T - G = \Delta S \tag{1.8}$$

The hydrologic equation is a useful tool. It can be employed in various ways to estimate the magnitude and time distribution of hydrologic variables. Following few examples are presented to describe some of the applications in day today use.

Sample Calculation 1.1

A reservoir of average surface area 1200 ha has water surface elevation of 105.0 m above datum at the beginning of a month. In that month, there were 350 mm rainfall and 200 mm evaporation from the reservoir. The runoff entering into the reservoir from a catchment area of 4500 ha in that month was measured to be 3.2 M m³. The outflow volume in that month from the reservoir of average surface area of 1200 ha was measured as 1.55 M m³. Write the water budget equation for the reservoir and compute the water surface elevation of the reservoir at the end of the month assuming there is no change in ground water storage. If during the period 1550 mm of water had entered into the groundwater reservoir and 650 mm of groundwater were exploited, then what would have been the water surface elevation of the reservoir?

Solution

(i) It is given that the reservoir has average surface area = 1200 ha = 12000000 m²

The outflow of water from the reservoir in a month was = 1.55 M m³ = 1550000 m³

In terms of depth unit, the outflow from the reservoir = 1550000/12000000 = 0.129 m = 129 mm

Runoff volume entering into the reservoir = 3.2 M m³ = 3200000 m³

Catchment area that contributes runoff to the reservoir = 4500 ha = 45000000 m².

The depth of water entering into the reservoir = 3200000/45000000 = 0.071 m = 71 mm

Rainfall in storage of the reservoir = 350 mm and evaporation = 200 mm

There is no groundwater contribution.

The water budget equation for the reservoir is:

Rainfall + runoff entering into the reservoir –evaporation – outflow from reservoir = change in reservoir storage

Change in storage, ΔS = 350 +71 – 200 -129 = 92 mm= 0.092m

Water surface elevation of the reservoir at the end of the month = 105.0 + 0.092 = 105.09 m $\approx$ = 105.1 m above the datum.

(ii) If we consider the groundwater flow then the water budget equation is written as:

Rainfall + runoff entering into the reservoir +groundwater inflow – groundwater outflow – evaporation – outflow from reservoir = change in reservoir storage

Change in storage, ΔS = 350 +71 +1550 – 650 – 200 -129 = 992 mm = 0.992 m

Water surface elevation of the reservoir at the end of the month = 105.0 + 0.992 = 105.99 m $\approx$ = 106 m above the datum.

Sample Calculation 1.2

A reservoir has an average surface area of 5000 ha. There is a rainfall of 1200 mm and evaporation of1050 mm occurring in the reservoir over a year. The average inflow to and outflow from the reservoir is measured as 5.5 and 5.8 m³/sec, respectively. The initial water level reading of the reservoir at the beginning of the year is 103.5 m above datum. Write the water budget equation and compute the final water level reading of the reservoir at the end of the year. Assume the groundwater inflow and outflow of the reservoir is same.

Solution

It is given that the reservoir has average surface area = A = 5000 ha = 50 M m²

Rainfall in the reservoir = P = 1200 mm =1.2 m

Volume of rainfall over the reservoir of area 5000 ha = $P.A$ = 50 x 1.2=60 M m³

Evaporation in the reservoir = E = 1050 mm = 1.050 m

Volume of evaporation in the reservoir of area 5000 ha = $E.A$ = 50 x 1.05 = 52.5 M m³

The average inflow to the reservoir = I = 5.5 m³/sec and average outflow = Q = 5.8 m³/sec

In time Δt, the volume of inflow and outflow of the reservoir are I. Δt and Q. Δt, respectively.

Now Δt =1 year = 365 x 24 x 60 x 60 = 31.536 M sec

Now I. Δt = 5.5 x 31.536 = 173.448 M m³ and

Q. Δt = 5.8 x 31.536 = 182.909 M m³

Since groundwater inflow and outflow of the reservoir is same, net groundwater flow component of the water budget equation is 0.

The water budget equation for the reservoir is:

I. Δt + P. A - Q. Δt - E. A= ΔS

Putting the values of I. Δt , P. A, Q. Δt and E. A as calculated above in the water budget equation, change in reservoir storage, ΔS is computed as:

173.448 + 60 – 182.909 – 52.5 = -1.961 M m³

Change in elevation = $\Delta S/A$ = -1.961 / 50 = - 0.039 m = - 39 mm.

Negative sign indicates that there is decrease in final water level reading in the reservoir.

The final water level reading in the reservoir at the end of the year = 103.5 – 0.039 = 103.461 m above the datum.

Sample Calculation 1.3

A micro watershed has an area of 100 ha and receives a rainfall of 12.5 cm in 2 hours. The outlet of the catchment has a stream gauging station. The gauging station measures no flow before the commencement of the rain in the watershed. The average outflow of the gauging station is measured as 2.8 m³/sec and the runoff is measured for 10 hours. After 10 hours, there is no flow as measured by the gauging station. Find out the total loss of water including infiltration, interception, evaporation and transpiration that do not contribute to the runoff. What percentage of rainfall is measured as runoff?

Solution

(i) The area of the watershed = 100 ha = 1 M m²

There is a rainfall of 10 cm in 2 hour and in rest 8 hours there is no rainfall.

The volume of rainfall in 10 hours over the watershed of area 100 ha = $P.A$ = (12.5 /100) x 1 = 0.125 M m^3

The average outflow of the gauging station = R = 2.8 m^3/sec and it occurs for 10 hours

Hence, the volume of outflow/runoff in 10 hours = $R.A$ = 2.8 x 10 x 60 x 60 = 0.1008 M m^3

Change in storage = 0 since the gauging station measures no flow both before and after the runoff measurement.

Hence, using water budget equation, the volume of water not contributing to the runoff due to losses including infiltration, interception, evaporation and transpiration is given as:

Losses = $P.A - R.A$

$= 0.125 - 0.1008 = 0.0242$ M m^3

In terms of depth unit, the losses = 0.0242 / 1.0 = 0.0242 m = 24.2 mm.

(ii) Volume of rainfall = $P.A$ = 0.125 M m^3

Volume runoff = $R.A$ = 0.1008 M m^3

Hence, percentage of rainfall that occurs as runoff = 0.1008/0.125 = 0.8064 = 80.64 percent. The ratio of runoff to rainfall is also called as runoff coefficient. Details of runoff coefficient are described in subsequent chapter in this book.

1.7 World Water Balance

Total quantity of water in the world is estimated to be about1.386 x 10^{18} m^3 out of which about 96.5% is contained in the ocean as saline water. Groundwater contains 1.69% of the total water resources out of which 0.93% is saline. Thus only 0.035 x 10^{18} m^3 of water is fresh (2.5% of total water resources) which can be used for various uses by plants, animals and human beings. However not all the 100% of this fresh water is readily available for different uses since majority of this amount (0.024 x 10^{18} m^3) is contained in frozen state as ice in the polar regions and on mountain tops and glaciers. So in practical sense only 0.76% of the total world water resources of 1.386 x 10^{18} m^3 is readily available for various forms of uses by animals and plants. Out of the total available groundwater on the earth, the quantity which is easily available up to about 700 m depth is only about 11% of all the fresh water (0.00385 x 10^{18} m^3) whereas the balance quantity is available below 700 m depth which is very difficult for exploitation (Garg, 1987). Table 1.1 represents the volume of different sources of water available in the world along with the percentage of storage in them (Subramanya, 2003).

Total area covered by the ocean and land surfaces in the world is 3.6130 x 10^8 km^2 and 1.488 x 10^8 km^2, respectively. Total amount of precipitation on the ocean and land surface are estimated as 1270 and 800 mm/year, respectively. Total amount of evaporation on the ocean and land surface are estimated as 1400 and 484 mm/year, respectively. Total runoff over the land surface is estimated as 316 mm/year (Subramanya, 2003). Considering the area covered by the ocean and land surfaces as reported above, the annual evaporation from the ocean and land surfaces come to be 5.05 x 10^5 km^3/year and 0.72 x 10^5 km^3/year, respectively. Similarly, the annual precipitation over the ocean and land surfaces come to be 4.58 x 10^5 and 1.19 x 10^5 km^3/year, respectively. Thus, there is more water evaporation than the precipitation over the oceans whereas the precipitation exceeds evaporation on land surfaces. The differential amount of about 0.47 x 10^5 km^3 flows as runoff from the land surfaces and groundwater to the oceans. It is to be noted that the above mentioned data are approximate only and they may vary as per the different studies. The main problems of this variation are due to the difficulty in obtaining reliable data on global scale.

Table 1.1 Estimated world water resources

Item	Volume of water (x 10^{15} m^3)	Percent of total water
Ocean	1338.00	96.50 (0.0)
Groundwater	23.40	1.69 (30.1)
Soil moisture	0.0165	0.0012 (0.05)
Ice and snow	24.3641	1.725 (69.6)
Lakes	0.1764	0.013 (0.26)
Marshes	0.01147	0.0008 (0.03)
Rivers	0.00212	0.0002 (0.006)
Biological water	0.00112	0.0001 (0.003)
Atmospheric water	0.01290	0.001 (0.04)
Total	1386.0	100.00 (100.00)

N.B. Values inside parenthesis in column 3 represents the percent fresh water.

Annual water balance studies of various continental land surfaces of the world are presented in Table 1.2. It is observed from this Table 1.2 that the annual precipitation over South America is the highest with a value of 1648 mm whereas that in North America is the lowest with a value of 670 mm. Similarly annual evaporation over South America is the highest with a value of 1065 mm whereas that in North America is the lowest with a value of 383 mm. Africa is the driest continent with a total annual precipitation of 686 mm out of which only

Table 1.2: Estimated water balances of different continents of the world (Subramanya, 2003)

Continent	Precipitation, mm/year	Evaporation, mm/year	Runoff, mm/year
Asia	726	433	293 (40)
Africa	686	547	139 (20)
Australia	736	510	226 (30)
Europe	734	415	319 (43)
North America	670	383	287 (43)
South America	1648	1065	583 (35)

N.B. Values in parenthesis in column 4 refers to the runoff as percentage of precipitation.

139 mm (20% of precipitation) is produced as the runoff whereas North America and Europe are the continents having highest amount of runoff with each having 43% of precipitation as runoff. Table 1.3 gives the annual water balance studies of various water masses (oceans) of the world. Table 1.3 indicates that the annual values of precipitation as well as evaporation varies from ocean to ocean. For example, the precipitation over the Pacific ocean is the highest with a value of 1210 mm whereas that of Arctic ocean is the lowest (240 mm). Similarly, evaporation is the highest for the Indian ocean (1380 mm) whereas that of the Arctic ocean is the lowest (120 mm).

Table 1.3: Estimated water balances of different oceans of the world (Subramanya, 2003)

Ocean	Precipitation, mm/year	Evaporation, mm/year
Atlantic	780	1040
Arctic	240	120
Indian	1010	1380
Pacific	1210	1140

1.8 Water Resources of India

The Ministry of Water Resources, Government of India (1993) has divided the country into 20 river basins (Fig. 1.3) out of which 12 are major river basins and eight are medium/small river basins. The 12 major river basins are:

a) Indus, (b) Mahanadi,(c) Pennar, (d) Brahmani-Baitarani, (e) Godavari, (f) Krishna, (g) Ganga-Brahmaputra-Meghna, (h) Cauvery, (i) Sabarmati, (j) Mahi, (k) Narmada, (l) Tapi

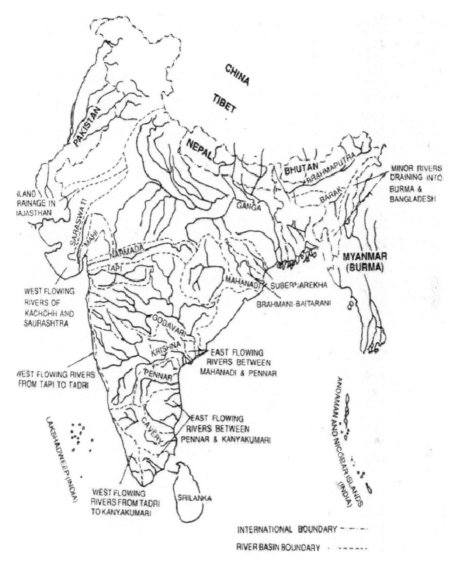

Fig. 1.3: River basins of India

The eight medium/small river basins are:

a)   Subarnarekha-combining small rivers between Subarnarekha and Baitarani

b)   East flowing rivers between Mahanadi and Pennar

c)   East flowing rivers between Pennar and Kanyakumari

d)   West flowing rivers of Kutch and Saurashtra including Luni

e)   West flowing rivers from Tapi to Dadri

f)   Inland drainage area in Rajasthan desert

g)   West flowing rivers from Dadri to Kanyakumari

h)   Minor rivers draining into Myanmar and Bangladesh.

The Central water Commission is maintaining 500 gauges and discharge observation stations in almost all inter-states-river systems in the country. On the basis of data available, the water resources potential of 20 river basins of India is presented in Table 1.4. Fig. 1.4 represents the basin wise distribution of estimated utilizable water resources of the country.

Table 1.4 Water resources of river basins in India (Central Ground Water Board, 1993)

Sl.No.	River basin	Catchment area (km^2)	Water resources potential (M m^3)		Groundwater potential (M m^3)
			Average	Dependable	
1.	Indus	321289*	7305	25543	
2.	Ganga-Brahmaputra-Meghna				
	Ganga	861452*	525023	436322	171725
	Brahmaputra	194413*	537240	441736	27857
	Barak and others	41723	48357	-	1795
3.	Godavari	312812	110540	80545	46762
4.	Krishna	258948	78124	69411	26646
5.	Cauvery	81155	21358	19375	13598
6.	Subarnerekha	29196	12368	9855	2185
7.	Brahmani-Baitarani	51822	28477	20051	5879
8.	Mahanadi	141589	66879	53786	21293
9.	Pennar	55213	6316	4393	5047
10.	Mahi	34842	11020	5713	4875
11.	Sabarmati	21674	3809	3146	3033
12.	Narmada	98796	45639	30829	11890
13.	Tapi	65145	14879	8860	8173
14.	West flowing rivers (Tapi to Dadri)	44940	87411	65663	9479
15.	West flowing rivers (Dadri to Kanykumari)	56177	113532	85285	8810
16.	East flowing rivers (Mahanadi - Pennar)	86643	22520	18768	22788
17.	East flowing rivers (Pennar – Kanykumari)	100139	16458	13930	20907
18.	West flowing rivers (Kutch-Saurastra-Luni)	321851	15098	-	13948

Contd.

Sl.No.	River basin	Catchment area (km²)	Water resources potential (M m³)		Groundwater potential (M m³)
			Average	Dependable	
19.	Inland drainage area in Rajasthan district	-	-	-	-
20.	Minor rivers draining into Myanmar and Bangladesh	36302	31000	-	-

* Area in Indian Territories

Table 1.4 and Fig. 1.4 indicates that the Ganga, Brahmaputra, Barak and others constitutes the largest share of water resources in the country with 47% of total water resources of all the basins. The river basin of Godavari also has large share of water resources with 11% of total water resources of all the basins. The total surface water potential of the country flowing through the above mentioned river basin in India is about 18,81,340 M m³. The gross command area of the country is 328 M ha. Thus the total annual surface runoff when is divided with the gross command area, the annual runoff comes as 0.574 m which is 574 mm. Considering the world average runoff of 260 mm, it is much higher and in terms of percentage of precipitation (average annual precipitation in India is 1250 mm), it is 46% which is the highest in the world and also amongst all the countries in Asia. Thus, India is called as the richest in its surface water resources.

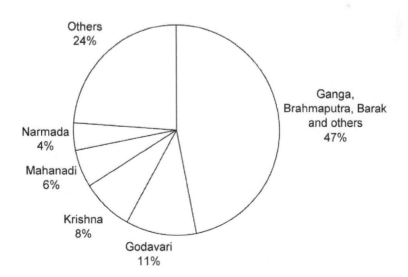

Fig. 1.4: Basin wise distribution of estimated utilisable water resources of the country

1.9 Hydrologic Data

Hydrologic data are needed to describe precipitation, stream flow, evaporation, sedimentation, soil moisture content, infiltration, transpiration, water quality etc. Depending on the problems, a hydrologist will require particular type of data relating to the various relevant phases. Various types of hydrologic data are classified into the following groups:

Field Data: These are the data measured in the field. Generally, groundwater depth, sediment load along the river flow etc. are some examples of it. These data can be generated at any time by carrying out field experiments and using some equipment.

Historic Data: These are the hydrologic data which are highly time dominant. If these data are not collected in time, then they are lost forever. Examples of historic data are rainfall, runoff, river discharge, evaporation, transpiration etc.

Laboratory and Field Experimental Data: The laboratory and field experimental data related to hydrology are generated in the same way as the data in hydraulic engineering. These data are seldom used in basic and applied researches.

There are some specific agencies that record and maintain the hydrologic data. For example, in India, India Meteorological Department and some state government agencies including Irrigation and Agricultural departments collect and maintain the meteorological data including rainfall, temperature, relative humidity, pan evaporation, sun shine hours etc. Stream flow data of major rivers are monitored by the Central Water Commission whereas stream flow data of small rivers and streams are monitored by the State Irrigation department. Similarly, groundwater data (both quality and quantity) are recorded by Central Groundwater Board as well as State Groundwater boards.

1.10 Hydrologic Process

A characteristic feature of hydrologic process is that it is not stationary. Rather it undergoes continuous changes with respect to time. The hydrologic processes may be classified as follows:

- **Deterministic Process:** In this process, change of occurrence of variables is ignored and the process is considered to follow a definite law of certainty but not the law of probability.

- **Stochastic Process:** In this case the change of occurrence of variables is taken into consideration and the process follows the law of probability. Therefore, this process is also called as probabilistic process. However, in real sense, stochastic process depends on time whereas the probabilistic process does not depend on time. In a time dependent stochastic process, the sequence of occurrence of the variables is considered and the variables are either pure random or non-pure random. If the variables of time series are independent among them, and constitutes a random sequence, then the hydrologic process is called as pure random stochastic process. However, in a non-pure random stochastic process, the variables of time series are dependent among them, and constitute a non-random sequence. Both pure random and non-pure random time dependent stochastic process may be classified as (i) stationary and (ii) non-stationary process. In a stationary process, probability distributions of variables remain constant throughout the process. But in non-stationary process, probability distributions of variables changes throughout the process.

1.11 Hydrologic System

Literally a system refers to the congregation of objects interlinked by some specific form of regular interaction. It may be classified as:

- **Stochastic System**: It is the system which contains probabilistic or stochastic processes.

- **Deterministic System:** It is the system which has deterministic processes.

- **Sequential System:** In a sequential system, there are inputs, outputs and some control volume or working mediums. For example, in a hydrologic cycle or watershed, water is the working medium.

Question Banks

Q1. What are the different forms in which water exhibits in nature?

Q2. Why is water important for all plants and human beings?

Q3. Explain why water is called as life.

Q4. Define hydrology. What are the scopes and importance of study of hydrology?

Q5. What are the classifications of hydrology? Differentiate between the scientific hydrology and engineering hydrology.

Q6. What are the different applications of engineering hydrology?

Q7. The basic knowledge of applied hydrology is a must for all civil and agricultural engineers. Enumerate the statement with suitable examples.

Q8. What are the various sub-divisions of hydrology?

Q9. Describe in details about how hydrology has developed in different time periods starting from the pre-historic periods till date.

Q10. With neat sketches, explain the hydrologic cycle. Explain how does water recirculate in the atmosphere?

Q11. What are the important influences of the hydrologic cycle in nature?

Q12. What is the most important component of the hydrologic cycle? Explain the significance of this component in day to day life.

Q13. Differentiate between precipitation and rainfall. What are the various forms of precipitation?

Q14. Write short notes on the following components of the hydrologic cycle:

(i) Evaporation (ii) Runoff (iii) Transpiration (iv) Infiltration and (v) Percolation

Q15. What do you mean by the term water budget? Write down the equation relating to various components of the hydrologic cycle.

Q16. Write down the water budget equation both above and below the earth surface and hence deduce the total water budget equation of the earth.

Q17. A surface reservoir has average surface area of 1500 ha. It has water surface elevation of 105.5 m above datum at the beginning of a week. In that week, there were 300 mm rainfall and 210 mm evaporation from the reservoir. The runoff entering into the reservoir from a catchment area of 4550 ha in that week was measured to be 3.3 M m³. The outflow volume in that month from the reservoir of average surface area of 1500 ha was measured as 1.58 M m³. Write the water budget equation for the reservoir and compute the water surface elevation of the reservoir at the end of the week assuming during the period 1580 mm of water had entered into the groundwater reservoir and 680 mm of groundwater were exploited.

Q18. A small drainage basin has an area of 250 ha and receives a rainfall of 12.0 cm in 2 hours. The outlet of the basin has a gauging station. The gauging station is dry before the commencement of the rain in the basin. The average outflow of the gauging station is measured as 2.9 m³/sec and the runoff is measured for 10 hours. After 10 hours, again the gauging station is dry without any flow. Find out the total loss of water that does not contribute to the runoff. What percentage of rainfall is measured as runoff?

Q19. The average inflow to and outflow from a reservoir of average surface area of 4800 ha is measured as 5.5 and 5.8 m³/sec, respectively. There is a rainfall of 1100 mm and evaporation of 1000 mm occurring in the reservoir over a year. The initial water level reading of the reservoir at the beginning of the year is 103.0 m above datum. Write the water budget equation and compute the final water level reading of the reservoir at the end of the year assume the groundwater inflow and outflow of the reservoir as same.

Q20. Narrate the world water balance from different sources.

Q21. Write notes on the estimated annual water balance of various continental land surfaces of the world.

Q22. Briefly describe the estimated water balances of different oceans of the world.

Q23. Describe in details the available water resources potential of different major, medium/small river basins of India.

Q24. With figure, describe the basin wise distribution of estimated utilizable water resources of the country.

Q25. India is considered to be the richest in its surface water resources compared to other nations in the world. Justify how far the statement is true.

Q26. What is the use of hydrologic data? Classify the various types of hydrologic data into different groups.

Q27. In India, who records and maintains the following hydrologic data:

(i) Meteorological data (ii) Stream flow data of major rivers (iii) Stream flow data of small rivers and streams (iv) Groundwater data

Q28. Write notes on the followings:

(i) Deterministic process and (ii) Stochastic process

Q29. Why is the stochastic process called as probabilistic process?

Q30. What is the difference between:

(i) Pure random stochastic process and non-pure random stochastic process.

(ii) Stationary process and non-stationary process.

Q31. What do you mean by hydrologic system? What are the various types of hydrologic systems?

References

Central Ground Water Board.1993. Annual Report. Government of India, New Delhi.

Eagleson, , P.S. 1970. Dynamic Hydrology. McGraw-Hill Book Company, New York.

Garg, S.K. 1987. Irrigation Engineering and Hydraulic Structures. Khanna Publishers, Delhi.

Government of India .1993. Annual Document. The Ministry of Water Resources, New Delhi.

Hansen, V.E., Israelsen, O.W. and Stringham, G.E. 1979. Irrigation Principles and Practices. John Wiley and Sons Inc., New York.

Meinzer,O.E., 1949. Hydrology, Vol. 9 of Physics of the Earth (New York McGaw-Hill Book Company, 1942). Reprinted by Dover Publications, Inc. New York.

Murty, V.V.N. 1998. Land and Water Management Engineering. 2nd Edition, Kalyani Publishers, New Delhi, pp. 586.

Subramanya, K. 2003. Engineering Hydrology. Tata McGraw-Hill Publishing Company Limited, New Delhi, pp. 392.

Suresh, R. 2008. Watershed Hydrology. Standard Publishers Distributors, Delhi, pp. 692.

Tripathy, L.K. 2008. Sustainable Water Management. Technical Annual of Institution of Engineers (India), Orissa State Centre: 55-60.

Viessman, W. Jr., Knapp, J.W., Lewis, G.L. and Harbaugh, T.E. 1977. Introduction to Hydrology. Thomas Y. Crowell Harper and Row Publishers, New York.

CHAPTER 2

Precipitation

2.1 Introduction

Precipitation is the general term for all forms of moisture emanating from clouds and then ultimately falling on the land surface. It is the total supply of water derived from the atmosphere in the forms of rain, snow, hail, mist, dew, sleet, fog etc. The liquid form of precipitation is called as rainfall whereas the frozen forms of precipitations are snow, hail and sleet. Of all the different forms of precipitation, rain and snow are the two dominant factors that contribute maximum amount of water. In Indian condition, rainfall is the most important form of precipitation that contributes the highest amount of water causing stream flow. Unless otherwise stated, the term rainfall will be used in this book synonymously with precipitation. It is the primary input vector of the hydrologic cycle. Generally it is expressed as the depth water on a horizontal surface over any specified time interval. If the specified time interval is day, then the amount of precipitation is called as daily precipitation. Similarly, the precipitation being contributed over time interval of a month, season and year is termed as monthly, seasonal or annual precipitation. Its form and quantity are influenced by the action of different climatic factors such as wind, temperature and atmospheric pressure. The study of precipitation forms a major portion of the subject of the hydrometeorology.

2.2 Forms of Precipitation

Some of the common forms of precipitation are (i) rain (ii) snow (iii) drizzle (iv) sleet (v) glaze(vi) hail (vii) dew (viii) frost (ix) fog and (x) mist.

Rain: In India, the principal form of precipitation is rain. The size of water droplets in rain is more than 0.5 mm. The maximum size of water droplets in rain can go up to 6 mm. The rain can be classified in to three categories based on its intensity. If the intensity is greater than 7.5 mm/hr, it is called as heavy rain. In moderate rain, the intensity is in the range 2.5 to 7.5 mm/hr. If the intensity is less than 2.5 mm/hr, then it is called as light rain.

Snow: It is another important form of precipitation. It consists of ice crystal which may be branched, hexagonal or stars form. When the ice crystals fuse together during fall, then it forms snow flake. When new, it has an initial density ranging from 0.06 to 0.15 gm/cc. The usual value of the density is considered as 0.1 gm/cc. Melting of snow produces enormous quantity of water in summer in the northern regions especially in the Himalayan regions of India. Because of this cause, most of the rivers originating from the Himalayan regions get water all the time in year and so these rivers are called as perennial rivers.

Drizzle: It is a liquid form of precipitation. The size of the water droplet in drizzle is less than 0.5 mm. The intensity is less than 1 mm/hr. The droplet is so small in size that it appears to float in air.

Sleet: It is frozen form of precipitation. It is produced when the rain falls through air at subfreezing temperature.

Glaze: When rain or drizzle comes in contact with cold ground surface at around 0°C, the water drops freeze to form an ice coating which is called as glaze or freezing rain.

Hail: It is a showery precipitation in the form of irregular pallets of ice of size larger than 8 mm. Hails are generally formed by alternate freezing and melting process, when ice pallets are falling down by violent thunderstorms in which vertical currents are very high.

Dew: It is the condensed water vapor on the cold surfaces. When, the temperature of the atmosphere becomes very low, especially in the winter season, condensed water vapours get deposited on the cold surfaces. With the sun light, they get evaporated soon.

Frost: It is the featuring deposit of ice, formed by dew or water vapour that has been frozen on the ground.

Fog: Fog is the thin form of cloud of varying density which is formed near the earth surface by condensation of atmospheric moisture.

Mist: When the fog is thin in density, it is called as mist.

2. 3 Occurrence and Causes of Precipitation

There are three essential requirements that cause precipitation to occur. In the first requirement, it is necessary for the air to get sufficiently cooled by some mechanisms so that condensation occurs and growth of the droplet occurs. In the second requirement, condensation nuclei are necessary which are present in the atmosphere in adequate quantities. In the third requirement, large scale cooling is needed for significant amounts of precipitation to occur. The phenomenon of occurrence of precipitation by the above three mentioned requirements are discussed as follows:

Because of evaporation, the water vapours from the water bodies are lifted up. The amount of water vapours in the atmosphere varies with altitude. Generally at higher altitude, the presence of water vapour in the atmosphere is more. The amount of water vapour hold by the air also varies with the temperature. Warm air contains relatively more amount of water vapour than the cold air. At the higher altitudes because of lower temperature, the water vapours get cooled which in turn saturates and sufficiently cools the surrounding air causing condensation. Condensation results in the formation of small droplets and hence cloud formation. These droplets will not in themselves form into rain drops and may remain indefinitely as droplets in a condition of colloidal suspension. When the minute droplets of moisture collect on particles in the air called as hygroscopic nuclei, rain drops or other forms of precipitation occurs. The very common hygroscopic nuclei are salt particles from ocean and products of combustion in atmosphere (Eagleson, 1970). It is to be remembered that during condensation, the air is allowed to cool below the dew point during which the available moisture in the air changes into liquid form. When the dew point falls below the freezing point, then water vapours are directly converted in to ice form by the process of sublimation. Under this condition, condensation starts after lowering of its temperature. Condensation in the atmosphere depends on relative humidity of air and its degree of cooling. In normal weather condition when condensation starts after cooling the air much below the dew point, the air becomes in super saturated condition. This occurs when air contains small size dust particles which are freezing nuclei. The process of cooling for condensation may occur through adiabatic or non-adiabatic process. In non-adiabatic process, the cooling is performed by radiation, conduction or mixing the saturated hot air mass to the cold air mass. There are some limitations in non-adiabatic process of cooling i.e. it produces small amount of precipitation by only fog, dew and frost. In adiabatic cooling process, large amount of precipitation in the form of rainfall occurs since in this process warm air mass is lifted upward in the atmosphere where it is expanded and cooled adiabatically forming conducive clouds and hence generating intense rainfall.

2.4 Process of Precipitation

There are two processes through which precipitation is released from the clouds. They are (i) coalescence process and (ii) ice-crystal process.

Coalescence Process: There are two forces which act up on the water droplets in the atmosphere. These forces are gravity force and frictional drag force. When these two forces are balanced, water droplets gain maximum velocity called as terminal velocity. Terminal velocity is directly proportional to the square of the radius of the droplet. Because of this terminal velocity, bigger size droplets fall down in the air and while falling it collects the smaller droplets on their forward path. While the droplets falls with a high speed in the air, it creates turbulence behind it as a result of which, smaller droplets also falls at higher speed and overtakes the heading droplets. In this way the droplets get embedded together and forms bigger size droplets. The droplets may sometimes grow to a size as large as 7 mm and the velocity can go up to 10 m/s (Suresh, 2008). When the velocity increases, sometimes the bigger droplets get flatten and break up into several mall sizes called as drizzles. In this way, the precipitation is released from the clouds by coalescence process.

Ice-Crystal Process: The water droplets can remain in the cloud mass at sub-freezing temperature down to about -40ºC. At this temperature, solidification of freezing nuclei occurs. During solidification, when ice elements are formed in the cooled cloud masses, an imbalance is caused because the equilibrium vapour pressure over the water droplets is greater in size than the ice crystal. This way some cloud mass continually grow in size whereas others shrink which cause uneven size distribution of cloud elements within the cloud. This phenomenon again favours for further growth of water droplets through collision and coalescence actions. In case of warm clouds, (clouds above sub-freezing temperature) the giant salt nuclei if present, serve to develop unusual large sizes water droplets which cause precipitation through fragmentation process.

2.5 Types of Precipitation

The type of precipitation to occur depends on the temperature of the air in which precipitation is formed and the temperature of the lower air layers through which it falls. The main driving force to create precipitation in the atmosphere is the lifting of air mass and subsequently followed by cooling mechanism. There are three different methods by which the air mass gets lifted. Depending on the way in which the air mass gets cooled so as to cause precipitation, precipitation may be convective, orographic and cyclonic types.

2.5.1 Cyclonic Precipitation

This type of precipitation is caused by lifting of an air mass due to pressure difference caused by unequal heating of earth surface. Unequal heating of air mass over the land and water surface cause pressure gradient which causes the air to flow from high pressure zone to the low pressure zone. This creates a front. Air from the surrounding zone flows horizontally to the low pressure zone casing air in the low pressure zone to move upward or get lifted upward. This phenomenon is called as cyclone. In the cyclonic zone the air is lifted upward and moves circularly upward. In the cyclonic zone, there is large drop of barometric pressure. There are two types of cyclones. They are (i) tropical cyclone and (ii) extra-tropical cyclone.

2.5.1.1 Tropical Cyclone

The tropical cyclone is generally called as cyclone in India. It is called as hurricane in USA and typhoon in South-East Asia. It is a wind system with an intensely strong depression with mean sea level (MSL) pressure going below 915 mbars sometimes. Normally, the cyclone may extend aerially from 100 to 200 km in diameter. The isobars are closely spaced. Wind blows in anticlock wise direction in northern hemisphere. Eye of the storm (centre of the cyclone is called as eye) extends to about 10 -50 km in diameter. The wind velocity remains almost quiet in the eye. At the right outside portion of the eye, there is very strong wind reaching to a speed of as much as 200 km/hr. The wind speed gradually decreases towards the outer edge whereas pressure gradually increases (Fig. 2.1). In the cyclone affected areas, there is generally high rainfall of high intensity. Cyclones are responsible for most of the rains in the central art of USA and for most of winter rains in Haryana and Punjab (Garg, 1987).

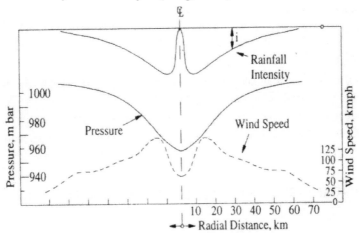

Fig. 2.1: Schematic view of a section of tropical cyclone

During the summer months, tropical cyclone originates in the ocean at around the latitudes of 5 to 10^0 and move at speed of about 0 to 30 km/hr towards the higher latitudes in an irregular path. During this process, they gain energy due to latent heat of condensation of ocean water vapour and increases in size as they move on oceans. When they strike the land masses, the source of energy is cut off and the cyclone dissipates its energy very fast. The intensity of the storm decreases rapidly. Due to the tropical cyclone, the coastal areas experience moderate to heavy rainfall for several days. High intense storms cause flood and heavy damage to domestic and national property. It also damages many buildings, loss to human and other animal lives and severe damage to crops.

2.5.1.2 Extra-tropical Cyclone

Extra-tropical cyclones are formed in locations outside the tropical zone. They possess a strong counter-clockwise wind circulation in the northern hemisphere being associated with a frontal system. The magnitude of precipitations and wind velocities are small as compared to the tropical cyclone. But in this case the duration of precipitation is longer and areal extent of coverage is also more.

2.5.1.3 Anticyclones

These are the regions of high pressure usually of large areal extent. The weather is almost calm at the centre. They cause clockwise wind circulations in the northern hemisphere and at the outer edge, cloudy and precipitation conditions exists.

Cyclonic precipitation may be divided into frontal and non-frontal types. The frontal precipitation can further be divided into warm-front and cold-front precipitation.

2.5.1.4 Frontal Precipitation

This type of precipitation occurs when lifting of air mass is carried out by fronts which are formed between two different kinds of air masses *i.e.* warm and cold air mass. The boundary between these two different kinds of air masses having different temperature, densities etc. is called as front or frontal surface. There are three types of fronts. They are

- Cold front

- Warm front

- Stationary or occluded front

Cold front is formed when cold air mass pushes the warm air mass back. The cold front moves faster than the warm front and usually overtakes the warm front. Warm front is formed when the warm air mass displaces the cold air mass during the passage of movement of the air masses. If the cold air mass pushes the warm air mass under and there is formation of front which does not move from its place then a stationary front is formed. Since the cold front moves faster than the warm front and usually overtakes the warm front, hence at this condition, the frontal surfaces of both fronts slide on each other. This phenomenon is called as occlusion and the resulting front is called as occluded front.

There are two types of frontal precipitations as:

- Warm front precipitation and

- Cold front precipitation

Warm front precipitation occurs when warm air mass moves gradually upward over a wedge of cold air mass. The warm air is replaced by the cold air. In this case rainfall of intensity light to moderate occurs over a large area. It is generally extended up to 500 km ahead of surface of the front. As discussed earlier, cold front occurs when the warm air mass is forced upward by an advancing wedge of cold air mass. During this process, precipitation in the from showery occurs at the place adjacent to the area of the surface front. It generally extends up to a distance of 150 km or more from the frontal surface. It consists of relatively narrow band of bad weather and passes quickly because of its high speed. The frontal precipitation is generally in the form of drizzle. It occurs mainly in the temperate zones.

2.5.1.5 Non- frontal Precipitation

This type of precipitation is produced due to lifting of air mass when there is a depression. Especially, when the moving cold air mass strikes the stationary warm air mass, the warm air is lifted upward over the cold air. This warm air at the higher altitudes gets cooled and condensed resulting formation of precipitation.

2.5.2 Convective Precipitation

Convective precipitation is generally tropical in nature and is mostly confined to the tropics. It is brought about by the heating of air mass at the interface with the ground. In this type of precipitation, the heated air mass is lifted by convective process. Solar energy is the main source of supplying heat for the development of the convective current in the air. The lifted air mass expands with a resultant reduction in its weight. During this period, increasing quantities of water vapour are taken up which sufficiently increases the relative humidity of the atmosphere. The warm moisture laden becomes unstable and the pronounced vertical currents are developed. Dynamic cooling takes place causing condensation and precipitation. Convective precipitation may be in the form of light showers or storms of extremely high intensity. Sometimes the convective precipitation is accompanied by thunder-lightening along with local winds. Sometimes hails are also appeared in this type of precipitation (Viessman *et al.*, 1977). This type of precipitation mostly occurs during the summer. The chief characteristics of this type of precipitations are:

- It occurs in the temperate air zones at low altitude.
- In this type of precipitation, intense rainfall occurs but for a short duration.
- Sometimes light showers may occur.
- Comulo-nimbus types of clouds cause precipitation.
- It occurs in the warm weather condition.
- Its areal extent of coverage is low.
- Dynamic cooling is the mechanism to form condensation and precipitation.
- It generally occurs in the afternoon or early evening of the day.

2.5.3 Orographic precipitation

It is the most important type of precipitation and is responsible for most of the heavy rains in India. It occurs due to lifting of moisture laden air along the orographic plane and cooling of the same adiabatically to form cloud. Orographic precipitation depends on the elevation and rate of rise of air mass. Mechanically lifted moist horizontal air currents when strikes some natural barriers such as mountain, it cannot advance further. Rather it moves upward, consequently undergo cooling, condensation and hence causing precipitation. This type of precipitation comprises (i) wind ward orographic precipitation and (ii) leeward orographic precipitation.

The side of the orographic plane from where the air is lifted and strikes the mountain range is called as wind ward side and precipitation occurring in this side is known as wind ward orographic precipitation. The opposite side of the mountain range is called as leeward side and the precipation in leeward side is called as leeward orographic precipitation. Leeward side precipitation is also called as rain-shadow. The wind ward side mountain range experiences higher rainfall than the leeward side mountain range. Fig. 2.2 shows the formation of orographic precipitation. The rainfall is composed of showers and steady rainfall.

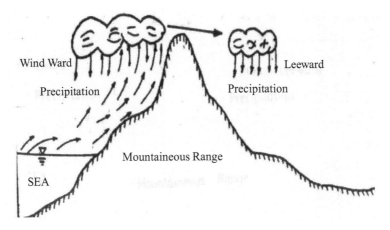

Fig. 2.2: Formation of orographic precipitation

The southern slope of the Himalayan range in India produces enormous amount of orographic precipitation which benefits agriculture and other allied sectors in India. Heavily moisture laden winds from the Bay of Bengal strike the southern slope of the Himalayan range causing intense rain at the foot hill region. Similarly the winds coming from the Pacific Ocean strike the western slope of the coastal ranges in Washington, USA causing heavy rain there (Linsley *et al.,* 1949).

2.6 Types of Cloud

Clouds are the condensed atmospheric moisture resulted from lifting of warm air masses in the sky. They are the main source of precipitation. The different forms of clouds are:

- High cloud
- Middle cloud and
- Low cloud

High cloud is formed at a mean attitude of 5 to 13 km above the land mass. Middle cloud is formed at an average altitude of 2 to 7 km whereas the low cloud is formed at 1 to 2 km above the land mass.

High clouds are of three types. They are:

- Cirrus
- Cirro-cumulus and
- Cirro-stratus

Middle clouds are of three types i.e.

- Alto-cumulus
- Alto-stratus and
- Nimbo-straus

Low clouds are divided into the following four types:

- Strato-cumulus
- Stratus
- Cumulus and
- Cumulo-nimbus

Cirrus clouds appear in the form of white, narrow bands or white patches. They contain ice crystal and do not form precipitation. These clouds produce beautiful colours at the time of sun rise and sun set in the sky. Cirro-cumulus clouds look like small globule or flakes. These are arranged in line or in group. Cirro-stratus clouds appear at greater height compared to other two high clouds. They are normally transparent in nature, whitish in colour and milky in appearance. They cover the sky fully or partly. They appear in the form of patches of cirrus or cirro-cumulus clouds. They contain ice crystals. Alto-cumulus clouds are found in white or grey or both grey and white in colours. They produce a dark shadow on the ground surface. These appear in the form of patches, sheets or layers and are arranged in parallel or wavy sequence. Alto-stratus clouds are grayish or bluish in colour and are found in the form of sheet or fiber. They cover the sky fully or partly in thin layers. Alto-stratus clouds produce precipitation in the form of drizzle or snow. Nimbo-stratus clouds are grey in colour with dark appearance. These are generally low clouds found at about 1000 feet high above the earth surface. Stratus clouds look like thick fog and so sometimes it is difficult to distinguish between them. These clouds result precipitation in the form of drizzle, ice prism or snow grain. Cumulus clouds are the detached clouds which appear in the form of sharp outlines. They appear in the sky during the day time and vanish in the night. They cause light precipitation. Cumulo-nimbus clouds are present at very low altitudes above the earth surface (1 to 2 km) and are heavy and dense. They appear in the shape of mountain or big tower in the sky. They produce heavy precipitation, thunder, lightning, hails etc. (Suresh, 2008).

2.7 Rainfall and its Variation

The total liquid product of precipitation or condensation from the atmosphere which is measured in a rain gauge is called as rainfall. A day receiving 2.5 mm or more rainfall is called as a rainy day. The amount of the rain that reaches the earth is usually expressed in millimeters of depth of water over an area. The daily rainfall pertains to the rainfall from 0830 hours of previous day to 0830 hours IST of date. The average annual rainfall in India is about 120 cm. The average annual rainfall is computed by adding the annual rainfall of several years with the total number of years. Since, the annual rainfall is different from year to year, the average annual rainfall will vary depending on the numbers of years whose annual rainfalls are taken into account for assessing average rainfall. For example, the average annual rainfall of 50 years may be different than the average annual rainfall of 60 years. Even for the same number of years, the average annual rainfall will be different for different periods of time. Average annual rainfall of 50 years from 1950 to 1999 may be different than the average annual rainfall of 50 years from 1958 to 2007. The term normal rainfall refers to the average annual rainfall of 30 years. The annual rainfall suffers from wide departure from the normal from year to year. Practically all parts of Assam, Meghalaya and the adjoining areas, the Western Ghats and the adjoining coastal strips and parts of the Himalayas receive heavy rainfall of more than 2000 mm. Cherrapunji of India records the highest rainfall in the world (about 1142 cm annual rainfall). On the other hand, Rajasthan, Kutch and the high peaks of Ladakh receives low annual rainfall ranging from 10 to 50 cm (Sharma, 1987).

Rainfall in India is seasonal in character. Most of the rainfall (about 75 to 85% of total annual rainfall) is contributed by South-West monsoon and the contribution is only from four months (June to September). The rest amount of annual rainfall is contributed by North-East monsoon (November and December). Some parts of the country receive hot weather rainfall between March and May due to thunderstorm activity called North-Westers. About 8 percent of total geographical area of India receives more than 250 cm annual rainfall, 50 percent area receives 100 cm or more, 35 percent area receives less than 75 cm and 12 percent areas receives annual rainfall less than 40 cm. The distribution of rainfall in India is very erratic. It has both spatial and temporal variation. The coefficient of variation of annual rainfall over large part of the country is about 30 percent. In some of the interior districts, the coefficient of variation may be as high as 100 percent. The coefficient of variation of rainfall is higher during the individual months compared to the annual values (Sharma, 1987). The variation of rainfall during different parts of the year is due to the difference in seasonal temperature between land and sea.

2.8 Characteristics of Rainfall in India

India is an agrarian country and the agriculture in the country is mostly dependent on the rainfall. As discussed above, rainfall variation in the country is very much erratic, uneven and mostly it is seasonal type. From the view of climate, the Indian subcontinent can be characterized to have two measure seasons and two transitional periods. These are:

- South-west monsoon
- Transition-I, post-monsoon
- Winter season and
- Transition-II, summer.

The chief precipitation characteristics of these seasons are described below (Subramanya, 2003).

2.8.1 South-West Monsoon

South west monsoon is commonly called as monsoon. It occurs for four months starting from June to September. It is the principal rainy season in India which gives about 70-80 percent total annual rainfall. Almost all parts of the country excepting some regions in the south-eastern part of peninsula and Jammu and Kashmir get maximum rainfall by this south west monsoon. July and sometimes August is the month that receives the highest rainfall due to the monsoon. The monsoon originates in the Indian Ocean and advances towards the land masses with high speed. It strikes Kerala in the month of May (last week) and sometimes in the first week of June. The onset of monsoon is accompanied by high south-westerly wind and low pressure regions at the advancing edge. The monsoon wind advances across the country in two branches: (i) The Arabian sea branch and (ii) the Bay of Bengal branch. The Arabian sea branch moves north wards over Karnataka, Maharashstra and Gujara whereas the Bay of Bengal branch moves over eastern and north-eastern parts of the country going up to and then turns west wards to advance Bihar and Uttar Pradesh etc. Both the branches reach Delhi by about last of June or first week of July. Monsoon trough originates between the two branches and most of the rainfall occurs due to this trough. Monsoon starts withdrawing from the country by end of September or first week of October. Monsoon withdraws first from the northern parts of the country and lastly from the southern parts. Depressions are formed frequently in the Bay of Bengal during the monsoon season which results heavy pour of rainfall which sometimes may go up to 100 to200 mm per day. During monsoon there may be several wet and dry spells. Dry spells of longer duration caused due to either no or very little rainfall cause drought which when coincides with the critical growth stage of the crops cause severe yield reduction. Similarly wet spells caused due to heavy shower for several days during the monsoon cause floods in several parts especially in the coastal regions including Odisha, West Bengal, Andhra Pradesh, Maharashtra etc. Monsoon results heavy annual rainfall to the tune of 2000 to 4000 mm in Assam and north eastern region whereas the lowest amount (about 1000 to 1200 mm) occurs annually in Uttar Pradesh, Punjab and Haryana. Details about the rainfall variation across the country are discussed as above.

2.8.2 Transition I, Post Monsoon

The period of October to November is called as the transition I, post monsoon. When the south west monsoon retreats from the Bay of Bengal, north easterly flow of air picks up moisture in the Bay of Bengal. This moisture laden air mass when strikes up the east coast of the southern peninsula, heavy rainfall occur. Severe tropical cyclones are formed during this period in the Bay of Bengal and the Arabian sea. The cyclones formed in the Bay of Bengal are more than that formed in the Arabian sea. These cyclones cause intense storms in the coastal areas and cause severe damage to the crops /properties and animal lives.

2.8.3 Winter Season

Winter season continues from December up to February. By mid-December, disturbances of extra tropical region travel east wards across Afghanistan and Pakistan. This disturbance is called as western disturbances which cause moderate to heavy rain and snow fall in the Himalayan region. Low pressures are also created in the Bay of Bengal during this period which causes moderate rain of about 100 to 150 mm in eastern and southern India.

2.8.4 Transition II, Summer

The period of March to May is called as the transition II, summer. Very little rainfall occurs during this period. At times, thunder storms accompanied by moderate to heavy rainfall are produced over Kerala, West Bengal, Odisha, Assam etc. Cyclonic activities may occur over the eastern parts of the country.

2.9 Measurement of Rainfall

For crop planning, design of different soil and water conservation structures, irrigation and drainage planning and/ or any other engineering studies, analysis of rainfall is required which necessitates to measure it. The instruments used to measure rainfall are called as rain gauges. Rain gauges are divided into two types i.e. (i) recording type gauges and (ii) non-recording type gauges.

2.9.1 Non-Recording Rain Gauge

It is also termed as non-automatic rain gauge. It is one of the most common type of non-recording rain gauge and is used to measure the total rainfall during a period. Observations are taken at the end of 24 hours period or at lesser intervals during rain. It cannot give the rainfall intensity during different time intervals during a day. The Symons rain gauge is a common type non-recording rain gauge (Fig. 2.3).

The rain gauge consists of a cylindrical vessel of 127 mm diameter with a suitable base and gun metal rim. It essentially has a collector to intercept the rainfall to be measured and a receiver consisting of a base and a bottle to collect and store the rainfall until measured. A funnel of exactly 127 mm diameter is inserted into the cylindrical vessel at the top and is used as the collector. The shank of the funnel is put into the receiving bottle having diameter 76 to 100 mm. The bottle has sufficient capacity to hold extremes of rainfall likely to occur in 24 hours at a location. The usual time to measure rainfall is 8.30 A.M. The rainfall collected in the receiving bottle is measured by a graduated measuring cylinder which can read to the nearest 0.1 mm. Recently Indian Meteorological Department has introduced fiberglass reinforced polyester rain gauge which is an improvement over Symon's gauge. In remote areas, storage gauge is used to measure total seasonal precipitation. It consists of a vertical 60 cm diameter steel pipe of sufficient length to place its 20 cm catch ring above maximum accumulated snow. To convert the snow which falls into the gauge to liquid

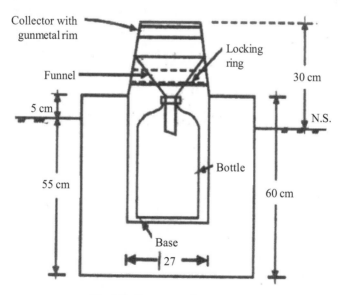

Fig. 2.3: Symons rain gauge

state, an antifreeze solution can be placed in the receiver. Following are the demerits of this rain gauge:

- It does not provide any information on intensity and duration of rainfall.

- The start and end time of rainfall cannot be known by it.

- Rainfall in any short duration out of total duration of occurrence cannot be known.

2.9.2 Recording Rain Gauge

Recording rain gauge also called as automatic rain gauge or self recording rain gauge is an improvement over the non-recording rain gauge. Since this rain gauge represents cumulative rainfall, it is also termed as integrated rain gauge or continuous rain gauge. It is especially suitable to inaccessible areas where it is used to give accumulated rainfall over an appreciable long period. In this rain gauge, the total rainfall received by a container over certain time period is recorded by a pen over a clock driven chart attached to the rain gauge. The graph of the cumulative rainfall is called as mass curve of the rainfall. The mass curve represents the total amount of rainfall received since the start of recoding at the station, duration of rainfall and intensity of rainfall. The rain gauge chart showing mass curve is illustrated in Fig 2.4 below.

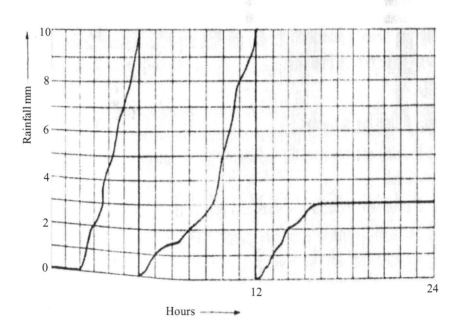

Fig. 2.4: Rain gauge chart showing mass curve

Following information can be known from the mass curve.

- Depth of rainfall at any time
- Starting and end time of rainfall of a day and
- Intensity of rainfall at any time during a day

Recording rain gauges are of following four types i.e. (i) weighing bucket type (ii) tipping bucket type (iii) float recording type and (iv) automatic radio reporting rain gauge.

2.9.2.1 Weighing Bucket Rain Gauge

Weighing type rain gauge is a common type rain gauge used for measuring rainfall and moderate snowfall. The snow fall is measured by collecting the snow in the receiver which is melted in the gauge itself by a heating device attached to the rain gauge or by using some chemicals such as calcium chloride, ethylene glycol etc. It gives the snow fall depth in terms of equivalent rainfall depth.

Weighing bucket rain gauge consists of a bucket which acts as the receiver. The bucket is supported on a weighing mechanism like a spring or liver mechanism. The weighing mechanism is attached to a pen arm through levers and links which moves over a chart mounted on a clock driven record drum (Fig. 2.5 a). Addition of water due to rainfall or snow in the bucket gives a trace of accumulated amount of rainfall with time and thus a mass curve is produced. It has the advantage that it is simple in mechanism. However, any rainfall less than the capacity of bucket compartments are not recorded.

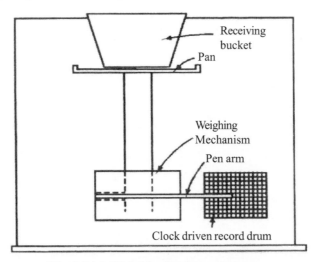

Fig. 2.5 (a): Weighing bucket rain gauge

2.9.2.2 Tipping Bucket Rain Gauge

It is an electrically operated rain gauge. It has the advantage that it can be used to record rainfall in remote areas. It consists of a sharp edged cylindrical receiver having a tipping bucket and funnel. The funnel is used to convey the rainwater into the bottle. Both the receiver and the bottle have diameter of 30 cm. Below the funnel, a tipping bucket with two compartments is pivoted on fulcrum. The capacity of each bucket is equivalent to 0.25 mm rainfall depth. When the rainwater gets filled up in the first compartment, water gets tipped up, empties and then the second compartment comes into position to receive water. When it is full, the bucket tips back into original position and thus process continue (Fig. 2.5 b). When bucket is tipped, it actuates an electrical circuit causing a pen to mark on a revolving drum attached with a graph. It has the advantage that it can be used for digitalizing of the output signal. However, the disadvantage is that the tipping of the bucket can be designed for particular rainfall intensity. At the time of occurrence of high intensity rainfall, tipping of the bucket is so fast that the recorded graphs will be much clung and thus accurate readings may not be obtained. Similarly during period of very low rainfall intensity, no lines may be found on the graph and thus indicating no occurrence of rainfall. By this rain gauge, it is difficult to characterize the starting and end time of rainfall. This rain gauge also requires some sophisticated mechanisms in repair and maintenance work.

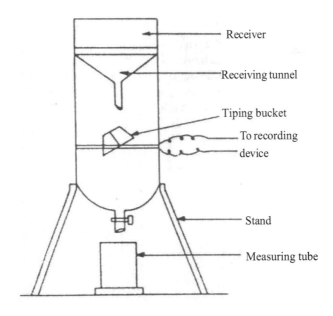

Fig. 2.5 (b): Tipping bucket rain gauge

2.9.2.3 Float Recording Rain Gauge

It consists of a rectangular float chamber carrying a funnel on one side and a drum on the other side. Both the funnel and the drum are connected to the float chamber at top. There is a light hollow float connected to the float chamber at its bottom (Fig. 2.5 c). When the rainwater enters into the float chamber, the float rises up. When the chamber is full of water, the float touches the top. At this time the accumulated water is drained out by siphoning mechanism. The float comes down and again the same process is continued during the occurrence of the rainfall. The movement of the float is transmitted through an arrangement to the pen which traces on a revolving chart. The revolving chart is actuated by a clock work. The clock mechanism revolves the drum once in 24 hours. By analysig the chart, it is possible to find out the total rainfall within any time interval. In most of the gauges, it is required to change the chart once in a day or in some, once in a week.

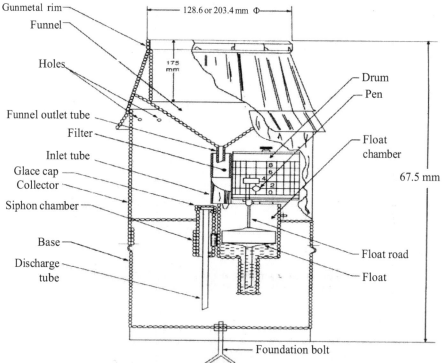

Fig. 2.5 (c): Float recording rain gauge

2.9.2.4 Automatic Radio Reporting Rain Gauge

This rain gauge signals rainfall data over radio at regular intervals. It is also called as telemetering rain gauges. The tipping bucket rain gauge is used to measure the rainfall. The tipping bucket gives electrical pulses equal in number to the millimeter of rainfall collected in the gauge. The pulses are converted into coded messages by recording unit and impressed on the transmitter during broadcast. A program clock switches the coding and programming units at predetermined time. The signals also indicate if there is any occurrence of rain at the time of broadcast and also the intensity of rainfall that occurs. These rain gauges are of utmost use in gathering information about rainfall data in mountainous and other inaccessible areas. It can cover distance about 100 km at a time. This rain gauge was developed by Indian Meteorological Department. The rain gauge helps to get information of rainfall and is also used for flood forecasting at the control centres. It provides to get instantaneous rainfall data of a place.

Advantages of Recording Rain Gauge

- It provides information about both amount and intensity of rainfall.

- It helps to collect the rainfall amount and intensity in remote areas.

- It gives comparatively more accurate result of amount and intensity of rainfall. Error in measurement of rainfall data due to human beings is eliminated.

- It provides information about start and end time of rainfall as well as duration of rainfall in a day.

Disadvantages of Recording Rain Gauge

- The cost of the recording rain gauge is higher than the non-recording type.

- Repair and maintenance cost of the instrument is higher than the non-recording type.

- It needs sophisticated mechanisms in repair and maintenance work.

- It does not give accurate measurement of rainfall for light rainfall or rainfall with low intensity.

- If there is any mechanical failure, then the instrument cannot measure any rainfall.

2.9.3 Radar Measurement of Rainfall

Meteorological radar is a powerful instrument for measuring the areal extent, orientation and movement of the rain storms. It can also be used to accurately measure the magnitude of rainfall over large areas. It works on the principle of echo-sounding effect. The radar transmits a regular succession of pulses of electromagnetic radiation in the sky in a narrow beam. When raindrops intercept a radar beam, some amount of the radiation is absorbed by them. It is observed that (Subramanya, 2003):

$$P_r = \frac{C\,Z}{r^2} \tag{2.1}$$

where, P_r = average echo power, Z = radar-echo factor, r = distance to target volume and C = a constant.

Factor Z is given as:

$$Z = a\,I^b \tag{2.2}$$

where, a and b are coefficients and I = intensity of rainfall, mm/hr. The values of a and b for a given radar station are determined by calibration with the help of recording rain gauges.

Meteorological radars operate with wave lengths ranging from 3 to 10 cm, the common value being 5 and 10 cm. A 10 cm radar is used for observing details about heavy flood producing rains. For light rains and snow a 5 cm radar is used. Radar can be considered as a remote sensing super gauge covering an areal extent of as much as 100,000 km². The data generated by the radar can be processed on-line by computers. Presently Doppler type radars are used for measuring the velocity and distribution of raindrops.

2.10 Selection of Site for Rain Gauge

Before installing rain gauges in any site, following considerations must be taken into account.

(i)   The site where rain gauge is installed should be true representative of the area to give rainfall. It should also be easily accessible.

(ii) The area should be open free from obstacles. If there is any obstacle like a building or any tree nearby, then the rain gauge should be located at a distance which is at least twice the height of the obstacle.

(iii) The site where the rain gauge is set up should be level. Generally the places such as terrace wall, sloppy ground are avoided for setting up the rain gauges.

(iv) Site should be shielded from strong wind.

(v)   In hilly areas, the rain gauge must be shield by fencing at least 5.5 x 5.5 m² area around it. Fencing controls the wind effect on precipitation catch.

(vi)   In forest areas, rain gauges must be installed at a distance which is at least twice the height of the tallest tree. The angle of elevation of the tree at the tip of the rain gauge should be within 20 to 30° and in extreme case may be 45°.

(vii)   Location of rain gauge should be avoided in the valley or at the ridge point of the area.

(viii) The surface over which the rain gauge is to be installed should be such that the receiver's height is about 75 cm above the ground level.

(ix)   Rain gauges must be installed in vertical position and any tilting should be avoided.

2.11 Common Errors in Rainfall Measurement

There may be errors in measurement of rainfall due to several factors as:

- Instrumental error
- Human error in taking the measurement
- Errors due to application of faulty recording or measuring procedures

In addition there are other errors in rainfall measurement and the following points must be taken into account while recoding the rainfall by a rain gauge.

The rainfall data from a non-recording rain gauge may have error due to displacement of water level by measuring scale which may range up to 2%. When the rainfall is collected in a dried rain gauge, there may be some amount lost in wetting the surface which gives less reading than the actual measurement. About 0.25 mm depth of rainfall is lost in wetting the dried surface of the rain gauge. There is also some loss of water from the rain gauge due to evaporation caused by high temperature which may be up to 2%. There may have some errors due to splashing of rain water from the collector. If there are some obstacles like tall building, tall tress etc. very near the rain gauges, then they may intercept the rainfall and the rainfall collected by the gauge is not true representative of the rainfall occurred in that place. The inclination of rain gauge affects the collection of the rainfall. When the angle of inclination is 10° from the vertical in the direction of wind movement, it catches about 1.5% less rainfall than the actual amount. But when it is inclined opposite to the wind direction, then the gauge catches more rainfall than the actual amount. Velocity of the wind also affects the rainfall collection. When the velocity of wind is 10 miles/hr, the rainfall collected is about 17% less and when the velocity is increased to 30 miles/hr, the reduction may rise up to 60%.

2.12 Rain Gauge Network

A rain gauge can measure rainfall up to certain areal extent. If the number of rain gauges in a place are less i.e. catching area of rain gauges is very small compared to the areal extent of a storm, then the rainfall recorded by the gauge is not true representative of the rainfall of that area. To get a representative figure of the rainfall over a catchment, the number of rain gauges should be as large as possible. On the other hand, if there will be adequate rain gauges in the catchment, then costs of setting the rain gauges may be high and hence not economical. Hence, there should be optimum number of rain gauges from which reasonably correct data about the rainfall information can be obtained. The World Meteorological Organisation (WMO) has recommended that:

- In flat regions of temperate, Mediterranean and tropical zones, there should be one station for 600 to 900 km² area which is ideal and one station for 900 to 3000 km² area which is acceptable.

- In mountainous regions of temperate, Mediterranean and tropical zones, there should be one station for 100 to 250 km² area which is ideal and one station for 250 to 1000 km² area which is acceptable.

- In arid and polar regions, there should be one station for 1500 to 10,000 km² area depending on feasibility.

From practical considerations of Indian conditions, the Indian Standard (IS: 4987-1968) recommends (IMD recommendations) that:

- In plains, there should be one station per 520 km² area.

- In regions of average elevation of 1 km, there should be one station per 260 to 390 km² area and

- In hilly areas, there should be one station per 130 km² area.

10% of the total number of rain gauges installed in a region should be recording type.

2.12.1 Adequacy of Raingauge Stations

McCulloch (1961) has developed a relationship for determining the adequacy of rain gauge. The formula is given as:

$$N = \left(\frac{C_v}{\varepsilon} \right)^2 \tag{2.3}$$

where, N = adequate/optimal number of rain gauge stations, C_v = coefficient of variation and ε = percent allowable error in estimate of the mean rainfall of the existing m number of rain gauges which is generally taken as 10%.

If there are m number of rain gauges in a catchment each recording rainfall of P_1, P_2, P_3........P_m in a given time, the coefficient of variation of these m number of rain gauges is given as:

$$C_v = \frac{\sigma_{m-1}}{\overline{P}} \times 100 \qquad (2.4)$$

where, σ_{m-1} = standard deviation of the rainfall data of m number of rain gauges and $\overline{P}$ = the mean rainfall of m number of rain gauges.

Mean rainfall is calculated as:

$$\overline{P} = \frac{\sum\limits_{i=1}^{i=m} P_i}{m} \qquad (2.5)$$

where, i = number of rain gauges which may be m and P_1, P_2, P_3........P_m are the rainfall in m number of rain gauges.

Standard deviation of the rainfall data of m number of rain gauges, σ_{m-1} is given as:

$$\sigma_{m-1} = \sqrt{\frac{\sum\limits_{i=1}^{i=m} (P_i - \overline{P})^2}{m-1}} \qquad (2.6)$$

where, all the terms are defined earlier.

If the value of N < m, then there is no need to take any more rain gauges. Otherwise $(N - m)$ is the extra number of rain gauges required to be taken.

Sample Calculation 2.1

Determine the optimum number of rain gauges required to be installed in a catchment of 600 km² area. The catchment has 6 number of rain gauges and the annual rainfall of these 6 rain gauges are recorded as 800, 1000, 750, 500, 670 and 400 mm, respectively. Assume the percent allowable error in estimate of the mean rainfall as 10%. Find out also the extra number of rain gauges required.

Solution

For the data of above example, $m = 6$

Mean rainfall $\bar{P} = \dfrac{\sum\limits_{i=1}^{i=m} P_i}{m} = (800 + 1000 + 750 + 500 + 670 + 400)/6 = 686.67 \text{ mm.}$

Standard deviation, σ_{m-1} is calculated as:

$$\sigma_{m-1} = \sqrt{\dfrac{\sum\limits_{i=1}^{i=m}\left(P_i - \bar{P}\right)^2}{m-1}}$$

$$= \sqrt{\dfrac{(800 - 686.67)^2 + (1000 - 686.67)^2 + (750 - 686.67)^2 + (500 - 686.67)^2 + (670 - 686.67)^2 + (400 - 686.67)^2}{6-1}}$$

$= 215.56 \text{ mm}$

Given percent allowable error in estimate of the mean rainfall $== 10$

Hence coefficient of variation is calculated as:

$$C_v = \dfrac{\sigma_{m-1}}{\bar{P}} \times 100 = (215.56 / 686.67) \times 100 = 31.39\%$$

Optimum number of rain gauges $N = \left(\dfrac{C_v}{\varepsilon}\right)^2 = \left(\dfrac{31.39}{10}\right)^2 = 9.85 = 10 \text{ rain gauges.}$

Extra rain gauges required $= 10 - 6 = 4$.

2.13 Estimating Missing Rainfall Data

Sometimes we may miss to record the rainfall data for a certain day or certain period because of some reasons due to absence of the data recorder, instrumental problems etc. Under such conditions, the missing data can be predicted with the help of the available data base of nearby stations by employing the following methods:

- Arithmetic mean method
- Normal ratio method
- Graphical method
- Long term mean rainfall of new station method

2.13.1 Arithmetic Mean Method

Following points must be considered before using this method:

- There should be availability of at least three rain gauge stations adjoining the station for which the rainfall data is missing.

- All these stations must be uniformly spaced from the station for which the rainfall data is missing.

- The percent variation of the normal annual rainfall data of the missing station and surrounding station should not exceed 10.

In such case arithmetic mean method can be employed to find out the missing rainfall data of a station. The procedure is explained as follows:

Let there be three stations 1, 2 and 3 very close to each other. There is another station X which has a missing rainfall record on a particular day in which the rainfall record of other three stations is available. The normal annual rainfall values at these four stations (i.e. 1, 2, 3 and X) should also be available. Now if the normal annual rainfall values at each of these three index stations differs within 10% of the normal annual rainfall value of the missing station X, then a simple arithmetic average of the rainfall (corresponding to the missing rainfall) at the three index stations will give the data of the missing station, X. If N_1, N_2, N_3 and N_x represents the normal rainfall values of stations 1, 2, 3 and X, respectively and P_1, P_2, and P_3 represents the rainfall data collected on a particular day in which the rainfall data of station X is missing, then the missing rainfall data of station X on that day (P_x) is:

$$P_x = \frac{(P_1 + P_2 + P_3)}{3} \qquad (2.7)$$

The above equation is valid provided N_1, N_2 and N_3 differs within 10% of N_x.

Sample Calculation 2.2

There are four rain gauge stations located in a catchment which are uniformly spaced. The rainfall data on a particular day in three stations A, B and C are 105, 90 and 120 mm, respectively. On that day the station X was inoperative and so no data could be collected. The normal annual rainfall values of stations A, B, C and X are 1015, 950, 1000 and 980 mm, respectively. Estimate the missing rainfall of station X.

Solution

Normal rainfall of stations X, $N_x = 980$ mm.

10% of $N_x = 98$ mm.

Thus, maximum permissible normal rainfall at any of the three stations for consideration of arithmetic average method is:

$= 980 \pm 98$

$= 882$ to 1078.

Thus, we find that normal rainfall of stations A,B and C differs within 10% of N_x and hence we can use arithmetic average to compute the missing rainfall of station X.

Hence, rainfall of station, X on the day when we have records of other three stations A, B and C is:

$$P_x = \frac{\left(P_A + P_B + P_C\right)}{3} = \frac{(105 + 90 + 120)}{3} = 105 \ mm$$

Hence, missing rainfall data is 105 mm.

2.13.2 Normal Ratio Method

The consideration of points 1 and 2 of arithmetic mean method as presented above remains same for normal ratio method also. However, normal ratio method is used when the percent variation of the normal annual rainfall data of the missing station and surrounding station exceeds 10. In this method, the missing rainfall is estimated by weighing the normal rainfall of each index station. The procedure of estimation of missing rainfall is as follows:

Let there be three stations A, B and C which have normal values N_a, N_b and N_c. There is a station X adjacent to stations A, B and C which has normal rainfall N_x. Let the rainfall of a particular day measured by the three index stations A, B and C are P_a, P_b and P_c and on that day the station X was inoperative so that no data could be collected. The missing rainfall of station X (P_x) is estimated as:

$$P_x = \frac{1}{3}\left[P_a \cdot \frac{N_x}{N_a} + P_b \cdot \frac{N_x}{N_b} + P_c \cdot \frac{N_x}{N_c} \right] \tag{2.8}$$

Sample Calculation 2.3

There are four rain gauge stations A, B, C and X located in a catchment which is uniformly spaced. The normal annual rainfall values of stations A, B, C and X

are 1115, 950, 1200 and 1000 mm, respectively. The rainfall data on a particular day in three stations A, B and C are 105, 100 and 120 mm, respectively. On that day the station X was inoperative and so no data could be collected. Estimate the missing rainfall of station X.

Solution

Normal rainfall of stations X, $N_x = 1000$ mm.

10% of $N_x = 100$ mm.

Thus, maximum permissible normal rainfall at any of the three stations for consideration of arithmetic average method is:

$= 1000 \pm 100$

$= 900$ to 1100.

Thus, we find that normal rainfall of stations A,B and C differs beyond 10% of N_x and hence we cannot use arithmetic average to compute the missing rainfall of station X. But we can use normal ratio method.

Hence, rainfall of station, X on the day when we have records of other three stations A, B and C is:

$$P_x = \frac{1}{3}\left[P_a \cdot \frac{N_x}{N_a} + P_b \cdot \frac{N_x}{N_b} + P_c \cdot \frac{N_x}{N_c} \right]$$

$$= \frac{1}{3}\left[\frac{1000}{1115} \cdot 105 + 100 \cdot \frac{1000}{950} + 120 \cdot \frac{1000}{1200} \right] = 99.7 \text{ mm}.$$

Hence, missing rainfall data is 99.7 mm.

2.13.3 Graphical Method

In this method, estimation of missing rainfall is based on the record of only two stations located near to each other. The data of rainfall of these two stations are availed for some years and in between the data of one station is missing in a year whereas the data of the other station is available for that year. We can estimate the missing rainfall of the station by graphical method as discussed below:

At first take a graph paper and plot the rainfall data of one station on ordinate of the graph paper whereas the rainfall data of the other station is plotted on abassica. Then a best fit line is drawn between the values of ordinate and abassica. With the help of this best fit line, the missing rainfall corresponding to the known rainfall amount of index station for which rainfall record is available is calculated by drawing a straight line which intersects the best fit line at a point and then corresponding value of rainfall of the missing station is read from the graph.

2.13.4 Long Term Mean Rainfall of New Station Method

In this method, long term mean rainfall of a newly established station which has only limited years of record can be estimated by comparing the long term rainfall of index stations which are adjacent to the newly established station. In this method, it is assumed that both the index stations and the newly established station for which data are missing are identical falling under the same hydrological conditions.

Let there be two raingauge stations A and B which are identical falling under the same hydrological conditions. The station B has limited years of rainfall record whereas station A has long term periods of record. In order to compute the long term mean rainfall of station B, following steps are to be considered:

- Estimate the mean value of rainfall of station A for the total long term periods of record available for station A. Let it be designated as A_{lt}.

- Estimate the mean value of rainfall of station B for the total limited/short term periods of record available for station B. Let it be designated as B_{st}.

- Estimate the mean value of rainfall of station A for the total limited/short term periods of record available for station B. Let it be designated as A_{st}.

- Now we can estimate the long term mean rainfall of station B (B_{lt}) as:

$$B_{lt} = \frac{A_{lt} \cdot B_{st}}{A_{st}} \tag{2.9}$$

Sample Calculation 2.4

There are two raingauge stations A and B. The station A has long term rainfall records but station B is a new station having limited periods of data being recorded. The rainfall data of these two stations as recorded are mentioned below. Estimate the long term mean rainfall of station B.

Year	1980	1981	1982	1983	1984	1985	1986	1987
Rainfall of station A (cm)	75	120	105	130	50	75	85	90
Rainfall of station B (cm)	-	-	-	-	-	-	-	75
Year	1988	1989	1990	1991	1992	1993	1994	1995
Rainfall of station A (cm)	75	65	85	105	135	65	100	105
Rainfall of station B (cm)	105	120	75	85	70	100	130	145

Solution

Raingauge station A has availability of long term rainfall data. Let us estimate the mean value of rainfall of station A for the total long term periods of record available for station A i.e. from 1980 to 1995 which is:

$A_{lt} = 91.60$ cm

Rainfall data of station B has records of 9 years i.e. from 1987 to 1995. The mean value of rainfall of station B for the short term periods of record available for station B i.e. from 1987 to 1995 is:

$B_{st} = 100.6$ cm

Now we estimate the mean value of rainfall of station A for the total limited/ short term periods of record available for station B i.e. we estimate the mean value of rainfall of station A for the periods 1987 to 1995 as:

$A_{st} = 91.67$ cm

The long term mean rainfall of station B (B_{lt}) is estimated as:

$$B_{lt} = \frac{A_{lt} \cdot B_{st}}{A_{st}} = \frac{91.6 \times 100.6}{91.67} = 100.53 \ cm.$$

2.14 Average Depth of Rainfall

As discussed, there is large variation of rainfall over a large area. The average depth of rainfall over the whole area is computed by recording the rainfall at different stations covering the area under consideration and employing any one of the following three methods i.e. (i) Arithmetic mean method, (ii) Thiessen method, and (iii) Isohyetal method.

2.14.1 Arithmetic Mean Method

In this method, the average rainfall is computed by adding the rainfall of individual stations and then dividing the addition with the number of stations.

2.14.2 Thiessen Method

This is also called as weighted mean method. In this method, location of rain gauges is plotted on a map and the rain gauge locations are joined by straight lines. Perpendicular bisectors are drawn on each of the straight lines forming polygons around each rain gauge (Fig. 2.6). Measuring the area of each polygon and with recorded rainfall of the rain gauge located inside each polygon, the average rainfall depth can be computed. The formula used for computation of average rainfall is:

$$P = \frac{\sum_{i=1}^{n} P_i A_i}{\sum_{i=1}^{n} A_i} \tag{2.10}$$

where P refers to the rainfall recorded by station, i and A refers to the area of the polygon enclosing the *station i* and n refers to the number of stations of the whole basin whose average depth of rainfall is to be computed..

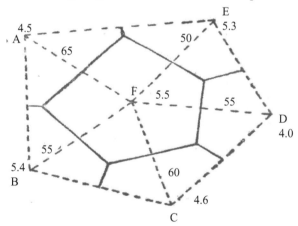

Fig. 2.6: Thiessen polygons to compute average depth of rainfall

2.14.3 Isohyetal Method

Isohyets are the lines joining places of equal rainfall. In this method, location of rain gauges with recorded amount of rainfalls are plotted on a map/graph paper and then isohyets are joined by interpolation (Fig. 2.7). The areas between the successive isohyets are measured by a planimeter.

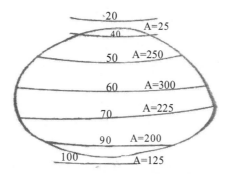

Fig. 2.7: Isohyetes of rainfall

The same formula as used in Theissen method (Eqn. 2.10) is used to compute the average rainfall. However, in this method, A_1, A_2 ... represents the areas between two successive isohyets and P_1, P_2... represent the average rainfall of the areas between two successive isohyets. In both Isohyetal and Theissen method, data of rainfall of stations located slightly beyond the coverage region can be used to compute the average rainfall.

Sample Calculation 2.5

Estimate the average depth of rainfall occurring in a basin of area 245 ha having 10 rain gauges. The rainfall recorded by each rain gauge and the polygon area enclosing each rain gauge are given as follows:

Station	A	B	C	D	E	F	G	H	I	J
Rainfall, cm	50	45	40	23	35	20	12	5	33	20
Polygon area, ha	10	13	18	15	26	25	20	23	40	55

Solution

Average depth of rainfall over a catchment by Thiessen polygon method is given as

$$P = \frac{\sum\limits_{i=1}^{n} P_i A_i}{\sum\limits_{i=1}^{n} A_i}$$

$$= \frac{50 \times 10 + 45 \times 13 + 40 \times 18 + 23 \times 15 + 35 \times 26 + 20 \times 25 + 12 \times 20 + 5 \times 23 + 33 \times 40 + 20 \times 55}{10 + 13 + 18 + 15 + 26 + 25 + 20 + 23 + 40 + 55}$$

$= 25.86$ cm

Sample Calculation 2.6

Estimate the average rainfall over an area with the following data:

Rainfall between isohyets, cm	Area, ha	Rainfall between isohyets, cm	Area, ha
125 - 130	20	140 – 145	22
130 - 135	25	145 – 150	35
135 - 140	28	150 - 155	40

Solution

Average rainfall by isohyetal method is:

Average rainfall between isohyets, P (cm)	Area, A (Ha)	Product ($P.A$)
$(125 + 130)/2 = 127.5$	20	2550
$(130 + 135)/2 = 132.5$	25	3312.5
$(135 + 140)/2 = 137.5$	28	3850
$(140 + 145)/2 = 142.5$	22	3135
$(145 + 150)/2 = 147.5$	35	5162.5
$(150 + 155)/2 = 152.5$	40	6100
Total	170	24110

$$P = \frac{\sum_{i=1}^{n} P_i A_i}{\sum_{i=1}^{n} A_i} = 24110 / 170 = 141.82 \text{ cm}$$

2.15 Selection of Suitable Methods for Computing Mean Areal Depth of Rainfall

Selection of suitable method among the three as outlined above depends on the following three factors:

- Raingauge network of the catchment
- Catchment size and
- Topographical features of the catchment

2.15.1 Raingauge Network of the Catchment

For the catchment with dense raingauge network, the isohyetal and Theissen polygon methods work the best to compute the mean areal depth of rainfall. However, with limited raingauge networks, arithmetic mean method and Theissen polygon methods perform the best to compute the mean areal depth of rainfall.

2.15.2 Catchment Size

If the area of the catchment is less than 500 sq. km, arithmetic mean method works the best to compute the mean areal depth of rainfall. For catchment of area 500 to 5000 sq. km, Theissen polygon method and for area greater than 5000 sq. km, isohyetal method works the best to compute the mean areal depth of rainfall.

2.15.3 Topography of the Area

For mountainous regions, Theissen polygon method is suitable. For plain land, arithmetic mean method and for hilly and rugged terrain lands, isohyetal method is suitable.

2.16 Test of Consistency of Record

Inconsistency in recoding of rainfall in a raingauge station would occur if there are some significant changes that have undergone during the periods of data collection. This inconsistency would be felt from the time the significant change took place. Some of the common causes of inconsistency of record are:

- Shifting of a raingauge station to a new place
- There are some marked changes in the adjacent stations
- Change in the ecosystem due to calamities like landslides, forest fire etc.
- Errors in collection of data from a certain date

The checking for inconsistency of a record is done by the *double-mass curve technique*. This technique is based on the principle that when each recorded data comes from the same parent population, they are consistent.

In this technique, a group of 5 to 10 base stations in the neighbourhood of the station which has inconsistency (say this station be X) is selected. Data of annual rainfall or mean monthly rainfall of the station X and also the average rainfall of the group of neighbouring base stations covering a long period is arranged in the reverse chronological order i.e. the latest record as the first entry and the oldest record as the last entry in the list. The accumulated rainfall of the station X (i.e.) and the accumulated values of the average of the group

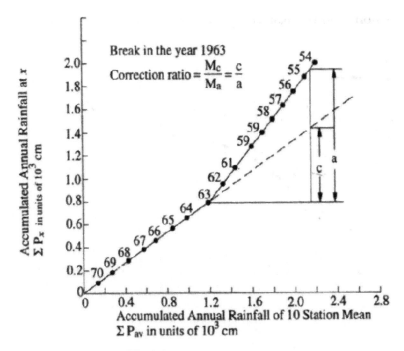

Fig. 2.8: Double- mass curve

of base stations (i.e.) are computed starting from the latest record. Values of are plotted against for various consecutive time periods (Fig. 2.8).

A marked break in the slope of the resulting plot indicates a change in the rainfall regime of station X. The rainfall values at station X beyond the period of change of regime (point 63 in Fig. 2.8) is corrected by using the relation:

$$P_{cx} = P_x \cdot \frac{M_c}{M_a} \qquad (2.11)$$

where, P_{cx} = corrected rainfall at any time period t_I at station X, P_x = original recorded rainfall at time period t_I at station X, M_c = corrected slope of the double mass curve and M_a = original slope of the double mass curve.

In this way, the older records are brought up to the new regime of the station. It is clear that the more homogeneous the base station records are the more accurate will be the corrected values at station X. A change in slope is normally taken as significant only where it persists for more than 5 years. The double mass curve is also helpful in checking arithmetical errors in transferring rainfall data from one record to another.

2.17 Presentation of Rainfall Data

A few commonly used methods to present the rainfall data that are helpful in interpretation and analysis are given below:

2.17.1 Mass Curve of Rainfall

The mass curve of rainfall is a graphical presentation of cumulative rainfall against time plotted in chronological order. Record of floating type and weighing bucket type gauges are of this form. Fig. 2.9 represents the view of a typical mass curve. The curve rises steeply in the beginning and then tends to be constant after some time. The mass curves are helpful in extracting information regarding duration of the storm and magnitude of the storm. Further, the intensity of the storm at any time interval in a storm can be obtained from the slope of the curve.

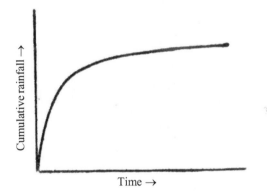

Fig. 2.9: Mass curve of rainfall

2.17.2 Hyetograph

A hyetograph is a graphical presentation of rainfall intensities against time. The hyetograph is derived from the mass curve and is commonly represented as a bar chart as shown in Fig. 2.10. It is a convenient way to represent the characteristics of storm and is particularly important in the development of design storms to predict extreme floods. The area under a hyetograph represents the total amount of rainfall received during the period. Sometimes the depth of rainfall is plotted against time (depth being taken as ordinate and time in abscissa) in bar chart form which is demonstrated in Fig. 2.12 with a sample problem (Sample calculation 2.7) as outlined below.

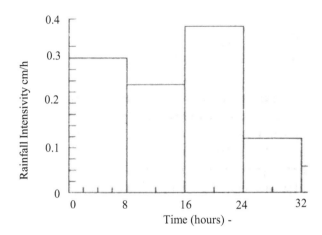

Fig. 2.10: Hyetograph of a rainfall event

2.17.3 Point Rainfall

Point rainfall is also called as station rainfall. It refers to the rainfall data collected at a station. It may be daily, weekly, monthly, seasonal and annual form. It is represented in a graph in point form taking rainfall in ordinate and time in abscissa. Such a presentation is also called as line diagram. A sample problem as presented below (Sample calculation 2.7) represents the point or line form (Fig. 2.11) presentation of rainfall data of a station.

2.17.4 Ordinate Form

Sometimes, the rainfall data are also presented in ordinate form by taking rainfall in ordinate and time in abscissa. But in this case the rainfall data are not joined point to point to form a line graph, but they are joined as ordinate line: each ordinate line (line drawn perpendicular to abscissa) represents the rainfall amount recorded for a particular year.

Sample Calculation 2.7

Following rainfall data of a rain gauge station recorded in various years are presented below. Present the data in (i) point form and (ii) bar chart.

Year	2000	2001	2002	2003	2004	2005	2006	2007	2008
Rainfall, cm	109	112	100	98	125	130	110	120	115

Solution

The presentation of rainfall data in point form and bar chart are shown in Figs 2.11 and 2.12, respectively.

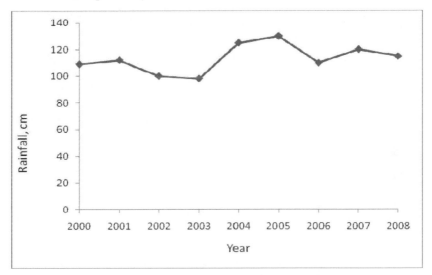

Fig. 2.11: Presentation of rainfall in point form

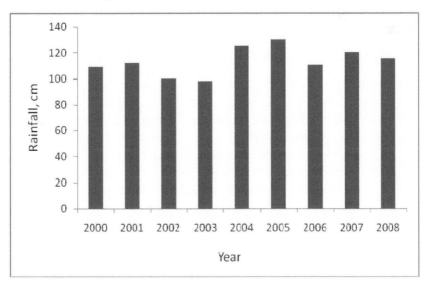

Fig. 2.12: Presentation of rainfall in bar chart

2.18 Depth-Area Relationship

Depth-area relationship is important for determining the variations of rainfall depth with respect to the variations of area of watershed during occurrence of a given storm. Horton developed a mathematical model which indicates that for a rainfall of given duration, the average depth of rainfall decreases with area of the watershed in exponential form as:

$$\overline{P} = P_o\, e^{-K\,A^n} \tag{2.12}$$

where, $\overline{P}$ = average depth of rainfall, cm; P_o = highest amount of rainfall, cm at the storm centre, A= area of the watershed, km^2 and K and n are constants for a given region.

Dhar and Bhattacharya (1975) determined the values of K and n for different durations for North India based on the observations of 42 sever most storms which is presented in Table 2.1 below.

Table 2.1 Values of K and n for different durations

Duration	Value of K	Value of n
1 day	8.256×10^{-4}	6.614×10^{-1}
2 days	9.877×10^{-4}	6.306×10^{-1}
3 days	1.745×10^{-3}	5.961×10^{-1}

In the analysis of large area storms, the highest station rainfall is taken as the average depth of rainfall over an area of 25 km^2.

2.19 Maximum Depth-Area-Duration Relationship

In many hydrological studies involving estimation of severe floods, it is necessary to have information on the maximum amount of rainfall of various durations occurring over various sizes of areas of the catchments. The development of relationship between maximum depth-area-duration for a given region is called as DAD analysis which is an important part in the field of hydro-meteorological studies. DAD curves prepared by DAD analysis of rainfall as described below are helpful in developing design storms for use in computing the design flood in the hydrological design of structures such as dam. Derivation of DAD relationship is carried out by plotting the progressively decreasing average rainfall depth over progressively increasing catchment area from the centre of the storm leading to its boundary. For this purpose, isohyetal maps and mass curves are required.

First, the severemost rain storms that have occurred in the region under question are considered. Isohyetal maps and mass curves of the storms are compiled. For a given duration, a depth-area curve of the storm is prepared. Then from a study of the mass curve of rainfall, various durations and maximum depth of rainfall in these durations are noted. The maximum depth-area-duration curve for a given duration D is prepared by assuming the area distribution of rainfall for a smaller duration to be similar to the total storm. The procedure is then repeated for different storms and the envelope curve of maximum depth-area for duration D is obtained. A similar procedure for various values of D results in a family of envelope curves of maximum depth vs. area with duration as the third parameter. These curves are called as DAD curves. Fig. 2.13 represents typical DAD curves of a catchment. In these curves of Fig. 2.13, the average depth denotes the depth averaged over the area under consideration. Study of the DAD curves reveals that maximum depth of rainfall for given duration of storm event decreases with increase in area of the catchment. Similarly for a given catchment with constant area, the maximum depth of rainfall increases with increase in the duration of the storms and vice versa.

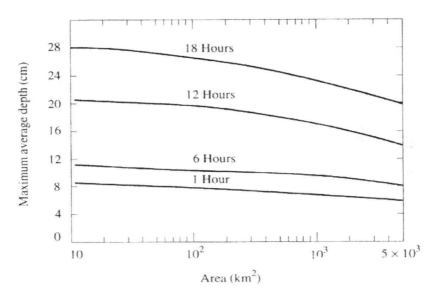

Fig. 2.13: Figure of a typical DAD curve

2.20 Intensity-Duration Relationship

Rainfall intensity is defined as the rate of fall of rainfall. Recording raingauge measures both rainfall and duration of rainfall in the form of a chart which is mass curve. Mass curve helps to find out the intensity of rainfall for any desired time period. In case of a non-recording raingauge, only total depth of rainfall over a particular time period, generally a day, is measured. From the known depth of rainfall recorded in a given time period, intensity is computed by dividing the depth of rainfall with duration. Thus, in a day if 100 cm rainfall is recorded by a non-recording raingauge, then the average intensity of rainfall is 100 cm/day which is equal to 4.7 cm/hr. This 100 cm rainfall might have occurred in 2 hours say 8 to 10 AM whereas in rest 22 hours no rainfall is recorded. Then, intensity of rainfall between 8 to 10 AM is 50 cm/hr and in rest of the time of that day intensity is zero. Thus, a non-recording raingauge gives vague value of intensity of the whole day. For analysis of hydrological problems, one should use the intensity taking rainfall data of a recording type raingauge.

Depending on the values of intensity there are three classifications:

- Light intensity – 2.5 mm/hr
- Moderate intensity – 2.5 to 7.5 mm/hr and
- Heavy intensity - > 7.5 mm/hr.

It is a general observation that the rainfall intensity is not same throughout the storm duration. Rather, it varies; may increase gradually and then decrease. For a longer storm duration, the intensity is less and vice-versa. For a short period, its value may be much higher than the average rainfall intensity of the storm. For computing the rainfall intensity at any time "t" during the storm, Richard (1944) developed a relationship between the intensity and duration of rainfall as:

$$\frac{i}{I} = \frac{T+K}{t+K} \tag{2.13}$$

where, i and I are the rainfall intensity for any time t and T, respectively; T is the storm duration and K is the constant. Except for extreme events, the value of K is taken as 1. Assuming value of K as 1, the intensity and duration of rainfall is expressed as:

$$\frac{i}{I} = \frac{T+1}{t+1} \tag{2.14}$$

Substituting intensity of rainfall, $i = P/T$ (P being depth of rainfall occurring in duration T) and rearranging the above equation we get:

$$i = \frac{P}{T}\left(\frac{T+1}{t+1}\right) \tag{2.15}$$

The above equation is used to compute intensity of rainfall (i) for any time (t) during the rainfall if the storm duration (T) and a average rainfall intensity ($I = P/T$) of the given storm are known. Tejwani et al. (1975) developed a graphical relationship between one hour rainfall intensity and other durations rainfall intensities which is discussed in Chapter 3.

2.21 Intensity-Duration-Frequency Relationship

As discussed earlier, the intensity of a storm decreases with increase in duration of the storm. Further, a storm of any given duration will have a larger intensity if its return period is large. In other words, for a storm of given duration, storms of higher intensity in that duration are rarer than storms of smaller intensity. The relationship amongst intensity (i in cm/hr), duration (t in hr) and return period (T in years) is commonly expressed as:

$$i = \frac{K\,T^a}{(t+b)^d} \qquad\qquad (2.16)$$

where, K, a, b and d are constants depending on the geographical locations of the area.

Variation of intensity i with duration t is shown in Fig. 2.14. Depth-duration-frequency curves are presented in Fig. 2.15.

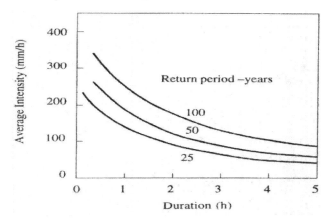

Fig. 2.14: Intensity-duration-frequency curves of rainfall event

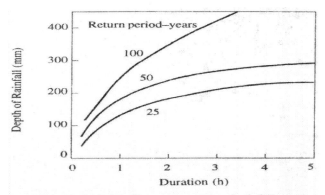

Fig. 2.15: Depth-duration-frequency curves of rainfall event

Tejwani *et al.* (1975) have determined values of K, a, b and d as 43.26, 0.17, 4 and 0.686, respectively for Oottacamund region of India and for other places, values of K, a, b and d are mentioned in Table 2.2.

2.22 Frequency of Point Rainfall

In many hydraulic engineering applications such as those concerned with floods, probability of occurrence of a particular extreme rainfall i.e. a 24-hr maximum rainfall, are study of importance. Such information is obtained by frequency analysis of point rainfall data. Rainfall is a spatial-temporal variable factor and hence is called stochastic event. At a given place, it varies from year to year constituting a time series. One of the common examples of time series is the annual series comprising of annual rainfall values over several years. Extreme events of the series in a year also constitute an annual series. For example if the maximum one day rainfall data of each year is chosen and data are collected for several years, then the one day maximum rainfall of each year constitute an annual series. The probability of occurrence of an event in this series is studied by frequency analysis of this annual data series. In this section, a brief description of the terminologies and a simple method of estimation of the frequency of an event is discussed.

Table 2.2: Values of K, a, b and n for different places of India to compute intensity-duration-frequency

Zone	Station	K	a	b	n
Northern Zone	Agra	4.911	0.1667	0.25	0.6293
	Allahabad	8.570	0.1692	0.50	1.0190
	Amritsar	14.41	0.1304	1.40	1.2963
	Dehradun	6.00	0.22	0.50	0.8000
	Jaipur	6.291	0.1026	0.50	1.1172
	Jodhpur	4.098	0.1677	0.50	1.0369
	Lucknow	6.074	0.1813	0.50	1.0331
	New Delhi	5.208	0.1574	0.50	1.1072
	Srinagar	1.503	0.2730	0.25	1.0636
	Northern Zone	5.914	0.1623	0.50	1.0127
Southern Zone	Bangalore	6.275	0.1262	0.50	1.1280
	Hyderabad	5.250	0.1354	0.50	1.0295
	Kodaikanal	5.914	0.1711	0.50	1.0086
	Madras	6.126	0.1664	0.50	0.8027
	Mangalore	6.744	0.1395	0.50	0.9374
	Tiruchirapalli	7.135	0.1638	0.50	0.9624
	Trivandrum	6.762	0.1536	0.50	0.8158

Contd.

Zone	Station	K	a	b	n
	Visakhapatnam	6.646	0.1692	0.50	0.9963
	Southern Zone	6.311	0.1523	0.50	0.9465
Eastern Zone	Agarthala	8.097	0.1177	0.50	0.8191
	Dumdum	5.940	0.1150	0.15	0.9241
	Gauhati	7.206	0.1557	0.75	0.9401
	Gaya	7.176	0.1483	0.50	0.9459
	Imphal	4.939	0.1340	0.50	0.9719
	Jamshedpur	6.930	0.1307	0.50	9.8737
	Jharsuguda	8.596	0.1392	0.75	0.8740
	North Lakhimpur	14.070	0.1256	1.25	1.0730
	Sagar Island	16.524	0.1402	1.50	0.9635
	Shillong	6.728	0.1502	0.75	0.9575
	Eastern Zone	6.974	0.1353	0.50	0.8801

The relation between the return period or called as recurrence interval or simply called as frequency *(T)* and probability of exceedance *(P)* is given as:

$$T = \frac{1}{P}$$
(2.17)

where, *P* is the probability of occurrence of an event i.e. rainfall or stream flow or any random hydrological variable whose magnitude is equal to or exceed a specified magnitude say *X*. This represents the average interval between the occurrence of a hydrologic event of magnitude equal to or greater than *X*. If it is stated that the return period of rainfall of 10 cm in a day is 20 years at any station, it implies that on an average rainfall magnitudes equal to or greater than 10 cm in a day occur once in 20 years. In a long period of say 100 years, 5 such events are likely to occur. However, it does not mean that every 20 years, one such event is likely to occur. The probability of occurrence of rainfall of 10 cm in a day in any one year at the station is $1/T = 1/20 = 0.05 = 5\%$.

It is to be noted that probability of an event has maximum value 1 i.e. 100%. If probability of exceedance or occurrence of an event is *P* then its probability of non-exceedance or non-occurrence is given by symbol *q* which is equal to *1 − P*. The binomial distribution used to find the probability of occurrence of the event *r* times in n successive years is given as:

$$P_{r,n} = {}^{n}C_{r} \, P^{r} \, q^{n-r} = \frac{n!}{(n-r)!\,r!} \, P^{r} \, q^{n-r}$$
(2.18)

where, $P_{r,n}$ = probability of a random hydrologic event (rainfall) of given magnitude and exceedence probability *P* occurring *r* times in *n* successive years. Thus, for example,

(i)   Probability of an event of exceedence probability P occurring 3 times in 10 successive years is:

$$P\ 3,10 = \frac{10!}{(10-3)!3!} P^3\ q^{10-3} = \frac{10!}{7!3!} P^3\ q^7$$

(ii)   Probability of an event not occurring any time in 10 successive years is:

$$P_{0,\ 10} = q^{10} = (1-P)^{10}$$

(iii)   Probability of an event occurring at least once in 10 successive years is:

$$P_1 = 1 - q^{10} = 1 - (1 - P)^{10}$$

Sample Calculation 2.8

One day maximum rainfall data of a raingauge station at Bhubaneswar was analysed and it was found that a value of 250 mm of one day maximum rainfall occurred at return period of 50 years. Calculate the probability of one day maximum rainfall equal to or greater than 250 mm occurring (i) once in 20 successive years (ii) 3 times in 15 successive years and (iii) at least once in 20 successive years.

Solution

Given return period = 50 years and so probability, $P = 1/50 = 0.02$

(i) Given $n = 20$ and $r = 1$

Hence, using Eqn. (2.18) we get probability of one day maximum rainfall equal to or greater than 250 mm occurring once in 20 successive years as:

$$P_{1,\ 20} = \frac{20!}{19!1!} x\ 0.02\ x\ (0.98)^{19} = 0.272$$

(ii) Given $n = 15$ and $r = 3$

Using Eqn. (2.18) we get probability of one day maximum rainfall equal to or greater than 250 mm occurring 3 times in 15 successive years as:

$$P_{3,\ 15} = \frac{15!}{12!3!} x\ 0.02^3\ x\ (0.98)^{12} = 0.0029$$

(iii)   Probability of one day maximum rainfall equal to or greater than 250 mm occurring at least once in 20 successive years is:

$$P_1 = 1 - q^{20} = 1 - (1 - P)^{20} = 1 - (1-0.02)^{20} = 0.332$$

2.22.1 Plotting Position

The purpose of frequency analysis of an annual series is to obtain a relation between the magnitude of the event and its probability of exceedenec. The probability analysis may be made either by empirical or by analytical method. A simple empirical method is given by plotting position formula given by Weibull called as also Weibull's formula described as:

$$P\ (\%) = \frac{m}{N+1} \times 100 \tag{2.19}$$

where, P = probability of an event equal to or exceeded in percent, m = rank number of the event after arranging the events in descending order of their magnitudes and N is the total number of years of recorded data.

The recurrence interval (T) now can be written as

$$T = \frac{N+1}{m} \tag{2.20}$$

It is to be remembered that Eqn. (2.20) is an empirical formula given by Weibull and there are other empirical formulae or plotting position formulae also available to compute P. Table 7.2 (Chapter 7) gives list of different plotting formulae to compute P.

Procedure for determining the probability of an event (maximum value of an event) by Weibull's formula is explained as below:

- At first the rainfall or any hydrologic data are arranged in decreasing order of their magnitudes.

- Rank number (m) is then allotted to each data with rank number 1 for the first data which is the highest value.

- Compute plotting position in percent of each rainfall event by using Eqn. (2.19). Eqn. (2.20) can also be used to compute the rainfall at different return periods.

- When two or more events have the same magnitude, then plotting position is calculated for the largest m value of the series.

- Plot the plotting positions or return periods in abscissa and rainfall in ordinate in a semi-log graph paper with rainfall in arithmetic scale and plotting position and/or return period in log scale.

- Now join all the plotted points by a smooth line. The obtained curve is called as rainfall frequency curve

By using the rainfall frequency curve, one can calculate the expected rainfall at any probability or return period by suitable extrapolation. The above mentioned empirical procedure can give good results for small extrapolations and the errors increase with the amount of extrapolation. For accurate result analytical methods like Gumbel's extreme value distribution and Log Pearson Type III method are commonly used and are discussed in subsequent chapters.

Sample Calculation 2.9

For a station A, annual 24 hour maximum rainfall are mentioned below. Develop a rainfall frequency curve and (i) Estimate the 24 hour maximum rainfall with return period of 15 and 50 years. (ii) What would be the probability of a rainfall of magnitude equal to or exceeding 10 cm occurring in 24 hour at a station A?

Year	1980	1981	1982	1983	1984	1985	1986	1987	1988	1989	1990
Rainfall, cm	13.0	14.2	16.0	12.1	12.4	10.8	11.2	10.2	10.6	9.6	9.5
Year	1991	1992	1993	1994	1995	1996	1997	1998	1999	2000	2001
Rainfall, cm	8.5	8.9	8.9	9.0	8.4	6.0	8.0	7.5	7.7	7.8	8.3

Solution: The data are arranged in decreasing order of their magnitude and probability and recurrence interval/return period of various events are computed by formulae (Eqns. 2.19 and 2.20, respectively) as indicted in Table 2.3 below.

There are 24 years of rainfall data which shows $N = 24$.

For rank number (m) 13 and 14, the two events have same magnitude of 8.9 cm. So the plotting position, P and return period, T are calculated for the largest rank number $i.e.$ rank number 14.

Now a graph is plotted between the rainfall (taken in ordinate) and return period, T (taken in log scale) in a semi-log graph paper as shown in Fig. 2.16. A smooth curve is drawn through the plotted points and then it is extended by judgment.

(i)   From the curve we get at rerun period of 15 year and 50 year, the 24 hour maximum rainfall as 14.95 and 18.00 cm, respectively.

(ii) From the curve for a rainfall of 10 cm, we get the return period as 2.4years which gives probability (P) value as 0.417= 41.7%.

It is to be remembered that to predict the probability of occurrence of a minimum value of an event by Weibull's formula, the various steps mentioned as above for prediction of a maximum value of the event by Weibull's formula remain same except that in this case the rainfall or any hydrologic data are arranged in increasing order of their magnitudes.

Table 2.3 Calculation of probability (plotting position) and return period for data of Sample calculation 2.9

Rank number, m	Rainfall, cm	Probability, P, %	Return period, T, years	Rank number, m	Rainfall, cm	Probability, P, %	Return period, T, years
1	16.0	4.3	23.00	12	9.0	52.2	1.92
2	14.2	8.7	11.50	13	8.9	-	-
3	13.0	13.0	7.67	14	8.9	60.9	1.64
4	12.4	17.4	5.75	15	8.5	65.2	1.53
5	12.1	21.7	4.60	16	8.4	69.6	1.44
6	11.2	26.1	3.83	17	8.3	73.9	1.35
7	10.8	30.4	3.29	18	8.0	78.3	1.28
8	10.6	34.8	2.88	19	7.8	82.6	1.21
9	10.2	39.1	2.56	20	7.7	87.0	1.15
10	9.6	43.5	2.30	21	7.5	91.3	1.10
11	9.5	47.8	2.09	22	6.0	95.7	1.05

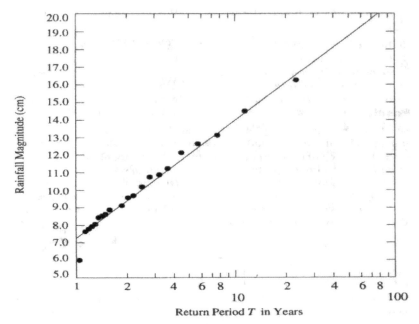

Fig. 2.16: Rainfall frequency curve

2.23 Probable Maximum Precipitation

In the design of major hydraulic structures such as spillway in large dams, it is the usual practice to keep the failure probability as low as possible *i.e.* virtually zero. This is done to avoid the failure of the structure to avert the heavy loss caused to life, property and economy. In such context, maximum possible precipitation that is probably to occur at a given location is to be considered for the design of the hydraulic structures. The probable maximum precipitation (*PMP*) is defined as the greatest or extreme rainfall for a given duration that is physically possible over a station or basin. From the operational point of view, *PMP* can be defined as that rainfall over a basin which would produce a flood flow with virtually no risk of being exceeded.

Computation of *PMP* is done based on two methods *i.e.* meteorological method and statistical study of rainfall data. Details of meteorological method that use storm models to compute *PMP* are available in reference Weisner (1970).

Computation of *PMP* based on the statistical study of rainfall data is done by the formula:

$$PMP = \overline{P} + K\sigma \tag{2.21}$$

where, $\overline{P}$ = mean of annual maximum rainfall series, σ = standard deviation of the series, and K = frequency factor which depends on the statistical distribution of the series, number of years of record and the value of return period. Value of *PMP* for North-Indian plains varies from 37 to 100 cm for 1 day rainfall (Suresh, 2008). The Indian Institute of Tropical Meteorology (1989) has published *PMP* atlas for India which shows 1 day *PMP* for various parts of India.

2.24 Normal Rainfall

Sometimes normal rainfall is of importance to the hydrologist. Normal rainfall of a year refers to the mean rainfall of consecutive last 30 years rainfall just preceding to it. Suppose we want normal annual rainfall of the year 2000. Then we have to find out the mean of 30 years annual rainfall starting from 1999 to 1970 which will give the normal annual rainfall of 2000.

2.25 Snow Contribution

Snow remaining on the surface of the earth contributes large amount of water. A snowfall of 10 cm with 10 cm over an area of 10 square kilometer will provide storage of 10,000 hectare-centimetre of water which will fill a surface reservoir of 100 hectare to a depth of 1 m. In cold countries, snow contributes major share to total water supply. In India, the rivers originating from the Himalayas including the Ganges carry large amount of water in summer because of melting of snow. About 90 percent of the yearly water supply in the high elevations of the Colorado Rockies of USA is derived from snowfall. It is important that the hydrologist understand the nature and distribution of snowfall and the mechanisms involved in the snow melt process. Estimation of snow melt water increases the total water availability and hence accordingly the stream flow is enhanced which helps in deciding the cropping plan and regional scale irrigation and drainage planning. The water yield from snowfall can be increased by increasing the evaporation of snow. An adequate understanding of meteorological factors is as much a prerequisite in considering the snow melt process as it is in dealing with evapotranspiration. Geographic, geologic, topographic, and vegetative factors affect the snowmelt runoff process. As in rainfall-runoff relationships, point rainfall measures are used in estimating aerial and temporal distribution over a basin, similar approach is considered in snow hydrology although the point areal relationships are more complex. Mathematical equations can be used to determine the various components of snowmelt at a given location.

Under the usual conditions encountered in regions with heavy snowfall, runoff from the snowmelt is the last occurrence of series of events beginning when snowfall reaches the earth. The time interval from the start to the end of the process may vary from a day or less to a period over a month. The temperature in a deep layer of accumulated snow is often below the freezing point after prolonged cold periods. When the summer approaches, melting occurs first at the snow pack surface. This initial melt water moves slightly below the surface and again freezes when comes in contact with the colder underlying snow. During the refreezing process, the heat fusion released from the melt water raises the snow pack temperature. Sometimes the heat of the overlying atmosphere and ground surface is transmitted to the snow pack resulting rise of temperature and hence enhancing snow melt process. When the melt rate exceeds the infiltration capacity of the soil, the water from snow melt enters into the soil at a rate of infiltration capacity. Excess water after infiltration flows over the soil surface as runoff.

2.25.1 Snow Measurement and Surveys

Snow measurements are done through the use of standard and recording rain gauges, seasonal storage precipitation gauges, and snow stakes. Rain gauges are equipped with shields to reduce the effect of wind. Snow stakes are calibrated wooden posts driven into the ground for periodic observation of the snow depth or are inserted into the snow pack to determine its depth. While estimating the average depth of snow, it is advisable to measure the snow depths at several locations in a large area so that erroneous measurement due to drifting or blowing can be avoided. Snow surveys help to overcome this problem. Snow surveys help to provide information on the snow depth, water equivalent, density and quality at various points along a snow course. A snow course includes a series of sampling locations, normaly more than 10 in number. The various stations are spaced about 50 to 100 ft apart in a geometric pattern designed in advance. Points are permanently marked so that the same locations will be surveyed each year. Survey data are obtained directly by foresters and others, by areal photographs and observations and also by the automatic recording stations that sends information by telemetry to a central processing location.

2.25.2 Estimates of Areal Distribution of Snowfall

Because of orographic and topographic effects, arithmetic average and Thiessen polygons do not provide reliable information for estimating areal distribution of snowfall. However, regional orographic effects are relatively constant for tracts small when compared with the arreal extent of general storms occurring in the region. This factor helps in estimating the areal snow distribution once the basin pattern has been found for a region. One method used to estimate basin precipitation (P_b) from point precipitation observations (P_s) assumes that the ratio of station precipitation to basin precipitation is approximately constant for storms. This is stated as:

$$\frac{P_b}{P_s} \equiv \frac{N_b}{N_s} \qquad\qquad (2.22)$$

where P_b is basin precipitation, Ps is the observed precipitation measured at any station, N_b is normal annual precipitation over the basin, and N_s is the normal annual precipitation for the station.

The normal annual precipitation is determined from a map displaying the mean annual isohyets for the region. The precipitation is determined by planimetering areas between the isohyets. It is to be kept in my that if the stations are not uniformly distributed in a basin, then weighting coefficients based on the percentage of the basin area portrayed by a gauge are sometimes used in determining N_s for the group.

2.25.3 Generalised Basin Snowmelt Equations

Extensive studies by the U.S. Army Corps of Engineering at various laboratories in the West have produced several equations for snowmelt during (1) rain-free periods, and (2) periods of rain. When rain is falling, heat transfer by convection and condensation is of prime importance. Solar radiation is slight and long wave radiation can be readily determined from theoretical considerations. When rain free periods prevail, both solar and terrestrial radiation become significant and may required direct equation. Convection and condensation are usually less critical during rainless intervals. The equations are summarized as follows (Viessman *et al.,* 1977):

1) Equations for periods with rainfall

a) For open (cover below 10%) or partly forested (cover from 10 to 60%) watersheds,

$$M = (0.029 + 0.0084\ ku + 0.007\ P_r)\ (T_a - 32) + 0.09 \qquad (2.23)$$

b) For heavily forested areas (over 80% cover)

$$M = (0.074 + 0.007\ P_r)\ (Ta - 32) + 0.05 \qquad (2.24)$$

where, M is the daily snowmelt (in. /day), P_r is the rainfall intensity (in. /day), T_a is the temperature of saturated air at 10 ft level (°F), u is the average wind at 50 ft level (mph), and k is the basin constant, which includes forest and topographic effects, and represents average exposure of the area to wind: values decrease from 1.0 for clear plains areas to about 0.3 for dense forest.

2) Equations for periods with no rainfall

a) For heavy forested areas

$$M = 0.074 \; (0.53 \; T'_a + 0.47T'_d \tag{2.25}$$

b) For forested areas (cover of 60 to 80%):

$$M = k \; (0.0084u)(0.22T'_a + 0.78 \; T'_d) + 0.029 \; T_a \tag{2.26}$$

c) For partly forested areas:

$$M = k' \; (1\text{-}F) \; (0.0040 \; I) \; (1\text{-} \; a) + k \; (0.0084u) \; (0.22T'_a + 0.78T'_{d)} + F \; (0.029 \; T'_d) \tag{2.27}$$

d) For open areas

$$M = k' \; (0.00508I) \; (1\text{-}a) + (1\text{-}N) \; (0.0212 \; T'_a - 0.84) + N \; (0.029T'_c) + k \; (0.0084u) \; (0.22 \; T'_a + 0.78T'_d) \tag{2.28}$$

where, T'_a is the difference between the 10 ft air and the snow surface (°F) temperatures, T'_d is the difference between the 10 ft dew point and the snow surface (°F) temperatures, I_i is the observed or estimated insolation (langleys), a is the observed or estimated mean snow surface albedo, k' is the basin short wave radiation melt factor (0.9-1.1) which is related to mean exposure of open areas compare to an unshielded horizontal surface, F is the mean basin forest canopy cover(decimal fraction), T'_c is the difference between cloud base and snow surface temperature (°F), and N is the estimated cloud cover (decimal fraction).

Question Banks

Q1.   What is hydrologic cycle? With a labeled diagram, describe the different parameters of the hydrologic cycle.

Q2.   What is engineering hydrology? What forms the basic part of study of engineering hydrology?

Q3.   How is the hydrologic cycle helpful for the study of hydrology?

Q4.   Define the followings:

   (i) Precipitation (ii) Evaporation (iii) Transpiration (iv) Runoff (v) Infiltration (vi) Percolation

Q5.   What are the differences between precipitation and rainfall?

Q6.   What are the different forms of precipitation? Which form of precipitation is important and why?

Q7.   Define the followings:

(i) drizzle (ii) sleet (iii) glaze (iv) hail (v) dew (vi) frost (ix) fog and (x) mist

Q8.   What is interflow? How does it help as a source to irrigation water?

Q9.   Write a mathematical equation for hydrologic cycle and explain the various terms of it.

Q10. Which component of the hydrologic cycle is very important in study of water resources and hydrology? How is it helpful in the fields of hydrolygy and water resources engineering?

Q11. Describe the mechanism for the occurrence of precipitation.

Q12. What are the different types of precipitation?

Q13. What is average annual rainfall? Describe the variation of average annual rainfall across the country.

Q14. Rainfall in India is seasonal. Justify the statement.

Q15. Write notes on (i) Recording type rain gauges and (ii) Non-recording type rain gauges.

Q16. What do you mean by mass curve of rainfall? How is it helpful in analysis of rainfall records? With a diagram explain how is it prepared?

Q17. What are the different types of recording rain gauges? Write short notes on each of them with labeled diagram.

Q18. What are the points to be considered while selecting sites for setting rain gauges?

Q19. Describe the different methods to compute the average rainfall of a place.

Q20. The annual rainfall of a gauging site is measured for 9 years and the data are presented in the tabular from as given below. The rainfall data for one year is missing. Find the missing data if the average annual rainfall of the place is 1250 mm.

Year	1999	2000	2001	2002	2003	2004	2005	2006	2007
Annual rainfal Rainfall, mm	1200	1340	1155	1205	1430	1250	-	1320	1190

Q21. Data of average depth of rainfall occurring in a basin of area 250 ha having 10 rain gauges are measured and given in table as below. The polygon area enclosing each rain gauge is also given in the table. However, rainfall, data of one gauge is missing. Find the missing data if the average depth of rainfall estimated by Thiessen method is 26 cm.

Station	A	B	C	D	E	F	G	H	I	J
Rainfall, cm	50	45	40	23	30	-	12	15	30	20
Polygon area, ha	13	15	18	15	26	25	20	23	40	55

Q22. Estimate average rainfall of a site by isohyetal method with the help of following recorded data.

Rainfall between isohyets, cm	120-130	130-140	140-150	150-160	160-170	170-180	180-190	190-200
Area enclosed between the isohyets, ha	22	24	30	23	25	33	37	18

Q23. Write notes on different characteristic features of rainfall in India.

Q24. Which type of monsoon gives maximum rainfall in India? Explain how it contributes rainfall in India.

Q25. What are the different advantages and disadvantages of recording raingauges?

Q26. How is the radar measurement of rainfall done?

Q27. Explain the working principle of automatic radio reporting raingauges.

Q28. What are the common errors in rainfall measurement? How can they be eliminated?

Q29. Describe how can the adequacy of raingauge station be calculated.

Q30. Determine the optimum number of rain gauges required to be installed in a catchment of 800 km² area. The catchment has 7 number of rain gauges and the annual rainfall of these 7 rain gauges are recorded as 850, 1000, 759, 500, 690, 550 and 400 mm, respectively. Assume the percent allowable error in estimate of the mean rainfall as 10%. Find out also the extra number of rain gauges required.

Q31. Enumerate the different procedures to estimate the missing rainfall data of a station.

Q32. There are four rain gauge stations located in a catchment which are uniformly spaced. The rainfall data on a particular day in three stations A, B and C are 108, 92 and 123 mm, respectively. On that day the station X was inoperative and so no data could be collected. The normal annual rainfall values of stations A, B, C and X are 1013, 945, 1000 and 986 mm, respectively. Estimate the missing rainfall of station X.

Q33. There are four rain gauge stations A, B, C and X located in a catchment which is uniformly spaced. The normal annual rainfall values of stations A, B, C and X are 1110, 955, 1202 and 1000 mm, respectively. The rainfall data on a particular day in three stations A, B and C are 108, 100 and 122 mm, respectively. On that day the station X was inoperative and so no data could be collected. Estimate the missing rainfall of station X.

Q34. There are two raingauge stations A and B . The station A has long term rainfall records but station B is a new station having limited periods of data being recorded. The rainfall data of these two stations as recorded are mentioned below. Estimate the long term mean rainfall of station B.

Year	1990	1991	1992	1993	1994	1995	1996	1997
Rainfall of station A (cm)	80	127	108	130	55	85	85	90
Rainfall of station B (cm)	-	-	-	-	-	-	-	75

Year	1998	1999	2000	2001	2002	2003	2004	2005
Rainfall of station A (cm)	75	65	85	105	135	65	100	105
Rainfall of station B (cm)	105	120	75	85	75	100	133	155

Q35. Describe how the selection of suitable site for computation of means areal depth of rainfall is done.

Q36. How can the consistency of rainfall record be checked?

Q37. With neat diagram, enumerate the different ways of presentation of ainfalldata.

Q38. Following rainfall data of a raingauge station recorded in various years are presented below. Present the data in (i) point form and (ii) bar chart.

Year	2000	2001	2002	2003	2004	2005	2006	2007	2008	2009
Rainfall, cm	117	115	100	96	125	130	110	120	115	122

Q39. With neat diagram explain the followings:

    (i)   Depth area relationship

    (ii) Maximum depth-area-duration relationship

    (iii) Intensity duration relationship and

    (iv) Intensity-duration-frequency relationship

Q40. Define probability of exceedence and recurrence interval. What is the relationship between them?

Q41. One day maximum rainfall data of a raingauge station at Sambalpur was analysed and it was found that a value of 220 mm of one day maximum rainfall occurred at return period of 50 years. Calculate the probability of one day maximum rainfall equal to or greater than 220 mm occurring (i) once in 15 successive years (ii) 2 times in 15 successive years and (iii) at least once in 15 successive years.

Q42. Describe with different steps how the probability of an event by Weibull's formula is determined.

Q43. For a station X, annual 24 hour maximum rainfall are mentioned below. Develop a rainfall frequency curve and (i) Estimate the 24 hour maximum rainfall with return period of 20 and 50 years. (ii) What would be the probability of a rainfall of magnitude equal to or exceeding 14 cm occurring in 24 hour at a station X?

Year	1980	1981	1982	1983	1984	1985	1986	1987	1988	1989
Rainfall, cm	13.5	14.9	16.5	12.1	12.4	10.8	11.2	10.2	10.6	9.6
Year	1990	1991	1992	1993	1994	1995	1996	1997	1998	1999
Rainfall, cm	8.5	8.9	8.9	9.0	8.4	6.0	8.0	7.5	7.7	7.9

Q44. Define probable maximum precipitation. How can it be estimated?

Q45. How does the snow contribute as a source to irrigation water?

Q46. Write notes on snow measurement and snow survey.

Q47. Explain the procedures to estimate areal distribution of snow fall.

Q48. Write down the different generalized basin snow melt equations and explain the terminologies used in these equations.

References

Dhar, O.N. and Bhattacharya, B.K. 1975. A study of depth area duration statistics of the severe most storms over different meteorological divisions of North India. Proceedings of National Symposium on Hydrology, Roorkee, India.

Eagleson, , P.S. 1970. Dynamic Hydrology. McGraw-Hill Book Company, New York.

Garg, S.K. 1987. Irrigation Engineering and Hydraulic Structures. Khanna Publishers, Delhi.

Indian Institute of Tropical Meteorology 1989. Probable Maximum Precipitation Atlas, IITM, Pune, India.

Linsley, R.K., Kohler, M.A. and Pauhus, J.L.H. 1949. Applied Hydrology. McGraw-Hill Publishing Co. Ltd., New York.

McCulloch, J.S.G. 1961. Statistical Assessment of Rainfall. COTA Publication No. 66, International African Conference on Hydrology, Nairobi.

Richard. 1944. The Flood Estimation and Control. Chapman and Hall, London.

Sharma, R.K., 1987. A Text book of Hydrology and Water Resources. Dhanpat Rai and Sons, Delhi.

Subramanya, K. 2003. Engineering Hydrology, Tata McGraw-Hill Publishing Co. Ltd., New Delhi.

Suresh, R. 2008. Watershed Hydrology. Standard Publishers Distributors, Delhi, pp. 692.

Tejwan, K.G. Gupta, S.K and Mohan, H.N. 1975. Soil and Water Conservation Research (1957 – 71), ICAR, New Delhi.

Viessman, W. Jr., Knapp, J.W., Lewis, G.L. and Harbaugh, T.E. 1977. Introduction to Hydrology. Thomas Y. Crowell Harper and Row Publishers, New York.

Weisner, C.J. 1970. Hydrometeorology, Chapman and Hall, London.

CHAPTER 3

Runoff

3.1 Runoff

Runoff is one of the important parameter of hydrologic cycle. It is defined as that portion of precipitation which is not absorbed by the deep soil strata but finds its way into the stream after meeting the demands of evapotranspiration, interception, infiltration, surface storage, and surface as well as channel detention. It includes surface runoff received into the channels after precipitation occurs, delayed runoff that enters the stream after passing through the land surface, and other delayed runoff from the temporarily detained pits, depressions, swamps or detained as snow fall. It is thus the flow collected from the drainage basin and appearing at the outlet of the basin. The runoff process is demonstrated in Fig. 3.1 as flow chart. Details of the runoff process are discussed as below.

The excess runoff moving over a basin to reach smaller channels is called as *overland flow*. Usually the length and depth of overland flow are small and the flow is in laminar range. Flows from several small channels join bigger channels and flows from these in turn combine to form a bigger stream. The bigger stream and thus the process continue till all the flows reach the catchment outlet. This overland flow together with the open channel flow passing in the streams of various grades constitutes the surface runoff.

A part of precipitation that infiltrates moves laterally through the upper crust of the soil is called as *interflow*. The interflow returns to the surface at some location away from the point of entry into the soil. *Interflow* is also called as *delayed runoff, through flow, storm seepage, subsurface storm flow,* or *quick return flow.* The contribution of *interflow* to runoff depends on the geologic strata in which it moves. If the strata are fairly pervious soil overlying

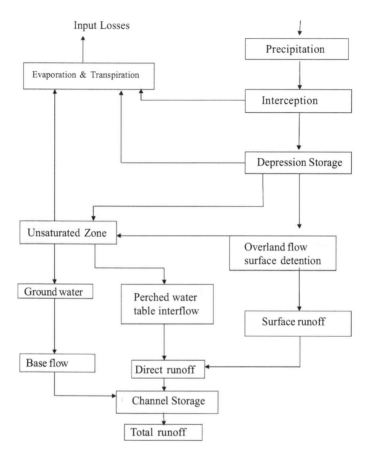

Fig. 3.1: Flow chart for computation of runoff

a hard impermeable surface, then the contribution of *interflow* to runoff will be significant.

Some portion of the infiltrated water after precipitation strikes the ground surface moves vertically down as deep percolation which enters the groundwater storage. The groundwater below the ground surface also moves depending on the available hydraulic gradient and ultimately reaches the stream. This is called as *base flow or simply groundwater flow*. Movement of groundwater is rather very slow and moves over a large and complicated path depending on the pores space of the subsoil strata. The time *lag i.e.* time difference between the entry of water into the soil and its appearance in the stream after moving as groundwater flow through a large complicated path is very large, being of the order of months and years. Groundwater flow provides the dry weather flow in perennial streams. Fig. 3.2 represents the different routes of runoff that reach a stream. It is to be noted that direct runoff is a term which refers to the part of the runoff that enters into the stream and includes surface runoff immediately after precipitation prompt interflow and precipitation on the channel surface. It may include snow melt water if it is there. Direct runoff is also called as *direct storm runoff* and *storm runoff.*

Runoff generated from a catchment is intervened frequently by human beings. Thus, it is very hard to get true runoff in the natural condition. A stream flow which is not intervened by works of human beings is called as *virgin flow.* Different storage works and diversion on a stream affects the virgin flow. The discharge diverted for different uses like industrial and agricultural uses and return flow from irrigation and other sources upstream of the gauging station need to be taken care in order to compute the true/virgin flow of the gauging station located at downstream. The *virgin flow* is given by Eqn. (3.1) (Subramanya, 2003) as:

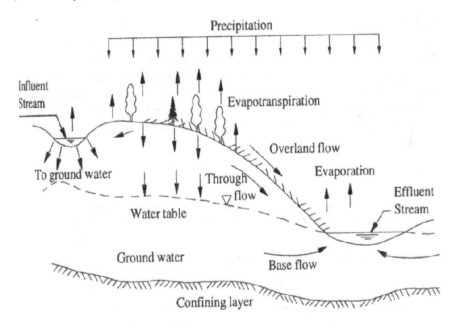

Fig. 3.2: Different components of runoff

$$R_v = V_s + V_d - V_r \qquad (3.1)$$

where, R_v = volume of *virgin flow,* V_s = volume of flow measured at the gauging station, V_d = volume of flow diverted from the stream at upstream of the gauging station and V_r = volume of return flow reaching the stream at upstream of the gauging station.

Sample Calculation 3.1 as given below shows computation of virgin flow of a stream gauging station.

Sample Calculation 3.1

The following table gives values of measured discharges at a stream gauging station in a year. At the upstream side of the gauging station there is weir through which 0.60 Mm³ of water is diverted per month for industrial uses and there is a return flow of 0.2 Mm³ per month. Estimate the *virgin flow.*

Month	Jan	Feb	Mar	Apr	May	Jun	Jul	Aug	Sep	Oct	Nov	Dec
Measured flow, Mm³	2.0	1.6	1.9	1.1	0.9	5.5	8.5	10.2	12.0	7.8	5.5	3.2

Solution

The *virgin flow* in a month is given by En. (3.1). The calculation is performed in the following tabular form:

Month	Volume measured at gauging station, Mm³	Volume diverted Mm³	Returned flow, Mm³	Virgin flow, Mm³
Jan	2.0	0.60	0.2	2.4
Feb	1.6	0.60	0.2	2.0
Mar	1.9	0.60	0.2	2.3
Apr	1.1	0.60	0.2	1.5
May	0.9	0.60	0.2	1.3
Jun	5.5	0.60	0.2	5.9
Jul	8.5	0.60	0.2	8.9
Aug	10.2	0.60	0.2	10.6
Sep	12.0	0.60	0.2	12.4
Oct	7.8	0.60	0.2	8.2
Nov	5.5	0.60	0.2	5.9
Dec	3.2	0.60	0.2	3.6
Total	60.2	7.2	2.4	65.0

Hence, total annual virgin flow is 65.0 Mm³.

3.2 Factors Affecting Runoff

Rate and volume of runoff from an area are influenced mainly by two factors i.e. (i) climatic factors and (ii) physiographic factors.

3.2.1 Climatic Factors

The various climatic factors that affect the runoff process are (Panigrahi, 2011):

(i)   Nature of precipitation

(ii)   Forms of precipitation

(iii) Interception

(iv)   Evaporation and

(v)   Transpiration

Nature of precipitation includes:

(i)   Rainfall intensity

(ii)   Frequency of intense rainfall

(iii) Duration of rainfall

(iv)   Distribution of rainfall and

(v)   Direction of storm movement

Surface runoff is produced when the rainfall intensity exceeds the infiltration capacity of the soil. Rainfall intensity influences both the rate and volume of runoff. Even though two storms have same total precipitation, an intense storm of short duration that exceeds the infiltration capacity of a soil at a faster rate produces greater volume of runoff than does a lesser intense storm of longer duration. The intensity-frequency relationship of a storm affects the runoff process in a way that an intense storm occurring frequently produces higher volume of runoff. Similarly a storm of given intensity produces higher amount of runoff if it occurs for a longer duration. In the initial stage of occurrence of a storm, the infiltration capacity may decrease with time and thus a storm of short duration may produce little runoff whereas a storm of longer duration but having same intensity will result in producing larger runoff. Distribution of rainfall also affects the rate and volume of runoff. When there is a higher intense storm occurring uniformly over the whole catchment, the entire watershed contributes and hence, the rate and volume of runoff becomes more. Further if there is an intense storm on one portion of the watershed, then it may produce more runoff than a moderate storm occurring over the entire watershed.

Forms of precipitation decide the magnitude of runoff. If there is a precipitation in the form of fog or dew etc having very low intensity, then the magnitude of runoff will be less compared to rainfall of high intensity. If the topography, soil and other conditions in the watershed are uniform, then for any given total volume of rain falling on the watershed, the more non-uniform the distribution is the greater is the peak runoff. The minimum peak is produced by a uniform distribution. Direction of storm movement also affects the peak rate of runoff. A storm moving in the direction of a stream produces a higher peak in a shorter period of runoff than a storm moving upstream. Interception, evaporation and transpiration depend on the meteorological parameters like temperature, humidity wind velocity and solar radiation and thus influence the runoff. More is the magnitude of these meteorological parameters, more will be losses due to interception, evaporation and transpiration and thus less will be the amount of runoff.

3.2.2 Physiographic Factors

The physiographic factors consist of both the watershed characteristics and channel characteristics. The watershed characteristics include:

(i)   Size of watershed

(ii)   Shape of watershed

(iii) Slope of watershed

(iv)   Orientation of watershed

(v)   Land use

(vi)   Soil moisture

(vii) Infiltration characteristics

(viii) Soil type and

(ix)   Topography

Channel characteristics relate mostly to hydraulic properties of the channel that governs the water flow. These factors include:

(i)   Size

(ii)   Slope

(iii) Cross section and

(iv)   Roughness coefficient of channel

Some of the factors mentioned as above are interrelated to each other which makes it difficult to quantify the affect of each individual factor on runoff. However, some dominant factors that affect the runoff of an area are discussed below.

If the climatic factors including depth and intensity of rainfall remain constant, then two watersheds/catchments irrespective of their sizes will produce the same amount of total runoff. However, for a watershed with larger area, it takes more time for the runoff to pass through the outlet hence the peak flow expressed as depth will be smaller. Runoff is influenced by the shape of the catchment. Two catchments having same area, fern-leaf shaped catchment gives lesser runoff because the tributaries of different lengths meet the main stream at regular intervals thereby the discharge are likely to be distributed over a large period of time (Fig. 3.3, a). On the contrary, fan shaped compact catchments tend to produce higher peak runoff rate than long and narrow catchments as in the former case, all parts of the catchment contributes to runoff at the outlet simultaneously and thus, the time of concentration to peak runoff is less (Fig. 3.3, b).

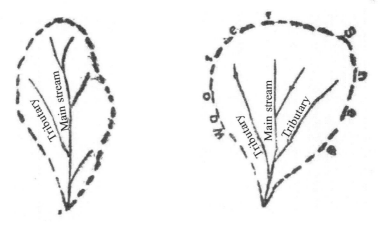

Fig. 3.3 (a): Fern shaped catchment.   Fig. 3.3 (b): Fan shaped catchment

In addition to the above mentioned two types of catchments, the catchments may be classified depending on the shape as tree like, rectangular type, radial type, parallel type or trellis type.

Generally the shape of the watershed is expressed by two factors. They are (i) form factor and (ii) compactness factor.

Form Factor: It is defined as ratio of average width of the watershed to axial length of the watershed and is expressed as:

Form Factor (F_f) = Average width of the watershed (B) / Axial length of the watershed (L)

Axial length of the watershed is the distance between outlet and the remotest point of the watershed. Width of the watershed is determined by dividing the area of the watershed (A) with the axial length (L).

Compactness Factor: It is the ratio of perimeter of the watershed to the circumference of a circle whose area is equal to the area of the watershed. It is expressed as:

Compactness Factor (C_f) = Perimeter of the watershed / Circumference of circle whose area is equal to the area of the watershed.

The slope of the catchment has pronounced affect on infiltration, surface runoff, soil moisture and ground water contribution to stream flow. It affects the period of over land flow. A watershed with flat slope having inadequate drainage systems results in getting more infiltration and hence less runoff than the watershed with steep slope and having well defined drainage systems. Orientation i.e. general direction of the slope of a catchment, with respect to prevailing winds, storm movement, and sun's position has considerable affect on the runoff. Soil texture and its permeability influences runoff greatly. Runoff generated from a watershed depends on the antecedent moisture condition of the soil in the watershed. If the soil moisture condition prior to occurrence of the rainfall in the watershed is high, then more runoff will be produced and vice-versa.

Shallow pervious soil and coarse textured soil like sandy soil tend to have more infiltration and thereby reducing runoff. Runoff is more for hard compacted subsoil since infiltration is less. Fault zones, fissures and cracks result in enhancing infiltration of in-*situ* rainfall and hence reducing runoff. However, the infiltrated water may contribute to the stream as base flow which is a delayed flow. Both land use and land treatment measures greatly affect the runoff of watershed. Land treatment measures like contour ploughing, strip cropping, bench and terrace cultivation and other soil and water conservation measures increase the water holding capacity of the soil and thus reduces runoff and soil loss in a watershed. Vegetated watersheds, due to, surface retention, and storage produces less runoff than bare watersheds. Storage characteristics like lakes, reservoirs, check dams, pits etc. act as flood modulator and reduce the runoff of a watershed. A channel having rough surface with higher values of roughness coefficient will retard the velocity of flow and thereby reduces the surface runoff contributed to a stream. Topographical factors like altitude and drainage density also affect the runoff. Higher catchments usually have steep slopes and produce greater total runoff. At higher altitudes, though the evaporation of the atmosphere is less, the rainfall, under certain condition, is more and hence total runoff may be more. Finally, a watershed with higher drainage density will produce more runoff than a watershed with less drainage density. Drainage density is defined as the ratio of total channel length prevailing in the watershed to the total area of the watershed and is given by the equation as:

$$\text{Drainage density } (D_d) = \frac{Total\ channel\ length\ (L)}{Total\ area\ of\ watershed\ (A)} \qquad (3.2)$$

3.3 Runoff Estimation

Runoff can be estimated from two sources i.e. due to rainfall and due to snow melt. Various methods used to estimate runoff due to rainfall (Sharma and Sharma, 2002) are:

(i)   Runoff coefficient

(ii)   Runoff percentage

(iii)   Strange's tables and curves

(iv)   Empirical formula

(v)   Water budget method

(vi)   Watershed routing technique

(vii)   Synthetic unit hydrograph method and

(viii) Infiltration method

3.3.1 Runoff Coefficient

Runoff is a function of rainfall. More is the amount of rainfall, more is the runoff. Thus, the runoff (R) and rainfall (P) can be correlated by a coefficient called as runoff coefficient or runoff factor (K). Runoff coefficient is defined as the ratio of volume of stream flow above the base flow to the volume of rainfall. The rainfall-runoff relation by this approach is considered as linear as given below:

$$R = K\,P \qquad (3.3)$$

where R is runoff (cm) and P is rainfall (cm). Runoff coefficient (K) depends on all those factors that affect the runoff. Values of K are presented in Table 3.1.

Table 3.1: Values of runoff coefficient (K) for various soil and vegetation condition

Type of soil/vegetation condition	Values of K	Type of soil/vegetation condition	Values of K
Plateau lightly covered	0.70	Loam largely cultivated and suburbs with garden, lawns	0.30
Clayey soils stiff and clayey soil lightly covered	0.55	Sandy soils, light growth	0.20
Loam lightly cultivated or covered	0.40	Parks, lawn, meadows, gardens	0.05-0.20

3.3.2 Runoff Percentage

3.3.2.1 Binnie's Percentage: Sir Alexander Binnie has suggested the following percentages of runoff to rainfall based on the data of two rivers of Madhya Pradesh.

Annual rainfall (cm)	Runoff (%)	Annual rainfall (cm)	Runoff (%)
50	15	90	35
60	21	100	38
70	25	110	40
80	29		

3.3.2.2 Barlow's Percentage: T.G. Barlow's runoff percentage, based on studies of catchments mostly under 130 km^2 in Uttar Pradesh is given below.

Class	Description of catchment	Runoff (%)		
		Season 1	Season 2	Season 3
A	Flat, cultivated and black cotton soils	7	10	15
B	Flat, partly cultivated various soils	12	15	18
C	Average	16	20	32
D	Hills and plains with little cultivation	28	35	60
E	Very hilly and steep, with hardly any cultivation	36	45	81

N.B.   Season 1: Light rain, no heavy downpour
        Season 2: Average or varying rainfall, no continuous downpour
        Season 3: Continuous downpour

The above values of runoff percentages are for average type of monsoon. They are modified by the application of following coefficients according to the nature of season.

Nature of season	Class of catchment				
	A	B	C	D	E
Light rain	0.70	0.80	0.80	0.80	0.80
Average or varying rainfall with no continuous downpour	1.0	1.0	1.0	1.0	1.0
Continuous downpour	1.50	1.50	1.60	1.70	1.80

3.3.3 Strange's Tables and Curves

Strange (1928), based on the rainfall and runoff data of the border areas of the present-day Karnataka and Maharashtra, has provided following tables and

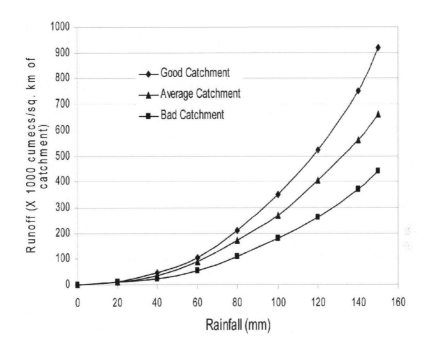

Fig. 3.4: Strange's curves to compute runoff

curves between rainfall and runoff. Strange's table (Table 3.2) and curves (Fig. 3.4) give runoff for daily rainfall for good, average and bad catchments and surface conditions being dry, damp and wet prior to rain.

3.3.4 Empirical Formulae

There are large numbers of empirical formulae developed for different regions and countries which are used for estimation of runoff. Most of these formulae relate rainfall with runoff. There are some formulae which relate runoff with other parameters in addition to rainfall like temperature and duration of rainfall. The empirical formulae have limitations that they are suitable for the particular regions for which they are developed and cannot be used confidently for other regions. Some of the important empirical formulae are mentioned below (Sharma and Sharma, 2002).

3.3.4.1 Lacey's Formula

$$R = \frac{P}{1 + \dfrac{304.8}{P}\left(\dfrac{F}{S}\right)}$$ (3.4)

Table 3.2: Daily runoff and percentage of runoff according to Strange for average catchment

Daily rainfall (mm)	Original stage of ground surface					
	Dry		Damp		Wet	
	Percentage of runoff	Runoff (mm)	Percentage of runoff	Runoff (mm)	Percentage of runoff	Runoff (mm)
5	-	-	4	0.20	7	0.35
10	1	0.10	5	0.50	10	1.0
20	2	0.40	9	1.80	15	3.0
25	3	0.75	11	2.75	18	4.5
30	4	1.20	13	3.90	20	6.0
40	7	2.80	18	7.20	28	11.20
50	10	5.00	22	11.0	34	17.0
60	14	8.46	28	16.8	41	24.6
75	20	15.0	37	27.75	55	41.25
100	30	30.0	50	50.0	70	70.0

N. B. Strange has suggested to add or deduct 25 per cent of runoff values for good or bad catchments.

where, R is runoff, P is rainfall, F is monsoon duration factor varying between 0.5 – 1.5 and S is catchment factor depending upon the slope; values of which vary from 0.25 for flat cultivated areas to 3.45 for very hilly areas.

3.3.4.2 Khosla's Formula: Khosla (1960) analysed the rainfall, runoff, and temperature data of various catchments in India and USA and derived a relation between rainfall and runoff as (Subramanya, 2003).

$$R_m = P_m - L_m$$

$$(3.5)$$

where, $L_m = 0.48 \ T_m$ for $T_m > 4.5°C$, $R_m =$ monthly runoff in cm and $P_m =$ monthly rainfall in cm. If in the above equation, $L_m > P_m$, then $R_m = 0$. In this equation, $L_m =$ monthly losses in cm.

For $T_m \leq 4.5°C$, loss L_m may provisionally be assumed as 2.17; 1.78 for $T_m = -1.0°C$ and 1.52 for $T_m = -6.5°C$.

Khosla's formula is used to estimate runoff on monthly basis. It is to be noted that Khosla's formula is indirectly based on the water balance concept and the mean monthly catchment temperature is used to reflect the losses due to evapotranspiration. Khosla's formula can be used to generate synthetic runoff data from historical rainfall and temperature data.

Sample Calculation 3.2

The mean monthly rainfall and temperature data of a station in U.P. as recorded are given below. Estimate the monthly and annual runoff by Khosla's method.

Month	Jan	Feb	Mar	Apr	May	Jun	Jul	Aug	Sep	Oct	Nov	Dec
Temp., °C	15	18	20	25	34	42	40	38	35	26	19	14
Rainfall, cm	4	3.5	4.7	2.5	9.2	12.3	20.1	24.2	18.9	2.1	1.0	0.0

Solution: In Khosla's formula, $R_m = P_m - L_m$ where $L_m = 0.48 \ T_m$ for $T_m > 4.5°C$

If the loss $L_m > P_m$, then $R_m = 0$. Calculation of monthly runoff is given as:

Month	Jan	Feb	Mar	Apr	May	Jun	Jul	Aug	Sep	Oct	Nov	Dec
Runoff, cm	0	0	0	0	0	0	0.9	5.96	2.1	0	0	0

Annual runoff = Summation of all monthly runoff = 8.96 cm.

3.3.4.3 Inglis and Desouza's Formula: Inglis and Desouza (1929) conducted experiment in 53 gauging sites of Western India and proposed the following two formulae to estimate annual value of runoff:

$R = 0.85\ P\ -30.5$ for ghat areas of western India   (3.6)

$$R = \frac{(P-17.8)P}{254} \text{ for Deccan plateau}$$ (3.7)

where, R is runoff (cm) and P is rainfall (cm).

3.3.4.4 Mayer's Formula: It is popular in USA for estimation of runoff. The formula has a limitation that it requires an area more than 10 km². The formula (Mayer, 1926) is:

$Q = 177.05\ A^{0.50}$ (3.8)

3.3.4.5 Coutagne Formula: This formula was developed for use in France and has limitations that it is to be used for watershed having area 400 to 3000 km². The formula is given as:

$Q = 150\ A^{0.50}$ (3.9)

3.3.4.6 Dicken Formula: The equation is given as:

$Q = C\ A^{3/4}$ (3.10)

where, Q is run off rate in m³/sec, C is a constant and A is watershed area in ha. This formula holds good for central and northern parts of India. Value of C varies from 2.8 to 5.6 for plain lands and 14 to 28 for mountainous regions.

3.3.4.7 Ryve's Formula: Ryve's formula is a modified form of Dicken's formula. It is suitable for coastal areas lying 25 km within the periphery of the sea. The formula is:

$Q = C\ A^{2/3}$ (3.11)

The terms in this formula are same as given in Dicken's formula but the value of the constant C is assumed as 6.8 for flat tracks and 42.40 for western ghat areas.

3.3.5 Water Budget Method

The hydrologic water budget equation for estimation of runoff for a given period is as:

$R = R_s + G_o = P - E_{at} - \Delta S$ (3.12) where, R = total runoff/streamflow, R_s = surface runoff, P = precipitation, E_{at} = actual evapotranspiration, G_o = net groundwater outflow and ΔS = change in the soil moisture storage.

The procedure to estimate R by the above equation is to start assuming a set of initial values by knowing the values of P, and functional dependence of E_{at}, ΔS and infiltration rates with catchment and climate conditions. For accurate results, the functional dependence of various parameters governing the runoff in the catchment and values of P at short time interval is needed. Calculations can then be done sequentially to get runoff at any time. This technique of predicting runoff, which is the catchment response to a given amount of rainfall, is called *deterministic watershed simulation*. For estimation of runoff, a model relating various parameters of rainfall and runoff is developed first. The calibrated model is then validated and after that the developed model is used to predict the runoff. Several watershed simulation models are available. Some important ones are

- Stanford Watershed Model (SWM) developed by Crawford and Linsely (1959)

- Hydrocomp Simulation Program

- Streamflow Synthesis and Reservoir Regulation Model (SSRRM) developed by Rockwood (1968)

- Kentucky Watershed Model (KWM)

3.3.6 Watershed Routing Technique

The watershed routing model is based on the assumption that the outflow from the watershed varies nonlinearly with the storage in the watershed. An equation as van den Akker and Boomgaard (1996) is written as:

$$S = KQ^n \tag{3.13}$$

where S = volume of stored water in the watershed, Q = volume of outflow from the watershed, and K and n are parameters.

For simplicity in solution of the above nonlinear equation, Beven (2001) assumed it to be linear. By doing so, the calculation was made much simpler while the accuracy was still acceptable. The linear form of the equation is expressed as:

$$S = KQ \tag{3.14}$$

By converting volume into depth of water spreading throughout the watershed, we get

$$s_1 = kq_1 \tag{3.15}$$

$$s_2 = kq_2 \tag{3.16}$$

where s_1 and s_2 = depths of water storage at time steps 1 and 2, respectively; q_1, q_2 = discharges as depth per unit time at time steps 1 and 2, respectively and k is coefficient. From continuity equation,

$$s_2 = s_1 + (i - 0.5(q_1 - q_2)) \Delta t \tag{3.17}$$

where i = intensity of excess rainfall, and Δt = time interval. By substituting Eqns. (3.15) and (3.16) in Eq. (3.17) and rearranging,

$$q_2 = \frac{k - 0.5\, \Delta t}{k + 0.5\, \Delta t} \cdot q_1 + \frac{\Delta t}{k + 0.5\, \Delta t} \tag{3.18}$$

Eqn. (3.18) is the watershed routing model. It can be used to predict the next step flow rate from knowing the present flow rate and intensity of rainfall. The depth flow rate, q, can be converted to volume flow rate, Q, by multiplying q with the watershed area, A.

Watershed routing technique starts with the assumption of the k value. The initial value of depth runoff, q, may be set to zero. The equation (3.18) was used consequently to obtain q of all time steps. The depth runoff value q were converted to the volumetric runoff Q by multiplying q with the watershed area, A. The k value is adjusted until the most suitable hydrograph was approached.

3.3.7 Synthetic Unit Hydrograph Method

The unit hydrograph is a direct runoff hydrograph resulting from a unit rainfall (1 cm depth) of specific rainfall duration (Shaw, 1994). The unit hydrograph is normally derived from records of rainfall and runoff data. When dealing with small watersheds, coupled rainfall and runoff data are hardly available, we therefore resort to the synthetic unit hydrograph. The unit hydrograph that is synthesized from topographic and climatic features is called a synthetic unit hydrograph. Essentially, the idea is that the lag time of each watershed is constant and can be evaluated from the watershed characteristics (Shaw, 1994). The lag time is the time lapse between the middle of the rainfall duration and the hydrograph peak. For a small watershed, the lag time can be assumed to be about 0.6 of concentration time. Time of concentration of a watershed can be computed by Kirpich (1940) formula. Details regarding estimation of runoff from rainfall by synthetic unit hydrograph are presented in separate chapter (Chapter-6).

3.3.8 Infiltration Method

Furthermore, infiltration method is also used to estimate the runoff from the record value of rainfall and infiltration loss. It is the most reliable method of computation of runoff. The rainfall rate in excess of infiltration capacity gives rise to surface runoff. Runoff from large areas can be determined by measuring the infiltration indices.

Infiltration is defined as the entry of water into the soil surface. It is the first process of hydrologic cycle after water strikes the ground surface. The capacity of absorption of water by the soil surface depends on many factors including the texture of the soil, antecedent moisture condition, management and cultural practice of the soil, groundwater table position etc. However, the general tendency of infiltration is that it goes on decreasing as time elapses for any soil. The infiltration rate is high initially and exponentially deceases as time increases. After some time, the infiltration rate remains constant and is the minimum. This minimum and constant infiltration rate may be zero at infinite time. Maximum infiltration rate is called as infiltration capacity. The trend of this infiltration rate with time is called infiltration capacity curve which is shown in Fig. 3.5. If the infiltration capacity curve is superimposed over the rainfall intensity which is in

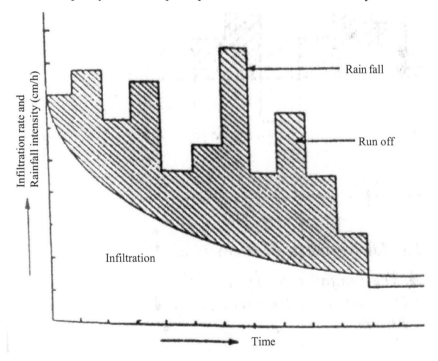

Fig. 3.5: Infiltration capacity curve

the form of hyetograph, the resultant amount will represent the runoff. The above method of computation of runoff can be used if the rainfall rate or intensity is more than the infiltration rate. When both the rainfall intensity and infiltration are plotted in the same axis (as ordinate) with time being taken as abscissa, then the area between the rainfall and infiltration capacity curve represents runoff. The area below the infiltration capacity curve represents loss of rainfall due to infiltration. This loss of water due to infiltration is more at the beginning resulting low values of runoff and subsequently decreases with the advancement of rainfall duration.

The infiltration capacity curve already determined on the test plots cannot be applied to a large basin having heterogeneous areas. In a large basin, at any instant of time, both the rainfall intensity and infiltration capacity will vary from point to point. Moreover, in case of large, the interflow will be predominant. Since the interflow is a part of infiltration, it will not be normally included in the runoff computation by the infiltration capacity curve determined on a test plot. In large basin, runoff is generally computed by using infiltration indices.

There are two types of infiltration indices. They are:

- ϕ-index and
- W – index

ϕ**-Index:** It is the rate of loss of water above which the volume of rainfall is equal to the volume of runoff as shown in Fig.3.6. ϕ index is given as:

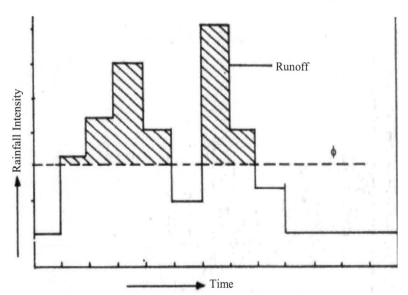

Fig. 3.6 Relationship between runoff and ϕ-index

$$\phi\text{-index} = \frac{Total\ basin\ recharge}{Rainfall\ discharge} \qquad (3.19)$$

The above formula is suitable when rainfall intensity exceeds the value of ϕ-index continuously. It assumes that at the beginning of a storm, ϕ-index is more producing less runoff and at the end, runoff becomes more with a reduced value of ϕ-index.

On the basis of rainfall and runoff correlation, CWC (1973) has found the following relation for the estimation of ϕ-index for flood producing storms and soil conditions prevalent in India:

$$R = \alpha\ I^{1.2} \qquad (3.20)$$

and

$$\varphi = \frac{I - R}{24} \qquad (3.21)$$

where, R is runoff, cm from a 24 hr rainfall of intensity I, cm/day and α is a coefficient which depends on soil texture as below:

Sl. No.	Type of soil	Value of α
1.	Sandy soil and sandy loam	0.17 – 0.25
2.	Coastal alluvium and silty loam	0.25 – 0.34
3.	Red soils, clay loam, gray and brown alluvium	0.42
4.	Black cotton and clayey soil	0.42 – 0.46
5.	Hilly soils	0.46 – 0.50

In estimating the maximum floods for design purposes, in the absence of any other data, a ϕ-index value of 0.10 cm/hr can be assumed.

W-Index: W – index is defined as the average water loss during a storm when rainfall intensity exceeds the infiltration capacity i.e. maximum infiltration rate. It is defined as:

$$W\text{–index} = \frac{F}{t_r} = \frac{P - Q}{t_r} \qquad (3.22)$$

where, F = total amount of water lost through infiltration including basin recharge, P = total precipitation occurring during time t_r, t_r = duration of rainfall in hr and Q = total runoff.

ϕ-index and W– index will be same for a uniform rain and heavy rain. Both the indices will be equal when infiltration rate of the soil becomes equal to the infiltration capacity and the surface retention is minimum. For a moderate rain of non-uniform intensities, ϕ-index is more than the W– index. It is to note that these two indices are not the actual infiltration rates but merely measures of potential basin recharge. It is to be noted that accurate estimation of W-index is difficult to determine since it is difficult to find out the true value of initial losses causing basin recharge. The minimum values of the W-index obtained under very wet soil conditions, representing the constant minimum rate of infiltration of the catchment, are called as W_{min}.

Following sample calculation illustrates the uses of W– index and ϕ-index.

Sample Calculation 3.3

The rainfall rates for successive 30 min intervals up to 3 hrs of a storm are: 0, 1.6, 2.5, 5.0, 3.0, 2.2 and 1.0 cm/hr. Assuming that the total runoff produced by the storm in 3 hr is 3.5 cm, calculate the value of ϕ-index. Also calculate the value of W– index.

Solution

Let I be the rainfall rate and ϕ be the infiltration value. Then in a time, t, total runoff (R) is given as $\Sigma(I - \varphi) \cdot t$

Substituting the value of I and t and R, we have,

$3.5 = 30/60 \{(1.6 - \varphi) + (2.5 - \varphi) + (5.0 - \varphi) + (3.0 - \varphi) + (2.2 - \varphi) + (1.0 - \varphi)\}$

$7.0 = (1.6 + 2.5 + 5.0 + 3.0 + 2.2 + 1.0) - 6\varphi$

The solution gives φ as 1.38 cm/hr.

Total precipitation in 3 hrs $= 0 \times 30/60 + 1.6 \times 30/60 + 2.5 \times 30/60 + 5.0 \times 30/60 + 3.0 \times 30/60 + 2.2 \times 30/60 + 1.0 \times 30/60 = 30/60 (0 + 1.6 + 2.5 + 5.0 + 3.0 + 2.2 + 1.0) = 7.65$ cm

Now $W\text{–}index = \dfrac{F}{t_r} = \dfrac{P-Q}{t_r}$

$= (7.65 - 3.5)/3 = 1.38$ cm/hr.

The above solution reveals that both the indices are same and hence it is a uniform rainfall.

Sample Calculation 3.4

The rainfall rates for successive 20 min intervals up to 140 min of a storm are: 2.5, 2.5, 10.0, 7.5, 2.0, 2.0 and 5.0 cm/hr. Assuming ϕ-index as 3.2 cm/hr find out the total runoff and also compute W-index.

Solution

Let I be the rainfall rate and ϕ be the infiltration value. Then in a time, t, total runoff (R) is given as $\Sigma(I - \phi) . t$

In the above sample problem, we have runoff in 3 cases when rainfall rate is more than the ϕ value.

These are the I values of 10.0, 7.5 and 5.0 cm/hr.
Substituting the value of I and t and ϕ, we have,

Runoff $= Q = (10 - 3.2).20/60 + (7.5 - 3.2). 20/60 + (5.0 - 3.2).20/60 = 4.3$ cm.

Total precipitation, $P = 20/60 \times (2.5 + 2.5 + 10.0 + 7.5 + 2.0 + 2.0 + 5.0) = 10.5$ cm

$$\text{Now } W\text{-}index = \frac{F}{t_r} = \frac{P-Q}{t_r}$$

$= (10.5 - 4.3)/ (140/60) = 2.66$ cm/hr.

3.4 Methods for Estimation of Runoff Rate

Following methods are used for estimation of runoff rate from a catchment.

(i)   Rational Method

(ii) Cook's Method

(iii) Curve Number Method

Out of the above mentioned three methods, the rational and US curve number method are mostly used for computation of runoff rate and hence are discussed in this book.

3.4.1 Rational Method

Rational method is widely used in computation of the peak runoff rate in small watersheds. In this method, the peak rate of runoff is given by the equation as:

$$Q = \frac{CIA}{360} \tag{3.23}$$

where, Q is peak runoff rate, m³/sec; I is rainfall intensity, mm/hr; C is runoff coefficient, and A is catchment area generating runoff, hectares.

Values of runoff coefficient (C) for different slopes and land use conditions as determined from the field experiments are given in Table 3.3. It is a dimensionless term and varies from 0 to 1.

For catchment with different values of C the weighted values of C as estimated below are considered for the whose catchment

$$C = \frac{\sum_{i=1}^{n} C_i A_i}{A} \tag{3.24}$$

where, C is the runoff coefficient of the whole catchment, A is the area of the whole catchment, C_i is the runoff coefficient for catchment i and A_i is the area of catchment i for which runoff coefficient is C_i, and i is the index for catchment ranging from number 1 to n.

The value of the intensity of rainfall to be used in Eq. 3.23 should be calculated for the period equal to the time of concentration of the catchment. While using the rainfall intensity in the rational method, the rainfall records at different places are analysed and the 1 hour rainfall intensity are calculated. For small water storage structures generally 10 years return periods is considered and hence, the rainfall intensity at 10 years frequency is considered for computation of runoff in rational formula. The values of maximum rainfall at 1 hour duration at 10 years frequency is given in Fig. 3.7. This 1 hour intensity is converted to

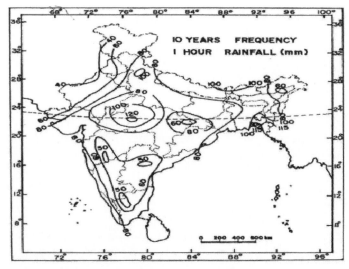

Fig. 3.7: One hour maximum rainfall at 10 years frequency

Table 3.3: Values of runoff coefficient, C used in rational method

Soil Types	Land Use		
	Cultivation	Pasture	Forest
Sandy or gravelly soil having high infiltration rate	0.29	0.15	0.10
Loamy soil and soils with no clay pans near the surface having average infiltration rate	0.40	0.35	0.30
Heavy clay soil or soils with clay pan near the surface or shallow soil above impervious rock having below average infiltration rate	0.50	0.45	0.40

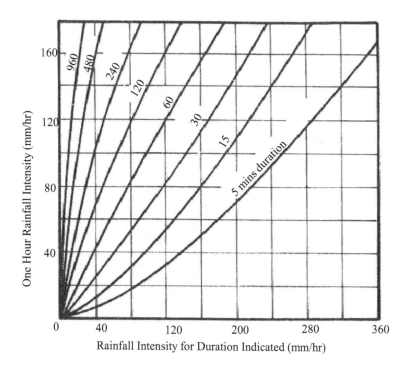

Fig. 3.8: Curves to convert one hour rainfall intensity to intensity of other duration

the intensity for the duration of time of concentration by using the graphs (Fig. 3.8) as proposed by Tejwani *et al.* (1975).

The time of concentration of the catchment is defined as the time required for water to flow from the remotest point of the catchment to the outlet of the catchment. When the duration of the rainfall equals to the time of concentration, all parts of the catchment will take part in contributing to the runoff and hence the discharge would be maximum.

Kirpich (1940) has given the following empirical formula to compute the time of concentration.

$$T_c = 0.02 \, L^{0.77} \, S^{-0.385} \tag{3.25}$$

where, T_c is time of concentration, min; L is the length of the channel reach, m and S is the average slope of the channel reach, m/m.

For small catchment with area less than 5 sq. km having less developed network of channels and where overland flow dominates the channel flow, the time of concentration is given by the Eq. 3.26 (Haan *et al.*, 1982) as:

$$T_c = 0.02 \; L^{\,0.77} \; S^{\,-0.385} + \left(\frac{2L_o n}{\sqrt{S_o}} \right)^{0.467} \tag{3.26}$$

where, L_o is the length of the overland flow, m; S_o is the slope along the flow path, m/m and n is the average roughness coefficient of the channels contributing overland flow.

The second term of Eq. 3.26 accounts for the overland flow. The value of this term has a maximum value regardless of watershed size. The values of n suggested by Haan *et al.* (1982) are given in Table 3.4.

Table 3.4 Values of roughness coefficient n used in overland flow

Nature of the surface	Values of n
Smooth, impervious surface	0.02
Smooth, bare packed surface	0.10
Cultivated row crops of moderately rough bare soil	0.20
Pasture or average grass	0.40
Deciduous forest areas	0.60
Timber land with deep forest litter or dense grass	0.80

Time of concentration can also be computed as:

$$T_p = T_c^{\,0.5} + 0.6 \; T_c \tag{3.27}$$

where, T_p is time from beginning of rise of hydrograph to the peak.

As reported by Suresh (2008), Morgali and Linsley (1965) and Aron and Erborge (1973) proposed kinematic wave formula as given below for computation of time of concentration:

$$T_c = \frac{0.94 \; L^{0.6} \; n^{0.6}}{i^{0.4} \; S^{0.3}} \tag{3.28}$$

where, L is the length of the overland flow(ft), n is Manning's roughness coefficient, i is rainfall intensity (inch/hr) and S is the average slope of the land surface (ft/ft).

Time of concentration can also be computed by SCS lag equation as:

$$T_c = \frac{100 \; L^{0.8} \left[\left(\frac{100}{CN} \right) - 9 \right]^{0.7}}{1900 \; S^{0.5}} \tag{3.29}$$

where, L is the length of the longest flow path of the watershed (ft), CN is the curve number and S is the average slope of the land surface (%). The above formula is applicable for agricultural watershed. However, it can be used in urban watershed with area less than 2000 acres. This formula gives satisfied result where area is completely paved. For heterogeneous soil surfaces in a watershed, it overestimates the value of time of concentration.

As reported by Garg (1987), the following formula can also be used to predict time of concentration for a small plot where there is no well defined flow channels from which runoff can occur. The formula is:

$$T_c = \frac{111\, b\left(L_0\right)^{1/3}}{(K.p)^{2/3}} \tag{3.30}$$

where, p is rainfall intensity factor, cm/hr, K is runoff coefficient, L_o is length of the overland flow, m and b is coefficient given as:

$$b = \frac{2.75 \times 10^{-4}\, p + Cr}{S_0^{1/3}} \tag{3.31}$$

where, p is the rainfall intensity, S_0 is the slope of overland flow and Cr is the retardance coefficient. These equations (Eqns. 3.30 and 3.31) are applicable when the product $p.L_0 < 387$. Time of concentration in Eqn. (3.30) is given in minute.

Values of Cr of Eqn. (3.31) depend on roughness of the surface and are given in Table 3.5.

Table 3.5 Values of retardance coefficient

Sl.No.	Types of surface	Values of retardance coefficient, Cr
1.	Dense blue grass turf	0.060
2.	Closely clipped soil	0.046
3.	Concrete pitched surface	0.012
4.	Tar and gravel pitched surface	0.017
5.	Smooth asphalt surface	0.007

Another formula for computation of time of concentration for small watershed has been developed by Linsely et al. (1979) which is:

$$T_c = C_t \left(L.\, L_{ca} / S\right)^n \tag{3.32}$$

where, C_t and n are constants, S is slope of the basin, L_{ca} is distance of the gauging point from the point opposite to the centre of the watershed and L is basin length measured along the main water course from the basin divide to the gauging point. This formula assumes that the time of concentration is approximately equal to the lag time.

3.4.1.1 Limitations of Rational Method: The limitations of rational method as proposed by (Singh, 1964) are:

(i)   Rational method assumes that the rainfall intensity over the whole catchment is uniform during the duration of the storm which is rarely satisfied.

(i)   Before the occurrence of runoff, the losses due to depression storage and infiltration must be fulfilled which is not taken care by the rational method.

(ii)   When there is flow in channels, the retardance affect caused due to surface detention and channel storage is not taken care in rational method.

(iii) This method is to be used for watershed having area up to 50 km^2.

(iv)   The runoff coefficient is computed only on the basis of the watershed characteristics and does not take into account the season of runoff generation and other atmospheric conditions.

3.4.1.2 Valley Storage/ Channel Storage: As the rate of overland flow and that of channel flow increases, the depth of flow also increases which cause a certain amount of water to be temporarily stored until runoff rate again decreases. This water which is stored temporarily within the channel in the basin is called as Channel storage or Valley storage. Equilibrium condition will occur only when a sufficient quantity of water has occurred to supply this valley storage. This storage is not taken care in Rational formula and hence is a limitation for use of the Rational formula.

Sample Calculation 3.5

A watershed of area 50 hectares having 20 hectares under cultivation, 10 hectares under pasture and 20 hectares under forest. The cultivated area has sandy soil, whereas the pasture and forest area have heavy clay soil with very low infiltration rate. The watershed has a fall of 10 m over the longest reach of length 1 km; the reach being connecting the remotest point to the outlet point of the watershed. Estimate the peak runoff rate if the 1 hour rainfall intensity for 10 years frequency of the watershed is 100 mm/hr. Assume that the ratio of 1 hour rainfall intensity to rainfall intensity of any other duration is 1.2,

Solution

Using Table 3.4, the values of C for cultivated, pasture and forest areas are calculated as 0.29, 0.45 and 0.40, respectively. The weighted value of C for the whole watershed is calculated as

$$C = \frac{20 \times 0.29 + 10 \times 0.45 + 20 \times 0.40}{50} = 0.366$$

For $L = 1$ km $= 1000$ m, slope $S = 10/1000 = 0.01$

Using Eq. 3.25, time of concentration, $T_c = 24$ min.

Given the 1 hour rainfall intensity of 10 year frequency is 100 mm/hr and the ratio of 1 hour rainfall intensity to that of any other duration is 1.2, the rainfall intensity for 24 minutes of rainfall is 100 x 1.2 = 120 mm/hr.

Hence peak runoff rate $= Q = \dfrac{CIA}{360} = \dfrac{0.366 \times 120 \times 50}{360} = 6.1 cumecs$

3.4.1.3 Modified Rational Method

When the rainfall occurs for a time greater than the time of concentration, rational formula as mentioned above does not hold good. In that case the modified rational method as mentioned by Suresh (2008) is used and is described below.

$$Q_p = C\,A\left(\frac{a}{b+T_d}\right) \tag{3.33}$$

where, Q_p = peak discharge, C is the runoff coefficient used in rational method, A is the area of the watershed, T_d is the duration of rainfall and a and b are constants.

Duration of rainfall (T_d) is given as:

$$T_d = \left(\frac{a.b.C.A}{Q_A - \dfrac{Q_A^{\,2}.T_p}{2.a.C.A}}\right) \tag{3.34}$$

where, T_p is time to peak or time concentration of the detention basin, Q_A is allowable discharge and other terms are defined as above.

Volume of runoff, V_r is given as:

$$V_r = Q_p \cdot T_d \tag{3.35}$$

Modified rational formula was developed to derive the hydrograph for storage design rather than peak discharge computation. It can be used for preliminary design of detention storage for the watershed area ranging from 20 to 30 acres.

3.4.2 Cook's Method

In this method, runoff is computed based on four watershed characteristic features. These features are (i) relief, (ii) infiltration rate, (iii) vegetal cover and (iv) surface storage. Numerical values to each of these four features are assigned based on hydro-meteorologically similar conditions and the values are mentioned in Table 3.6.

The computational steps for peak discharge are presented below.

Step I: Evaluate the degree of watershed characteristics like relief, infiltration rate, vegetal cover and surface storage by comparing with hydro-meteorologically similar conditions of other watersheds and assign numerical values (W) to each of these characters with the help of data of Table 3.6.

Step II: Add all the numerical values (ΣW) assigned to each character. Compute the weighed value of ΣW for heterogeneous basin having different values of ΣW.

Step III: Estimate the runoff rate against ΣW by using runoff curve (Fig. 3.9) with known values of watershed area. It is to be noted that Fig. 3.9 is valid for estimation of peak rate of runoff for a recurrence interval of 10 years and for a given geographic location. For periods of other recurrence interval and any geographic location, the estimated runoff rate which is also called as unadjusted runoff rate is to be modified.

Step IV: Calculate the adjusted runoff rate for desired recurrence interval and watershed location by formula:

Table 3.6: Numerical values of watershed characteristic features used in Cook's formula

Sl.No.	Numerical values of watershed characteristic features used to produce runoff				
	Range	Relief	Infiltration rate	Vegetal cover	Surface storage
1.	Normal	Land is in rolling shape having slope 5 – 10% (10 -20)	Varies from 0.75 – 2 cm/hr; soil is normal and deep permeable in nature (10)	Forest or equivalent; about 50% area is under good grass cover or other equivalent cover (10)	Considerable depressions, lakes, ponds and marshy land covering less than 2% of total drainage area (10)
2.	High	Hilly lands with average slopes ranging from 10 – 30% (20 – 30)	Varies from 0.25 – 0.75 cm/hr; soil is relatively hard such as clay (15)	Vegetal cover varies from poor to fair; less than 10% area is under grass cover or other equivalent cover (15)	Area well drained with less surface depressions (15)
3.	Extreme	Steep lands having rugged terrain with slope up to 30% (30 – 40)	Infiltration rate is very low (20)	Bare land having no effective grass cover (20)	Area very well drained with negligible surface depression; area having no ponds/tanks (10)

N.B. Values in parenthesis represents the numerical values assigned to each watershed characteristic feature.

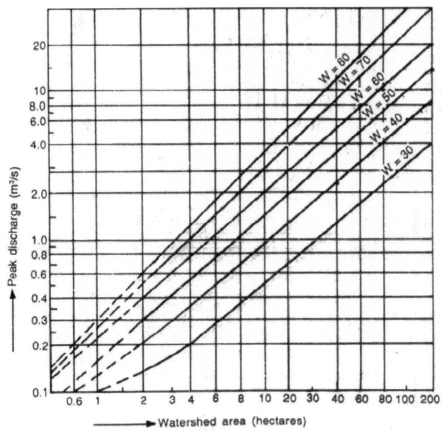

Fig. 3.9: Computation of peak rate of runoff by Cook's method

$$Q_{peak} = p.r.f.s \tag{3.36}$$

where, Q_{peak} is the adjusted runoff rate for any geographical location and any recurrence interval, p is the uncorrelated runoff rate as obtained in Step III above, r is geographic rainfall factor (Fig. 3.10), f is recurrence interval/frequency factor (Fig. 3.11) and s is the shape factor (Table 3. 7).

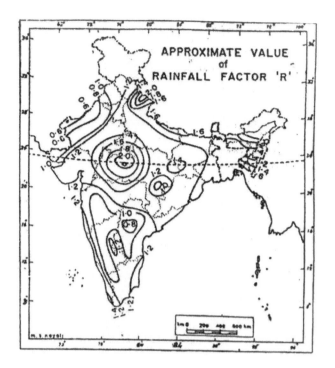

Fig. 3.10: Geographical rainfall factor (*r*) for various places

Fig. 3.11: Recurrence interval/frequency factor (f) for various places

Table 3.7 Shape factor for use in Cook's formula

Lengh : width	Area of watershed (ha)				
	20	40	80	200	240
1	1.00	1.00	1.00	1.00	1.00
1 – 1.5	0.92	0.92	0.91	0.90	0.90
2	0.88	0.87	0.86	0.84	0.83
2 – 2.5	0.85	0.84	0.82	0.80	0.78
3	0.81	0.80	0.78	0.76	0.74
4	0.76	0.72	0.70	0.68	0.66
5	0.74	0.70	0.68	0.66	0.64
6	0.72	0.68	0.66	0.64	0.62
7	0.68	0.66	0.64	0.61	0.59

Sample Calculation 3.6

A watershed has an area of 200 ha having following characteristics. Estimate the runoff rate considering 25 years recurrence interval.

Land use with area	Average slope,%	Infiltration rate, cm/hr	Vegetal cover	Surface storage
Cultivated, 140 ha	1	0.70	Less than 10% area grass cover	Ponds/tanks cover less is under good than 2% drainage area and area is not well drained
Pasture, 60 ha	5	1.00	About 95% area is under good grass cover	Negligible surface storage

The watershed has length to width ratio 3. Assume the rainfall factor as 1.2 and frequency factor as 1.0. Assume the relief factor for the cultivated land as 10 and that for the pasture land as 15.

Solution

Using Table 3.7 and considering the values of length to width ratio 3 and watershed area 200 ha, the value of shape factor is obtained as 0.76.

The relief factor, R for the cultivated and pasture lands are given 10 and 15, respectively. From Table 3.6 and using the data given in the problem, the soil infiltration factor (I) for cultivated and pasture lands are found to be 15 and 10, respectively. Similarly using Table 3.6 and using the data given in the problem, the vegetal cover factor (V) for cultivated and pasture lands are found to be 15 and 10, respectively. Moreover using Table 3.6 and using the data given in the problem, the surface storage/depression factor (D) for cultivated and pasture lands are found to be 10 and 10, respectively.

The sum of all the numerical values (ΣW) for the cultivated and pasture lands are now worked out to be 50 and 45, respectively.

The weighted value of ΣW for the entire watershed is $(50 \times 140 + 45 \times 60)/200 = 48.5$

From Fig. 3.9, with the values of ΣW as 48.5 and watershed area of 200 ha, we get uncorrelated peak runoff rate as $14.5 \text{ m}^3/\text{sec}$.

Given the rainfall factor as 1.2 and frequency factor as 1.0. Also the shape factor is found to be 0.76.

Using Eqn. (3.36), we get the actual peak runoff rate for the watershed as:

$Q_{peak} = 14.5 \times 1.2 \times 1.0 \times 0.76 = 13.22 \text{ m}^3/\text{sec}$.

3.4.3 Curve Number Method

This method is also called as Hydrologic Soil Cover Complex Number method or US Soil Conservation Service (SCS) method. It is based on the recharge capacity of the watershed. The recharge capacity is determined by the antecedent moisture condition (AMC) and by the physical characteristics of the watershed. Let I_a be the initial quantity of interception, depression storage and infiltration that must be satisfied by any rainfall before runoff can occur. It is assumed that the ratio of the direct runoff Q and the rainfall minus initial losses $(P - I_a)$ and the storage capacity Sr are related by:

$$\frac{Q}{P-I_a} = \frac{P-Q-I_a}{Sr} \tag{3.37}$$

I_a is assumed to be certain percentage of Sr. Assuming it to be 0.2 Sr, the runoff is given as

$$Q = \frac{(P-0.2Sr)^2}{(P+0.8Sr)} \text{ if } P > 0.2Sr \tag{3.38}$$

$$Q = 0 \qquad \text{if } P \le 0.2Sr \tag{3.39}$$

where, Q = surface runoff, mm; P = rainfall, mm and Sr = potential maximum retention, mm.

For Indian condition, the value of I_a is assumed as 0.3 Sr.

The parameter Sr (value in mm) is related to the curve number (CN) as:

$$Sr = 254\left(\frac{100}{CN} - 1\right) \tag{3.40}$$

The curve number for antecedent moisture condition (CN for AMC II or CN = CN_2) for different land use conditions and hydrologic soil groups are given by USDA (1972) (Table 3.8). Fig.3.12 helps in the computation of runoff from rainfall for different values of curve number (CN) for AMC II. A heterogeneous basin may be divided into sub areas with different values of curve numbers and in such case, a weighted value of CN for AMC II is calculated for the whole watershed. The corresponding values of CN for dry (CN for AMC I = CN_1) and wet (CN for AMC III = CN_3) conditions are given by Sharpley and Williams (1990) as:

$$CN_1 = CN_2 - \frac{20\left(100 - CN_2\right)}{100 - CN_2 + \exp\left\{2.533 - 0.0636\left(100 - CN_2\right)\right\}} \tag{3.41}$$

$$CN_3 = CN_2 + 10\exp\left\{0.00673\left(100 - CN_2\right)\right\} \tag{3.42}$$

and Sr given by them is:

$$Sr = Sr_1\left[1 - \frac{FFC}{FFC + \exp\left(W_1 - W_2.FFC\right)}\right] \tag{3.43}$$

where, Sr_1 is the value of Sr associated with CN_1 and FFC is the availability of soil water expressed as fraction of field capacity given as:

$$FFC = \frac{SMC_i - WP}{FC - WP} \tag{3.44}$$

where SMC_i is the soil moisture content on day i, FC and WP are the soil moisture content at field capacity and wilting point, respectively. W_1 and W_2 are given as:

$$W_1 = \ln\frac{Sr_3}{Sr_1 - Sr_3} + W_2 \tag{3.45}$$

$$W_2 = \ln\left(\frac{0.5Sr_2\left(Sr_1 - Sr_3\right)}{Sr_3\left(Sr_1 - Sr_2\right)}\right)^2 \tag{3.46}$$

In which Sr_2 and Sr_3 are the values of Sr when $FFC = 0.5$ and 1, respectively. Sr_1, Sr_2 and Sr_3 are the retention parameters corresponding to CN_1, CN_2 and CN_3, respectively. Once the parameters W_1, W_2 and FFC are estimated, value of Sr can be obtained from equation 3.43. The estimates of peak runoff rate from total runoff computed by Eq. 3.38 as suggested by US Soil Conservation Service (1964) is:

Table 3.8 Curve numbers for hydrologic soil cover complexes

(For watershed condition $I_1,I^a=0.2S$) Land Use or cover	Treatment or Practice	Hydrologic Condition	Hydrologic Soil Group			
			A	B	C	D
Fallow- Row Crops	Straight row		77	86	91	94
	Straight row	Poor	72	81	88	91
	Straight row	Good	67	78	85	89
	Contoured	Poor	70	79	84	88
	Contoured	Good	65	75	82	86
	Contour & terraced	Poor	66	74	80	82
	Contour & terraced	Good	62	71	78	81
Small Grain	Straight Row	Poor	65	76	84	88
	Straight	Good	63	75	83	87
	Contour	Poor	63	74	82	85
	Contour	Good	63	75	83	87
	Contoure & terraced	Poor	61	72	79	82
	Contour & terraced	Good	59	70	78	81
Seeded Legumes or Rotation Meadow	Straight row	Poor	66	77	85	89
	Straight row	Good	58	72	81	85
	Contour	Poor	64	75	83	85
	Contour	Good	55	69	78	83
	Contour & terraced	Poor	63	73	80	83
	Contour & terraced	Good	61	67	76	80
Pasture or range		Poor	68	79	86	89
		Fair	49	69	79	84
		Good	39	61	74	80
	Contour	Poor	47	67	81	88

(For watershed condition 11,I^a=0.2S) Land Use or cover	Treatment or Practice	Hydrologic Condition	Hydrologic Soil Group			
			A	B	C	D
	Contour	Fair	25	59	75	83
	Contour	Good	06	35	70	79
Meadow (permanent)		Good	30	58	71	78
Woodlands		Poor	45	66	77	83
(farm woodlots)		Fair	36	60	73	79
Farmsteads			59	74	82	86
Roads (dirt)			72	82	87	89
Roads (Hard surface)			74	84	90	92

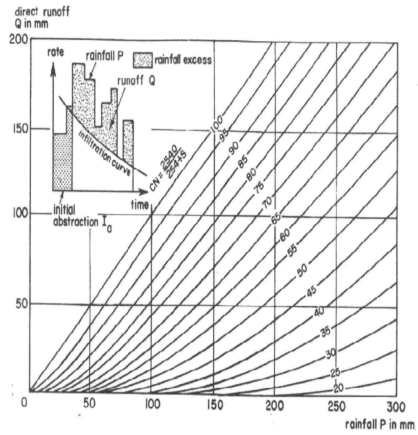

Fig. 3.12: Curve number to convert rainfall to runoff

$$Q_{peak} = \frac{0.0208AQ}{T_p} \tag{3.47}$$

where, Q_{peak} is peak rate of runoff, m³/sec; A is area, ha; T_p is time to peak, hr; T_c is time of concentration, hr as discussed earlier in Eqs. 3.25 and 3.26 and Q is runoff, cm. T_p is given as:

$$T_p = \sqrt{T_c} + 0.6T_c \tag{3.48}$$

Antecedent moisture condition (AMC) is an index for watershed wetness. The three levels of AMC are:

AMC I: Lowest runoff potential with dry condition of watershed.

AMC II: Average condition.

AMC III: Highest runoff potential with wet condition of watershed.

The AMC class is determined using 5 days antecedent rainfall in dormant or crop growing season as given below (Table 3.9).

Table 3.9: Values of 5 days antecedent rainfall in deciding AMC class

AMC	5 days antecedent rainfall, cm	
	Dormant season	Growing season
I	< 1.25	< 3.5
II	1.25 - 2.75	3.5 – 5.25
III	> 2.75	> 5.25

The hydrologic soil groups are divided into 4 classes depending on runoff potential as :

Hydrologic Soil Group A (Low runoff potential): Soil having high infiltration rate and consisting chiefly of deep well to excessively drained sands or gravels.

Hydrologic Soil group B (Moderately low runoff potential): Soil having moderate infiltration rate and consisting chiefly of moderately deep to deep, moderately well to well drained soils having moderate coarse texture.

Hydrologic Soil group C (Moderately high runoff potential): Soil having low infiltration rate and consisting chiefly of moderately deep to deep, moderately well to well drained soils having moderate coarse texture.

Hydrologic Soil group D (High runoff potential): Soil having very low infiltration rate and consisting of clay soils with high swelling potential and soil with clay layer near the surface.

Hydrologic soil groups as mentioned above is given by Gupta et al. (1970) and is given in Table 3.10.

Table 3.10 Hydrological grouping of different soil types

Sl. No.	Soil type	Tentative hydrological groups
1.	Alluvial soil non calcareous to moderately calcareous	B
2.	Alluvial soil highly calcareous	B
3.	Alluvial soil developed on coast as alluvium	A
4.	Alluvial soil developed on deltaic alluvium	B
5.	Alluvial soil reverine affected by salinity and alkalinity	C
6.	Pedocal sicrozem of alluvial origin	C
7.	Pedocal brown soil of alluvial origin	C
8.	Grey and brown (desert soil)	D

Contd.

Sl. No.	Soil type	Tentative hydrological groups
9.	Desert soil	A
10.	Deep black soil	D
11.	Medium black soil	D
12.	Shallow black soil	D
13.	Black soil affected by salinity and alkalinity	D
14.	Black soil undifferentiated	D
15.	Mixed red and black soil	C
16.	Ferruginous red soil	B
17.	Ferruginous red gravely soil	B
18.	Red and yellow soil	C
19.	Laterite	B
20.	Laterite and black soil	D
21.	Brown soil under deciduous forest	B
22.	Forest soil	B
23.	Podsolic soil	B
24.	Foothill soil	C
25.	Mountain and hill soil	B
26.	Mountain meadow soil	B
27.	Peat	B

Sample Calculation 3.7

A watershed of area 50 hectares consist of 20 hectare with row crops cultivated in good terrace land and 30 hectares of good pasture land. The soil of the watershed belongs to hydrologic soil group C. Values of field capacity and wilting point soil moisture contents are 20 and 10%, respectively and soil moisture content of the soil on a particular day is 16%. Find the runoff amount on that day when a rainfall amount of 100 mm occurs in the watershed.

Solution

Using the data of land use, treatment conditions, hydrologic soil group and hydrologic condition, values of CN (AMC II) for 20 and 30 hectares lands are found from Table 3.8 as 78 and 74, respectively. The weighted CN value is given as

$CN_2 = (78 \times 20 + 74 \times 30)/50 = 75.6$

Using Eq. 3.41, $CN_1 = 57.68$ and Eq. 3.42, $CN_3 = 87.40$

Value of Sr_1 for $CN_1 = 57.68$ using Eq. 3.40 (replacing CN as CN_1) = 186.36. Similarly using Eq. 3.40 and replacing CN as CN_3, value of Sr_3 is obtained as 36.62.

Value of FFC (Eq. 3.44) = (16-10)/(20-10) = 0.6

Using Eq. 3.46, W_2 is obtained as 0.944 and using Eq. 3.45, W_1 is obtained as -0.461.

Using Eq. 3.43, value of Sr is obtained as 69.63 mm

Using Eq. 3.38, Q is obtained as

$$Q = \frac{(100 - 0.2 \times 69.63)^2}{100 + 0.8 \times 69.63} = 47.6 \ mm.$$

3.5 Water Year

Water year is the time duration from start of rainfall to the end of rainfall in a year. In general it starts from the time when precipitation exceeds the average evapotranspiration losses. In India it starts from 1st of June and ends on 31st May.

3.5.1 Runoff Characteristics of Streams

Streams can be classified into the following three categories based on the flow duration. They are:

- Perennial stream
- Intermittent stream and
- Ephemeral stream

3.5.1.1 Perennial Stream: A stream that carries flow throughout the year is called as perennial stream (Fig. 3.13). In such streams, there is considerable contribution of ground water flow throughout the year. Here water level remains above the stream bed always. Streams in humid zones are such type. In India, the streams in the Himalayan region get considerable flow due to snow melt in summer and these streams carry water almost throughout the year. Hence these streams are perennial type.

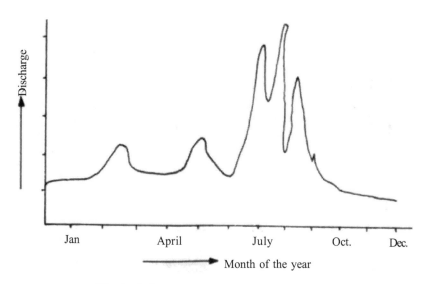

Fig. 3.13: Hydrograph of a perrineal stream

3.5.1.2 Intermittent Stream: An intermittent stream has limited contribution of groundwater. These streams do not carry water throughout the year. In rainy season, the groundwater table rises and feeds considerable amount of water to the streams such that water level in the stream remains above the stream bed. However, in summer (dry) season, the groundwater table goes down and does not feed to the stream. Further, there is no or scanty rainfall in summer. Hence in summer, the water level in the stream goes below the stream bed. Stream in southern and eastern part of India are mostly intermittent type. Fig. 3.14 shows the hydrograph of an intermittent stream.

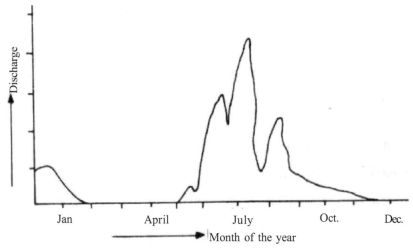

Fig. 3.14: Hydrograph of a intermittent stream

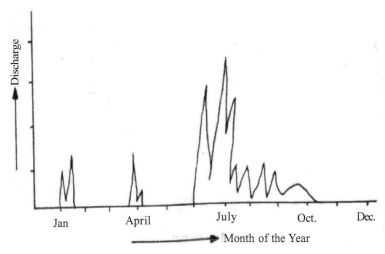

Fig. 3.15: Hydrograph of an ephemeral stream

3.5.1.3 Ephemeral Stream: An ephemeral stream does not have any base flow/groundwater contribution. They are found in arid zones. They carry runoff due to rainfall only. The hydrographs of such streams show series of short duration spikes marking flash flows in reference to storms (Fig. 3.15). The streams become dry soon after the cessation of rainfall. Typically an ephemeral stream does not have any well defined channels.

3.6 Flow Duration Curve

It is a fact that the stream flows varies over a water year. One of the popular methods to study the variability of stream flow is through flow-duration curve. A flow duration curve of a stream is the graphical presentation of discharge data against the per cent of time the flow was equaled or exceeded which is also called as discharge-frequency curve. Plotting position formula such as a one given by Weibul is used to find out the per cent of time flow was equaled or exceeded to a given value. Plotting position is described briefly as given below.

The stream flow data which may be daily, weekly, monthly or annually are arranged in descending order with the highest valued data being given rank number 1. If the data are too large in number, a class interval may be used for finding out the plotting position. The plotting position for any discharge or class value is given as:

$$P \ (\%) = \frac{m}{N+1} \ x \ 100 \tag{3.49}$$

where, P = probability of an event equal to or exceeded in percent, m = rank number of the event (discharge or class value) after arranging the events in descending order of their magnitudes and N = total number of years of recorded data. A plot of discharge against probability (given by Eqn. 3.49) is the flow duration curve. In case of grouped data having class intervals, lowest value of discharge may be plotted as ordinate and corresponding probability may be plotted in abscissa of a log-log paper. Details about estimation of the plotting position for a hydrological event are discussed in Chapter 2. From the flow duration curve, one can calculate the discharge at any desired probability level. Fig. 3.16 represents the flow duration curve of a perennial, intermittent and ephemeral stream. Flow duration curve represents the cumulative frequency distribution and can be considered to represent the stream flow variation of an average year. The ordinate of the discharge (Q_p) at any percentage probability *(P)* represents the flow magnitude in an average year that can be expected to equal or exceed P per cent of time and is termed as *P%* dependable flow. In case of perennial stream, Q_{100} = 100% dependable flow is considered as finite value whereas in intermittent and ephemeral streams, Q_{100} is zero i.e. in intermittent and ephemeral streams the stream flow is zero for a finite part of the year.

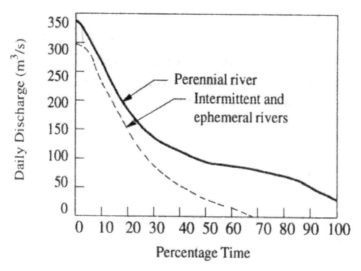

Fig. 3.16: Flow duration curve of a stream

3.6.1 Characteristics of Flow Duration Curve

The characteristics of flow duration curves are:

- The slope of a flow duration curve for a given stream may be steep, flat, may be flat at top of the curve or may be flat at the lower end of the curve depending on the interval of the data selected. For daily stream flow data the slope is steep whereas for monthly data there is smoothening of small peaks which makes the slope curve type. The steep slope indicates highly variable flow. Flat slope which is curve type represents there is little variation in stream flow. Flat portion at the top of the curve represents that the stream has large flood plain and there is no rain in winter season. Flat portion at the lower end of the curve indicates that there is considerable groundwater flow in the stream.

- The presence of a reservoir in a stream considerably modifies the virgin flow duration curve depending on the nature of flow regulation. Fig. 3.17 represents the effect of a typical flow regulation.

- The virgin flow duration curve when plotted on a log probability paper plots as a straight line at least over the central region. From this property, various coefficients expressing the variability of the flow in a stream can be developed for the description and comparison of various streams.

- The chronological sequence of occurrence of the flow is masked in the flow-duration curve. It does not specify the period of occurrence of the flow. For example a discharge of 100 m^3/sec in a stream will have the same percentage P whether it occurred in February or July. This aspect must be taken into account while interpreting a flow-duration curve.

The flow characteristics of different streams can be studied well when the flow-duration curve is plotted in a log-log paper. The steep slope of such a curve indicates highly variable discharge whereas the flat slope indicates low variability.

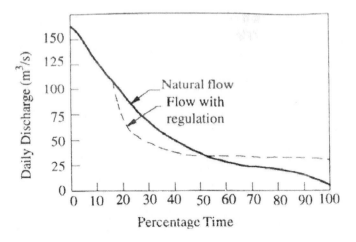

Fig. 3.17: Reservoir regulation effect

3.6.2 Use of Flow Duration Curve

Flow duration curve is used for the following purposes.

- It is used for planning of water resources projects.
- It helps in design of drainage system.
- It can be helpful in deciding the hydro-power potential of a stream.
- It is used in flood control studies.
- It is used in estimating the sediment load and dissolved solids loads in a stream.
- It is used in comparing the adjacent catchments with a view to extend the stream flow data.

Sample Calculation 3.8

The daily flows (m³/sec) of a river measured for three consecutive years are grouped into the various class intervals and are given below along-with their total frequency of occurrence. Out of these data of three years, one year was leap year. Estimate the flow at 50, 75 and 100 percent probability level.

Discharge, m³/sec	5.1-10.0	10.1-15	15.1-20	20.1-25	25.1-30	30.1-40	40.1-50	50.1-60	60.1-80	80.1-100	00.1-1120	120.1-140
Total frequency of occurrence	1096	1091	1046	963	837	665	430	236	132	70	25	6

Solution

The number of data collected in three consecutive years including one leap year is 365 x2 + 366 =1096. Hence, N = 1096. Total frequency of occurrence here represents the number of days the flow is equal to or greater than the class interval which gives the rank number, m. Computation of plotting position i.e. probability of exceedenec (given by Eqn. 3.49) is shown in tabular form as below.

The smallest value of discharge in each class interval is plotted against probability, P on a log-log paper (Fig. 3.18).Using this plot, the flow at 50, 75 and 100 percent probability level i.e. Q_{50}, Q_{75}, and Q_{100} are estimated as 35, 26 and 4.9 m³/sec, respectively.

Discharge, m³/sec	140-120.1	120-100.1	100-80.1	80-60.1	60-50.1	50-40.1	40-30.1	30-25.1	25-20.1	20-15.1	15-10.1	10-5.1
m	6	25	70	132	236	430	665	837	963	1046	1091	1096
P (%) = $\dfrac{m}{N+1} \times 100$	0.55	2.28	6.38	12.03	21.51	39.19	60.62	76.30	87.78	95.35	99.45	99.91

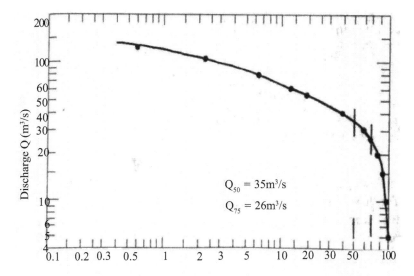

Po = Percentage time indicated discharge is equalled or exceeded

Fig. 3.18: Flow-duration curve of sample calculates as 3.8

3.7 Flow – Mass Curve

The flow-mass curve is a plot of accumulated discharge volume or accumulated total flow against time plotted in chronological order. The ordinate of mass curve, V at any time t is thus given as:

$$V = \int_{t_0}^{t} Q \, dt \qquad (3.50)$$

Where, t_0 = time at the beginning of the curve and Q = discharge rate. The flow-mass curve or simply called as mass curve is an integral or summation curve of the hydrograph. The flow-mass curve is also known as Rippl mass-curve, after the name of Rippl (1982), who first suggested its use. The ordinate of the mass-curve is in units of volume in million m^3. Other units of ordinate may be m^3/sec day, ha. m etc. Similarly the abscissa represents time (taken in chronological order) which may be month, week, day depending on the data used. The slope of the mass-curve represents discharge rate (i.e. $dV/dt = Q$). Fig. 3.19 shows a typical flow-mass curve. If two points say M and N are connected by a straight line, the slope of this line represents the average rate of flow that can be maintained between time t_m and t_n provided the reservoir has adequate storage. The average discharge over the whole period of plotted record is represented by the slope of the line AB joining the first and last point of the mass curve.

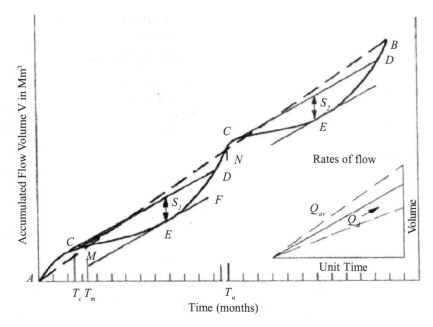

Fig. 3.19: A typical flow-mass curve

Mass-curve method can be used to compute the storage capacity of reservoir.

3.8 Yield of a Stream

The total amount of water that can be expected from a stream in a given period say a year is called as annual yield/yield of the stream. It is used to calculate the mean annual runoff volume of the stream. The calculation of yield is of fundamental importance in all the water resources planning and development studies. Various methods used to compute the yield of a stream are:

- Rainfall-runoff correlation
- Empirical formulae and
- Watershed simulation technique

3.8.1 Rainfall-Runoff Correlation

Runoff generated from a catchment/watershed due to rainfall is a complex phenomenon depending on several factors including watershed and climatic factors. These factors have been discussed earlier in this chapter. There are some problems in the gauging site to measure the runoff. The paucity of the runoff data creates problem in planning of water resources projects and in the study of hydrology. However, if a relation between the rainfall and resulting runoff in a watershed can be developed, then this will help to simulate the runoff for known values of rainfall. For this purpose, several years of data of rainfall and the resulting runoff of a watershed need to be collected and the data need to be plotted in a graph taking rainfall in abscissa and runoff as ordinate of the graph. Then a best fit line may be drawn between the data points which gives a rough estimate to compute the runoff from known values of rainfall. A liner regression line may be fitted between rainfall, P and runoff, R and if the value of correlation coefficient, is close to unity, then the regression line can be used to estimate R for given value of P. Equation of the linear regression line between P and R is given as:

$$R = a P + b \tag{3.51}$$

Where, a and b are the coefficients of regression given as:

$$a = \frac{N\left(\sum P R\right) - \left(\sum P\right)\left(\sum R\right)}{N\left(\sum P^2\right) - \left(\sum P\right)^2} \tag{3.52}$$

$$\text{and } b = \frac{\sum R - a \sum P}{N} \tag{3.53}$$

in which N = number of data sets of P and R. The value of correlation coefficient (r) is computed as:

$$r = \frac{N\left(\sum P R\right) - \left(\sum P\right)\left(\sum R\right)}{\sqrt{\left[N\left(\sum P^2\right) - \left(\sum P\right)^2\right] \cdot \left[N\left(\sum R^2\right) - \left(\sum R\right)^2\right]}} \tag{3.54}$$

The value of r lies between 0 to +1. A value close to +1 gives good correlation between P and R. However a value 0.6 to +1.0 indicates satisfactory correlation.

Sometimes other correlations except linear may exist between P and R. For large catchments, power form relationship may exist. This power form relationship may be studied in the same way as the linear relationship by transforming the data of P and R into logarithmic form and then computing coefficient a and b by the Eqns. (3.52) and (3.53). Finally the value of coefficient a and b so calculated is transformed into anti-logarithmic form to produce the power form relation between P and R.

Sample Calculation 3.9

Following table gives the rainfall and runoff recorded in a basin. Develope a correlation between them.

Rainfall, P, cm	100	120	45	50	65	23	30	112	80	85	34	20	12	25
Runoff, R, cm	45	52	23	25	37	12	14	55	43	42	18	11	7	13

Solution:

For the given data we have N = 14

$\Sigma P = 801$, $\Sigma R = 397$, $\Sigma P^2 = 63073$, $\Sigma R^2 = 14893$,

$\Sigma PR = 30537$, $\left(\sum P\right)^2 = 641601$, $\left(\sum R\right)^2 = 157609$

For the correlation equation, $R = aP + b$ by (Eqn. 3.51) we have

$$a = \frac{N\left(\sum PR\right) - \left(\sum P\right)\left(\sum R\right)}{N\left(\sum P^2\right) - \left(\sum P\right)^2}$$

Putting the values of data so computed as above for the problem in the above equation, we get $a = 0.453$.

Thus, the value of b given by Eqn. (3.53) is:

$$b = \frac{\sum R - a \sum P}{N}$$

Putting the values of data so computed as above for the problem in the above equation, and value of a as 0.453, we get $b = 2.401$.

Hence the correlation between rainfall, P and runoff, R is:

$R = 0.453\, P + 2.401$

In this equation both P and R are in cm.

Correlation coefficient r is given by Eqn. (3.54) as:

$$r = \frac{N\left(\sum PR\right) - \left(\sum P\right)\left(\sum R\right)}{\sqrt{\left[N\left(\sum P^2\right) - \left(\sum P\right)^2\right]\cdot\left[N\left(\sum R^2\right) - \left(\sum R\right)^2\right]}}$$

Putting the values of data so computed as above for the problem in the above equation we get $r = 0.988$ which indicates good correlation between rainfall and runoff.

3.8.2 Empirical Formulae

Different empirical formulae to compute the runoff from rainfall have already been discussed earlier in this Chapter.

3.8.3 Watershed Simulation Technique

Recall the hydrologic budget equation (Eqn. 1.7) of Chapter 1 which is:

$P - R - E - T - G = \Delta S$

where, P = precipitation, R = runoff, E = evaporation, T = transpiration, G = groundwater flow and and ΔS = change in storage.

The above equation can be rearranged to get R and the value of R can be estimated by knowing other values of the parameters used in the hydrologic budget equation.

Now-a-days digital computers are available for use to compute runoff by water budget equation. This technique of predicting runoff from a known value of rainfall in a catchment is called as *deterministic watershed simulation*. In this the mathematical relationships describing the interdependence of various parameters in the system are first prepared which is called as a model. The model is then calibrated by simulating the known values of rainfall and runoff records. The calibrated model is further verified which is called as model validation by using recorded rainfall data and the simulated runoff is compared with the observed ones. Various statistical tests may be undertaken for model validation. Several scientists have proposed several watershed simulation models to compute runoff. Some of the widely used hydrological models are Stanford Watershed Model (SWM), Streamflow Synthesis and Reservoir Regulation Model (SSARR model), Kentucky Watershed Model (KWM) etc. It is to be remembered that each model has got its own limitations and before using the model to predict runoff these limitations are to be taken care. Further, the model has to be calibrated and validated using sufficient numbers of years of data.

Question Banks

Q1.   Define the followings:

(i) Interflow (ii) Overland flow (iii) storm runoff (iv) Groundwater flow (v) Virgin flow (vi) Yield of a stream (vii) Hydrograph (viii) Hyetograph (ix) Flow-duration curve (x) Flow-mass curve (xi) probability of exceedenace (xii) -index (xiii) W – index (xiv) Valley Storage (xv) Drainage density (xvi) Form factor (xvii) Compactness factor.

Q2. With a flow chart explain the formation of runoff produced in a catchment due to rainfall.

Q3. What do you mean by runoff? What are the factors on which the runoff of catchments depends on?

Q4. What is virgin flow? The following table gives values of measured discharges at a stream gauging station in a year. At the upstream side of the gauging station there is weir through which 0.62 Mm^3 of water is diverted per month for industrial uses and there is a return flow of 0.23 Mm^3 per month. Estimate the *virgin flow.*

Month	Jan	Feb	Mar	Apr	May	Jun	Jul	Aug	Sep	Oct	Nov	Dec
Measured flow, Mm^3	2.2	1.65	1.96	1.13	0.95	5.1	8.5	10.2	12.0	7.8	5.5	3.5

Q5. How is the nature of precipitation affect the runoff of a catchment?

Q6. Explain the different physiographic factors that influence the runoff of a catchment?

Q7. Which one of the following will produce more runoff and explain why?

(i) Fan shaped catchment and (ii) Fern shaped catchment.

Q8. What do you mean by rainfall-runoff modeling? Explain the watershed routing method to estimate the runoff from a given rainfall mount.

Q9. Briefly explain the different methods used to compute the runoff produced from a catchment.

Q10. Write notes on the following empirical formulae to estimate the runoff from a catchment.

(i) Lacey's formula (ii) Khosla's formula (iii) Inglis and Desouza's formula (iv) Mayer's formula (v) Coutagne formula (vi) Dicken formula (vii) Ryve's formula.

Q11. Describe how is the runoff estimated by the following methods:

(i) Binnie's percentage (ii) Barlow's percentage (iii) Strange's tables and curves

Q12. Describe the rational method to estimate the runoff rate of a catchment.

Q13. What are the limitations to use the rational method for computation of runoff rate of a catchment?

Q14. A watershed has an area of 100 ha under pasture, 120 ha under cultivation and 50 ha under the forest. The cultivated area has loamy soil whereas the pasture and forest lands have heavy clay soil with very low infiltration rate. The average longitudinal slope of the watershed with the longest reach contributing runoff is0.1 percent. The one hour rainfall intensity for 10 years frequency interval of the watershed is 60 mm/hr. Estimate the runoff rate of the watershed. Assume the ratio of one hour rainfall intensity to rainfall intensity of any other duration as 1.3.

Q15.   What is modified Rational formula? How is it used to compute runoff?

Q16.   The mean monthly rainfall and temperature data of a station in Odisha as recorded are given below. Estimate the monthly and annual runoff by Khosla's method.

Month	Jan	Feb	Mar	Apr	May	Jun	Jul	Aug	Sep	Oct	Nov	Dec
Temp., °C	16	18	22	25	34	42	40	38	35	26	19	15
Rainfall, cm	4.3	3.2	4.9	3.5	9.7	12.3	20.9	24.2	18.9	2.1	1.3	0.0

Q17.   Explain how the runoff rate of a catchment is estimated by the curve number method.

Q18.   Write about the four hydrologic soil groups. What are their characteristic features?

Q19.   Which one of the following will produce more runoff rate and explain why?

(i) Hydrologic soil group A and (ii) Hydrologic soil group B.

Q20.   A watershed has an area of 100 ha out of which 60 ha is under the row crop with good terrace and the rest is under good pasture land. Soil of the watershed belongs to the hydrologic soil group B. Values of field capacity and wilting point soil moisture contents are 22 and 10%, respectively and soil moisture content of the soil on a particular day is 18%. Find the runoff amount on that day when a rainfall amount of 120 mm occurs in the watershed.

Q 21.   What do you mean by the term AMC? What are the factors on which the AMC depends on?

Q22.   What do you mean by the term CN? Write in short the factors on which the CN depends on?

Q 23.   What is lag time and how is it determined?

Q24.   Write short notes on the followings:

(i) Hydrologic water budget equation (ii) Watershed simulation technique

(iii) Synthetic unit hydrograph (iv) Models used in runoff estimation.

Q25.   Explain in details how Cook's formula is used to estimate peak rate of runoff? A watershed has an area of 200 ha having following characteristics. Estimate the runoff rate considering 50 years recurrence interval.

Land use with area	Average slope, %	Infiltration rate, cm/hr	Vegetal cover	Surface storage
Cultivated, 150 ha	2	0.75	Less than 10% area is under good grass cover	Ponds/tanks cover less than 2% drainage area and area is not well drained
Pasture, 50 ha	5	1.00	About 95% area is under good grass cover	Negligible surface storage

The watershed has length to width ratio 3. Assume the rainfall factor as 1.3 and frequency factor as 1.0. Assume the relief factor for the cultivated land as 10 and that for the pasture land as 15.

Discharge, m³/sec	5.1 – 10.0.	10.1 - 15	15.1 - 20	20.1 - 25	25.1 -30	30.1 -40	40.1 - 50	50.1 - 60	60.1 - 80	80.1 - 100	100.1 - 120	120.1 - 140
Total frequency of occurrence	1096	1091	1046	963	837	665	430	236	132	70	25	6

Q26. With diagram describe the characteristic features of the following streams:

(i) Perennial stream (ii) Intermittent stream and (iii) Ephemeral stream

Q27. How is the variability of stream flow studied by flow-duration curve? How is the curve prepared? Explain its characteristics and use in hydrology.

Q28. The daily flows (m³/sec) of a river measured for three consecutive years are grouped into the various class intervals and are given below along-with their total frequency of occurrence. Out of these data of three years, one year was leap year. Estimate the flow at 25, 50 and 100 percent probability level.

Q29. What is a flow-mass curve? How is it prepared? Describe its use in the field of Hydrology.

Q30. Describe the linear relationship between the rainfall and runoff in a basin. How are the coefficients of the correlation determined? Also write the equation to compute the coefficient of correlation between the rainfall and runoff.

Q31. Following table gives the rainfall and runoff recorded in a basin. Develop a correlation between them.

Rainfall, P, cm	110	120	43	57	68	23	30	112	80	85	34	22	12	30
Runoff, R, cm	49	55	23	25	39	14	14	55	43	42	18	11	6	15

Q32. The rainfall rates for successive 30 min intervals up to 3 hrs of a storm are: 0, 1.8, 2.5, 5.2, 3.0, 2.2 and 1.0 cm/hr. Assuming that the total runoff produced by the storm in 3 hr is 3.4 cm, calculate the value of -index. Also calculate the value of W – index.

Q33. The rainfall rates for successive 20 min intervals up to 140 min of a storm are: 2.3, 2.8, 10.2, 7.5, 2.2, 2.2 and 5.1 cm/hr. Assuming -index as 3.3 cm/hr find out the total runoff and also compute W-index.

References

Beven, K.J. 2001. Rainfall-Runoff Modelling: The Primer. Wiley.

Central Water Commission, India. 1973. Estimation of Design Flood Peaks, Flood Estimation Directorate, Report No. 1/73, New Delhi.

Garg, S.K. 1987. Irrigation Engineering and Hydraulic Structures. Khanna Publishers, Delhi.

Gupta, S.K., Ram Babu and Tejwani, K.G. 1970. Nomograph and important parameters for estimation of peak rate of runoff from small watershed. Indian Journal of Agricultural Engineering, ISAE, New Delhi, Vol. 8(3): 25-32.

Hann, C.T. , Johnson, H.P. and Brakensick, D.L. 1982. Hydrologic Modelling of Small Watersheds. (9 edition) American Society of Agricultuural Engineers, St. Joseph.

Kirpich, Z.P. 1940. Time of concentration of small agricultural watersheds. Civil Engg., 10:362.

Linsley, R.K., Kohler, M.A. and Paulhus, J.L.H. 1979. Applied Hydrology. Tata McGraw-Hill Publishing Co. Ltd., New Delhi.

Mayer, L.D. 1926. Transactions of ASCE, Vol. 89, 985.

Panigrahi, B. 2011. Irrigation Systems Engineering. New India Publishing Agency, New Delhi. 325 pp.

Sharma, R.K. and Sharma, T.K. 2002. Irrigation Engineering including Hydrology, S. Chand and Company Ltd. New Delhi.

Sharpley, A.N. and Williams, J.R. (Edn.) 1990. EPIC-Erosion/Productivity Impact Calculator: (1) Model Documentation, pp.235.

Shaw, E.M. 1994. Hydrology in Practice. 3rd ed., Chapman and Hall.

Singh, J. 1964. Irrational Use of the Rational formula. Er presented at the third annual meeting of the Indian Society of Agricultural Engineers, Allahabad.

Subramanya, K. 2003. Engineering Hydrology, Tata McGraw-Hill Publishing Co. Ltd., New Delhi, 392 pp.

Suresh, R. 2008. Watershed Hydrology. Standard Publishers Distributors, Delhi, pp. 692.

Tejwan, K.G. Gupta, S.K and Mohan, H.N. 1975. Soil and Water Conservation Research (1957 – 71), ICAR, New Delhi.

US Soil Conservation Service.1972. National Engineering Handbook, Section 4, Hydrology, Department of Agriculture, Washington, D.C.

van den Akker, C. and Boomgaard, M.E. 1996. Hydrologie. Technische Universiteit Delft.

CHAPTER 4

Abstractions from Precipitation

4.1 Introduction

Precipitation is a major component of *hydrological cycle*. It occurs because of condensation of water vapour molecules in the atmosphere. When the precipitation falls on the grounds, it undergoes several phases before it appears as runoff in the streams. Certain amount of precipitations is intercepted by the foliage of plants and roofs and other parts of the building and other hydraulic structures. This intercepted water is called as interception loss which is evaporated to the atmosphere. Sometimes the dried surface of the buildings and other structures soak certain amount of precipitation while falling on them which also adds to interception loss. After striking the land masses, certain amount of the precipitated water gets infiltrated into the soil surface which then moves both in lateral as well as in vertical direction. This loss is called as infiltration loss. For the study of groundwater, infiltration loss is not a loss but rather it is a gain in terms of groundwater recharge. However for a surface water hydrologist, it is a loss since it decreases the amount of runoff contributing to the stream or any other water bodies.

There are other losses also like evaporation from soil and transpiration by the plants. The combined loss due to evaporation of soil mass and transpiration by plants is called as evapotranspiration. Both evaporation and transpiration form important links in the hydrologic cycle. Before the precipitation reaches the outlet of a basin as runoff, the depression storage over the uneven land masses is to be satisfied. The combined loss due to interception and depression storage is called as *initial loss*. The rest of the precipitation after satisfying the demands of losses like evaporation, transpiration, interception, infiltration and depression storage moves over the land mass which may be called as overland flow or surface runoff. It is to be remembered here that surface runoff is a part of total runoff or simply called as runoff or stream flow. Sometimes, some portion of the infiltrated water in the form of interflow and base flow also contributes to the runoff. Thus for a hydrologist the difference of precipitation and total runoff reaching a stream/waterbody is refereed as water loss. Various water losses as discussed above which are also called as abstractions from precipitation are very important from the point of view of engineering hydrology which are dealt separately in this chapter.

4.2 Evaporation

Conversion of water molecules into the vapour forms due to the effect of heat flux and temperature is termed as evaporation. It is a natural process and continues because of falling of incident rays of the solar radiation on the water bodies and also on wet soil masses. It is observed that the evaporation loss in hot months of May and June is 2-5 times more than that in winter months of December and January. Estimate of evaporation across the country indicates that it is as high as 2000 mm in the semi-arid tropics. Annual average value of the evaporation loss varies from 1400 to 1800 mm across major parts of the country: the value being highest in west Rajasthan, parts of Saurasthtra and Tamilnadu and lowest in coastal Mysore, Bihar plateau and east M.P. Reliable statistics reveal that about 70 M ha m of water evaporates from the water bodies and land surfaces out of the total annual precipitation of about 400 M ha m received by the country. The total evaporation losses from the water surfaces only are about 5 M ha m from the total storage of 15 M ha m in the reservoirs including ponds, lakes spread all over the country (Belgaumi *et al.,* 1997).

4.3 Evaporation Losses

Evaporation is the process of conversion of water from liquid state to gaseous/ vapour state at the free surface below the boiling point through the transfer of heat energy. It is a natural process and occurs when solar radiation falls on a water body and energy so transmitted is acquired by water molecules which in turn gets separated and move upward causing evaporation. It is a cooling process in that the latent heat of vapourisation must be provided by the water body. In the hydrologic cycle, evaporation plays a vital role in causing precipitation. Rate of evaporation depends on the atmospheric demand like air and water temperature, vapour pressure at the water surface and air above, wind velocity, atmospheric pressure, quality of water as well as nature and size of the water body. The factors affecting the rate of evaporation loss are discussed below.

Temperature: Rate of evaporation is directly proportional to both air as well as water temperature i.e. it increases with temperature and vice-versa. This is because when temperature increases the saturated vapour pressure also increases which accelerates the rate of evaporation. However reports reveal that there is a good correlation between water temperature and rate of evaporation whereas the correlation is not strong with air temperature. Because of this reason a water body may have different degrees of evaporation rate in various months having same mean monthly air temperature in theses months.

Vapour Pressure: The rate of evaporation is directly proportional to the difference between the saturation vapour pressure at the water temperature (e_w) and the actual vapour pressure in the air (e_a). This difference of vapour pressure is called vapour pressure deficit. The above law is called as *Dalton's law of evaporation* and is given as:

$$E_L = C (e_w - ea) \tag{4.1}$$

where, E_L = rate of evaporation, mm/day and C = a constant. The other terms are defined as above. The terms e_w and e_a are in mm of mercury. It is clear from Eqn. (4.1) that the rate of evaporation will continue till e_w and e_a are equal. At higher humidity, the vapour pressure deficit will be less which decreases the rate of evaporation.

Wind Velocity: Wind affects the evaporation rate to a great extent. Wind aids in removing the evaporated water vapour from the zone of evaporation and ultimately creates higher scope for evaporation. However, if the wind velocity is large enough to remove all the evaporated water vapour, then any further increase in wind velocity does not influence the evaporation. Hence, the rate of evaporation increases with the wind velocity up to a thresh hold speed beyond which any further increase in the wind speed has little influence on rate of evaporation. This thresh hold wind speed value is a function of size of the water surface. For example for large water bodies, high speed turbulent winds are required to cause maximum rate of evaporation.

Atmospheric Pressure: Other factors remaining same, a decline of atmospheric pressure at the high altitudes, cause evaporation to increase.

Water Quality: The vapour pressure of a solution containing dissolved salts is less than that of pure water and hence the solution containing the salt retards the evaporation rate. The percent rate of reduction of evaporation approximately corresponds to the percent rate increases of the specific gravity. This is the reason why the rate of evaporation in sea water is less (about 2 to 3%) than that of the surface water which is fresh.

Nature and Size of Water Body: Water spread area of a particular size and shape of storage structure with a given side slope varies with depth of water in the structure and hence the evaporation loss of the storage structure varies daily. Deep water bodies have more heat storage than the shallow ones and thus affects the rate of evaporation. For example, the deep storage in large water bodies may store radiation energy in summer and may release it in winter causing less evaporation in summer and more in winter as compared to a shallow storage in the same water body in similar situation. However, this affect seldom affects the annual evaporation rate.

4.4 Estimation of Evaporation

Estimation of evaporation is of utmost importance in many hydrologic problems associated with planning, operation and management of reservoirs and irrigation systems. In arid regions, this is very important to conserve the scarce and precious water resources. The various methods of estimating evaporation are (i) Analytical method (ii) Empirical method and (iii) Using evaporimeter data by Pan evaporation method.

4.4.1 Analytical Method

There are three analytical methods to compute evaporation losses. They are water budget method, mass transfer method and energy budget method.

4.4.1.1 Water Budget Method: This is based on water balance approach in which various inflows and outflows to and from the reservoir are considered. The equation used to evaluate evaporation is:

$$E = S_1 + R + I - O \pm U - S_2 \tag{4.2}$$

where, E is evaporation, S_1 and S_2 are initial and final storage R is rain-fall, I is surface inflow to the reservoir, O is the surface outflow from the reservoir and U is the underground inflow or outflow into or from the reservoir. The various terms used in above equation are expressed in terms of depth unit, generally in cm.

4.4.1.2 Mass Transfer Method: The evaporation in this method is computed as:

$$E = \frac{-10K^2\rho(q_2 - q_1)(u_2 - u_{1)}}{\left(\ln\left\{\frac{z_2}{z_1}\right\}\right)^2} \tag{4.3}$$

where, E = evaporation, mm; K = von Karman constant = 0.41; ñ = density of air, (g/cc); q_2 and q_1 are specific humidity at heights z_2 and z_1, respectively that are taken as 2 m and 1 m, respectively above the water surface; and U_2 and U_1 are wind speeds (cm/s) at heights z_2 and z_1, respectively.

4.4.1.3 Energy Budget Method: In this method, evaporation is given as:

$$E = \frac{R - G}{L}\left(\frac{1}{1 + \beta}\right)$$ (4.4)

where, $R = H + LE + G$ and β is given as

$$\beta = \frac{T_s - T}{e_s - e} \times 0.61 \times 0.001\, P$$

In the above equations, E = evaporation (mm); R = net radiation flux received at the surface (cal/cm²); H = sensible heat flux (cal/cm²); LE = latent heat flux, where L = latent heat of vapourisation (cal / g); E as defined above; G = heat flux in the soil or water (cal/cm²); β = Bowen ratio; T_s = temperature of water surface (⁰C); T = air temperature (⁰C); e_s = saturation vapour pressure at T_s in hectopascal (hpa); e = vapour pressure of air at T in hpa; and P = station level pressure hpa.

Based on Penman's equation (1948), the Kohler-Nordenson-Fox equation (Pochop *et al.*, 1985) describes evaporation from water bodies as the combination of water loss due to radiation heat energy and the aerodynamic removal of water vapor from a saturated surface.

The general form for the combination equation is

$$E = \frac{\Delta}{\Delta + \gamma} Rn + \frac{\gamma}{\gamma + \Delta} Ea$$ (4.5)

where, E is the evaporation in inches per day; Δ is the slope of the saturation vapor pressure curve at air temperature in inches of mercury per degree F; γ is the psychrometric constant in inches of mercury per degree F; R_n is the net radiation exchange expressed in equivalent inches of water evaporated; and E_a is an empirically derived bulk transfer term. E_a is expressed as

$$E_a = f\,(u)\,(e_s - e_d)$$

where, $f\,(u)$ is a wind function; and $(e_s - e_d)$ is the vapour pressure deficit.

So, the evaporative water loss from water body can be checked by obstructing the radiation heat energy coming on to the water surface and further reduction can be achieved by lessening the vapour pressure deficit or the wind speed in the microclimate. Solar radiation is considered as the vital source for causing evaporation from water bodies.

4.4.2 Empirical Method

There are several empirical methods available for computing evaporation loss. Most of these methods take into account the evaporation loss of the reservoir which in turn depends on temperature of air –water interface, wind velocity, relative humidity and saturated vapour pressure of air. A few of the popular empirical methods are mentioned below.

4.4.2.1 Meyer's Formula: According to this formula evaporation loss is estimated as:

$$E_L = K_M \left(e_w - e_a \right) \left(1 + \frac{u_9}{u_1} \right) \tag{4.6}$$

where, E_L is lake evaporation, mm/day; e_w is saturated vapour pressure at the water surface temperature in mm of mercury; e_a is actual vapour pressure of overlying air at a specified height in mm of mercury u_9 is monthly mean wind velocity in km/hr at about 9 m above the ground, u_1 is monthly mean wind velocity at 1 m above ground level in km/hr and K_M is coefficient accounting for various other factors with a value of 0.36 for large water bodies and 0.50 for small and shallow water bodies (Subramanya, 2003).

4.4.2.2 Rohwer's Formula: This formula considers a correction for the effect of pressure in addition to the wind speed effect and is given as:

$$E_L = 0.771 \, (1.465 - 0.000732 \, p_a) \, (0.44 + 0.\,0733 \, u_0) \, (e_w - e_a) \tag{4.7}$$

where, E_L, e_w and e_a are as defined above; p_a is mean barometric reading in mm of mercury, u_0 is mean wind velocity in km/hr at ground level.

It is to be noted that the above mentioned empirical formula are based on several assumptions and are region based and so cannot be used universally to estimate evaporation loss. However, they can predict approximate values and so should be used cautiously at other locations.

In the use of estimation of evaporation loss by Eq. (4.6) and (4.7), saturation vapour pressure at a given temperature (e_w) can be calculated as:

$$e_w = 4.584 \exp\left(\frac{17.27\, t}{237.3 + t}\right) \text{ mm of mercury} \tag{4.8}$$

where t = temperature in °C.

Alternatively one can use Table 4.1 to find out the value of e_w for any temperature, t in °C.

Table 4.1 Saturation vapour pressure of water at different temperature

Temperature, °C	Saturation vapour pressure, e_w, mm of Hg	A, mm/°C
0	4.58	0.30
5.0	6.54	0.45
7.5	7.78	0.54
10.0	9.21	0.60
12.5	10.87	0.71
15.0	12.79	0.80
17.5	15.00	0.95
20.0	17.54	1.05
22.5	20.44	1.24
25.0	23.76	1.40
27.5	27.54	1.61
30.0	31.82	1.85
32.5	36.68	2.07
35.0	42.81	2.35
37.5	48.36	2.62
40.0	55.32	2.95
45.0	71.20	3.66

The wind velocity data at any desired elevation, h within 500 m above the ground level , (u_h) can be calculated by $1/7^{th}$ power law as:

$$u_h = u_1\, h^{1/7} \tag{4.9}$$

where u_1 is wind velocity at 1 m above the ground level in km/hr, u_h is wind velocity at h m above the ground level and h is height above ground level in m.

Sample Calculation 4.1

Estimate the evaporation loss in a reservoir in a week which has a surface area of 300 ha. The average values of the reservoir parameters within the week are as follows:

Water temperature is 20°C, relative humidity is 30%, wind velocity at 1 m above the ground level is 15 km/hr.

Solution

The reservoir has surface area = 300 ha = 3×10^6 m².

Water surface temperature as given is 20°C.

Using Eq. (4.8) or using Table (4.1) at temperature 20 °C, e_w = 17.54 mm of mercury.

Since relative humidity is given as 30%, we have e_a = 0.30 x 17.54 = 5.262.

The wind velocity at 1 m above the ground (u_1) is given as 15 km/hr.

Wind velocity at 9 m above ground level (u_9) can be estimated by Eq. (4.9) as:

u_9 = 15 x $(9)^{1/7}$ = 20.53 km/hr.

Assuming the reservoir having area 300 ha as large water body, the value of coefficient, K_M is taken as 0.36.

Average daily evaporation loss by Meyer's formula (Eq 4.6) is estimated as:

E_L = 0.36 x (17.54 – 5.262) x (1 + 20.53/15) = 10.47 mm/day.

Evaporation loss in a week = 7 x 10.47 = 73.29 mm.

Volume of evaporation in a week = 73.29 mm x 3×10^6 m² = 73.29 x 3×10^6 x 10^{-3} = 219870 m³.

4.4.3 Pan Evaporation Method

In this method, the evaporation loss measured by an evaporimeter located near the reservoir is measured. The ratio of the evaporation loss measured by the evaporimeter to the reservoir evaporation loss on annual basis is a constant and hardly varies from region to region. However, on short term period basis, the ratio depends on a number of factors like heat storage of the evaporimeter/pan, size of the pan, water depth in the pan, colour of pan etc. The evaporation loss from a pan is greater than that of the reservoir. As the size of the pan increases, the difference of the evaporation loss of both pan and reservoir decreases. The rate of evaporation, in general, is observed to be more for deep buried pan with higher depth of water than shallow pans with less depth of water. Similarly, the evaporation from dark pans is greater than that from unpainted, galvanized pans. The white coloured pan shows the lowest evaporation. Covering the pan with a screen helps to reduce the pan evaporation to the equivalent of that from a reservoir. The screen helps to maintain the pan temperature uniform throughout day and night and reduces turbulence at the water surface so as to maintain uniform rate of evaporation (Lenka, 2001). .

4.4.3.1 Selection of Site of Evaporimeter/Evaporation Pan: While selecting the site of evaporimeter near the reservoir, care should be taken so that:

(i)   It is very near to the reservoir.

(ii)   The evaporimeter must be placed on a level ground.

(iii) There should not have any obstruction for evaporation of water from the pan. If there are any tree or building near by, then the pan should be placed at the location which is at least 10 times more then the height of the tree or building.

(iv)   The pan should not get submerged due to heavy rain or flood.

(v)   It should be located upwind of the reservoir.

(vi)   The highest ground water level must be more than 2 m from the ground surface.

(vii) Moreover the site of evaporimeter must be easily accessible.

4.4.3.2 Types of Evaporimeters/Pan Evaporations: Pan evaporation is an instrument for measuring the amount of water lost by evaporation per unit area in a given time interval from a shallow container. The pan evaporation is used to interpret the evaporation loss from the surface if the reservoir. The commonly used evaporation pans are as described below (Sharma and Sharma, 2002)

4.4.3.3 B.I.S. Class A Pan (Modified): This pan evaporimeter is standardized by Bureau of Indian Standards (B.I.S.) (IS: 5973-1970) and used in India. It is also called as modified Class A Pan and ISI Standard Pan. It consists of a large cylindrical pan of 1220 mm diameter and 255 mm deep (Fig. 4.1). It has a fixed point gauge which indicates the level of water. It is made up of 0.9 mm thick copper sheet tinned inside and painted white outside. It is covered with a wire mesh. The pan rests on a white painted wooden grill of 100 mm above the ground so that air circulation becomes free. The wooden grill is square in size having width of 1225 mm. A stilling well is provided near the fixed point gauge for providing undisturbed water surface near the point gauge. A thermometer is fixed to the side of the pan to measure surface temperature of water. Fixed point gauge is used as a reference point for free water surface in the pan. A graduated cylinder is used to measure the volume of water to be added to the pan that is lost by evaporation in a given time so that the water level reaches the reference point. Dividing the volume of water so added with the pan surface area gives depth of evaporation loss from the pan. If there is any rainfall during

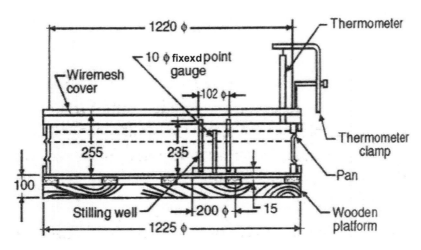

Fig. 4.1: B.I.S. Class A pan (modified)

the time period in which evaporation loss is measured, then due allowance is made from the catch of nearby raingauge. Dividing the depth of evaporation loss with the time period during which evaporation measurement is taken, evaporation rate can be found out. The coefficient used to obtain evaporation loss from free water surface area of the reservoir from pan evaporation may vary between 1.10 - 0.90 for reservoir evaporation of the order of 4-5 mm/day, between 0.75 - 0.65 for reservoir evaporation over 10 mm/day and 0.80 for evaporation between 5 -10 mm/day.

4.4.3.4 Class A U.S. Weather Bureau Land Pan

It is a circular pan made of galvanized iron or monel metal. It is 1210 mm in diameter and 255 mm in depth and is used by US Weather Bureau and is shortly called as Class A Land Pan. The pan rests over a wooden open platform 150 mm above the ground. The platform allows free wind circulation below the pan. The pan is filled with water within 50 mm from the rim of the pan (Fig. 4.2). A hook gauge is used to measure the evaporation loss of water from the pan. The coefficient of 0.7 is used to convert the pan evaporation to the reservoir evaporation.

4.4.3.5 Colorado Sunken Pan

It is a unpainted galvanized iron of 920 mm square and 460 mm deep. The pan is sunk in the ground till the rim remains 50 mm above the ground surface (Fig. 4.3). The water level is maintained at or slightly below the ground level in the pan. A hook gauge is used to measure depth of water lost as evaporation from the pan. The measurement can also be made with the help of fixed point gauge.

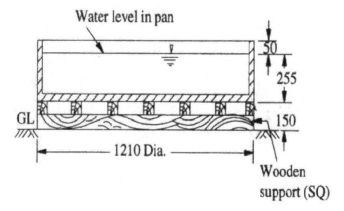

Fig. 4.2: Class A, U.S. Weather Bureau Land Pan

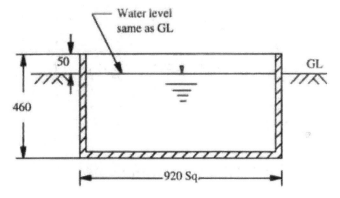

Fig. 4.3: Colorado Sunken Pan

The hook can read up to 0.02 mm. The coefficient used to convert the pan evaporation to reservoir evaporation is 0.89.

4.4.3.6 US Geological Survey Floating Pan

It is developed to simulate the conditions similar to those of the surrounding water in the reservoir. The pan is made of galvanized iron. It is 900 mm in square and 450 mm deep. It is supported on drum floats in the centre of a raft 4.25 m x 4.87 m and kept afloat in the reservoir. The water level in the pan is kept at the same level in the surrounding water body leaving a rim of 75 mm. This pan is hardly used. If it is used then, Class A U.S. Weather Bureau Land Pan is also installed near the floating pan to supplement the data during the missing period. The floating pan is less accessible for taking readings. Sometimes wind affects the accuracy in measurement since the waves formed by wind may cause splashing of water. The coefficient used to convert the pan evaporation to reservoir evaporation is 0.8.

4.4.3.7 Class A U.S. Weather Bureau Floating Pan

It is a land pan (Fig.4.4) but is not rested over wooden open support. The water level inside the pan is 75 mm below the top of the pan. The water level both inside and outside are kept at same level. A fixed index point in the centre of the pan is marked. The quantity of water added to the pan to bring the water level reading up to this fixed index point gives the amount of water lost as evaporation from the pan.

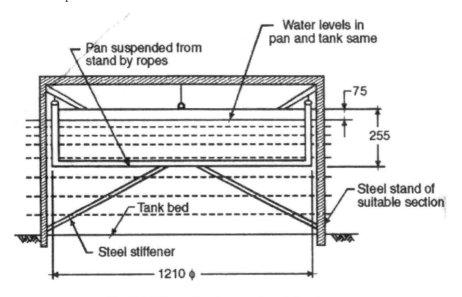

Fig. 4.4: US weather bureau class A floating pan

Sample Calculation 4.2

An evaporation pan is 1.5 m in diameter and is used to measure evaporation loss of reservoir. The initial and final water level reading in the pan are found to be 10 cm and 11 cm, respectively. A rainfall of 4 cm has occurred during the period the observation was taken. To keep the water level same in the pan, 2 cm depth of water has been removed from the pan. Find out the pan evaporation. If the pan coefficient used to convert the pan evaporation to reservoir evaporation is 0.8, find out the reservoir evaporation also.

Solution

Initial water level in the pan = 10 cm.

Rainfall during the observation period = 4 cm.

Water removed from the pan = 2 cm.

Net addition of water in the pan = 4 – 2 = 2 cm.

Had there been no loss, the water level in the pan would have been = 10 + 2 = 12 cm.

Final water level in pan = 11 cm.

Hence, evaporation loss of the pan = 12 – 11 = 1 cm.

Pan coefficient given = 0.8

Hence, reservoir evaporation loss = 0.8 x 1 = 0.8 cm.

4.5 Soil Evaporation

It is the evaporation that takes place from soil surface. If the soil is wet, then the rate of soil evaporation is high than when the soil is dry. Soil evaporation is also expressed as "evaporation opportunity" (EO) which is defined as (Garg, 1987):

$$EO = \frac{Actual\ evaporation\ from\ land\ surface\ at\ a\ given\ time}{Equivalent\ evaporation\ from\ free\ water\ surface} \times 100 \qquad (4.10)$$

4.5.1 Factors Affecting Soil Evaporation

There are various factors that affect soil evaporation. They are:

- Soil type and soil cover
- Moisture content of the soil
- Humidity and wind velocity
- Management practice

It is to be noted that soil evaporation is directly proportional to the moisture content of the soil and for a given soil, the rate decreases as time increases. The rate of evaporation is more when the soil is barren without any crop cover. In a barren soil, evaporation is very much dependent on moisture content of the soil. Evaporation mainly takes place from the soil layer up to 15 cm. However, if there is any crop or mulching is present over the soil surface, then the rate of soil evaporation decreases. Soil cover greatly affects the evaporation by reducing the effect of wind velocity, solar radiation etc. on the soil surface. Again a soil which has less water holding capacity like sandy soil has more evaporation rate compared to clay soil having more water holding capacity. But when both the soil has saturation moisture content, then they have same evaporation rate. Even colour of the soil affects soil evaporation. For example a dark coloured soil has higher rate of evaporation since it absorbs more radiant energy.

4.5.2 Reduction of Evaporation Losses

Storage of water in a reservoir is a very valuable commodity. Appreciable evaporation loss from large water spread area of a reservoir has a consequential loss on power generation, irrigation supply and many other associated activities. Some of the methods used for reducing evaporation losses are (Panigrahi, 2011):

Reduction of Reservoir Surface Area: Evaporation is a surface phenomenon and hence if the surface area of the reservoir is reduced, then their evaporation loss is considerably reduced. A reservoir with greater depth of stored water having smaller surface area has less evaporation loss compared to a shallow reservoir with large surface area; both the reservoirs having same volume of stored water.

Covering the reservoir with shading materials: Shading materials like plastic films, thatches, paddy straw, sugarcane trash etc. reduces the radiation affects on water surface and decreases the evaporation loss. Biological shading like creepers over the small reservoir water surface can help to reduce the evaporation loss of the reservoir. Some other shading materials are dye mixed with pond water, coloured pan, plastic mesh and sheet, polystyrene beads and sheet, white spheres, white butyl sheet etc. Out of these measures, polystyrene raft, plastic sheet and foamed butyl rubber are the best measures since they are reported to reduce the evaporation loss by more than 90% (Fig. 4.5).

Monomolecular Film

Alcohol such as cetyl alcohol, also called as hexadecanol and stearyl alcohol (octo decanon) are known to form a monomolecular film on contact with water which is sufficiently enduring during field conditions. The invisible film is non-toxic and reduces evaporation by about 50-60% and at the same time helps as a free passage for rain, oxygen and sun light (Cooley, 1974). However, the monomolecular films have the disadvantages that they get diluted in water quickly and then become ineffective to reduce the evaporation of water.

Wind Breaks

It is a well known fact that the turbulence affect of the water surface caused due to wind action increases the evaporation loss. Hence, wind breaks like growing tall trees around reservoir can minimize the evaporation loss to some extent. But, these wind breaks are useful only for small reservoirs. It is found out that a reduction of wind velocity by 25% reduces the reservoir evaporation loss by only 5% and thus is not significant on sustained basis.

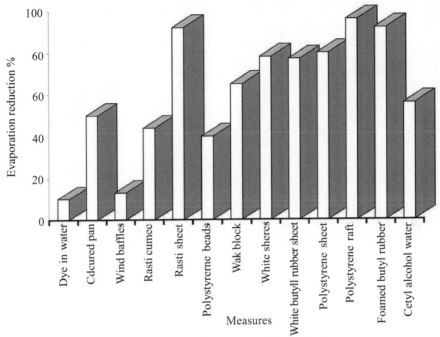

Fig. 4.5: Effectiveness of various evaporation reduction measures

4.6 Evapotranspiration

4.6.1 Transpiration

The process by which water leaves the body of a living plant and reaches the atmosphere as water vapour is called as transpiration. Plants take out water from the soil through root systems and send it to the leaves for photosynthesis. It is the leaves through which the transpired water leaves the plant body and goes to atmosphere. This transpiration is a very important process for the growth and living of plants. It is affected by (i) atmospheric vapour pressure, (ii) wind velocity, (iii) temperature, (iv) intensity of light, (v) root and leaf systems of the plants etc. Transpiration is mainly confined to the day time when there is sun light. Almost 90 to 95% of the total transpiration occurs during the day time. The rate is maximum around the noon time. During morning and afternoon, the rate is less. Evaporation is the process which continues throughout day and night although the rate of evaporation is more in day than in night. Moreover, transpiration depends on the growth period of the crop whereas evaporation does not depend. A typical relationship between evapotranspiration (ET), transpiration (T) and evaporation (E) is shown in Fig. 4.6.

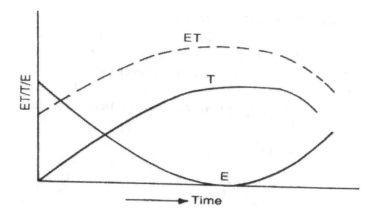

Fig. 4.6: Relationship between evapotranspiration, transpiration and evaporation

4.6.1.1 Measurement of Transpiration: Transpiration can be measured by an instrument called as phytometer. Phytometer consists of a closed tank which is filled with soil. Plants are transplanted in the tank. The tank is so constructed that only the plants inside the tank are exposed to the atmosphere. Manually water is applied to the plants. Initial weight of the phytometer (W_1) is noted before applying any water. After application of water to the plants in tank, weight of water (W_2) is taken. Finally the weight of tank with plants and applied water is taken and let it to be W_3. The weight of water consumed by the plant as transpiration, W_e is given as:

$$W_e = W_1 + W_2 - W_3 \qquad (4.11)$$

The weight of transpired water so calculated as above gives unadjusted value which is corrected by a correction factor. Sometimes transpiration loss is also determined in terms of transpiration ratio (TR) defined as the ratio of weight of absorbed water by the root system of plants during respiration process throughout the growing season to the weight of dry matter produced exclusive of roots. The value of the transpiration ratio for different crops is generally found in the range of 300 to 800. For paddy crop, the value is about 700 (Suresh, 2008).

4.6.2 Evapotranspiration

Total loss of water through evaporation from the soil surface and transpiration by the plants is called as evapotranspiration. It is to be noted that evpotranspiration occurs only from the cropped land. It is difficult to separate evaporation and transpiration and so commonly a single term evpotranspiration is used as water lost from soil-plant system. Evapotranspiration is sometimes called as water requirement of crops. The term consumptive use is the sum of water required for metabolic activity and demand to meet evapotranspiration. Since the metabolic requirement is less (about 1%), the term consumptive use is sometimes called simply evapotranspiration. Estimation of evapotranspiration is very important from several points such as planning and design of irrigation systems, irrigation scheduling etc. From hydrologic point of view, it is important to consider as a loss which needs to be assessed so that runoff can be determined.

4.6.2.1 Factors Affecting Evapotranspiration: Evapotranspiration is affected by the following factors:

- Soil factor
- Plant factor
- Atmospheric factor
- Cultural practice factor

Soil Factor: Soil plays a dominant role in evapotranspiration in the following ways:

(i) It supplies water as soil moisture content in controlling the rate of evaporation and evapotranspiration.

(ii) Altering the absorption of radiant energy.

(iii) Altering the heat or vapour transmission for evaporation.

Plant factor: Evapotranspiration depends on the type of plants and their species. The root and leaves systems of the plants affect the rate of evapotranspiration. Plants with short height and less vegetative coverage have less evapotranspiration than tall and bushy plants. The plants which have deeper roots and more root density, they extract moisture from the deeper layers and supplies to the plants. Hence they have higher evapotranspiration. The growth stages of the crops also decide the rate of evapotranspiration. Mid season and reproductive stage of any crop is associated with higher rate of evpotranspiration than the initial/ crop establishement and maturity/late season/harvesting stage. The plant factor is associated with crop evapotranspiration in the form of crop coefficient which is discussed subsequently in this chapter.

Atmospheric factor: Evapotranspiration is dependent on atmospheric conditions like temperature, relative humidity, wind velocity, sun shine hours and its intensity of radiation etc. Thus, the rate of evapotranspiration is not same at all places and even in the same place, it varies from seasons to seasons. Rate of evapotranspiration is more in summer than in winter. Again the rate of evapotranspiration is more in equatorial region than the polar region.

Cultural practice factor: The cultural practice factors like irrigation methods, irrigation frequencies, mulching, shading, timing of irrigation etc. affect evapotranspiration. If irrigation interval is more and soil surface is dried, then evapotranpiration is reduced. In flooding method of irrigation,more amount of water is applied to the crops and hence, evapotranspiration is generally more for flooding irrigation methods. Shading also reduces evapotranspiration in the way that it reduces the radiation affect. Mulching has positive affect in reducing rate of evapotranspiration.

4.6.2.2 Potential Evapotranspiration: For a given set of atmospheric conditions, evapotranspiration depends on the availability of soil moisture. If sufficient moisture is available such that there is no moisture stress affect on growth of plants and the plants actively and uniformly grows covering the whole area, than the evapotranspiration is called as potential evapotranspiration (PET). It is to be noted that PET depends entirely on the atmospheric conditions and not on soil and plant factors. It is the maximum rate of evapotranspiration and may be more or equal to actual evapotranspiration (AET). It is important to note that except in a few specialized studies in hydrology, all applied studies in hydrology use PET for various estimation purposes. It is generally agreed that PET is a good approximation for lake evaporation.

4.6.2.3 Actual Evapotranspiration: The real evapotranspiration occurring in a specific situation is called as actual rate of evapotranspiration (AET). In practice, AET is less than PET since the availability of soil moisture may be a binding factor. If water supply to plant is adequate such that soil moisture content is always at or more than the field capacity, then there will be no moisture stress and under such condition, plants actively grows at potential rate of evapotranspiration. Under unsaturated phase when moisture content is lower than field capacity, AET < PET and the ratio of these two factors is less than 1. The decrease of AET/PET with available moisture (available moisture = field capacity – wilting point) depends on the type of soil and rate of drying. Generally for clay soil, AET/PET ≈ 1.0 for nearly 50% drop in the available moisture. When soil moisture reaches permanent wilting point, AET reduces to zero (Fig. 4.7).

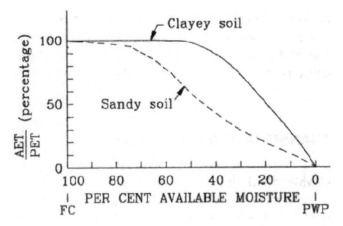

Fig. 4.7: Variation of actual evapotranspiration (Subramanya, 2003)

4.6.3 Methods to Determine Evapotranspiration

The most important component of water requirement of crop is its evapotranspiration. There are several methods used to determine the crop evapotranspiration. Broadly they can be classified as (Panigrahi, 2013):

(i)   Direct measurements and

(ii)   Estimation method

The direct measurement method comprises the followings:

(i)   Lysimeter experiment

(ii)   Field experimental plots

(iii) Soil moisture depletion studies and

(iv)   Water balance method.

Estimation of evapotranspiration can be achieved by:

(i)   Aerodynamic method

(ii)   Energy balance method

(iii) Combination method and

(iv)   Empirical method

4.6.3.1 Lysimeter Method: Lysimeter studies involve the growing of crop in large containers (lysimeters) and measuring their water loss and gains. Mainly there are two types of lysimeters (Fig. 4.8). They are (i) weighing type and (ii) drainage type. Weighing type lysimeters give more accurate results compared to the drainage type. To get more accurate and short period results, weighing type lysimeter can be used. The soil and crop conditions of the lysimeters should be closed to the natural condition. From the irrigation point of view, weighing type lysimeter is set up to enable the operator to measure the water balances: water added, water retained by the soil, and water lost through all sources-evaporation, transpiration and deep percolation. These measurements require weighing that may be made with scales or by floating the lysimeters in water in which case the change of liquid displacement is computed against water loss from the tank. The techniques yield measurement of total water loss and is useful as an indicator of field water loss. The precautions needed in this case is that the tank must be permanently buried in the ground and surrounded by a large area of crop with same height as crop. The water table is maintained as specific depth in the tank by connecting it to a supply tank provided with a float mechanism which has an arrangement for receiving excess water that tends to

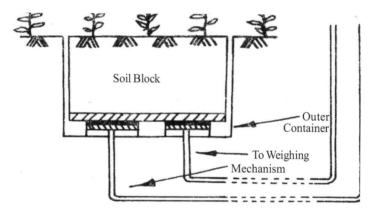

(a) Weighing type lysimeter (Panigrahi, 2013)

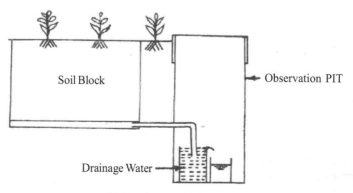

(b) Drainage type lysimeter

Fig.4.8: Lysimeters to measure water requirement of crops (Panigrahi, 2013)

build up in the tank. Water is applied in measured amount to the lysimeter as irrigation is applied to the crops in the fields where lysimeter is set up. The water received either from the reservoir or precipitation excluding the outflow, constitutes the water used by the crops (Lenka, 2001).

4.6.3.2 Soil Moisture Depletion Studies: This method is usually suitable for irrigated areas where soil is fairly uniform and the depth to the groundwater is such that it will not influence the soil moisture fluctuations within the crop root zone.

In this method, the soil moisture within the crop effective root zone before and just after irrigation and also within each irrigation cycle is measured. The soil moisture content from different profile layers within the crop effective root zone are measured and then these values are used to compute the total soil moisture content in the entire crop root zone. There are a number of methods used to measure the soil moisture content i.e. gravimetric method, neutron probe method, infra red, tensiometer, gypsum block, pressure membrane and pressure plate technique.

The water use from the crop root zone is given (Mishra and Ahmed, 1987) as:

$$u = \sum_{i=1}^{n} \frac{(M_{1i} - M_{2i})}{100} \cdot A_i \cdot D_i + ER - Dp \tag{4.12}$$

where, u = water use from the crop root zone for successive sampling periods or within one irrigation cycle, mm; n = number of soil profile layers in the crop effective root zone D; M_{1i} = soil moisture percentage at the time of the first sampling in the ith layer; M_{2i} = soil moisture percentage at the time of the second sampling in the ith layer; A_i = apparent specific gravity of the ith layer of the soil and D_i = depth of the ith layer of the soil profile, mm, ER = effective rainfall, mm and Dp = percolation from the crop effective root zone, mm. The seasonal value of water use by the crop is calculated by adding the water use of each sampling interval. A correction is made by adding the potential evapotranspiration value(s) for the accelerated water loss for the interval(s) just after irrigation(s) and before soil moisture sampling.

4.6.3.3 Field Experimental Plots: In this method, measurement of water supplies to the fields and change in soil moisture contents of field plots are made. The seasonal water requirement is computed by adding measured quantities of irrigation water, effective rainfall received during the season and contribution of soil moisture from the crop effective root zone. This is expressed by the relationship (Mishra and Ahmed, 1987):

$$WR = IR + ER + \sum_{i=1}^{n} \frac{(M_{bi} - M_{ei})}{100} \cdot A_i \cdot D_i \qquad (4.13)$$

where, WR = seasonal water requirement, mm; IR = total irrigation water supplied, mm ; ER = seasonal effective rainfall, mm; M_{bi} and M_{ei} = soil moisture percentage at the beginning and end of the season in the ith layer of the soil, respectively and other terms of equation are as discussed above. This method has the limitation that it does not provide information on intermediate soil moisture conditions, short term use, profile use, deep percolation and peak use rate of the crop.

Sample Calculation 4.3

Profile soil moisture content of a potato field in 60 cm effective root zone depth are measured as follows:

Profile layer, cm	Soil moisture content,%		Apparent Specific gravity, gm/cc
	Beginning of irrigation cycle	End of irrigation cycle	
0-15	18	10	1.60
15-30	20	11	1.58
30-45	25	15	1.59
45-60	25	15	1.60

If effective rainfall within the irrigation cycle is 10 mm, and percolation from the effective root zone of the crop is 10 mm, estimate the water use from the crop root zone within the irrigation cycle.

Solution

Water use of the crop is given by Eq. (4.12) as

$$u = \sum_{i=1}^{n} \frac{(M_{1i} - M_{2i})}{100} \cdot A_i \cdot D_i + ER - Dp \qquad (4.14)$$

Given effective rainfall, $ER = 10$ mm and deep percolation loss $Dp = 10$ mm.

The soil has 4 layers of each 15 cm thick ($Di = 15$ cm).

The apparent specific gravity (Ai) is given in the question for each layer.

The values of soil moisture content at beginning and end of each irrigation cycle (M_{1i} and M_{2i}) are also mentioned in the question for each layer.

Using the given data, water use is calculated as

= (18-10)/100 x 1.60 x 150 + (20-11)/100 x 1.58 x 150 + (25-15)/100 x 1.59 x 150 + (25-15)/100 x 1.60 x 150 + 10 -10 = 88.5 mm.

Sample Calculation 4.4

A field is sown with ground nut. The profile soil moisture content of the field at sowing and at harvest time are measured and presented in table as below.

Profile layer, cm	Soil moisture content,%		Apparent specific gravity, gm/cc
	At sowing	At harvest	
0-15	20	10	1.62
15-30	22	11	1.60
30-45	25	13	1.59
45-60	25	15	1.60

If the crop is supplied with 10 cm irrigation, effective rainfall in the crop growing season is 12 cm and there is no percolation loss from the crop effective root zone in the entire season, then estimate the seasonal water requirement of the crop.

Solution

Water requirement, WR of the crop is given by Eq. (4.13) as:

$$WR = IR + ER + \sum_{i=1}^{n} \frac{(M_{bi} - M_{ei})}{100} \cdot A_i \cdot D_i$$

Given effective rainfall, ER = 12 cm = 120 mm and irrigation = IR = 10 cm = 100 mm.

The soil has 4 layers of each 15 cm thick (D_i = 15 cm).

The apparent specific gravity (A_i) is given in the question for each layer.

The values of soil moisture content at beginning and end of crop growing season (M_{bi} and M_{ei}) are also mentioned in the question for each layer.

Using the given data, water requirement is calculated as:

= (20-10)/100 x 1.62 x 150 + (22-11)/100 x 1..60 x 150 + (25-13)/100 x 1.59 x
150 + (25-15)/100 x 1.60 x 150 + 100 -120 = 323 mm.

4.6.3.4 Water Balance Method: The method is suitable for large fields over
long periods. It is based on inflow and outflow of water to the soil control
volume. The water balance model of a crop field (generally non-paddy) under
unsaturated phase is expressed (Panigrahi, 2001) as:

$$S_{w,i} = S_{w,i-1} + R_i + I_{s,i} - ETcrop_i - SR_i - S_i - D_i \qquad (4.15)$$

where, R = rainfall; I_s = supplemental irrigation, $ETcrop_i$ = evapotranspiration
of the crop; SR = runoff from the crop field; D = percolation loss in the crop
field, S = seepage loss from the crop filed and S_{wi} and S_{wi-1} = soil moisture
content on ith and $(i-1)$th day, respectively. The term i of the above equation
refer to the day which may be taken as 1 day.

Water balance model for paddy field (which is generally under saturated / ponding
phase) is given (Panigrahi, 2001) as follows:

$$D_{p,i} = D_{p,i-1} + R_i + I_{s,i} - ETcrop_i - SR_i - S_i - D_i \qquad (4.16)$$

where, D_{pi} and D_{pi-1} are the ponding depths occurred at the end of day i and $i-1$, respectively. Other terms of the equation are described as above. The unit of
various terms in the above equation is mm. It is to be remembered that the
capillary rise due to groundwater table is ignored (Odhiambo and Murty, 1996).

Measuring all the parameters of the water balance model (Eqs.4.15 and 4.16)
except crop evapotranpiration, we can calculate evapoptranspiration. The method
has the advantage that we can compute the crop evapotranspiration on short
term basis like daily basis. Details of estimation of the various parameters of
the above water balance model are available in the book (Panigrahi, 2011).

4.6.3.5 Estimation of Evapotranspiration: All methods for computing
evapotranspiration ($ETcrop$) involve the following equation:

$$ETcrop = Kc\ ET_o \qquad (4.17)$$

where, $ETcrop$ = crop evapotranspiration or simply evapotranspiration; Kc =
crop coefficient and ET_o = reference crop evapotranspiration or potential
evapotranspiration.

Potential evapotranspiration is the maximum rate at which water can be removed
from the soil and plant surfaces. Potential evapotranspiration depends on the
amount of energy available for evaporation and from day to day. Reference

crop evapotranspiration, simply called as reference evapotranspiration, is the potential evapotranspiration for a specific crop (usually either grass or alfalfa) and set of surrounding (advective) conditions. Reference crop evapotranspiration is defined as the evapotranspiration from an extensive surface of 8 to 15 cm tall green grass cover of uniform height, actively growing, completely shading the ground and having no shortage of water (Doorenbos and Pruitt, 1977). Reference evapotranspiration is preferred over the potential evapotranspiration, since the later may vary from crop to crop due to the differences in aerodynamic roughness and surface reflectance (albedo) and from location to location. Reference evapotranspiration, on the other hand, is defined for a specific crop and set of advective conditions.

Many methods with differing data requirements and levels of sophistications have been developed for calculating ET_o. Some are physically based while others are determined empirically. Details of estimation of ET_o by various methods are discussed below.

4.6.3.6 Estimation by Aerodynamic Method: In the aerodynamic methods, vapour flux is proportional to mean wind speed and the vapour pressure difference between the evaporating surface and the surrounding air.

Dalton proposed the following equation for estimating evaporation from a water surface (James, 1988).

$$ET_o = (e_s - e) f (u) \tag{4.18}$$

where, e_s = vapour pressure at the plant surface (within the boundary layer surrounding the leaf); e = vapour pressure at some height above the plant and $f(u)$ = function of the horizontal wind velocity.

Dalton equation is not widely used to estimate ET_o because of the difficulty of determining e_s within the boundary layer surrounding the plant. This is due to primarily the very large vapour pressure and temperature gradients that often exist within the boundary layer.

A more common aerodynamic method of determining ET_o involves determining the vapour pressure and wind speed differences between measurement heights z_1 and z_2. The evaporation in this method is computed as:

$$ET_o = \frac{-10K^2\rho(q_2 - q_1)(u_2 - u_1)}{\left(\ln\left\{\frac{z_2}{z_1}\right\}\right)^2} \tag{4.19}$$

where, E = evaporation, mm; K = von Karman constant = 0.41; ρ = density of water vapour, (g/cc); q_2 and q_1 are specific humidity at heights z_2 and z_1, respectively that are taken as 2 m and 1 m, respectively above the water surface; and u_2 and u_1 are wind speeds (cm/s) at heights z_2 and z_1, respectively.

4.6.3.7 Estimation by Energy Balance Method: Based on the principles of conservation of energy and considering a soil column extending from the surface to some depth where vertical heat exchange is negligible, the energy balance can be written (Michael, 1987) as:

$$(Q + q)(1 - a) + I\downarrow - I\uparrow + H + LE + G + M = 0 \tag{4.20}$$

where, Q = direct solar radiation incident on earth; q = diffuse solar radiation incident on the earth; a = surface resistance (albedo) which is a fraction; $I\downarrow$ = incoming long wave radiation; $I\downarrow$ = outgoing long wave radiation from the land; H = sensible heat transfer; LE = latent heat flux due to evaporation; G = net rate at which heat content of soil column is changing and M = miscellaneous heat terms.

Eq. (4.20) can be used for any time period. The units are generally expressed in cal/cm²/unit time. It is to be noted that 1 cal/cm²/min ≈ 1 mm of water evaporating per hour. In Eq. (4.20) $(Q + q)(1 - a) + I\downarrow - I\uparrow = R_n$ where R_n = net solar radiation which is the difference between the total incoming and outgoing solar radiation.

Neglecting M, Eq. (4.20) can be written as:

$$R_n = H + LE + G \tag{4.21}$$

Based on Penman's equation (1948), the Kohler-Nordenson-Fox equation (Pochop et al., 1985) describes evaporation from water bodies as the combination of water loss due to radiation heat energy and the aerodynamic removal of water vapor from a saturated surface.

4.6.3.8 Estimation by Combination Method: Based on Penman's equation (1948), the Kohler-Nordenson-Fox equation (Pochop *et al.,* 1985) describes evaporation from water bodies as the combination of water loss due to radiation heat energy and the aerodynamic removal of water vapor from a saturated surface.

The general form for the combination equation is:

$$ET_o = \frac{\Delta}{\Delta + \gamma} Rn + \frac{\gamma}{\gamma + \Delta} Ea \tag{4.22}$$

where, ET_o = evaporation in mm/day; Δ = slope of the saturation vapor pressure curve at air temperature T_a in mbar/^{0}C; γ = psychrometric constant in mbar/^{0}C; R_n = net radiation expressed in mm/day; and E_a = an empirically derived bulk transfer term (aerodynamic term) in mm/day. E_a is expressed as:

$$E_a = f(u)(e_s - e_d) \tag{4.23}$$

where, $f(u)$ = wind function; and $(e_s - e_d)$ = vapour pressure deficit.

Detail calculations of various terms of the above equation are available in the book (James, 1988).

4.6.3.9 Estimation by Empirical Method: Owing to the difficulty in measuring the different water balance parameters including the data required for computation of crop evapotranspiration by lysimeters, estimation methods by using climatic data are preferred. There are several climatologically factors which will influence and decide the rate of evaporation. In fact when water is not a limiting factor, the rate of evaporation is decided by atmospheric demand of the water. Some of the important factors of climate influencing the evaporation are radiation, temperature, humidity, and wind speed. The estimation methods are empirical methods. Some of the estimation methods require only one or two climatic data and does not give accurate result of evapotranspiration. However, there are other estimation methods like FAO Penmain-Monteith model which require a lot of climatic data and give accurate result of evapotranspiration. The empirical method/estimation method involves estimating the reference crop evapotranspiration and then multiplying the reference crop evapotranspiration with crop coefficient we can estimate the crop evapotranspiration. Some of the more commonly used empirical methods to estimate reference crop evapotranspiration are given below. For detail estimation of various other empirical methods, readers are suggested to refer books (Doorenbos and Pruitt, 1977; Allen *et al.*, 1998).

4.6.3.10 Thornthwaite Method: Thornthwaite (1948) assumed that the amount of water lost through evapo-transpiration from soil surface covered with vegetation is governed by climatic factors and is independent of species when moisture supply is not limiting. He proposed the following formula:

$$e = 1.6 \left(\frac{10\,t}{I} \right)^a \tag{4.24}$$

where, e = unadjusted potential evapotranspiration, cm/month; t = mean air temperature, ^{0}C; I = annual or seasonal heat index, the summation of 12 values of monthly heat indices (i) where $I = (t/5)^{1.514}$ and a = an empirical exponent computed by the equation

$$a = 0.000000675 \ I^3 - 0.0000771 \ I^2 + 0.01792 \ I + 0.49239 \qquad (4.25)$$

The factor e is an unadjusted value based on a 12 hour day and 30 day month and is corrected by actual day length in hours and day in a month, to give the adjusted potential evapotranspiration *PET* (reference evapotranspiration) as.

$$PET = e \text{ x correction factor} \qquad (4.26)$$

The correction factor for different latitudes and months are provided in Table 4.2.

Sample Calculation 4.5

Kharagpur is situated at a latitude of 25^0N. Mean monthly temperature of Kharagpur for 12 months are given below:

Find out the reference crop evapotranspiration/*PET* for the region for a crop grown during June to September.

Month	Jan	Feb	Mar	Apr	May	June	Jul	Aug	Sept	Oct	Nov	Dec
Mean monthly temp,^{0}C	18	24	28	30	31	31	32	29	28	26	25	23

Solution

Thornthwaite proposed the formula to compute reference crop evapotranspiration /*PET* as:

$$e = 1.6 \left(\frac{10t}{I} \right)^a$$

where, e = unadjusted reference rop evapotranspiration/*(PET)*, cm/month; t = mean air temperature, ^{o}C; I = annual or seasonal heat index ,the summation of 12 values of monthly heat indices, where $I = \left(\dfrac{t}{5} \right)^{1.514}$ and a = an empirical exponent computed by the equation, a = 0.000000675 I^3 − 0.0000771 I^2 + 0.01792 I + 0.49239

Reference crop evapotranspiration/*(PET)* = e × correction factor

Table 4.2 Correction factors to be used in Thornthwaite formula (Mean possible duration of sunlight expressed in units of 30 days of 12 hours each)

North Latitude	Jan	Feb	Mar	Apr	May	June	Jul	Aug	Sept	Oct	Nov	Dec
0	1.04	0.94	1.04	1.01	1.04	1.04	1.04	1.04	1.01	1.04	1.01	1.04
5	1.02	0.93	1.03	1.02	1.06	1.03	1.06	1.05	1.01	1.03	0.99	1.02
10	1.00	0.91	1.03	1.03	1.08	1.06	1.08	1.07	1.02	1.02	0.98	0.99
15	0.97	0.91	1.03	1.04	1.11	1.08	1.12	1.08	1.02	1.01	0.95	0.97
20	0.95	0.9	1.03	1.05	1.13	1.11	1.14	1.11	1.02	1.00	0.93	0.94
25	0.93	0.89	1.03	1.06	1.15	1.14	1.17	1.12	1.02	0.991	0.9	0.91
26	0.92	0.88	1.03	1.06	1.15	1.15	1.17	1.12	1.02	0.99	0.91	0.91
27	0.92	0.88	1.03	1.07	1.16	1.15	1.18	1.13	1.02	0.99	0.90	0.90
28	0.91	0.88	1.03	1.07	1.16	1.16	1.18	1.13	1.02	0.98	0.90	0.90
29	0.91	0.87	1.03	1.07	1.17	1.16	1.19	1.13	1.03	0.98	0.90	0.89
30	0.9	0.87	1.03	1.08	1.18	1.17	1.20	1.14	1.03	0.98	0.89	0.88
31	0.9	0.87	1.03	1.08	1.18	1.18	1.20	1.14	1.03	0.98	0.89	0.88
32	0.89	0.86	1.03	1.08	1.19	1.19	1.21	1.15	1.03	0.98	0.88	0.87
33	0.88	0.86	1.03	1.09	1.19	1.20	1.22	1.15	1.03	0.97	0.88	0.86
34	0.88	0.85	1.03	1.09	1.20	1.20	1.22	1.16	1.03	0.97	0.87	0.86
35	0.87	0.85	1.03	1.09	1.21	1.21	1.23	1.16	1.03	0.97	0.86	0.85
36	0.87	0.85	1.03	1.10	1.21	1.22	1.24	1.16	1.03	0.97	0.86	0.84
37	0.86	0.84	1.03	1.10	1.22	1.23	1.25	1.17	1.03	0.97	0.85	0.83
38	0.85	0.84	1.03	1.10	1.23	1.24	1.25	1.17	1.04	0.96	0.84	0.83
39	0.85	0.84	1.03	1.11	1.23	1.24	1.26	1.10	1.04	0.96	0.84	0.82
40	0.84	0.83	1.03	1.11	1.24	1.25	1.27	1.18	1.04	0.96	0.83	0.81
41	0.83	0.83	1.03	1.11	1.25	1.26	1.27	1.19	1.04	0.96	0.82	0.80
42	0.82	0.83	1.03	1.12	1.26	1.27	1.28	1.19	1.04	0.95	0.82	0.79
43	0.81	0.82	1.02	1.12	1.26	1.28	1.29	1.20	1.04	0.95	0.81	0.77
44	0.81	0.82	1.02	1.13	1.27	1.29	1.30	1.20	1.04	0.95	0.80	0.76

Contd.

North Latitude	Jan	Feb	Mar	Apr	May	June	Jul	Aug	Sept	Oct	Nov	Dec
45	0.80	0.81	1.02	1.13	1.28	1.29	1.31	1.21	1.04	0.94	0.79	0.75
46	0.79	0.81	1.02	1.13	1.29	1.31	1.32	1.22	1.04	0.94	0.79	0.74
47	0.77	0.80	1.02	1.14	1.30	1.32	1.33	1.22	1.04	0.93	0.78	0.73
48	0.76	0.80	1.02	1.14	1.31	1.33	1.34	1.23	1.05	1.93	0.77	0.72
49	0.75	0.79	1.02	1.14	1.32	1.34	1.35	1.24	1.05	0.93	0.76	0.71
50	0.74	0.78	1.02	1.15	1.33	1.36	1.37	1.25	1.06	0.92	0.76	0.70
South Latitude												
5	1.06	0.95	1.04	1.00	1.02	0.99	1.02	1.03	1.00	1.05	1.03	1.06
10	1.08	0.97	1.05	0.99	1.01	0.96	1.00	1.01	1.00	1.06	1.05	1.10
15	1.12	0.98	1.05	0.98	0.98	0.94	0.97	1.00	1.00	1.07	1.07	1.12
20	1.14	1.00	1.05	0.97	0.96	0.91	0.95	0.99	1.00	1.08	1.09	1.15
25	1.17	1.01	1.05	0.96	0.94	0.88	0.93	0.98	1.00	1.10	1.11	1.18
30	1.20	1.03	1.06	0.95	0.92	0.85	0.90	0.96	1.00	1.12	1.14	1.21
35	1.23	1.01	1.06	0.94	0.89	0.82	0.87	0.94	1.00	1.13	1.17	1.25
40	1.27	1.06	1.07	0.93	0.86	0.78	0.84	0.92	1.00	1.15	1.20	1.29
42	1.28	1.07	1.07	0.92	0.85	0.76	0.82	0.92	1.00	1.16	1.22	1.31
44	1.30	1.06	1.07	0.92	0.83	0.74	0.81	0.91	0.99	1.17	1.23	1.33
46	1.32	1.10	1.07	0.91	0.82	0.72	0.79	0.90	0.99	1.17	1.25	1.35
48	1.34	1.11	1.08	0.90	0.80	0.70	0.76	0.89	0.99	1.18	1.27	1.37
50	1.37	1.12	1.08	0.89	0.77	0.67	0.74	0.88	0.99	1.19	1.29	1.41

Step 1: For determination of annual heat index I

Month	Mean temp °C	Annual heat index
Jan	18	6.954116887
Feb	24	10.74977151
Mar	28	13.57552622
Apr	30	15.07026861
May	31	15.83729542
Jun	31	15.83729542
July	32	16.61714772
Aug	29	14.31627336
Sept	28	13.57552622
Oct	26	12.13470051
Nov	25	11.43511632
Dec	23	10.07895154
Total		156.1819897

Step 2: For determination of a = (an empirical exponent)

a = 0.000000675 − 0.0000771 + 0.01792 + 0.49239

  = 0.000000675− 0.0000771 + 0.01792 + 0.49239

  = 3.982045

Step 3: For calculation of unadjusted reference crop evapotranspiration/(*PET)*, cm/month for four month from June to September:

$$e_{June} = 1.6\left(\frac{10\times31}{156.182}\right) = 24.52997$$

$$e_{July} = 1.6\left(\frac{10\times32}{156.182}\right) = 27.83573$$

$$e_{Aug} = 1.6\left(\frac{10\times29}{156.182}\right) = 18.80885$$

$$e_{sept} = 1.6\left(\frac{10\times28}{156.182}\right) = 16.35596$$

Step 4: For calculating adjusted reference crop evapotranspiration/$(PET) = e \times$ correction factor

Correction factor can be calculated from table for given month with known value of latitude

$$PET_{Jun} = e_{Jun} \times correction \ factor$$

$$= 24.53 \times 1.14$$

$$= 27.96417$$

$$PET_{July} = e_{July} \times correction \ factor$$

$$= 27.83573 \times 1.17 = 32.56781$$

$$PET_{Aug} = e_{Aug} \times correction \ factor$$

$$= 18.80885 \times 1.12 = 21.06592$$

$$PET_{Sept} = e_{Sept} \times correction \ factor$$

$$= 16.35596 \times 1.02 = 16.35596$$

Total *PET* from June to Sept = 27.96417+32.56781+21.06592+16.68308

= 98.28098 cm/month = 98.3 cm/month

4.6.3.11 Modified Blaney-Criddle Method: In the modified Blaney-Criddle method the relationship recommended, representing mean value over the given month is expressed (Doorenbos and Pruitt, 1977) as:

$$ET_0 = C\left[P\left(0.46\,T + 8\right)\right] \qquad (4.27)$$

where, ET_0 = reference crop evapotranspiration, mm/day for the month considered; T = mean daily temperature,^{0}C over the month considered; P = mean daily percentage of total annual daytime hours (Table 4.3) for a given month and latitude and C = adjustment factor which depends on minimum relative humidity, sunshine hours and day time wind estimates.

The adjustment factor is determined by using Fig. 4.9. Fig.4.9 can be used to estimate ET_0 graphically using the calculated values of.The values of is given on X-axis and the value of ET_0 can be read directly from the Y axis. Relationships are presented in Fig. 4.9 for (i) minimum relative humidity (RH_{min}) (ii) three levels of the ratio of actual to maximum possible sun shine hours (n/N) and (iii) three levels of day time wind velocity at 2 m height (U_{day}). Values of maximum possible sunshine hours for different latitudes and months are given in Table 4.4 (Doorenbos and Pruitt, 1977). The limitations of use of the method are also mentioned in the above mentioned reference.

1. U daytime 0 - 2 m/sec. (= 10)
2. U daytime 2 - 5 m/sec. (= 35)
3. U daytime 5 - 8 m/sec. (= 65)

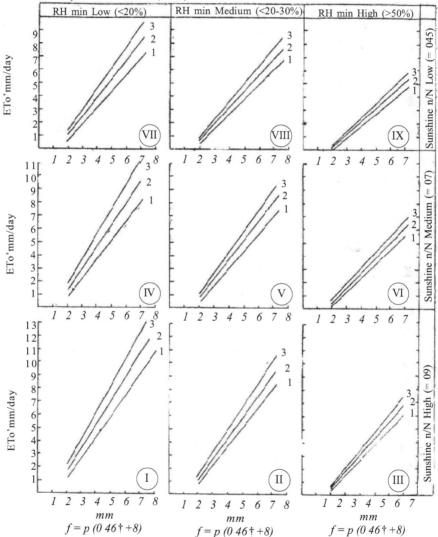

Fig.4.9: Correction factor to be used in adjustment of reference evapotranspiration by Blaney-Criddle method

Table 4.3: Mean daily percentage (P) of annual daytime hours for different latitudes

Latitude North South	Jan.-July	Feb.-Aug	March -Sept.	April-Oct.	May-Nov.	June-Dec.	July-Jan.	Aug.-Feb.	Sept.-March	Oct.-April	Nov.-May	Dec.-June
60°	0.15	0.20	0.26	0.32	0.38	0.41	0.40	0.34	0.28	0.22	0.17	0.13
58°	0.16	0.21	0.26	0.32	0.37	0.40	0.39	0.34	0.28	0.23	0.18	0.15
56°	0.17	0.21	0.26	0.32	0.36	0.39	0.38	0.33	0.28	0.23	0.18	0.16
54°	0.18	0.22	0.26	0.31	0.36	0.38	0.37	0.33	0.28	0.23	0.19	0.17
52°	0.19	0.22	0.27	0.31	0.35	0.37	0.36	0.33	0.28	0.24	0.20	0.17
50°	0.19	0.23	0.27	0.31	0.34	0.36	0.35	0.32	0.28	0.24	0.20	0.18
48°	0.20	0.23	0.27	0.31	0.34	0.36	0.35	0.32	0.28	0.24	0.21	0.19
46°	0.20	0.23	0.27	0.30	0.34	0.35	0.34	0.32	0.28	0.24	0.21	0.20
44°	0.21	0.24	0.27	0.30	0.33	0.35	0.34	0.31	0.28	0.25	0.22	0.20
42°	0.21	0.24	0.27	0.30	0.33	0.34	0.33	0.31	0.28	0.25	0.22	0.21
40°	0.22	0.24	0.27	0.30	0.32	0.34	0.33	0.31	0.28	0.25	0.22	0.21
35°	0.23	0.25	0.27	0.29	0.31	0.32	0.32	0.30	0.28	0.25	0.23	0.22
30°	0.24	0.25	0.27	0.29	0.31	0.32	0.31	0.30	0.28	0.26	0.24	0.23*
25°	0.24	0.26	0.27	0.29	0.30	0.31	0.31	0.29	0.28	0.26	0.25	0.24
20°	0.25	0.26	0.27	0.28	0.29	0.30	0.30	0.29	0.28	0.26	0.25	0.25
15°	0.26	0.26	0.27	0.28	0.29	0.29	0.29	0.28	0.28	0.27	0.26	0.25
10°	0.26	0.27	0.27	0.28	0.28	0.29	0.29	0.28	0.28	0.27	0.26	0.26
5°	0.27	0.27	0.27	0.28	0.28	0.28	0.28	0.28	0.28	0.27	0.27	0.27
0°	0.27	0.27	0.27	0.27	0.27	0.27	0.27	0.27	0.27	0.27	0.27	0.27

Table 4.4: Mean daily maximum duration of bright sunshine hours (N) for different months and latitudes

Latitude North South	Jan.-July	Feb.-Aug	March-Sept.	April-Oct.	May-Nov.	June-Dec.	July-Jan.	Aug.-Feb.	Sept.-March	Oct.-April	Nov.-May	Dec.-June
50	8.5	10.1	11.8	13.8	15.4	16.3	15.9	14.5	12.7	10.8	9.1	8.1
48	8.8	10.2	11.8	13.6	15.2	16.0	15.6	14.3	12.6	10.9	9.3	8.3
46	9.1	10.4	11.9	13.5	14.9	15.7	15.4	14.2	12.6	10.9	9.5	8.7
44	9.3	10.5	11.9	13.4	14.7	15.4	15.2	14.0	12.6	11.0	9.7	8.9
42	9.4	10.6	11.9	13.4	14.6	15.2	14.9	13.9	12.6	11.1	9.8	9.1
40	9.6	10.7	11.9	13.3	14.4	15.0	14.7	13.7	12.5	11.2	10.0	9.3
35	10.1	11.0	11.9	13.1	14.0	14.5	14.3	13.5	12.4	11.3	10.3	9.8
30	10.4	11.1	12.0	12.9	13.6	14.0	13.9	13.2	12.4	11.5	10.6	10.2
25	10.7	11.3	12.0	12.7	13.3	13.7	13.5	13.0	12.3	11.6	10.9	10.6
20	11.0	11.5	12.0	12.6	13.1	13.3	13.2	12.8	12.3	11.7	11.2	10.9
15	11.3	11.6	12.0	12.5	12.8	13.0	12.9	12.6	12.2	11.8	11.4	11.2
10	11.6	11.8	12.0	12.3	12.6	12.7	12.6	12.4	12.1	11.8	11.6	11.5
5	11.8	11.9	12.0	12.2	12.3	12.4	12.3	12.3	12.1	12.0	11.9	11.8
0	12.1	12.1	12.1	12.1	12.1	12.1	12.1	12.1	12.1	12.1	12.1	12.1

Calculation of mean daily ET_0 should be made for periods no shorter than one month. ET_0 should preferably be calculated for each calendar month for each year of record rather than by using mean temperature based on several years' record. Once ET_0 has been determined, the crop coefficients (Kc) can be used to determine $ETcrop$.

Sample Calculation 4.6

Calculate the reference evapotranspiration by modified Blaney-Criddle method using the following data for a place which has Latitude 29^0N: Altitude 283.89 m. The month is December. Mean daily maximum temperature = 22.96 ^{0}C and mean daily minimum temperature = 5.86 ^{0}C. $RHmin$ = 65%, day time wind velocity at 2 m height = 4 m/sec and actual sunshine hour = 9.4 hours.

Solution

T daily mean = (22. 96 + 5.86) /2 = 14.41 ^{o}C.

P from Table 4.3 for 29^0N, 0.23

$P (0.46\ T+8)$ = 0.23 (0.46 X 14.41 + 8) = 3.36 mm/day

$RHmin$ = high

Value of N is taken from Table 4.4 and using given value of n (9.4 hr) we get,

n/N = high

u_2 day time at 2 m height = moderate

ET_0 from Fig.4.9 Block III (line 2) =2.3 mm/day

4.6.3.12 Pan Evaporation Method: This method is a relatively inexpensive and simple method of assessing the evaporative demand of the atmosphere. It has been observed that a close relationship exists between the rate of consumptive use by crops and the rate of evaporation from a properly located evaporation pan. There are different types evaporation pans in use which have been discussed earlier. The standard USDA class A open pan evaporimeter may be used for the estimation (Fig. 4.10). In this case daily reading of water depth in the pan is taken. The difference of the two consecutive day's water depth reading in the pan (assuming that water loss due to wind action, animals, birds etc has been prevented or is negligible) gives the daily pan evaporation. If in between the two consecutive readings of taking water depth in the pan, there is some rainfall, then this amount of rainfall is to be subtracted from the difference of the two consecutive days water depth reading of the pan so that the actual pan evaporation data is obtained. The relationship between reference evapotranspiration *(ETo)* and pan evaporation *(Epan)* is given by the pan coefficient *(Kp)* as:

$$ETo = Epan \cdot Kp \qquad (4.28)$$

The value of *Kp* depends on environment and geographical location and also on the nature and characteristics of pan. The shape and colour of the pan also affects the value of *Kp*. For example, water loss from circular pans is independent of wind direction, while evaporation from square pans depends on wind direction. Colour difference between the pans affects the reflection of radiation and hence evaporation. Screens mounted above pans to prevent birds and animals to drink from the evaporation pan reduce pan evaporation by as much as 10 percent (Doorenbos and Pruitt, 1977). Conditions upwind of the evaporation pans also significantly affect *Kp*. Table 4.5 lists values of *Kp* for different winds, relative humidities and windward ground cover conditions as (illustrated in Fig.4.11) for class A evaporation galvanized pans annually painted with aluminum. There will be little difference in values of *Epan* when inside and outside surfaces of the pan are painted white. An increase of *Epan* value up to 10 percent may occur when the pan is painted black.

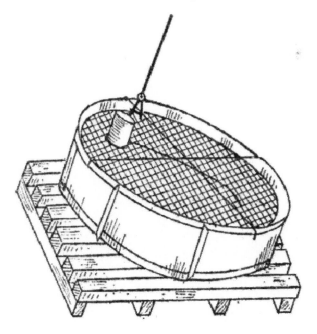

Fig. 4.10: Figure of U.S.D.A. class A open pan

Table 4.5 Values of K_p for class A open pan for different ground covers, 24 hour wind speed and levels of mean relative humidities (RH).

Class A Pan	Case A[a] Pan surrounded by Short Green Crop				Case B[b] Pan surrounded by Dry-Fallow Land			
RH mean, Percent	Upwind distance of Green Crop (m)	Low (<40)	Medium (40-70)	High (>70)	Upwind distance of Dry Fallow (m)	Low (<40)	Medium (40-70)	High (>70)
Average daily wind speed (m/sec)								
Light (<2)	1	0.55	0.65	0.75	1	0.7	0.8	0.85
	10	0.65	0.75	0.85	10	0.6	0.7	0.8
	100	0.70	0.8	0.85	100	0.55	0.65	0.75
	1000	0.75	0.85	0.85	1000	0.5	0.6	0.7
Moderate (2-5)	1	0.5	0.6	0.65	1	0.65	0.75	0.8
	10	0.6	0.7	0.75	10	0.55	0.65	0.7
	100	0.65	0.75	0.8	100	0.5	0.6	0.65
	1000	0.7	0.8	0.8	1000	0.45	0.55	0.6
High (5-8)	1	0.45	0.5	0.6	1	0.6	0.65	0.7
	10	0.55	0.6	0.65	10	0.5	0.55	0.65
	100	0.6	0.65	0.7	100	0.45	0.5	0.6
	1000	0.65	0.7	0.75	1000	0.4	0.45	0.55
Very High (> 8)	1	0.4	0.45	0.5	1	0.5	0.6	0.65
	10	0.45	0.55	0.6	10	0.45	0.5	0.55
	100	0.5	0.6	0.65	100	0.4	0.45	0.5
	1000	0.55	0.6	0.65	1000	0.35	0.4	0.45

Source: Doorenbos and Pruitt (1977)

[a] Cases A and B are defined in Fig. 4.11

[b] For extensive areas of bare fallow soils and areas without agricultural development, reduce K_p values by 20 percent under hot windy conditions and by 5 to 10 percent for moderate wind, temperature and humidity conditions.

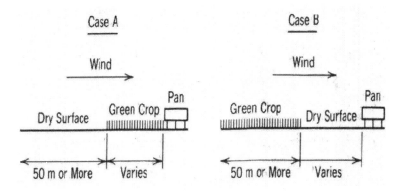

Fig. 4.11: Definition sketch for Cases A and B in Table 4.5 (from Doorenbos and Pruitt, 1977)

Sample Calculation 4.7

Calculate the reference crop evapotranspiration value (ET_o) as measured by class A open galvanized pan in a given site under the following conditions with data measured as:

Water level in the pan on previous day at 7 A.M. is 18.5 cm and that measured after 24 hours at the same time is 20. 5 cm. There is 7.0 cm rainfall in between the 24 hours. The pan is surrounded by green crop and is situated upwind distance of green crop of 100 m. Mean relative humidity is 50 percent and wind speed is 4 m/sec.

Solution

There is 7 cm rainfall within the two consecutive day's readings of water measurement.

Water level readings on previous day = 18.5 cm

Water level reading on day after 24 hours = 20.5 cm

Hence, the value of $Epan$ = 18.5 +7.0 – 20.5 = 5 mm/day.

Kp = 0.75 (referring Table 4.5).

Hence, $ETo = Epan\ Kp$

$$= 5 \times 0.75 = 3.85 \text{ mm/day.}$$

Other methods used for computation of ETo are Penmen method, Modified Penmen method, Radiation method, Christiansen method, Energy Balance method, Oliver method, Van Bavel method, FAO Penmen-Monteith method etc. Out of these methods, FAO Penmen-Monteith method gives reasonable correct value of ET_o since it considers a number of climatic data. The details of FAO Penmen-Monteith method is discussed in FAO Paper No. 56 (Allen *et al.*, 1998).

4.6.4 Approximate Range of Seasonal Crop Water Requirement of Some Crops

As discussed above, the values of crop water requirements (*ETcrop*) is a variable factor. The values mainly depend on climatic condition of the place and crop type including sowing/planting time, crop growing period, crop growth stages, cultural and water management practices etc. Because of these reasons, the same crop may have wide variation in the value of crop water requirement. Table 4.6 given below reports the seasonal variation of crop water requirement of some important crops grown in different places (Doorenbos and Pruitt, 1977).

Table 4.6: Approximate range of seasonal *ETcrop* of some crops

Crop	Seasonal *ETcrop*, mm	Crop	Seasonal *ETcrop*, mm
Alfalfa	600-1500	Potato	350-625
Banana	700-1700	Rice	500-950
Beans	250-500	Sorghum	300-650
Coffee	800-1200	Soybeans	450-825
Cotton	550-950	Sugarbeets	450-850
Deciduous trees	700-1050	Sugarcane	1000-1500
Grains (small)	300-450	Sweet potato	400-675
Grape	650-1000	Tobacco	300-500
Maize	400-750	Tomato	300-600
Oil seeds	300-600	Vegetables	250-500
Onions	350-600	Vineyards	450-900
Oranges	600-950	Walnuts	700-1000

The above mentioned values of *ETcrop* may be used for design of irrigation projects. However, it is recommended that local available values of *ETcrop* should be used in irrigation project design.

4.6.5 Potential Evapotranspiration over India

Annual PET varies from 1400 to 1800 mm over most part of the country. The annual PET is generally more in arid climate than humid one. It is the highest at Rajkot, Gujarat with a value about 2145 mm. Extreme south-east of Tamil Nadu also shows high average value of more than 1800 mm. The highest PET for southern Peninsula is at Tiruchirapalli, Tamil Nadu with a value of 2090 mm. Valuable PET data relevant to various part of the country are available in Reference of Rao *et al.* (1971).

4.7 Initial Loss

As discussed earlier, runoff resulting from precipitation results when the losses like infiltration, evapotranspiration, interception, depression storage and channel detention are fulfilled. Of these losses, interception and depression storage are two losses which are not of major abstraction but they affect the runoff in a catchment. These two losses together are called as initial loss. These losses must be satisfied before overland flow begins. These losses are briefly discussed below.

4.7.1 Interception

Interception is that part of precipitation which is caught or intercepted by the stems and canopy cover of plants, buildings and other structures. Some of the water intercepted by the plants fall back to the ground surface called as through fall which contribute to the runoff. Some amount of rain water may run along the leaves and branches and down the stems of the plants finally reaching the ground surface which is called as stem flow. The rest part of the intercepted water gets evaporated and mixes in the atmosphere as water vapour. Sometimes, when there is rainfall and the surfaces of the buildings or other structures above the ground are dry, some amount of rain water is used to wet these surfaces so that runoff begins then. This amount of water required to wet the dry surfaces of intercepted objects is called as interception storage which is to be satisfied before runoff begins. It is to be remembered that interception loss is solely due to evaporation and does not include transpiration, through fall and stem flow.

4.7.1.1 Factors Affecting Interception: Following factors affect interception.

- Storm factor
- Plant factor
- Prevailing wind and
- Season of the year

Storm Factor: Nature and duration of the storm affect interception. If the intensity of storm is very less i.e. in the form of drizzle, then most of the rain is intercepted by the leaves of plants. Field studies reveal that if the precipitation is less than 0.254 mm in depth then it is completely intercepted and if it is more than 1.02 mm, then interception varies from 10 to 40% of total precipitation. However, if the intensity and duration of precipitation or both increases, then interception decreases. The relationship amongst storm precipitation, duration of storm and interception is shown in Fig.4.12.

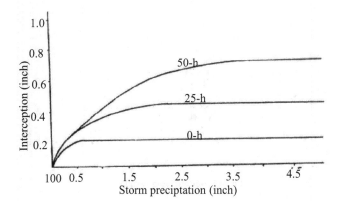

Fig. 4.12: Relationship amongst storm precipitation, storm duration and interception

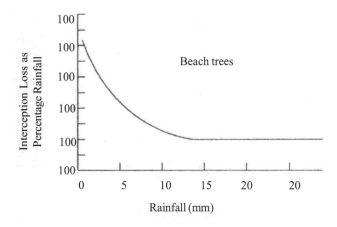

Fig. 4.13: Typical interception loss of a small storm

In general, interception loss is large for a small storm and leaves off to a constant value for large storms. Typical interception loss of a small storm for beech tress is presented in Fig. 4.13.

For a given storm, interception loss is estimated as:

$$I_i = S_i + K_i E\, t \tag{4.29}$$

where, I_i is interception loss, mm; S_i is interception storage whose value varies from 0.25 to 1.25 mm depending on the nature of vegetation; K_i is ratio of vegetal surface area to its projected area; E is evaporation rate during precipitation, mm/hr, t is duration of storm/precipitation, hr and "i" is the day.

Linsley *et al.* (1949) developed the following formula for computation of interception loss:

$$I_i = S_i \left(1 - e^{\frac{-p}{S_i}} \right) + K_i\, E\, t \tag{4.30}$$

where, I_i, S_i, K_i, E, and t are defined earlier and p is the rainfall. If rainfall is more, than the above Eqn. (4.30) is reduced to Eqn. (4.29).

Plant Factor: Phenological characteristics of the plants such as canopy density, canopy size and leaf characteristics affect interception. A plant with large size canopy or leaves like Sal tree holds more precipitation than plants with small size canopy or leaves. Similarly if the canopy is upward concave shaped, then it holds more precipitated water and hence, the interception loss is more. Further, if the density of canopy is more, then interception loss is also more. In general, coniferous trees have more interception than the deciduous trees. Also, dense grasses have nearly same interception losses as full grown plants and may account for nearly 20% of the total rainfall in the season. Leaf with rough surface holds more water than smooth surface and so interception loss in rough surface leaves is more. Variation of interception loss with types of vegetal cover for various plants is presented in Fig 4.14.

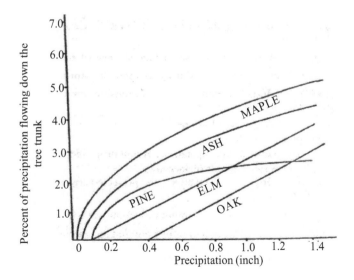

Fig. 4.14: Variation of interception loss with types of vegetal cover for various plants

Prevailing Wind: If wind velocity is very less and there is steady rainfall, interception loss is relatively more and it continues till storage capacity of the canopy is completely filled and then it starts decreasing at slower rate. However, if the wind velocity is more, then most of the falling precipitation is drifted away and so the interception loss is less. Wind also affects interception by increasing the evaporation rate from the stored water in the canopy of plants.

Season of the Year: Vegetative growth and phenological characteristics of plants are different for different seasons of a year. During the vegetative growth period, there is more canopy density and so interception loss during this period is more. During initial and maturity stage of the crop, the canopy coverage/density is less and so during this period, interception loss is less.

4.7.1.2 Measurement of Interception: Interception can be measured by an instrument called as interceptometer. The instrument is placed on the ground surface under canopy to collect the falling rainfall. The collected rainfall by the interceptometer is compared with the rainfall caught by the rain gauge installed in an open area. The difference in the amount of rainfall collected by interceptometer and the rain gauge is considered as interception.

4.7.2 Depression Storage

When the precipitation falls on the ground surface, it fills up the pits, depressions, ditches etc. present on the ground surface and then runoff flows over the surface. The volume of water required to fill up these depressions is called as depression storage. This amount is eventually lost to runoff through the process of infiltration and evaporation and thus forms a part of initial loss.

4.7.2.1 Factors Affecting Depression Storage: Following factors affect depression storage.

- Type of soil
- Condition of the surface reflecting the amount and nature of depression
- Slope of the catchment and
- Antecedent moisture condition

The antecedent moisture condition affects the runoff in the sense that it decreases loss to runoff in a storm due to depression. Values of 0.50 cm in sand, 0.4 cm in loam and 0.25 cm in clay can be taken as the representative value for the depression storage loss during the intensive storms.

4.8 Infiltration

Initially when some amount of water is poured on the soil surface, it is found that poured water is soaked by the soil surface which seeps and moves in different directions inside the soil. This movement of water through the soil surface is termed as infiltration. In general it is the entry of water through top soil surface. It was first defined by Horton (1935). In hydrology, it is considered as a loss and plays a vital role in deciding generation of runoff. Infiltration is the primary step in the natural groundwater recharge.

The mechanism of infiltration can be explained by the physiology of soil mass. The soil comprises a network of macro-pores extending laterally in all the directions and forms a continuous path. This path is like a gravity channel. Through this gravity channel, water moves downwards under gravity following the least resistant path. The soil also contains small size capillary pores which have comparatively greater affinity with water than air. Capillary pores have also grater capillary force in al directions than gravity force. Water when comes in contact with the oil surface, it tends to enter the soil profile and move downwards through gravity channels and capillary pores. The capillary water moves from the wet to dry zone inside the soil mass. As the soil moisture content goes on increasing with time, the rate of advancement of water into the soil called as infiltration goes on decreasing.

Infiltration Capacity: Infiltration occurs both prior to and during the occurrence of surface runoff but decreases with time until a minimum rate is reached. The rate at which infiltration takes place during a rainfall or when irrigation water is applied to crop fields, expressed in terms of depth of water per unit time (usually mm/hr, cm/hr) is called as infiltration rate. The maximum rate of infiltration is called as infiltration capacity. It is designated by symbol f_c. When the intensity of rainfall (i) is greater than equal to the infiltration capacity (f_c), then the actual infiltration rate (f) is equal to the infiltration capacity of the soil. Otherwise, actual infiltration rate (f) is equal to the rainfall intensity (i). The rate of infiltration is initially high and it decreases exponentially as time elapses (Fig. 4.15).

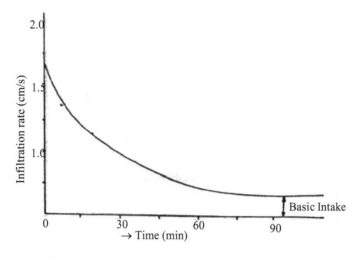

Fig. 4.15: Variation of infiltration rate with time

4.8.1 Factors Affecting Infiltration Rate

Following factors the infiltration rate:

- Rainfall factor
- Soil factor
- Climate factor
- Surface cover factor and
- Fluid characteristics

Rainfall Factor: Various rainfall factors that affect infiltration rate are (i) rainfall intensity, (ii) forms of precipitation and (iii) duration of rainfall.

When the intensity of rainfall is less than the maximum infiltration rate i.e. infiltration capacity of the soil, then whole amount of rainfall falling on the soil surface is absorbed by the soil. The infiltration rate then becomes equal to the intensity of rainfall. However, if the intensity of rainfall is greater than the infiltration capacity of the soil, then infiltration rate is affected by the mechanisms:

- Increment in supply of moisture to the soil
- Mechanical effects and
- Inwashing of the soil particles

Addition of moisture to the soil reduces the infiltration rate due to decrease in intake capacity of the capillary pores. Mechanical effects are due to the striking impact of heavier size raindrops consequent upon higher intensity of rainfall which compacts the soil and creates a physical barrier on the soil surface that reduces the intake of water into the soil. A heavy intensity storm also causes inwashing effect. During the occurrence of intense storms, when a bigger drop falls over the soil surface, the coarser soil particles are broken into finer ones which enter into the pore spaces and thus seal them. This decreases the sizes of the ore spaces and hence reducing infiltration rate.

Soil Factor: Textural class of the soil i.e. sand, silt or clay, its structures, permeability, slopes, compaction, cultivation practice, depth of groundwater table, soil moisture content present in the soil are some of the dominant factors that affect soil infiltration rate. A light texture soil like sand has higher infiltration rate than heavy texture soil like clay. A heavily compacted soil has lower rate of infiltration than a light less compacted soil. Infiltration rate is affected by the underground water table. It is observed from various experiments that infiltration rate decreases when the underground water table is nearer to the ground surface. However, as the water table rises, an upward thrust is applied by the rising water table which tends to reduce the intake of infiltrated water. However, a good under drainage accelerates infiltration rate. Sometimes because of sun rays, some cracks are developed in the soil surface that enhances the infiltration rate. Water enters into the soil and sub-soil layers in the cracks at a greater rate. Soil slopes affects the infiltration arte in the sense that as the slope increases, water moves over the soil at a greater speed. There is less opportunity time for the moving water to come in contact with the soil and hence, infiltration rate decreases. Cultivation practice has a greater influence on infiltration. If the land is barren, then it is noticed that infiltration rate is less and runoff is more. However, a cover of vegetation on the soil surface reduces runoff by increasing infiltration. A close growing crop like groundnut and peas cover the soil to the maximum extent. As a result the falling raindrop is intercepted by the foliage of the plants which dissipates the effect of kinetic energy of the raindrop. This process reduces both runoff and soil loss. The cultural practices like tillage make soil porous and creates rough surface. When it rains, the flowing rain water is intercepted by these rough surfaces. Water cannot move forward since infiltration rate of the soil is more and depression storages are to be fulfilled before run begins. A dry soil absorbs more water than a wet soil which pore spaces are already full having higher moisture content. Finally the land use has a significant influence on infiltration rate. For example, a forest soil which is rich in organic content has a much higher value of infiltration rate than the soil in urban areas the soil is devoid of organic matter and is subjected to compaction.

Climate Factor: Climatic factors like temperature affects infiltration rate. Temperature reduces the viscosity of water as a result water enters a greater rate into the soil. Because of this reason, infiltration rate is more in summer than in winter. Moreover, the activities of the micro-organisms are activated more due to increased temperature which helps in swelling of the soil and reduces surface tension. All these affect the infiltration rate.

Surface Cover Factor: At the soil surface, the impact of raindrops causes the fines of the soils to be displaced and these in turn can clog the pore spaces in the upper layers. This affects the infiltration rate. Thus a surface covered by grass or any other vegetation can prevent the falling raindrop to come in direct contact with the soil and prevents clogging of the pore spaces which in turn decreases infiltration rate. It is very important to note that right choice of the crops can help in deciding the infiltration. For example, a busy cover having thick foliage and canopy density like ground nut or peas or a pasture land has greater influence on enhancing infiltration than row crops having less canopy cover like corn or maize or a bare land. Fig. 4.16 shows typical infiltration capacity curves of two crops i.e. pasture land and widely spaced row crop i.e. corn grown on the same pieces of land. This figure reveals how infiltration rate is more for pasture land than corn.

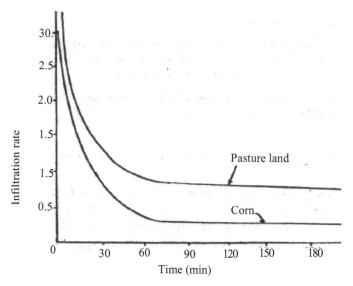

Fig. 4.16: Typical infiltration capacity curves of pasture land and corn

Presence of litter also affects infiltration rate. Penman (1963) reported that litter layers present on the soil surface has appreciable effect on infiltration rate as compared to the vegetative layer. Litter acts in two ways: (i) it dissipates the impact of kinetic energy of the falling rain drop over the ground surface and (ii) it filters the water trickling through it. These two factors combinedly protect the pore spaces against chocking from the finer particles and thus augmenting infiltration rate of the soil.

Fluid Factors: Water that enters into the soil may contain colloidal impurities, both in solution and suspension. The turbidity of water, especially the clay and colloid content may clog the pore spaces of the soil and reduce infiltration rate. As discussed earlier, increased temperature of water affect viscosity of water which in turn affects the infiltration rate of the soil. Further, contamination of water by the dissolved salts can affect the soil structures which in turn may affect the infiltration rate.

4.8.2 Infiltration Models

Several researchers have conducted a number of experiments on infiltration and have suggested a number f equations or models. Some of the famous models are described below.

Kostiakov's Model: Kostiakov (1932) developed the following model for estimateion of infiltration rate:

$$I = k\ t^n \tag{4.31}$$

where, I is infiltration rate; k and n are constants and t is time. The value of n in the above equation lies between 0 and 1.

In a small time, dt, the depth of infiltration is given as:

$$I\ dt = k\ t^n\ dt \tag{4.32}$$

Integrating the Eqn. (4.32) for time 0 to any time, t we get total depth of infiltration (Y),

$$Y = \frac{k}{n+1}\ t^{n+1} \tag{4.33}$$

Limitation of this model is that the above proposed equation predicts infiltration as zero at a large time interval which is not correct. In fact, at any time (large time interval), infiltration will not be zero which may be very small and continues up to infinity. However, it computes accurate value of infiltration depth for the time, which is generally used in irrigation practice.

Lewis and Milne Model: This model is also called as *Horton's model* and is expressed as:

$$Y = a\,t + b\left(1 - e^{-rt}\right) \tag{4.34}$$

where, Y is total depth of infiltration, t is time and a, b, r are constants.

Differentiating Eqn. (4.34), we get rate of infiltration, I as:

$$I = a + b\,r\,e^{rt} \tag{4.35}$$

The above equation holds good for both small and large time intervals.

Phillip's Model: Phillip's model is used to estimate cumulative infiltration depth which is given as:

$$Y = S.\ T^{1/2} + A.\ t \tag{4.36}$$

where, Y is cumulative infiltration; S is soil property known as sorptivity, t is time and A is constant.

Differentiating Eqn. (4.36), we get rate of infiltration, I as:

$$I = (S/2)\ t^{-1/2} + A \tag{4.37}$$

Michael et al. (1972) have proposed a model in the form as:

$$Y = a\,t^{\alpha} + b \tag{4.38}$$

where, Y is accumulated infiltration rate; t is time and a, b and α are constants. Values of a, b and α vary from 0 to 1.

Differentiating Eqn. (4.38), we get rate of infiltration, I as:

$$I = a.\alpha\ .\ t^{\alpha-1} \tag{4.39}$$

Horton's Model for Infiltration Capacity Values: Infiltration capacity for a given soil decreases with time with the time from start of rainfall. It decreases with saturation of soil and depends on type of soil. Horton expressed this decay of infiltration capacity with time as:

$$f_{ct} = f_{cf} + \left(f_{co} - f_{cf}\right)e^{-K_h t} \text{ for } 0 \geq t \leq t_d \tag{4.40}$$

where, f_{ct} is infiltration capacity at any time t from the start of rainfall; f_{co} is initial infiltration capacity at time $t = 0$; f_{cf} is final steady state infiltration; t_d is duration of rainfall and K_h is constant which depends on soil characteristics and vegetation cover. It is very difficult to find out the variation of the three parameters f_{co}, f_{cf} and K_h with soil characteristics and antecedent moisture conditions and so it is difficult to use the above equation in practice.

Eqn. (4.40) can be solved by rearranging it and taking natural logarithm in both the sides as:

$$\ln\left(f_{ct} - f_{cf}\right) = \ln\left(f_{co} - f_{cf}\right) - K_h\, t \qquad (4.41)$$

The above Eqn. (4.41) is a linear form equation where the y-intercept is $\ln\left(f_{co} - f_{cf}\right)$ and slope is $-K_h$. Negative sign of the slope indicates that as the time, t increases, infiltration capacity, f_{ct} decreases and therefore $\ln\left(f_{ct} - f_{cf}\right)$ decreases.

If two values of infiltration capacity at two different time, t i.e. f_{ct} are known and f_{cf} (final steady state infiltration) is also known, a straight line can be drawn through these two points and slope of the line can be measured. After knowing the slope, equations for the infiltration capacity can be written.

Following sample calculation helps in developing the model:

Sample Calculation 4.8

The infiltration capacity of an area at different intervals of time is recorded as below. Develop the Horton's exponential model using the data.

Time, hr	0	0.25	0.50	0.75	1.0	1.25	1.50	1.75	2.0
Infiltration capacity, cm/hr	10.0	5.5	3.0	2.0	1.5	1.2	1.0	0.9	0.9

Solution

Horton's infiltration capacity model is:

The recorded data as given in the problem represents that final steady state infiltration rate, f_{cf} reaches after 1.75 hr and the value is 0.9 cm/hr. In the problem, values of infiltration capacity at different time (f_{ct}) are given. We can tabulate the values of and also vs. time, t as shown below:

Time, t (hr)	0	0.25	0.50	0.75	1.0	1.25	1.5	1.75	2.0
f_{ct} (cm/hr)	10.0	5.5	3.0	2.0	1.5	1.2	1.0	0.9	0.9
$(f_{ct}-f_{cf})$ cm/hr	9.1	4.6	2.1	1.1	0.6	0.3	0.1	0	0
$ln\,(f_{ct}-f_{cf})$ cm/hr	2.208	1.526	0.742	0.095	-0.510	-1.204	-2.303		

Now plot in an arithmetic graph paper time, t vs. $ln\,(f_{ct}-f_{cf})$ taking time as abscissa and $ln\,(f_{ct}-f_{cf})$ as ordinate and the following graph is obtained.

From this figure when we fit a linear equation, y-intercept is found to be 2.283 and slope of the line is obtained as -2.924.

Hence, slope $-K_h$ = -2.924 which gives K_h = 2.924.

Hence Horton's infiltration model is:

$$f_{ct} = 0.9 + \left(10 - 0.9\right) e^{-2.924\,t}$$

$$f_{ct} = 0.9 + 9.1\, e^{-2.924\,t}$$

where, f_{ct} is in cm/hr and time, t is in hr.

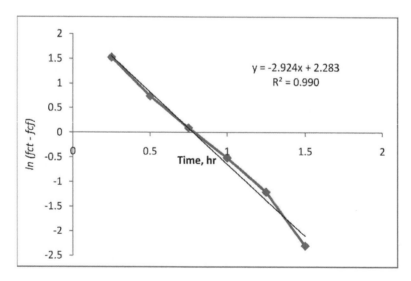

In addition to the above mentioned infiltration models, other models like Green and Ampt model, Lewis's model etc. are also available.

4.8.3 Measurement of Infiltration

Infiltration can be measured by an instrument called as infiltrometer. There are two kinds of infiltrometer. They are:

- Flooding type infiltrometer and

- Rainfall simulator

4.8.3.1 Flooding Type Infiltrometer: They are simple or single ring infiltrometer and double ring infiltrometer or simply called a ring infiltrometer. Single ring infiltrometer (Fig. 4.17) consists of a metal cylinder made of M.S. having 30 cm diameter and 60 cm height and is open at both the ends. The cylinder is driven into the ground upto a depth 50 cm. Water is poured into the cylinder up to a depth of 5 cm above the ground as shown in Fig. 4.18 and a pointer is set to the mark the water level. As time elapses, the level of water in the cylinder goes down as infiltration. The volume of water lost in the cylinder in a given time is noted from the pointer used to mark the water level. The volume of water measured in a graduated container is added to the cylinder to make up the loss of water in the cylinder in that given time. Knowing the volume of water added at different time intervals, infiltration rate is calculated as:

$$I = \frac{V}{a\,t} \tag{4.42}$$

where, I is the infiltration rate; V is the volume of water added to the ring in time t and 'a' is the cross sectional area of the cylinder/ring. After getting a number of readings, a plot of infiltration rate vs. time is drawn. Sample calculation 4.9 illustrates the procedure of computation of infiltration rate. The experiments are continued till a constant or uniform rate of infiltrations are obtained which may take 2-3 hrs. Initially, the time interval for measurement must be kept shorter, because infiltration rate is high at the beginning and as time increases, soil moisture content increases and infiltration rate then decreases. The surface of soil is usually protected by a perforated disk to prevent formation of turbidity and its setting on the soil surface.

There is a drawback in using single ring infiltrometer. When water is poured in to the single ring infiltrometer, it starts flowing out both latterly and vertically as shown in Fig. 4.17 and as such the tube area is not representative of the area in which infiltration takes place. To overcome this problem, a double ring infiltrometer or sometimes simply called as ring infiltriometer is used. The ring infiltrometer consists of two concentric rings as shown in Fig. 4.18. Both the rings are open at top and bottom and are inserted into the ground up to a depth of about 50 cm leaving a depth of 10 cm above the ground. Water is poured into the cylinder up to a depth of 5 cm above the ground Water is poured into the cylinder up to a depth of 5 cm above the ground as shown in Fig. 4.18 and a pointer is set to the mark the water level. Same water level is maintained in

Time, t (hr)	0	0.25	0.50	0.75	1.0	1.25	1.5	1.75	2.0
f_{ct} (cm/hr)	10.0	5.5	3.0	2.0	1.5	1.2	1.0	0.9	0.9
$(f_{ct}-f_{cf})$ cm/hr	9.1	4.6	2.1	1.1	0.6	0.3	0.1	0	0
$\ln (f_{ct}-f_{cf})$ cm/hr	2.208	1.526	0.742	0.095	-0.510	-1.204	-2.303		

Now plot in an arithmetic graph paper time, t vs. $\ln (f_{ct}-f_{cf})$ taking time as abscissa and $\ln (f_{ct}-f_{cf})$ as ordinate and the following graph is obtained.

From this figure when we fit a linear equation, y-intercept is found to be 2.283 and slope of the line is obtained as -2.924.

Hence, slope $-K_h$ = -2.924 which gives K_h = 2.924.

Hence Horton's infiltration model is:

$$f_{ct} = 0.9 + (10 - 0.9)\, e^{-2.924\,t}$$

$$f_{ct} = 0.9 + 9.1\, e^{-2.924\,t}$$

where, f_{ct} is in cm/hr and time, t is in hr.

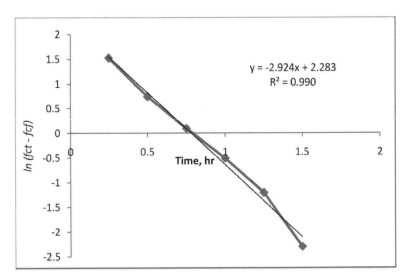

In addition to the above mentioned infiltration models, other models like Green and Ampt model, Lewis's model etc. are also available.

4.8.3 Measurement of Infiltration

Infiltration can be measured by an instrument called as infiltrometer. There are two kinds of infiltrometer. They are:

Solution

Calculations are shown in following table.

Accumulated time, min	Volume of water added, cc	Depth during time interval, cm	Infiltration rate, cm/min
0	0	0	0
1	90	0.13	0.13
2	80	0.11	0.11
5	75	0.10	0.04
10	65	0.09	0.018
15	54	0.076	0.015
30	107	0.151	0.010
60	110	0.156	0.005
90	91	0.129	0.004
120	88	0.124	0.004

Basic intake rate of soil is 0.004 cm/min.

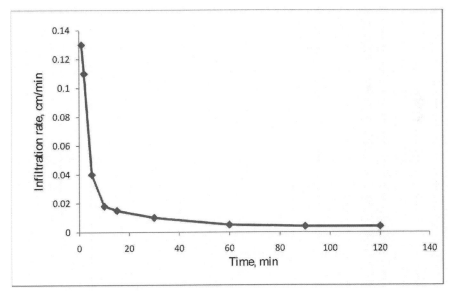

Fig. 4.19: Represents the view of infiltration capacity curve for the recorded data.

4.8.3.2 Rainfall Simulator: Rainfall simulator simulates rainfall which falls over some distance above the ground. The falling raindrop strikes the ground surface and some portion of it is infiltrated into the soil surface. United States Department of Soil Conservation Services has introduced two types of sprinkling infiltrometers. They are (i) F-type infiltrometer and (ii) FA type infiltrometer. Constructional features of both the types are same with differences in respect of size of the experimental plots where they are placed. For setting F- type infiltrometer, a small piece of land of about 2 m x 4 m in size is required. However, for FA type infiltrometer, large plot area of about 30 m x 76 m is required. The rainfall simulator is provided with a series of nozzles on the longer side with arrangements to collect and measure the surface runoff. The nozzles are specially designed to produce rain drops falling from a height of 2 m and are capable of producing rainfall of different intensities. However, for the experiment, controlled conditions are required for various combinations of intensities and durations and in each case, the surface runoff is measured. Knowing the amount of rainfall, surface runoff, duration of rainfall and ignoring the evaporation loss (since the experiment is conducted for a small interval of time) and finally employing the water budget equation, one can calculate the amount of infiltration in a given time. From this calculated values of time and amount of infiltration, one can compute the infiltration rate. Thus, infiltration rate over a wide range of time can be computed and an infiltration capacity curve can be drawn. It is important to remember that for obtaining the infiltration capacity values, the rainfall intensity must be greater than the infiltration rate of the soil. The rainfall simulator type infiltrometer gives lower values of infiltration rate than the flooding type infiltrometer. The reason is due to the effect of rainfall impact and turbidity of the surface water present in the former.

4.8.3.3 Measurement of Infiltration by Furrow Infiltrometer Method:
The cylinder infiltrometer method gives accurate measurement of infiltration rate from level surface of the soil. In corrugated condition like furrow, it does not give accurate result, because of variations in compactness of the soil at the sides and bottom of the furrow. Under such case, furrow infiltrometer works well. Furrow infiltrometer is of two types i.e. (i) blocking type and (ii) by-pass.

Blocking Type Furrow Infiltrometer: This type of infiltrometer was designed by Bordurant in 1957. It consists of two metal plates, which are placed in the furrow section to confine a certain length of the furrow. A supply tank fills water in this confined section of the furrow. A constant water level is maintained in the furrow section throughout the experiment. As time elapses, water level in the furrow goes down because of infiltration in the furrow. The lowering of water level in the supply tank at different time intervals is noted. The cumulative infiltration depth that has entered into the soil during a certain time period, t is given as:

$$F = (h - h_t) \, K \qquad\qquad (4.43)$$

where, F is the cumulative depth of infiltration; h is the initial height of water level in the tank, h_t is the final height of water level in the tank and K is the multiplying factor which is:

$$K = A/a \tag{4.44}$$

where, A is the cross sectional area of the supply tank and 'a' is the area of the furrow section.

Area of the furrow section assuming furrow to be trapezoidal in cross section is given as top width of the furrow section multiplied by the spacing of two plates in the furrow section and is expressed as:

$$A = (b + 2.D.Z)\ S \tag{4.45}$$

where, b is the bottom width of the furrow; D is the constant depth of water level maintained in the furrow, Z is the side slope of the trapezoidal furrow section (horizontal: vertical) and S is the spacing of plates in the furrow section.

Once, the accumulated depth of infiltration and the accumulated time in which the accumulated depth of infiltration water has entered into the soil surface is know, infiltration rate at different time intervals can be computed. The sample calculation as described below illustrates the use of blocking type infiltrometer for computation of depth of infiltration.

Sample Calculation 4.10

Compute the cumulative infiltration depth and infiltration rate by using the following information as recorded by a blocking type infiltrometer.

Elapsed time, min	0	5	10	20	40	60	120
Water level in supply tank, cm	32	28	25	22	20	18	17

The furrow cross section is trapezoidal in shape with bottom width 20 cm and side slope 2:1 (horizontal:vertical). Spacing of the plate in the furrow section is 60 cm. Constant water level in the furrow section is 5 cm and supply tank has a size 30 cm.

Solution

Bottom width of furrow = 20 cm

Side slope, $Z = 2{:}1$ (horizontal:vertical).

Depth of constant water level in furrow, $D = 5$ cm.

Hence, the top width of the furrow = $b + 2.D.Z = 20 + 2 \times 5 \times 2 = 40$ cm

Area of furrow cross section = area of top of the water surface in the furrow, 'a' = 40 x 60 = 2400 cm^2.

Cross sectional area of the tank, $A = \dfrac{\pi}{4} \cdot d^2 = \dfrac{\pi}{4} \cdot 30^2 = 706.85$ cm².

Multiplying factor, $K = A/a = 706.85/2400 = 0.29$

Computation of cumulative infiltration depth and infiltration arte is computed in the tabular form as below:

Elapsed time, min	Water level in the supply tank, cm	Decrease in water level in tank, cm	Accumulated depth of infiltration, cm	Infiltration rate, cm/min
0	32	0	0	0
5	28	4	1.16	0.232
10	25	7	2.03	0.174
20	22	10	2.90	0.087
40	20	12	3.48	0.029
60	18	14	4.06	0.029
120	17	15	4.35	0.0048

Accumulated depth of infiltration (f) is the product of decrease in water level in the tank and multiplying factor, K.

By-Pass Type Furrow Infiltrometer: This type of infiltrometer was designed by Shull in 1961. This type of infiltrometer simulates similar condition as in the furrow section as generally occurs during irrigation. In this case a small section of furrow is isolated by the infiltrometer and in the rest section, irrigation water is by-passed to flow. Under such circumstances, water does not flow in the enclosure of the infiltrometer. Other measurement procedures are same as discussed in blocking type furrow infiltrometer.

4.8.3.4 Infiltration Indices: In hydrological calculations involving floods, it is found convenient to use a constant value of infiltration rate for the duration of the storm. The average infiltration rate is called as infiltration index. There are two types of infiltration indices. They are (i) ϕ index and W- index. These two indices have been discussed earlier in Chapter 3.

Question Banks

Q1.   What is abstraction? What is evaporation loss? How does it vary with respect to time?

Q2.   What are the factors on which evaporation loss depend? Briefly enumerate them.

Q3.   How does the evaporation loss of water bodies vary from place to place in India?

Q4.   How does the depth of water in a water body affect the evaporation loss?

Q5.   Write down the water budget method to estimate the evaporation loss in a water body. Describe how it is helpful in estimation of evaporation loss in a water body.

Q6.   Describe the mass transfer and energy budget method to compute the evaporation rate in a water body.

Q7.   How does the open water surface area of a water body influence evaporation rate?

Q8.   Describe the combination equation for estimation of evaporation loss in a reservoir. Write the equation and explain the various terms of this.

Q9.   Draw the diagram of USDA Class A Pan evaporimeter and write about its working principles.

Q10.   What are the criteria followed in selecting site for a pan evaporimeter?

Q11.   What are the various types of evaporimeters? With diagram explain the construction and working principles of each type.

Q12.   Narrate the different procedures to estimate the evaporation loss in a reservoir.

Q13.   Describe different empirical methods used to estimate evaporation rate in a water body?

Q14.   Write short notes on the following empirical formulae to compute evaporation rate?

(i)   Meyer's formula (ii) Rohwer's formula

Q15.   Estimate the evaporation loss in a reservoir in a week which has a surface area of 250 ha. The average values of the reservoir parameters within the week are as follows:

Water temperature is 20 °C, relative humidity is 32%, and wind velocity at 1 m above the ground level is 15 km/hr.

Q16. Write brief notes on the following evaporimeters.

(i) B.I.S. Class A Pan (Modified)

(ii) Class A.U.S. Weather Bureau Land Pan:

(iii) Colorado Sunken Pan:

(iv) US Geological Survey Floating Pan

(v) Class A U.S. Weather Bureau Floating Pan

Q17. An evaporation pan is 1.6 m in diameter and is used to measure evaporation loss of reservoir. The initial and final water level reading in the pan are found to be 10.5 cm and 11.6 cm, respectively. A rainfall of 3.8 cm has occurred during the period the observation was taken. To keep the water level same in the pan, 2 cm depth of water has been removed from the pan. Find out the pan evaporation. If the pan coefficient used to convert the pan evaporation to reservoir evaporation is 0.8, find out the reservoir evaporation also.

Q18. What do you mean by evaporation opportunity time? Write the formula to compute it.

Q19. What are the factors that affect soil evaporation? Explain these factors.

Q20. Briefly explain the various methods to reduce evaporation loss in a water body.

Q21. How are the shading materials helpful in reducing evaporation loss in a water body?

Q22. What do you mean by biological shading? How are they used to decrease evaporation losses?

Q23. What is transpiration? What are the factors that affect transpiration?

Q24. What are the differences between evaporation and transpiration?

Q25. How can the transpiration be measured? Explain the working mechanisms of phyto meter to measure transpiration.

Q26. Define the followings:

(i) Consumptive use (ii) Evapotranspiration (iii) Water requirement

Q27. What do you mean by evapotranspiration? What are the various factors that affect evapotranspration?

Q28. Differentiate between actual crop evapotranspiration and potential evapotranspiration.

Q29. What do you mean by reference crop evapotranspiration? How does it vary from place to place?

Q30. How does the reference crop evapotranspiration vary from potential evapotranspiration?

   (a)   Briefly explain the energy balance method and aero-dynamic method to estimate the crop water requirement.

   (b)   What is a combined method for estimation of reference crop evapotranspiration?

Q31. With label diagram, describe how the evapotranspiration of crop is measured by the lysimeters.

Q32. Which type of lysimeter gives more correct measured values of crop evapotranspiration? Describe how this lysimeter is used to measure the crop evapotranspiration.

Q33. Describe how the soil moisture depletion study is helpful to compute the crop evapotranspiration.

Q34. Profile soil moisture content of a potato field in 60 cm effective root zone depth are measured as follows:

Profile layer, cm	Soil moisture content,%		Apparent Specific gravity, gm/cc
	Beginning of irrigation cycle	End of irrigation cycle	
0-15	20	12	1.60
15-30	22	13	1.58
30-45	24	15	1.59
45-60	25	15	1.60

If effective rainfall within the irrigation cycle is 20 mm, and percolation from the effective root zone of the crop is 11 mm, estimate the water use from the crop root zone within the irrigation cycle.

Q35. A field is sown with ground nut. The profile soil moisture content of the field at sowing and at harvest time are measured and presented in table as below.

Profile layer, cm	Soil moisture content,%		Apparent Specific gravity, gm/cc
	At sowing	At harvest	
0-15	22	10	1.62
15-30	24	11	1.60
30-45	25	13	1.59
45-60	25	15	1.60

If the crop is supplied with 20 cm irrigation, effective rainfall in the crop growing season is 12 cm and there is no percolation loss from the crop effective root zone in the entire season, then estimate the seasonal water requirement of the crop.

Q36. Write the general water balance model of crop field (both paddy and non-paddy) and explain the different terms. How can the model be used to predict the soil moisture content in different soil layers of the crop filed?

Q37. Write down the different steps involved in estimating the reference crop evapotranspiration by Thronwaite method.

Q38. Following data are recorded for a particular place having latitude 29°N. Estimate the daily reference crop evapotranspiration for the month of January using the data and using Thornwaite method.

Month	Jan	Feb	Mar	Apr	May	Jun	Jul	Aug	Sep	Oct	Nov	Dec
Mean temperature, °C	13	15.9	16	23	29.8	30.4	29.0	28.6	25.7	20.6	18.6	12

Q39. Write down the different steps used to compute the reference evapotranspiration by the modified Blaney-Criddlee method.

Q40. Calculate the reference evapotranspiration with using the following data.

Place is Chiplima having 22°N latitude and altitude is 205 m period is month December. The mean maximum temperature is 23°C and mean minimum temperature is 6°C. The place has actual sunshine hour (average monthly) is 9 hr. The day wind velocity at 2 m height is 3.5 m/s and the mean minimum relative humidity is 54%.

If the crop coefficient value is 1.05 then find out the actual crop evapotranspiration.

Q41. (a) The water level in a pan evaporimeter on a day at 7 AM is 20 cm below the rim of the pan and that on the next day is 19.2 cm. If there is no rainfall within the time both readings are taken then find out the pan evaporation. Had there been 5 mm rainfall then what would have been the value of pan evaporation? Using pan coefficient value as 0.8, find out the reference evapotranspiration in both the cases.

(b)   Briefly describe what the factors that influence the pan coefficient value.

Q42. How does the potential evapotranspiration vary over different places over the country?

Q43. What is interception loss? How is it calculated?

Q44. What are the factors on which interception loss depends? Briefly explain them.

Q45. Describe Linsley formula for computation of interception loss. Write down the equation and define the terms of this equation.

Q46. How do the storm factor and plant factor influence interception?

Q47. What is depression storage? How does it affect the runoff? Write down the factors that affect depression storage.

Q48. Define infiltration rate. Describe the mechanism of infiltration as explained by the physiology of soil mass.

Q49. Define infiltration capacity. What are the factors that affect infiltration capacity? Describe these factors how they affect it.

Q50. Enumerate different types of infiltration models, write the equations of these models and explain the terms used in these equations.

Q51. Write short notes on the following models:

(i)   Phillip's model (ii) Kostiakov's model (iii) Horton's model (iv) Lewis and Milne model

Q52. Write the Horton's infiltration capacity model and explain the various terms used in this model. How can you solve this model to find out the infiltration capacity of a soil?

Q53. The infiltration capacity of an area at different intervals of time is recorded as below. Develop the Horton's exponential model using the data.

Time, hr	0	0.25	0.50	0.75	1.0	1.25	1.50	1.75	2.0
Infiltration capacity, cm/hr	9.5	5.0	3.2	2.2	1.5	1.2	1.0	0.95	0.9

Q54. How is infiltration measured in field? Describe how the infiltration rate is measured by flooding type infiltrometer and rainfall simulator?

Q55. What are the disadvantages of flooding type infiltrometer?

Q56. Differentiate between single ring infiltrometer and double ring infiltrometer. Which one gives accurate measurement of infiltration arte and why?

Q57. A double ring infiltrometer was set up in a field for measurement of infiltration and the following readings are recorded.

Accumulated time, min	0	1	2	5	10	15	30	60	90	120
Volume of water added, at different time interval, cc	0	80	72	70	65	60	50	30	25	24

Diameter of the inner ring of the infiltrometer is 30 cm and constant depth of water maintained in the ring 4 cm. Compute the infiltration rate at different time intervals and find out the basic intake rate of the soil.

Q58. In corrugated condition like furrow, how can the infiltration rate be measured? Explain how does furrow infiltrometer work in corrugated soil in measuring infiltration.

Q59. Write brief notes on the following two types of furrow infiltrometers.

(i) Blocking type furrow infiltrometer and (ii) by-pass type furrow infiltrometer

Q60. Compute the cumulative infiltration depth and infiltration rate by using the following information as recorded by a blocking type infiltrometer.

Elapsed time, min	0	5	10	20	40	60	120
Water level in supply tank, cm	32	28	25	22	20	18	17

The furrow cross section is trapezoidal in shape with bottom width 20 cm and side slope 2:1 (horizontal:vertical). Spacing of the plate in the furrow section is 60 cm. Constant water level in the furrow section is 5 cm and supply tank has a size 30 cm.

References

Allen, R.G., Pereira, L.S., Raes, D. and Smith, M. 1998. Crop Evapotranspiration: Guidelines for Computing Water Requirements. Irrigation and Drainage Paper 56, FAO, Rome, pp 300.

Belgaumi, M.I., Itnal, C.J. and Radder, C.D. 1997. Farm Ponds. Technical Series 4, Publication Centre, University of Agricultural Sciences, Dharwad, pp. 47.

Cooley, K.R. 1974. Evaporation suppression for conserving water supplies. Proceedings of Water Harvesting Symposium, Phoenix, Arizona: 192-199.

Doorenbos, J. and Pruitt, W.O. 1977. Guidelines for Predicting Crop water Requirements. Irrigation and Drainage Paper 24, FAO, Rome

Garg, S.K. 1987. Hydrology and Water Resources Engineering. Khanna Publishers, Delhi, pp. 581.

James, L.G. 1988. Principles of Farm Irrigation System Design. John Wiley and Sons, New York, pp. 543.

Lenka, D. 2001. Irrigation and Drainage. Kalyani Publishers, New Delhi.

Linsley, R.K., Kohler, M.A. and Pauhus, J.L.H. 1949. Applied Hydrology. McGraw-Hill Publishing Co. Ltd., New York.

Michael, A.M., Mohan, S and Swaminathan, K.R. 1972. Design and Evaluation of Irrigation Methods. IARI Monograph No. 1, WTC (IARI), New Delhi.

Misra, R.D. and Ahmed, M. 1987. Manual on Irrigation Agronomy, Oxford & IBH Publishing Co. Pvt. Ltd., New Delhi, pp. 412.

Odhiambo, L.O. and Murty, V.V.N. 1996. Modeling water balance components in relation to field layouts in low land paddy fields. I. model development, Agricultural Water Management, Vol. 30(2): 185-199.

Panigrahi, B. 2001. Water Balance Simulation for Optimum Design of On-Farm Reservoir in Rainfed Farming System. Ph.D. Thesis, Indian Institute of Technology, Kharagpur.

Panigrahi, B. 2011. Irrigation Systems Engineering. New India Publishing Agency, New Delhi.

Panigrahi, B. 2013. A Handbook on Irrigation and Drainage. New India Publishing Agency, New Delhi.

Penman, H.L. 1948. Natural evaporation from open water, bare soil and grass. Proceedings of Royal Society of London, Vol. 193: 120-146.

Penman, H.L. 1963. Vegetation and Hydrology. Farnham Royal, Common Wealth Agricultural Bureaux.

Pochop, L.O., Borrelli, J. and Hasfurther, V. 1985. Design Characteristics for Evaporation Ponds in Wyoming. Final Report submitted to the Wyoming Water Research Centre, Laramie, WY: 1-13.

Rao, K.N., George, C.J. and Ramasastri, K.S. 1971. Potential evapotranspiration (PE) over India. Symposium on Water Resources, I.I.Sc., Bangalore, India.

Sharma, R.K. and Sharma, T.K. 2002. Irrigation Engineering including Hydrology, S. Chand and Company Ltd. New Delhi.

Subramanya, K. 2003. Engineering Hydrology, Tata McGraw-Hill Publishing Co. Ltd., New Delhi, 392 pp.

Suresh, R. 2008. Watershed Hydrology. Standard Publishers Distributors, Delhi, pp. 692.

Thornthwaite, C.W. 1948. An approach towards a rational classification of climate. Geographical Review, Vol. 38: 55-94.

CHAPTER 5

Streamflow Measurement

5.1 Introduction

A steam can be defined as a flow channel in to which the surface runoff flowing over the land surfaces and base flow/ground flow flowing below the ground surfaces meet. It receives the total flow from a specified basin. Streamflow representing the runoff phase of the hydrologic cycle is the most important basic data for hydrologic studies. Interestingly, stream flow is the only part of the hydrologic cycle that can be measured accurately. Stream flow is a historical data and like rainfall and runoff, it is a stochastic hydrologic variable. It is also called as discharge or rate of flow. Discharge is defined as the volume of water flowing through a cross section in a unit time. It can be expressed as cubic metre per second (cumec), cubic feet per second (cusec), litre per sec (lps) etc. one cumec is equivalent to 35.5 cusecs or 1000 lps. The measurement of discharge is very important in everyday life starting from deciding the amount of irrigation to be supplied to a crop at a particular growth stage, or amount of water supply for domestic and industrial uses, for flood control and power generation etc. It plays a vital role in water resources planning. The measurement of discharge in a stream forms an important branch of *hydrometry*, the art of water management. If a stage-discharge (also called as gauge-discharge) relationship for a particular gauging site is developed, then daily measurement of discharge in the gauging site is not required. Discharge can be estimated from the developed relationship of discharge versus stage if the measurement of stage is known.

5.2 Objectives of Streamflow Measurement

The various objectives of streamflow measurements are given as:

(i)   For accurate assessment of available and dependable water resources of river basins,

(ii) For flood forecasting and flood warning,

(iii) For comprehensive and coordinated planning for the optimum utilization of water resources in rivers and their tributaries,

(iv) For ascertaining annual and seasonal variation of runoff,

(v)   For estimation of water losses and regeneration in the various reaches of the river and

(vi) For investigation and design of multipurpose river valley projects for development of irrigation and power potential, flood control, water supply etc.

5.3 Methods of Streamflow Measurement

The various methods of streamflow measurement can be broadly classified into the following types (i) velocity-area methods (iii) slope-area method (iv) gauge-discharge relationship (v) flow measuring structures/hydraulic structures such as orifices, mouthpieces, flumes and weirs and (vi) chemical dilution method and (vii) radioisotope method (viii) electromagnetic method and (ix) ultrasonic method.

5.3.1 Velocity-Area Method

In this method the rate of flow passing in a pipe or open channel is determined by multiplying the cross sectional area of the flow at right angles to the direction of flow at a selected section called the *gauging site* by the velocity of water. The formula is:

Discharge (Q) = Area (A) x Velocity (V)   (5.1)

The gauging site must be selected with care to assure that the stage-discharge curve is reasonably constant over a long period of time of about a few years. The essential requirements of gauging site are discussed subsequently in this chapter. The discharge is generally expressed in the unit of m^3/sec, area in m^2 and velocity in m/sec. The velocity is usually measured by float method, tracer method or by a current meter. If the small irrigation channel is of regular cross section in the form of a rectangle or trapezium, then cross sectional area of flow can be measured by measuring the depth of flow, base width and side

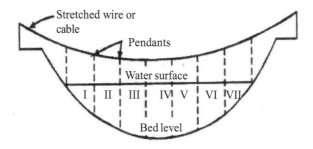

Fig. 5.1: Segmentation of cross-section

slope (in case of trapezoidal section) of the channel and using the formula to compute the area of rectangle or trapezium as the case may be. In case of large irrigation channels, the cross section is divided into suitable compartments or vertical sections and total discharge is computed by adding the discharges in various sections on the basis of the depth and velocity observed in each vertical section. The method of dividing the large section in to several small vertical sections called as segments is termed as segmentation (Fig.5.1). In discharge computation by velocity-area method, two approaches are followed. They are mid-section method and mean section method.

5.3.1.1 Mid-Section Method

In mid-section method, the cross section is assumed to be composed of several small vertical sections/compartments each bounded by two adjacent verticals (Fig. 5.2). Mean velocity of each compartment is measured at 0.6 m depth

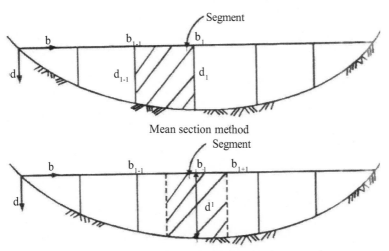

Fig. 5.2: Mid and mean section methods

below the top water surface if the depth of flow in channel is less or taking the arithmetic average of velocities taken at 0.2 and 0.8 depth below the top water surface if the depth of flow in channel is large. Total discharge flowing in the channel is computed by the formula as (Sharma and Sharma, 2002):

$$Q = V_1 d_1 \left[\frac{b_1 + b_2}{2} \right] + V_2 d_2 \left[\frac{b_2 + b_3}{2} \right] + \ldots\ldots\ldots \tag{5.2}$$

where $b_1, b_2, b_3 \ldots$ are width of the vertical sections 1, 2, 3... , $d_1, d_2 \ldots$ are the depths at the mid point of the vertical sections 1 and 2, vertical sections 2 and 3...., $V_1, V_2, \ldots\ldots$ are the mean velocity at the mid point of the vertical sections 1 and 2, vertical sections 2 and 3....where depths $d_1, d_2 \ldots$ are taken. In this method, flow at the edges is neglected.

5.3.1.2 Mean-Section Method

In this method, width of the strip of a section is taken as the distance between two verticals. Velocity is the average value of mean velocities in two adjacent verticals (Fig. 5.2). Discharge of each strip is given as (Sharma and Sharma, 2002):

$$q = b \left[\frac{d_1 + d_2}{2} \right] \left[\frac{V_1 + V_2}{2} \right]. \tag{5.3}$$

and total discharge in the channel is the sum of discharge of all the strips. The first and last verticals are sited as close to the banks as possible, since discharge in the end segments are neglected.

5.3.2 Slope Area Method

This method of discharge measurement is approximate and is used when the actual measurement by velocity-area method is not possible. In this method, the average cross sectional area is multiplied with the velocity derived from the observed water surface slope. Cross sectional area is measured at three locations, two at the ends and one at the centre of a straight reach of section. The mean cross sectional area, A is given as (Sharma, 1987):

$$A = \frac{\left[A_1 + 2 A_2 + \ldots\ldots.(2 A_{n-1}) + A_n \right]}{2 (n-1)} \tag{5.4}$$

where, $A_1, A_2 \ldots A_n$ are the cross sectional areas measured at various locations.

Mean wetted perimeter, P is similarly calculated as (Sharma, 1987):

$$P = \frac{\left[P_1 + 2\,P_2 + \ldots\ldots(2\,P_{n-1}) + P_n\right]}{2\,(n-1)} \tag{5.5}$$

where, P_1, P_2.....P_n are the wetted perimeter measured at various locations.

Mean velocity is determined from the observed water surface slope by either Chezy's formula or Manning's formula as written below.

Chezy's Formula: $V = C\,\sqrt{R\,S}$ (5.6)

Manning's Formula: $V = \dfrac{1}{n}\,R^{2/3}\,S^{1/2}$ (5.7)

where, V = mean velocity, m/sec; C = Chezy's constant depending on shape and surface the channel; R = hydraulic mean radius, m; n = rugosity coefficient (value of n varies from 0.020 to 0.060 depending on types and sizes of bed materials of the channel) and S = water surface slope between two ends of a canal section or reach.

Hydraulic mean radius is the ratio of area to wetted perimeter where area is given by Eqn. (5.4) and wetted perimeter is given by Eqn. (5.5). If the depth of water levels at two ends of a canal section of length, 'L' are y_1 and y_2, respectively and the two ends are at elevations of z_1 and z_2 above some datum, then water surface slope is given as,

$$S = \frac{\left(y_1 + z_1\right) - \left(y_2 + z_2\right)}{L} \tag{5.8}$$

It is to be noted that in case of flat slope or for a reach of sufficient length with uniform flow, kinetic energy difference between the two ends of the canal section where water surface slope is measured is neglected. But if the slope of the canal section is high then there may have some kinetic energy difference between the two ends of the section/reach and so in such case, energy slope instead of water surface slope is to be taken into account to compute mean velocity of flow either by Manning's or Chezy's formula as given above. Energy slope, S_f is computed as follows:

Consider a longitudinal canal section of length L between two sections 1 and 2. Let the bed slope of the section be S_0 (Fig. 5.3). Let the two sections 1 and 2 be at an elevations of z_1 and z_2 above some datum, have water depth y_1 and y_2, velocity of flow V_1 and V_2, respectively Let head loss in the reach be h_L. The head loss h_L is sum of frictional loss, h_f and eddy loss, h_e. Applying energy equation to sections 1 and 2, we get

$$z_1 + y_1 + \frac{V_1^2}{2g} = z_2 + y_2 + \frac{V_2^2}{2g} + h_e + h_f \tag{5.9}$$

Denoting $z + y = h$, we get frictional loss, h_f as:

$$h_f = (h_1 - h_2) + \left(\frac{V_1^2 - V_2^2}{2g}\right) - h_e \tag{5.10}$$

Now energy slope, S_f is the ratio of head loss in the reach (h_f) and length of the reach (L) and is given as:

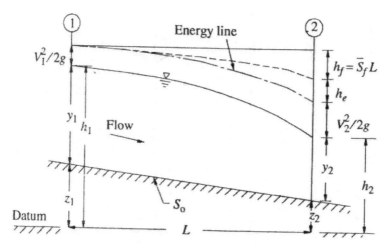

Fig. 5.3: Slope-area method

$$S_f = \frac{h_f}{L} \tag{5.11}$$

Using Manning's formula, we can write

$$S_f = \frac{Q^2}{K^2} \tag{5.12}$$

where, Q is the discharge and K is the conveyance of the channel given as:

$$K = \frac{1}{n} A R^{2/3} \tag{5.13}$$

If the flow is not uniform, an average conveyance may be used as:

$$K = \sqrt{K_1} \cdot \sqrt{K_2} \qquad (5.14)$$

where, $K_1 = \dfrac{1}{n} A_1 R_1^{2/3}$ and $K_2 = \dfrac{1}{n} A_2 R_2^{2/3}$

The eddy loss, h_e is expressed as:

$$h_e = K_e \left(\frac{V_1^2 - V_2^2}{2g} \right) \qquad (5.15)$$

where, K_e is eddy loss coefficient having values as below.

Characteristics of reach section	Value of K_e	
	Expansion	Contraction
Uniform	0	0
Gradual transition	0.3	0.1
Abrupt transition	0.8	0.6

Eqns. (5.10) to (5.15) together with continuity equation enables one to compute the discharge for known values of h, channel cross-sectional properties and Manning's roughness or rugosity coefficient, n.

The discharge is calculated by trial and order procedure using the following steps:

1. First assume velocity at two sections V_1 and V_2 be same. This gives, eddy loss zero. Now solving Eqn. (5.10), we get $h_f = h_1 - h_2$

2. Using Eqn. (5.12), find out the discharge, Q.

3. From continuity equation find out V_1 and V_2 using the formula $V_1 = Q/A_1$ and $V_2 = Q/A_2$. Now calculate velocity head, $V^2/2g$ and eddy loss from Eqn.(5.15).

4. Now recalculate h_f by Eqn. (5.10) and go to step 2 to find out discharge, Q. Repeat the calculations till two successive calculations give values of discharge nearly same.

Following sample calculation helps in estimating discharge by slope area method as discussed above.

Sample Calculation 5.1

A rectangular channel section is 10 m wide. The depth of flood water at two sections in this channel at 200 m apart is 3.1 and 3.0 m, respectively. The two sections have elevations of 15.5 and 15.0 m, respectively above datum. Assume Manning's coefficient at both the sections as same with a value of 0.025 and considering the flow to be uniform, compute the flood discharge through the channel section.

Solution

Let us use suffix 1 and 2 to denote the upstream and downstream sections, respectively. Now the cross-sectional properties of these two sections are as below:

Section 1

Depth of water $= y_1 = 3.1$ m

Width of the channel section $= 10$ m

Cross sectional area $= A_1 = 31$ m^2

Perimeter $= P_1 = 16.2$ m

Mean hydraulic radius $= R_1 = 1.914$ m

Conveyance of channel $= K_1$ (Eqn. 5.13) $= 1911.95$

Section 2

Depth of water $= y_2 = 3.0$ m

Width of the channel section $= 10$ m

Cross sectional area $= A_2 = 30$ m^2

Perimeter $= P_2 = 16.0$ m

Mean hydraulic radius $= R_2 = 1.875$ m

Conveyance of channel $= K_2$ (Eqn. 5.13) $= 1825.04$

Average conveyance for the reach $= K$ (by Eqn. 5.14) $= 1867.99$

Now calculate $h_1 = z_1 + y_1 = 15.5 + 3.1 = 18.6$ m and $h_2 = z_2 + y_2 = 15.0 + 3.0 = 18.0$ m

Hence $h_f = h_1 - h_2 = 18.6 - 18.0 = 0.6$ m

Assume in the first step $V_1 = V_2$.

The flow is uniform and so eddy loss = 0

From Eqn. (5.11), $S_f = \dfrac{h_f}{L} = h_f/200$

Discharge, $Q = K \sqrt{S_f} = 1867.99 \sqrt{S_f}$

Velocity head, $\dfrac{V_1^2}{2g} = (\dfrac{Q^2}{31^2})/19.62$ and $\dfrac{V_2^2}{2g} = (\dfrac{Q^2}{30^2})/19.62$

$h_f = \left(h_1 - h_2\right) + \left(\dfrac{V_1^2 - V_2^2}{2g}\right) - h_e = = \left(h_1 - h_2\right) + \left(\dfrac{V_1^2 - V_2^2}{2g}\right)$ (since eddy loss = 0).

$h_f = 0.6 + \left(\dfrac{V_1^2 - V_2^2}{2g}\right)$ (5.16)

The calculations are now shown in Tabular form as below.

Trial	h_f (trial)	$S_f x$ (10⁻³)	Q (m³/sec)	$V_1^2/2g$ (m)	$V_2^2/2g$ (m)	h_f (Eqn. 5.16) (m)
1.	0.6	3.0	102.314	0.555	0.593	0.562
2.	0.562	2.81	99.021	0.520	0.555	0.5647
3.	0.5647	2.82	99.263	0.523	0.558	0.5646

At the 3rd trial we find there is negligible difference between the values of h_f (see values of h_f in 1st and last column of trial number 3).

Hence, the discharge of the channel is 99.263 m³/sec.

5.3.3 Depth Measurement

In streamflow measurement, depth of flow is to be measured correctly. The depth of flow is measured directly by (i) suspension rod (ii) sounding rod (iii) lead lines (iv) reel line and cranes and indirectly by (i) echo sounder and (ii) theodolite. Suspension rod is used when the depth of flow is less (up to 1 m) and velocity of flow is of the order of 1 m/sec. The length of the suspension rod is 3 m. They are equipped with a base plate at the bottom which prevents the rod from sinking into the stream bed. The rods have graduations up to meter, decimeter and centimeter by using the enamel paint. The flow depth is directly measured by placing the suspension rod into the stream flow over the bed.

Sounding rods are used when the depth of flow is more (up to 6 m) and velocity is medium. The sounding rods may be made of wood or bamboo. The sounding rods are also graduated in meters, decimeters and centimeters and are painted white, black or red. The letter size is is usually kept 2.5 cm. The sounding rods are of two types i.e. (i) Wooden rod and (ii) Bamboo rod. Wooden sounding rods are used for streamflow measurement in the rivers or channels having low depth of flow of 2 to 3 m. They are made of strong and well seasoned wood having sizes ranging from 2 to 3 m and cross –sectional area is 80 x 20 mm². Just like suspension rod, they are equipped with a base plate at the bottom which prevents the equipment from sinking into the stream bed. The plate is made of metal sheet of 150 mm diameter and 5 mm thick. A circular flange having 80 mm diameter opening is also provided to the rod for connecting the nipple of the same diameter (80 mm). Bamboo type sounding rods are preferred to wooden type since they are light in weight and potable. The length varies from 3 to 6 mm and diameter can go up to 38 mm. A metal base plate of 100 mm diameter along with a flange is also provided at the base of the rod. Bamboo rods work well in measuring the streamflow in rivers of depth 3 to 6 m.

Lead lines also called as log lines are used when the depth of flow is more then 6 m but the velocity of flow is less. It consists of a cable of 2.5 mm diameter to the lower end of which is attached a hanger for placement of the current meter and sounding weight during observations. Weight of 7 to 14 kg is necessary to avoid to avoid deflection of cable from the true vertical. A core of insulated copper wire in the log line enables electrical circuit to be made for counting ticks of the revolution of the current meter when the current meter is operated by battery. When the depth of flow is high and velocity of flow is more, a cable line is lowered by means of a crane to measure the depth of flow. The arrangement, either mounted on a boat or used from a bridge parapet, consists of a hand-operated reel fixed to a pedestal and carrying 3 mm diameter hard steel flexible cable passing over pulley at the other end.

When the channel is of sufficient depth having high velocity of flow, echo sounder is used for accurate and rapid measurement of depth. Echo-sounder can also be used to measure velocities up to a small depth of 0.30 m, but these measurements are not so accurate. A recording type echo sounder records a continuous graph of the stream bed during the traverse on moving chart drawn by a stylus. An indicator type echo sounder records a single reading for each depth measurement. The working principle consists of transmitting the pressure wave by a transmeter from water surface to the stream bed for a short duration. The transmitted wave is reflected back which is received by a receiver. Time from start of transmission of the pressure wave to the receipt of the echo by the receiver is noted. By knowing the velocity of sound in water and time of travel, one can compute the depth of the stream bed from water surface.

5.3.4 Velocity Distribution in Open Channel

The velocity in a channel varies approximately as a parabola, from zero at the bottom of the channel to a maximum at or near the water surface at the top. The maximum velocity occurs at a depth ranging from 0.10 to 0.20 of the depth below top water surface. The ratio of the maximum to average velocity is nearly equal to 1.2. In shallow streams up to 3 m depth, the average velocity on any vertical section occurs at a depth 0.6 of the total depth below the top water surface. This procedure to measure the velocity is called as single-point observation method. In moderately deep stream, the average velocity at two depths i.e. 0.2 and 0.8 depth below the water surface equals the mean or average velocity. Velocity measurement using two points method is more correct than that done by one point (velocity taken at 0.6 depth) method. In rivers having flood flows, only the surface velocity is measured within a depth of about 0.5 m below the top water surface and then this value is multiplied with a reduction factor to get the average velocity in river. Values of the reduction factor vary from 0.85 to 0.95. In case of a channel section with abruptly changing velocities at deferent depths in a vertical section, several measurements of depth and velocity along the vertical section are taken and then using integration approach, the average velocity is calculated as (Sharma, 1987):

$$V = \frac{\displaystyle\int_{i=1}^{n} v_i d_i}{\displaystyle\int_{i=1}^{n} d_i} \tag{5.17}$$

where, V = mean velocity, m/sec; v_i = velocity (m/sec) of ith point in the vertical section at depth d_i (cm) and n = total number of points along the vertical sections where velocities and depths are measured.

5.3.5 Velocity Measurement

Velocity can be measured by the following methods.

5.3.5.1 Velocity Measurement by Float Method

Approximate values of velocity of flow in a channel can be measured by float method. In this method the velocity is measured by noting the rate of movement of a floating body. Float methods of velocity measurement can be of two types i.e. (i) Surface float and (ii) Sub-surface float.

Surface Float

A surface float usually a small piece of wooden block of 7 to 15 cm diameter or a long necked bottle partly filled with water is used to travel a known length of a straight channel section having uniform cross section. The time taken for the surface float to cover the known length of the channel section is noted. Dividing the time with the known length of the channel section, surface velocity of water called as float velocity is obtained. Several measurements of time to cover the same known length of the channel section are taken from which average float velocity is determined. To convert the average float velocity to average velocity of flow of the channel/stream, a conversion factor 0.85 is selected (Michael, 1987). Several measurements of depth and width are made within the trial section of known length in order to find out the average cross sectional area. Multiplying the average velocity of flow of the channel/stream with its average cross sectional area, rate of flow is determined. This method of measuring velocities is primitive. Still they find use in special circumstances such as in (i) small stream with flood and (ii) small stream with rapidly changing water surfaces.

While any floating object can be used, normally specially made leak proof and easily identifiable floats are used (Fig. 5.4). There are some pre-cautions for use of surface float. The stream section where surface float is used should be straight and uniform. Floats are generally affected by surface winds, eddy and water waves etc. Under such conditions, the surface floats do not give accurate velocity. To overcome these effects, a metallic hollow cylindrical float is also attached to the float by means of a thin string. Rod float sinks in water and only a small portion of it remains above water surface. This rod float is least affected by surface winds and eddies and work better in measurement of velocity.

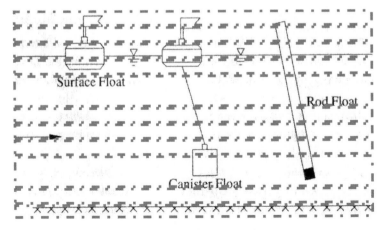

Fig. 5.4: Floats to measure velocity

Limitations of use of surface float: Followings are the limitations of use of surface float.

- Surface float performs better in streams having small depth up to 4.5 m only. It is not suitable for higher depths.

- To get the correct velocity, one has to take a large number of observations.

- Control of floats in stream flow is very difficult.

- It involves more timing to take the observations.

- At the vicinity of stream bed where there may have some irregularities in channel cross-sections which affect the float and there it does not give correct readings.

- If there is high surface winds and waves in water surface, the velocity measurement is not accurate.

Sub-surface float: Sub-surface floats are of two types. They are (i) double float and (ii) twin float. Double floats consist of two floats, in which one is kept over the water surface and the second below water surface. The one remaining below the water surface is called as sub-surface float. The sub-surface float is suspended from the surface float by a thin cord known as length. Cord length is decided based on the depth of flow. The sub-surface float is positioned in the channel at 0.2 times the depth of flow in the channel above the channel bed. The use of double float for measurement of flow velocity is shown in Fig. 5.5. Twin float consists of two similar spherical floats made of hollow metal. Both the floats are connected with a thin rod. The upper float floats over the water surface and the second one sinks in water and keeps the surface float in

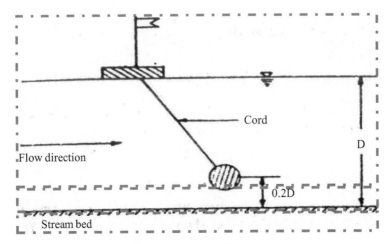

Fig. 5.5: Double float

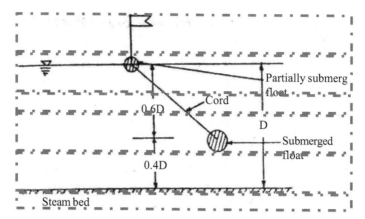

Fig. 5.6: Twin Float

vertical position. During measurements, the surface float is partially submerged in the flowing water while the lower one is completely submerged in the stream flow. The lower float is positioned in the channel at 0.4 times the depth of flow in the channel above the channel bed (Fig. 5.6).

5.3.5.2 Velocity Measurement by Current Meter

The most commonly used instrument in hydrometry to measure the velocity at a point in the flow cross-section is the current meter. Current meter is of two types: (i) Vertical axis current meter and (ii) Horizontal axis current meter. Vertical axis current meter is also called as differential meter, direct meter, Price or Gurley current meter. Horizontal axis current meter is called as propeller type current meter. Sometimes to measure the low velocity of water (velocity up to 1 m/sec) in small shallow streams/irrigation channels such as minors and sub minors, a miniature type current meter called as Pigmi current meter is preferred. Propeller type current meter was invented by Robert Hooke (1663) to measure the distance travelled by a ship. The present-day cup-type instrument and the electrical make-and-break mechanism were invented by Henry in 1868.

5.3.5.3 Vertical Axis Current Meter

Vertical axis current meter is also called as *differential meter, direct meter, Price or Gurley current meter*. The principal components of a vertical axis current meter are (i) head (ii) tail (iii) hanger (iv) recording or indicating device and (v) suspending device (Fig. 5.7). The head carries a wheel of 12.7 cm diameter with six conical cups attached to a horizontal frame. The wheel revolves in anti-clockwise direction on a shaft which passes into a commutator box that eclectically shows the revolutions. The tail consists of a stem with two vanes

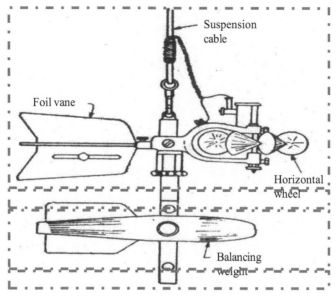

Fig. 5.7: Vertical axis current meter

out of which one is adjustable carrying adjustable weight and the other is fixed. The tail balances the head and retains the meter in the direction of flow. The hanger is a thin bar of steel passing through the meter frames having some weight at the bottom. This weight is called as balancing weight or sounding weight. The weight serves to keep the meter plumb when hung by a cable. The sounding weights enable the current meters to be positioned in a stable manner at the required locations in flowing water. These weights are of streamlined shape with a fin in the rear. Sounding weights come in different sizes and the minimum weight is estimated as:

$$W = 50 . \overline{v} . d \text{ s} \tag{5.18}$$

where, W is minimum weight, N; $\overline{v}$ is average stream velocity in vertical direction, m/sec and d is depth of flow in vertical direction, m.

The recording device consists of a headphone and battery equipment for conveying the ticks to the observer which indicates the number of revolutions. Each rotation of the rotor operates a contact mechanism which energizes a headphone as well as an electric flash counter. Velocity of water is determined by counting the total number of revolutions of the rotor in a given time interval and referring to the rating table (discussed below). The accuracy of this type of current meter is about 1.5% at threshold value and improves to about 0.3% at speeds in excess of 1.0 m/sec. However, if the velocity of water is less than 0.15 m/sec then the current meter does not give accurate reading. Vertical-axis current meter has the disadvantages that it cannot be used in situations where there are appreciable vertical components of velocities. The example is that the instrument gives positive velocity when it is lifted vertically in still water. In a deep water body the current meter is suspended by a cable suspension whereas in small rivers with shallow water body, it is suspended by a wadding rod. Sometimes railway or road bridges are employed as gauging stations for velocity measurement by current meter. The velocity measurement is done on the downstream portion of the bridge to minimize the instrument damage due to drift and knock against the bridge piers. For wide rivers, boats are used in current meter measurement. A cross-sectional line is marked by distinctive land markings and buoys. The position of the boat is determined by using two theodolites on the bank through an intersection method (Sharma and Sharma, 2002).

5.3.5.4 Horizontal Axis Current Meter

Horizontal axis current meter also called as propeller current meter consists of a propeller mounted at the end of a horizontal shaft (Fig. 5.8). The propeller diameter comes in the range of 6 to 12 cm and can register velocities in the range of 0.15 to 4.0 m/sec. The rotations about the horizontal axis are accomplished by means of screws or propeller shaped blades. Its accuracy is about 1% at threshold value and about 0.25% at a velocity of 0.3 m/sec and above. The advantage of this meter is that it measures velocity without being affected by oblique flows of as much as 15°. Ott, Neyrtec and Watt-type meters are typical instruments under this kind.

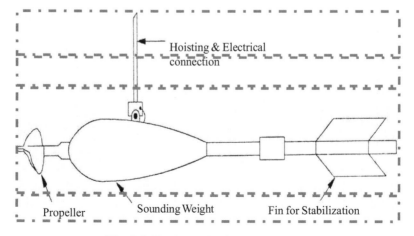

Fig. 5.8: Horizontal axis current meter

5.3.5.5 Pigmi Current Meter

Sometimes to measure the low velocity of water (velocity up to 1 m/sec) in small shallow streams/irrigation channels such as minors and sub minors, a miniature type current meter called as Pigmi current meter is preferred. It is similar to the vertical axis current meter with the size of all parts suitably diminished. The head carries a wheel of 5 cm diameter instead of 12.7 cm with a single revolution contact box. This current meter has no tail fins as is there in vertical axis current meter. The advantage of this current meter is that it can measure small velocities with great accuracy.

5.3.5.6 Rating of Current Meter

The current meter has to be carefully rated in still water before using it to measure velocity of water in a stream. It is mounted on a carriage which moves on rails along a straight channel. The rating is done by moving the carriage through a still water tank also called as towing tank. The rotating elements of the current meter are immersed to a specified depth in the water tank. The speed of the carriage is estimated from the time required to traverse a known distance in a straight channel and the corresponding average value of the revolutions per second of the instrument is determined. The instrument has a provision to count the number of revolutions in a known interval of time. This is usually accomplished by the making and breaking of an electric circuit either mechanically or electro-magnetically at each revolution of the shaft. In older model instruments, the breaking of the circuit is counted through an audible sharp signal ("tick") heard on a headphone. The revolutions per second are calculated by counting the number of such signals in a known interval of time, usually about 100 sec. The present-day model uses electro-magnetic counters with digital or analogue displays. Several readings of runs/revolutions of meter per second at various speeds of flow are taken and a curve is plotted called as rating curve.

The rating curve otherwise called as calibration curve is expressed as:

$$V = a N + b \tag{5.19}$$

where, V = velocity of flow, m/sec; N = number of revolutions of the meter per second; 'a' and 'b' = constants given by the manufacture or generally determined by the rating curve. Typical values of 'a' and 'b' for a standard size 12.5 cm diameter Price-type (cup-type) current meter is 'a' = 0.65 and 'b' = 0.03. For a current meter of small size cup diameter (Pigmy type having cup diameter = 5 cm), values of 'a' and 'b' are 0.30 and 0.003, respectively.

It is to be remembered that Eqn. (5.19) is used for certain threshold velocity and below this threshold value, the rating curve does not work.

Sample Calculation 5.2

Following velocities are recorded in irrigation channel with a current meter.

Depth above the bed of channel, m	0	1	2	3	4
Velocity, m/sec	0	0.4	0.75	0.8	1.0

Find the discharge per unit width of the channel near the point of measurement. Depth of flow at the point is 5 m.

Solution

Given depth of flow = 5 m. Velocity of flow at 0.2 depth below water surface (1m) i.e. 4 m above the channel bed is given as 1.0 m/sec. Similarly, the flow velocity at 0.8 depths below water surface (4 m) i.e. 1 m above the channel bed is given as 0.4 m/sec. Assuming velocity measurements by 2 points method i.e. at 0.2 and 0.8 depths below water surface, the mean velocity is estimated as $(1.0 + 0.4) / 2 = 0.70$ m/sec.

Hence, discharge per unit width of the section is computed as Q = area x mean velocity = (5 x 1) x 0.7 = 3.5 m/sec.

Sample Calculation 5.3

Calculate the discharge of a canal section from the following observations taken by the current meter whose rate curve is given by $V = 0.3 N + 0.05$

Distance from bank, m	2	4	6	8	10	12	14	16
Depth (d), m	0.4	0.8	1.0	1.2	0.9	0.7	0.5	0.5
Current meter revolution/sec (N)	0.8	1.1	1.5	2.0	1.5	1.2	0.8	1.0

Solution

Total discharge is computed in tabular form as given below.

Distance from bank, m	Segmented width (b), m	Depth (d), m	Current meter revolution (N)	Velocity, m/sec $V = 0.3$ $N + 0.05$	Discharge, m³/ sec $Q =$ $b \, x \, d \, x \, V$
2	2	0.4	0.8	0.29	0.232
4	2	0.8	1.1	0.38	0.608
6	2	1.0	1.5	0.50	1.000
8	2	1.2	2.0	0.65	1.560
10	2	0.9	1.5	0.50	0.900
12	2	0.7	1.2	0.41	0.574
14	2	0.5	0.8	0.29	0.290
16	2	0.5	1.0	0.35	0.350
Total discharge					5.514

Sample Calculation 5.4

Water is flowing in a river cross section. The velocity of a surface float of the river cross section as observed is given below. Estimate the total discharge of the river cross section passing at the gauging site. Assume the velocity factor to convert the surface float to mean velocity of water as 0.89.

Width of section (b), m	Mean depth of each section (d), m	Surface velocity observed (V_s), m/sec
25	1.5	0.8
30	1.5	1.0
30	2.0	1.0
28	2.5	0.8
25	2.0	0.8
25	2.2	1.2
30	1.8	1.1
30	1.8	1.2
30	2.0	0.8
30	2.0	0.8

Solution

Estimation of total discharge of river cross section is given in tabular form as below.

Width of section (b), m	Mean depth of each section (d), m	Surface velocity observed (V_s), m/sec	Mean velocity (V_m), m/sec $V_m = 0.89 \times V_s$	Area (A), m² $A = b \times d$	Discharge (q), m³/sec $q = A \times V_m$
25	1.5	0.8	0.712	37.50	26.70
30	1.5	1.0	0.890	45.00	40.05
30	2.0	1.0	0.890	60.00	53.40
28	2.5	0.8	0.712	70.00	49.84
25	2.0	0.8	0.712	50.00	35.60
25	2.2	1.2	1.068	55.00	58.74
30	1.8	1.1	0.979	54.00	52.87
30	1.8	1.2	1.068	54.00	57.67
30	2.0	0.8	0.712	60.00	42.72
30	2.0	0.8	0.712	60.00	42.72
Total Discharge					460.31

Sample Calculation 5.5

Following readings were taken by a current meter in a main canal section.

Find out the total discharge flowing in the canal section by mean-section method. If the common segment width in the canal section is 0.5 and depth and velocity of flow at 0.2 and 0.8 depth below the water surface are same as mentioned in the above problem, find out the total discharge in the canal by mid-section method.

Distance from one bank to the other, m	0	0.6	1.2	1.8	2.4	3.0	3.6	4.2	4.8
Depth (d), m	0	0.3	0.8	1.0	1.5	1.2	1.0	0.6	0
Velocity at 0.2 depth below water surface, $(V_{0.2})$, m/sec	0	0.4	0.6	0.8	1.0	0.9	0.8	0.5	0
Velocity at 0.8 depth below water surface, $(V_{0.8})$, m/sec	0	0.2	0.4	0.6	0.8	0.7	0.6	0.4	0

Solution

Computation by mean section method

Computation of total discharge is shown in the tabular form as given below.

Total discharge per unit segment width, q = 0.025 + 0.220 + 0.540 + 1.000 + 1.148 + 0.825 + 0.464 + 0.069 = 4.228 m²/sec

Common segment width = b = 0.6 m

Hence, total discharge = Q = q x b = 0.6 x 4.228 = 2.537 m³/sec.

$(V_{0.2})$, m/sec	0	0.4	0.6	0.8	1.0	0.9	0.8	0.5	0
$(V_{0.8})$, m/sec	0	0.2	0.4	0.6	0.8	0.7	0.6	0.4	0
Mean V, m/sec	0	0.3	0.5	0.7	0.9	0.8	0.7	0.45	0
Average velocity at two adjacent verticals, m/sec		0.15	0.4	0.6	0.8	0.85	0.75	0.58	0.23
Average depth, m		0.15	0.55	0.90	1.25	1.35	1.10	0.80	0.30
Discharge at the verticals per unit segment width, m (average depth x average velocity)		0.025	0.220	0.540	1.000	1.148	0.825	0.464	0.069

Computation by mid-section method

Computation of total discharge is shown in the tabular form as given below.

Total discharge, Q = 0.045 + 0.20 + 0.35 + 0.675 + 0.48 + 0.35 + 0.135 = 2.235 m³/sec.

$(V_{0.2})$, m/sec	0	0.4	0.6	0.8	1.0	0.9	0.8	0.5	0
$(V_{0.2})$, m/sec	0	0.4	0.6	0.8	1.0	0.9	0.8	0.5	0
$(V_{0.8})$, m/sec	0	0.2	0.4	0.6	0.8	0.7	0.6	0.4	0
Mean V, m/sec	0	0.3	0.5	0.7	0.9	0.8	0.7	0.45	0
Depth, m	0	0.3	0.8	1.0	1.5	1.2	1.0	0.6	0
Segment width, m	0.5	0.5	0.5	0.5	0.5	0.5	0.5	0.5	0.5
Area, m²		0.15	0.40	0.5	0.75	0.6	0.5	0.3	
Discharge, m³/sec (Area x mean velocity)		0.045	0.20	0.35	0.675	0.48	0.35	0.135	

5.3.5.7 Velocity Measurement by Pitot Tube

Pitot tube is generally used to measure velocity in laboratory. Pitot tube consists of a "L" shaped tube, the lower end of which is bent at 90⁰ angle and is fitted with a nozzle. It is inserted into the up-stream side of the stream so that flowing water enters into the tube through the nozzle up to some height say "h" above the centre of the nozzle as shown in Fig. 5.9. The flow velocity is computed as:

$$V = \sqrt{2 g h} \tag{5.20}$$

where, V is velocity, m/s; h is water height in the vertical tube, m and g is acceleration due to gravity, 9.81 m/sec².

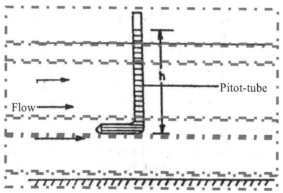

Fig. 5.9: Discharge measurement by pitot tube

Sample Calculation 5.6

A pitot tube is used to measure the flow velocity in a channel having depth of flow 2 m. The position of the pitot tube above the channel bed and the height of rising water in the tube are as follows. Compute the flow velocities at different positions of the tube.

Position of the pitot tube above channel bed, m	0.2	0.5	0.8	1.0	1.4	1.8
Depth of water in the pitot tube, m	0.75	0.80	0.92	1.0	1.1	1.15

Solution

The flow velocity in a pitot tube is given as

$$V = \sqrt{2\,g\,h}$$

The calculations of velocities at different positions of the pitot tube is given in tabular form as:

Position of the pitot tube above channel bed, m	0.2	0.5	0.8	1.0	1.4	1.8
Depth of water in the pitot tube, m (h)	0.75	0.80	0.92	1.0	1.1	1.15
Velocity, V	3.83	3.96	4.25	4.43	4.65	4.75

5.3.6 Gauge-Discharge Relationship

Continuous measurements of discharge at a site in stream are very difficult, costly and time consuming also. However, at such site indirectly the discharge can also be computed. Measurement of discharge by the indirect methods involves a two step procedure of which development of gauge – discharge otherwise called as stage-discharge relationship is very important. Once the gauge-discharge relationship (G-Q) is established, then in the subsequent steps, only the gauge is measured and this gauge is used to compute the discharge from the known G-Q relationship. The gauge-discharge relationship at a particular site in stream section is developed by using the current meter or any other direct stream flow measuring instrument. The gauge-discharge relationship is also called as *rating curve*. The procedure of developing a stage-discharge relationship is described as follows:

In particular gauge site of a canal section or river section, numerous observations of the gauge height or called as stage above certain datum and the corresponding discharge passing the gauge site are measured. A relationship in graphical form called as gauge-discharge or stage-discharge curve is developed. A typical gauge-discharge curve is shown in Fig. 5.10 below (Sharma and Sharma, 2002). This developed curve helps in predicting the discharge in periods when no discharge records are available but the stage records are available. The curve is

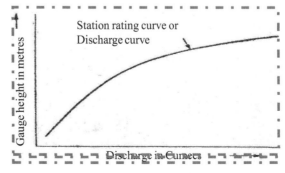

Fig. 5.10: A typical gauge-discharge relationship

prepared by taking a series of data of both stages and corresponding discharges and plotting the discharges in abscissa and the stages in ordinate. From this developed curve, discharge at any observed stage can be found out. The stage-discharge curve in case of low or medium flood stage is a single curve. In the gauge-discharge relationship, there is integrated effect of a wide range of channel and flow parameters. The combined effect of these parameters is termed as *control*. The controls are of two types. In the permanent control case, there is no change in the G-Q relationship for a gauging station over years. However, in shifting control system, the G-Q relationship changes.

5.3.6.1 Permanent Control

In non-alluvial rivers, the G-Q relationship remains constant over years and so in these rivers, the control is permanent type. The singled values relation between the gauge and discharge in such case is expressed as:

$$Q = C_r \left(G - a \right)^{\beta} \tag{5.21}$$

where, Q is stream discharge, G is gauge height (stage), 'a' is a constant which represents the gauge reading corresponding to zero stage, C_r and β are the constants of the rating curve. Eqn. (5.21) is called as rating equation. Values of discharge and gauges are plotted in an arithmetic graph paper and the relationship (gauge-discharge) in the form a graph (Fig. 5.10) is developed. In a logarithmic graph paper when the discharge is taken in abscissa which is in logarithmic scale and gauge $(G - a)$ is taken as ordinate which is also in logarithmic scale, a linear form of gauge-discharge relationship is developed. Values of the rating curve constants C_r and can be calculated as follows.

Taking logarithm of both sides of Eqn (5.21) we get

$$log\ Q = \beta\ log\ (G - a) + log\ C_r \tag{5.22}$$

Eqn. (5.22) is in linear form where slope of the line is β and $log\ C_r$ is y-intercept. A number of measured values of discharge and stage $(G - a)$ are plotted in arithmetic graph paper with X-axis taken as $log\ (G - a)$ and ordinate as $log\ Q$ and a best fit line is drawn. From this best fit line, the slope of the line and y-intercept are calculated. Taking anti logarithm of the y-intercept, we can get the value of rating constant C_r. The slope of the rating curve gives value of β. Values of β and $log\ C_r$ are computed as:

$$\beta = \frac{N \left(\sum \{ \log (G - a) . \log Q \} \right) - \sum \log (G - a) . \sum \log Q}{N . \sum \{ \log (G - a) \}^2 - \{ \sum \log (G - a) \}^2} \tag{5.23}$$

$$\log C_r = \frac{\sum \log Q - \beta \left(\sum \log \{G - a\}\right)}{N}$$ (5.24)

Taking anti-logarithm of *log* C_r, we can find out the value of C_r.

Stage for Zero Discharge, *a*: In Eqn. (5.21), the constant '*a*' represents the gauge/stage height for zero discharge in the stream. It is difficult to find out the value of '*a*'. Following alternative methods as proposed by Subramanya (2003) are used to determine the value of '*a*'.

Step I: First plot *Q vs G* on an arithmetic graph paper and draw a best fit line. Extrapolate the line till it intercepts the Y-axis. Take value of *G* at *Q* = 0. Using the value of '*a*, plot *log Q vs log (G – a)* and verify whether the data plots as a straight line. If not, then select another value in the neighbourhood of previously assumed value and redraw the curve and check whether straight line is obtained. Continue till a straight line relationship of *Q vs G* is obtained. That gives the value of '*a*' at zero discharge.

Step II: There is another method called as Running's method to find out the value of '*a*'. In this method, discharge, *Q* and gauge, *G* are plotted in an arithmetic graph paper and a smooth curve through the plotted points is drawn. On the smooth curve, select three points say A, B and C such that the discharges are in geometric progression i.e. $Q_a \cdot Q_c = Q_b^2$

Draw vertical lines at points A and B and horizontal lines at B and C to get D and E as the intersection points with the verticals (Fig. 5.11). Two straight lines ED and BA are drawn to intersect at F. The ordinate at F is the required value of '*a*'. In this method, the lower part of the gauge-discharge curve is assumed to be parabola.

Step III: Now-a-days, optimization methods are also available based on the use of computer which helps in quick determination of value of '*a*'.

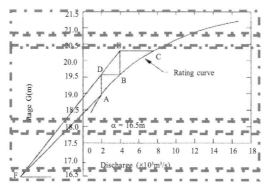

Fig. 5.11: Running's method to find out the value of '*a*'

Sample Calculation 5.7

Following gauge and discharge data of a river site is collected by stream gauging operation. Develop a gauge-discharge relationship for the stream at this section. Assume the value of gauge at zero discharge as 7.0 m. Using the developed relation, estimate the value of discharge at gauge height of 10 m at this stream section.

Gauge reading, m	Discharge, m³/sec	Gauge reading, m	Discharge, m³/sec
7.6	15.1	8.4	175.0
7.7	28.8	8.5	260.8
7.8	55.5	8.6	312.8
7.9	60.8	8.7	355.8
8.0	75.9	8.8	432.0
8.1	100.5	9.2	518.5
8.2	140.5	9.5	600.0

Solution

The gauge-discharge relationship can be written as (Eqn. 5.22):

$$log\ Q = \beta\ log\ (G - a) + log\ C_r$$

Values of gauge reading (G) and discharge (Q) are given in problem as above. Value of 'a' is given as 7.0. Now taking the logarithmic values of $(G - a)$ and Q we get the calculations as follows:

Sl. No.	Gauge, G (m)	Discharge, Q (m³/sec)	$log\ (G\text{-}a)$	$log\ Q$	$log\ Q.\ log\ (G\text{-}a)$
1.	7.6	15.1	-0.221	1.179	-0.262
2.	7.7	28.8	-0.155	1.459	-0.226
3.	7.8	55.5	-0.097	1.744	-0.169
4.	7.9	60.8	-0.046	1.783	-0.082
5.	8.0	75.9	0	1.880	0
6.	8.1	100.5	0.041	2.002	0.083
7.	8.2	140.5	0.079	2.148	0.170
8.	8.4	175.0	0.146	2.243	0.328
9.	8.5	260.8	0.176	2.416	0.425
10.	8.6	312.8	0.204	2.495	0.509
11.	8.7	355.8	0.230	2.551	0.588
12.	8.8	432.0	0.255	2.635	0.673
13.	9.2	518.5	0.342	2.714	0.930
14.	9.5	600.0	0.398	2.778	1.105
Total			1.354	30.031	4.073

N = 14

Putting the above values in Eqn. (5.23), we get $\beta = 3.432$ and using Eqn. (5.24) we get $log\ C_r = 1.813$ and hence $C_r = 6.130$.

Hence the gauge-discharge relationship is

$Q = 6.130\ (G - a)^{3.432}$

For gauge height, $G = 10$ m, discharge from the above developed relation comes as:

$Q = 6.130\ x\ (10 - 7)^{3.432} = 266.03$ m³/sec.

5.3.6.2 Shifting Control

There may be some change in the gauge-discharge relationship at a site if there is some weed growth or any other channel encroachment, aggradations and degradations phenomenon in an alluvial channel, back water effects affecting he station or gauging section or due to any unsteady flow effects of a rapidly changing stage. If the effects as mentioned above are due to reasons first or second, then there is no permanent corrective measures to tackle the shifting controls. However, with frequent current meter readings rating curve is to be updated which is the only solution in such case. Shifting controls due to the causes three and four are discussed as below.

Unsteady Flow Effect: Unsteady flow effect is also called as rapidly changing flow effect. When a flood wave passes a gauging station in the advancing portion of the wave, the approaching velocities are larger than in the steady flow at corresponding stage. Thus, for the same stage, more discharge than in a steady uniform flow occurs. On the other hand in the retreating phase, approaching velocities are smaller than in the equivalent steady flow case. Hence, in case of unsteady flow case, the gauge-discharge relation will not be a straight line as is the case in steady flow case. Rather, it becomes a looped curve as shown in Fig. 5.12.

The looping curve has two stages i.e. falling stage and rising stage. For the same gauge, in the falling stage less discharge passes through the river whereas in the rising stage, more discharge passes. If Q_n is the normal discharge at a given stage under steady uniform flow case and Q_m is the actual measured unsteady flow then the relationship between Q_n and Q_m is written as:

$$\frac{Q_m}{Q_n} = \sqrt{(1 + \frac{1}{V_w\ S_0} \cdot \frac{dh}{dt})a} \tag{5.25}$$

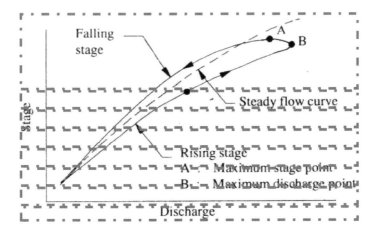

Fig. 5.12: Loop rating curve

where, S_o is channel slope (water surface slope for uniform flow), dh/dt is the rate of change of stage and V_w is the velocity of flood wave. If the channel is natural, then V_w is 1.4 times average velocity (V) for a given stage estimated by applying Manning's equation and energy slope. Moreover, in case of unsteady flow case, we use energy slope instead of S_o in the denominator of Eqn. (5.25). If measurements on discharge and dh/dt are taken under changing flow stages, then the term $(1/V_w S_0)$ can be computed as:

$$\frac{1}{V_w \, S_0} \frac{dh}{dt} = \left(\frac{Q_m}{Q_n}\right)^2 - 1 \tag{5.26}$$

Backwater Effect: If the shifting control is due to the effects of backwater, then the same stage will give different discharges. To correct this effect, another gauge called as auxiliary gauge or secondary gauge is installed at some distance downstream of the gauging station and readings of both the gauges are taken. Difference between the readings of both the gauges is called as fall (F) of the water surface in the reach. For a given main stage reading, the discharge under variable backwater condition is a function of the fall, F and is expressed as:

$Q = f\,(G,\, F)$

Adjustment of G - Q curve for constant fall rate is done in the following way:

Constant Fall Rating: Under this condition, the mean value of the fall (F_0) is taken into consideration. The mean fall value is also called as normalized fall value which is assumed as constant for all the stages. Let the actual fall for a discharge, Q be F. For adjusting the rating curve, all observed discharge data are adjusted for the condition of constant fall given as:

$$\frac{Q}{Q_0} = \left(\frac{F}{F_0}\right)^p \tag{5.27}$$

where, Q_o is the nomalised discharge when fall is F_0, and p is an exponent which has a value close to 0.5.

Step I: From the observed data, a convenient value of F_0 is taken.

Step II: Now plot the curve between Q/Q_0 and F/F_0 and draw the best fit curve. The obtained curve is called as adjustment curve.

Step III: Find the adjusted value of Q/Q_0 for each given value of F/F_0.

Step IV: Calculate the adjusted value of Q_0 by dividing the measured Q by Q/Q_0.

This curve is called as constant fall curve.

Limiting Fall Rating: Normally, the gauge-discharge relationship is not affected by the existing downstream conditions till a limiting or constant fall is created in the stream flow. Back water effect is produced when there is reduction in the fall. Reductions in fall consequences reductions in discharge for same gauge or stage also. The limiting fall depends on the flow stage. A relationship between the ratio of measured fall to limiting fall and measured discharge to back water free discharge is developed as shown in Fig. 5.13 (Suresh, 2008). Each discharge measurement is adjusted according to this ratio and a normal stage-discharge

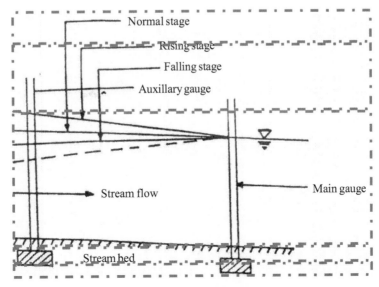

Fig. 5.13: Auxiliary gauge for flow measurement

relationship is derived. The derived relationship is called as limiting fall or normal fall rating.

5.3.6.3 Extrapolation of Rating Curve

Extreme flood flows are considered in most of the hydrological designs. In the design of hydraulic structures like bridge, barrages etc., maximum flood discharges and maximum flood levels are required. The stage discharge relationship at a site will help in predicting the stage corresponding to design flood discharge. However, the available stage-discharge data hardly covers design flood range and hence, there is a need to extrapolate the rating curve. The curve is extended backwards for low stage-discharge and forward for high stage-discharge. Before extrapolation is done, it is necessary to examine the site and collect the relevant data on changes in the river cross section due to flood plains, roughness and backwater effects. The common methods of rating curve extrapolation are:

- Logarithm method
- Steven's method and
- Conveyance method

Logarithmic Method: In this method, the stage-discharge relationship given by Eqn. (5.21) is used. On a logarithmic graph paper stage and discharge are plotted and best fit line is then drawn through the plotted points lying on the high stage range. The best fit line is then extended to cover the range of extrapolation. In another way, coefficients of Eqn. (5.21) are obtained by the least square error method and the coefficients are computed as:

$$\beta = \frac{N\left(\sum\{\log(G-a).\log Q\}\right) - \sum\log(G-a).\sum\log Q}{N.\sum\{\log(G-a)\}^2 - \{\sum\log(G-a)\}^2}$$

and $\log C_r = \dfrac{\sum\log Q - \beta\left(\sum\log\{G-a\}\right)}{N}$ as discussed earlier. Taking anti-logarithm ic of $\log C_r$ one can get the value of the coefficient C_r.

The relationship governing the stage and discharge is now written as:

$$G - a = C_r\,Q\,\beta \tag{5.28}$$

Using Eqn. (5.28), value of stage corresponding to a design flood discharge is estimated.

Steven's Method: In this method, Chezy's formula to compute the discharge is used. Chezy's formula is:

$$Q = C \cdot A \cdot \sqrt{R \cdot S} \qquad (5.29)$$

where, Q is discharge, A is cross-sectional area, S is energy slope, R is mean hydraulic radius $(R = A/P; P$ is wetted perimeter) and C is a constant depending on roughness coefficient. For a wide and shallow stream one can assume $R H$" D (D is depth of flow) and so Eqn. (5.29) can be written as:

$$Q = C \cdot A \cdot \sqrt{D \cdot S} \qquad (5.30)$$

For a shallow wide stream it can be assumed that $C \sqrt{S}$ is constant and hence as per Eqn. (5.30),

$$Q = f(A \sqrt{D}) \qquad (5.31)$$

Since A and D are functions of gauge height, hence a curve can be plotted between Q and based on the observation of stage and discharge which is in the form of a straight line provided value of is constant. By extending this curve in forward direction, discharge for any value of during high flow stage can be obtained. Extension of rating curve using Steven's method is shown in Fig. 5.14.

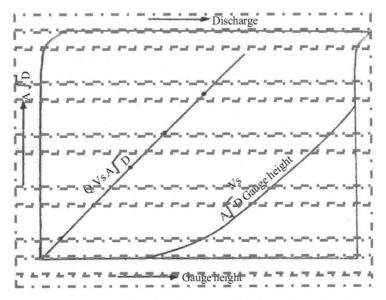

Fig.5.14: Extension of rating curve using Steven's method

Conveyance Method: This method is similar to Steven's method. But in this case, instead of Chezy's formula, Manning's formula is used to compute the discharge at both lowest and highest flow stages.

As mentioned earlier, Manning's formula for discharge computation in case of non-uniform flow can be written as:

$$S_f = \frac{Q^2}{K^2} \tag{5.32}$$

where, Q is the discharge, S_f is the energy slope and K is the conveyance of the channel given as:

$$K = \frac{1}{n} A R^{2/3} \tag{5.33}$$

where, n is Manning's roughness constant, A is cross-sectional area of flow and R is mean hydraulic radius. Both A and R are functions of stages and hence, the values of conveyance, K for various values of stages are calculated by using Eqn. (5.33). These values of calculated K are plotted against stages. The range of the stages should include values beyond the level up to which extrapolation is desired. Then a smooth curve is drawn through the plotted points (Fig. 5.15).

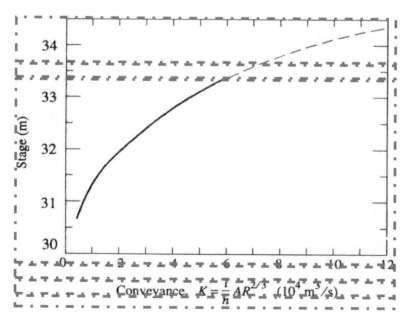

Fig. 5.15: Conveyance method of rating curve extension: K vs Stage

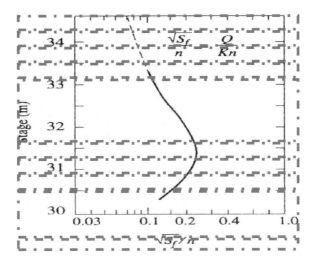

Fig. 5.16: Conveyance method of rating curve extension: S_f vs Stage

Using the available stage and discharge data, values of S_f are calculated by using Eqn. (5.32) and are plotted against the stages. A smooth curve is drawn through the plotted points (Fig. 5.16). This curve is then extrapolated keeping in mind that S_f approaches a constant value at high stages.

Using Figs. (5.15) and (5.16), the discharge at any stage is calculated as $Q = K\sqrt{S_f}$ and a stage-discharge curve covering the desired range of extrapolation is constructed. This extrapolated curve helps in determining the stage at any design flood discharge.

5.3.7 Selection of Gauge Site

Selection of gauging site along with the essential requirement of gauge site are very important to know before data on stage and discharge are recorded and required relationship between the recorded data is developed.

A gauge site should be selected after careful study of different features of the available topographical map of the site. These sites are later inspected and surveyed to ensure that the physical and hydraulic features of the proposed site confirm to the requirements for the application of the method of flow measurement on a river/stream to furnish the gauge data.

5.3.7.1 Essential Requirements of a Gauge Site

The essential requirements of a gauging site are:

(i)   The gauging site should be permanent and should not be shifted frequently because of silting or souring problem.

(ii)   The site should be clearly visible and easily accessible.

(iii) The site should always register the same elevation of water surface corresponding to the given discharge.

(iv)   The site should be so located that it can record the highest and lowest stage data corresponding to the high and low floods.

5.3.8 Types of Gauges

Gauges are generally divided into two types. They are direct gauges and indirect gauges. The direct gauges include staff gauges, hook gauges and float gauges. The indirect gauges include crest gauges and self record type gauges. Again the direct gauges may be permanent type or temporary type. A permanent type gauge has a fixed position to gauge the discharge of a site where as in an indirect type gauge, the gauge site changes time to time. The temporary gauges may be wooden or made of steel plates but the latter is preferred for its longevity. Followings are different types of gauges commonly used.

5.3.8.1 Staff Gauges

A staff gauge is a graduated scale on a staff or metal plate against which elevation of water surface is read. It is economical in construction and has long lasting in use. It resists to alternate wetting and drying. It is also simple in construction. Depending on the material of construction, it may be wooden type or vertical enameled type. Similarly depending on installation, it may be vertical staff gauge or inclined staff gauge.

5.3.8.2 Vertical Staff Gauge

Vertical staff gauge preferably of enameled plate gauge is graduated to millimeter. The graduation marks are clear and have bold line indications at every 20 mm interval. The graduations are made either by etching or by painting or enameling so as to read centimeters, decimeters, and metres clearly. The smallest graduation mark is the centimeter. If there are large fluctuations in flow, then several gauges running normal to the flow direction may be used. The staff gauge is generally fixed at or near the water in the river which facilitates in taking reading easily. It also helps in easy accessibility to the site for taking gauge treading.

5.3.8.3 Inclined Gauge

The inclined staff gauge is fixed on a sloping river bank. It is calibrated on the site by precise leveling. Inclined gauge may be constructed to one continuous slope or may be a compound of two or more slopes. It is convenient to construct a flight of steps along the side of inclined gauge to facilitate in taking readings.

5.3.8.4 Hook Gauge

It a gauge made of brass rod pointing upwards with its top coinciding with the water surface in the river or stream. If the rod points downward, then it is called as a point gauge. Hook gauge is used in a vertical well.

5.3.8.5 Crest Stage Gauge

It is a gauge used to obtain the records of flood crests at sites where recording gauges are not installed. Different types of crest stage gauges are used. In one type, a small float is used which rises with the stage but are restrained at the maximum level. In another type, water soluble paints are used at bridge pier side shielded from rain. The portion of the paint coming in contact with water is washed out to indicate high water level.

5.3.8.6 Self Recording Gauge

Self recording gauge is also called as stage recorder. Like the self recording type rain gauge, it is an instrument which continuously records the rise and fall of water level of the stream or river in which water is flowing with respect to time on a graph paper. The recorder has two essential elements i.e. (i) time element consisting of a clock mechanism actuated by a spring, a weight and electricity and (ii) a gauge height element consisting of non-corrosive metallic floats, float cables, a counter weight and a gear reduction mechanism. There are various types of record. Commonly the time element operates parallel to the axis of the drum while the gauge height element rotates the drum in proportion to the rise or fall in water level. The recorder is usually operated by an 8-day spring driven clock. The height element activates a pen stylus which moves parallel to the axis of the paper roll so that its travel represents a change in water stage. Self recording gauge gives continuous record of gauges from which maximum and minimum gauges as also time of their occurrences can be known. From the continuous records, daily average gauge height can be estimated.

5.3.9 Discharge Site

Generally the site selected for gauge installation is the ideal site for discharge observations. If there are some limitations for the two sites to be distinct, then they should be as close as possible. The discharge measurement site should consider all the essential requirements of gauge site. In addition, it should consider additional requirements like (Sharma and Sharma, 2002):

- At or as close as possible to the gauge site in the case of velocity measurement by current meter or near the middle of a straight reach where float is used to measure the velocity,

- Located in a straight reach of length which is the least of 3 to 4 times the river width or 0.75 km during high flood both upstream and downstream of the normal cross section where velocity is to be measured,

- The flow in the straight reach is streamlined, steady and showing less departure from the normal flow at all stages of the rivers,

- The water flow is preferably in a single channel,

- Site is free from aggradations and degradation,

- Located where greatest range of fluctuations in river stage occurs,

- Site relatively free from any obstructions,

- Site is free from the tendency of formation of backward flow,

- Located preferably away from any hydraulic structures,

- Reasonably stable bed and banks for affording a regular gauge-discharge relationship, and

- Easily accessible all the time.

In case of a hilly area where discharge measurement site is selected, including the above requirements, some additional requirements are needed. They are

- Straight reach of minimum length of 150 m above section line,

- Free from projection in the beds and sides,

- Minimum 150 m removed from incoming torrent upstream and downstream from the section line,

- A bridge without piers and

- Nearest to a proposed dam site.

5.3.10 Weir

Weir is a notch of regular form through which irrigation streams are allowed to pass. It is built across the stream or channel. They may be built as stationary structures or may be portable. They consist of a weir wall of concrete, timber, strong tin plates, galvanized iron sheet or metal with a sheet metal weir plate fixed to it. They are generally divided into three types as (i) rectangular (ii) trapezoidal and (iii) triangular. The weir should be so placed across a channel section that the flow is under free flow condition. This necessitates the nappe of the freely falling water not to be submerged by tail end water of the channel section.

According to crest, the weir may be sharp crested weir or broad crested weir. In a sharp crested weir, water passing over the crest of the weir just touches only a line. In case of broad crested weir, there is a rounded upstream edge, or a crest so broad that water flowing over the weir touches the crest with a surface. In practice, the sharp crested weir is more accurate than a broad crested weir.

5.3.10.1 Rectangular Weir

It takes the name from the shape of the notch. It is used to measure large discharge. It has a horizontal crest and vertical sides. As described above, it may be sharp crested or broad crested type. It is also dived into two more types i.e. contracted rectangular or suppressed rectangular weir. In a contracted weir, the crest length is less than the width of the up stream channel. In a suppressed rectangular weir, the crest length is equal to or more than the width of up stream channel or called as approach channel. The discharge formula through a rectangular weir can be calculated empirically by Francis' formula, Bazin's formula and Rehbock's formula as stated below:

Francis' Formula

It is one of the most commonly used formula for calculating discharge over sharp or narrow crested weirs with or without end contraction. The formula was proposed by J.B. Francis. In this formula, the length of the crest varied from 1.07 to 5.19 m, but in most of the experiments, the crest length was 3.05 m. The head varied from 0.18 m to 0.49 m. He conducted more than eighty experiments and from these experiments, he proposed the following formula for computation of discharge over rectangular weir. For a suppressed rectangular weir (Fig. 5.17), discharge is given by (Murty, 1998):

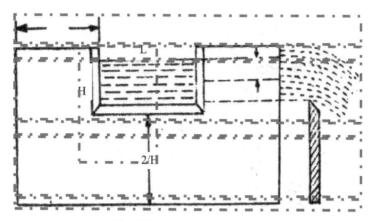

Fig. 5.17: Rectangular weir

$$Q = C L H^{3/2} \qquad (5.34)$$

where Q = discharge in lit/sec; L = length of crest in cm; H = height of water called as head above the crest in cm and C = coefficient of discharge which is 0.0184.

In a contracted rectangular weir (both sides contracted), the discharge is given as:

$$Q = C (L-0.2 \, H) \, H^{3/2} \qquad (5.35)$$

For one side contraction, the formula is:

$$Q = C (L-0.1 \, H) \, H^{3/2} \qquad (5.36)$$

where the terms are defined as above.

The value of C is 0.0184. The terms inside the bracket i.e. $(L- 0.1 \, H)$ in Eq. (5.36) and $(L-0.2H)$ in Eqn. (5.35) are called as effective length of the weir. Fig. 5.17 (as given above) shows the view of a rectangular weir without end contraction.

It is to be noted that in above equations (Eqns. 5.34 to 5.36), the velocity of approach is neglected. When the velocity of approach is taken into account, the discharge formulae for Eqn. (5.34), Eqn. (5.35) and Eqn. (5.36) becomes as

$$Q = C L (H_I^{3/2} - h_a^{3/2}) \qquad (5.37)$$

Similarly for one side end contraction, discharge is given as:

$$Q = C (L-0.1 \, H_I) \, (H_I^{3/2} - h_a^{3/2}) \qquad (5.38)$$

For both sides end contraction, discharge is:

$$Q = C (L-0.2 \, H_I) \, (H_I^{3/2} - h_a^{3/2}) \qquad (5.39)$$

where, $H_1 = (H + h_a) = (H + V_a^2/2g)$; H = height of water surface above the crest of the weir and V_a = velocity of flow in the upstream side of the weir in the channel/stream.

Bazin's Formula

In 1886, H. Bazin of France conducted a number of experiments with rectangular weirs. From these experiments, he proposed the following formula for computation of discharge over the rectangular weir as follows:

Without velocity of approach, the discharge formula is:

$$Q = m (2 g)^{1/2} L H^{3/2} \tag{5.40}$$

In which according to Bazin, the value of the coefficient m is:

$$m = (0.405 + 0.003/H) \tag{5.41}$$

With consideration of velocity of approach, discharge is given as:

$$Q = m_1 (2 g)^{1/2} L H_1^{3/2} \tag{5.42}$$

In which the coefficient m_1 is:

$$m_1 = (0.405 + 0.003/H_1) \tag{5.43}$$

and H_1 is (H_1 is still water head) given as:

$$H_1 = (H + \alpha V_a^2/2g) \tag{5.44}$$

where α is a constant, the mean value of which is proposed by Bazin as 1.6.

Rehbock's Formula

In 1929, T. Rehbock on the basis of experiments with suppressed rectangular weir, proposed the following empirical formula for computation of discharge as:

$$Q = 0.667 (0.605 + 0.08 H/Z + 0.001/H) (2 g)^{1/2} L H^{3/2} \tag{5.45}$$

where, H = head, m, and Z = crest height, m.

5.3.10.2 Trapezoidal Weir

It is also called as cipoletti weir. It is a contracted weir in which each side of the notch has a slope of 1 horizontal to 4 vertical. The weir has sharp crest and sharp sides which is beveled from downstream side only. It is commonly used to measure medium discharge.

The discharge through a trapezoidal weir is given as:

$$Q = C\,L\,H^{3/2} \tag{5.46}$$

where, Q = discharge, lit/sec; L = crest length, cm; H = height of water called as head over the crest, cm and C = coefficient of discharge which is 0.0186.

The crest length is given as

$$L = (L_1 + L_2)\,/2 \tag{5.47}$$

where L_1 = length of crest at bottom and L_2 = length of crest at top. Fig. 5.18 represents a trapezoidal weir.

5.3.10.3 Triangular Weir

It is also called as V notch. The 90^0 triangular or V notch has greater practical utility than any other weir. It is ideal to measure discharge under medium or low flow condition. It is popular since it is easy construct and install. It has both sides sharp, beveled from downstream side only. The discharge equation is given as (Lenka, 2001):

$$Q = 0.0138\ H^{5/2} \tag{5.48}$$

where Q = discharge, lit/sec and H = height of water above the cres, cm.

Fig. 5.19: Represents a 90^0 triangular or V notch.

In the above equations used to measure discharge, the coefficient of discharge values are used when they are not available. In practice the values of the coefficient of discharge for each type of weir should be found out by calibration in laboratory. The calibrated values of the coefficient can be used provided similar condition in the field exists. To get correct flow, the devices should be correctly designed and installed. The flow quantity should be checked from time to time.

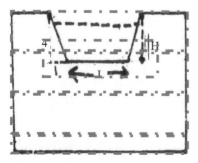

Fig. 5.18: Trapezoidal weir

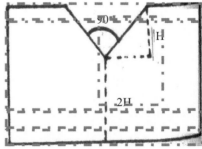

Fig. 5.19: 90^0 Triangular weir/V notch

5.3.11 Spillway and Siphon Spillway

A spillway is a portion of the dam over which excess water that cannot be stored in the reservoir, flows to the downstream end of the dam. In general the shape of the spillway profile is made to follow the profiles of a lower nappe of a well ventilated sharp crested weir. The spillway is formed by filling the space between the sharp crested weir and the lower nappe with concrete or masonry as shown in Fig. 5.20 (a). Such a spillway is called as *ogee spillway or ogee weir.* The discharge over an ogee spillway is given as (Modi and Seth, 2005):

$$Q = C\,L\,H^{3/2} \qquad\qquad\qquad (5.49)$$

where, C = coefficient of the spillway, L = length of the spillway and H = head above the crest of the spillway. The value of the coefficient C is determined by calibration.

Siphon spillway consists of an ogee weir which is provided with an air-tight cover as shown in Fig. 5.20 (b). This converts the discharge face of the spillway into a large rectangular sectioned pipe connecting the upstream and downstream water surfaces. In the case of a siphon spillway the head H under which the water flows is equal to the difference between the water surface on the upstream and downstream sides. As such as compared with an ordinary spillway, the head in the case of a siphon spillway is more, on account of which the siphon spillway has a much greater discharge for a given length than an ordinary open spillway. The working of a siphon spillway is automatic.

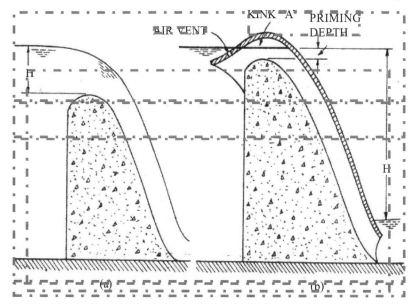

Fig. 5.20 (a): Ogee spillway (b) Siphon spillway

When the water level in the upstream side of the reservoir rises above the crest level of the spillway, water starts flowing over the crest through the siphon. The jet of the flowing water strikes the inside of the cover, thus enclosing a small space A called kink in the upstream portion of the cover, in which air is trapped. As more and more water flows down, it sucks the trapped air from the kink, thus creating a partial vacuum in this portion which sucks up the water from the reservoir and completely fills the pipe. The symphonic action is started and the effective head causing the flow of water becomes equal to H, the difference between the water surface in the upstream and downstream sides. When the level of water in the reservoir decreases and becomes almost equal to the crest level of the spillway, the symphonic action stops. Air vents are provided in the cover at a level slightly above the crest of the spillway to facilitate this action (Garg, 1978).

The discharge through a siphon spillway is given same as ogee spillway which is

$$Q = C L H^{3/2} \tag{5.50}$$

A siphon spillway has the following advantages over the ogee spillway.

(i)   The operating head is increased considerably allowing higher discharge to flow.

(ii)   The depth of water (priming depth) needed for the starting of the siphonic action is only a few centimeter above the crest of the spillway. As such the crest of the siphon spillway can be raised, thereby allowing more amount of water to get stored in the reservoir.

5.3.12 Proportional Weir (Sutro Weir)

In case of the rectangular or trapezoidal weir as discussed above, the discharge, Q varies directly with the head, H not linearly but the relation is in the form of power function i.e. $Q \, \alpha \, H^n$, where $n = 3/2$ for a rectangular or trapezoidal weir and $n = 5/2$ for a triangular weir. It is possible to design a proportional weir or called as Sutro weir in which the discharge, Q is linearly proportional to the head, H over the crest of the weir i.e. $Q \, \alpha \, H$.

The analytical relationship for the shape of the proportional weir profile having $Q \, \alpha \, H$, is

$$x \, \alpha \, y^{-1/2} \tag{5.51}$$

However, it may be noted from the above relationship that as $y \to 0$, $x \to \infty$ that means the width of the weir becomes infinitely at the crest. The infinitely tending profile is however not practicable.

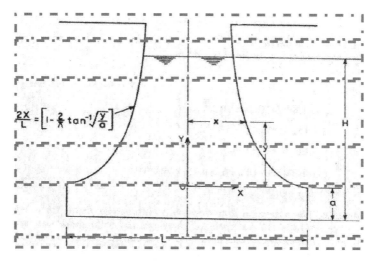

Fig. 5.21: Proportional weir

In order to overcome this problem, the shape of the weir profile was modified so that a finite width at the weir crest may be provided. Fig. 5.21 shows the view of the modified proportional weir profile that has its sides diverging downward in the form of hyperbolic curves having the equation as (Modi and Seth, 2005):

$$2 \ x/L = [1 - 2/\prod tan^{-1}(y/a)^{1/2}] \tag{5.52}$$

where, a and L are respectively the height and width of the small rectangular shaped aperture that forms the base of the weir. The discharge through this weir is given by

$$Q = k \ (H - a/3) \tag{5.53}$$

where k is given as:

$$k = C_d L \ (2 \ g \ a)^{1/2} \tag{5.54}$$

In the above equation C_d is the coefficient of discharge which varies from 0.60 to 0.65.

Sample Calculation 5.8

Find the discharge over a suppressed rectangular weir having length 5 m and head over the crest being 0.60 m.

Solution

The discharge through a rectangular weir as given by Francis formula is:

$Q = 0.0184 \ L \ H^{3/2}$

Given L = 5 m = 500 cm and H = 0.6 m = 60 cm

Hence Q = 0.0184 x 500 x (60)$^{3/2}$ = 4276 lit/sec.

5.3.13 Flow through Orifices and Mouthpieces

Water flows to the crop fields both through open channel as well as through pipes. The flow in pipes may be through orifices or mouthpieces. An orifice is an opening having a closed perimeter, made in the walls or bottom of a tank containing a fluid, through which fluid may be discharged. A mouthpiece is a short tube of length not more than two to three times its diameter, which is fitted to a circular opening or orifice of the same diameter, provided in a tank containing the fluid, such that it is an extension of the orifice and through which also the fluid may be discharged. According to the discharge condition, the orifice may be classified as orifice discharging free and submerged orifice. The submerged orifice may be further divided as completely or partially submerged orifice. The mouthpiece may be classified on the basis of their shape, position and discharge conditions. According to the shape the mouthpiece may be classified as cylindrical, convergent, divergent and convergent-divergent (Modi and Seth, 2005). According to the position, the mouthpiece may be classified as external and internal mouthpieces. According to the discharge condition, the mouthpiece may be classified as running full and running free mouthpieces. Formulae for calculation of discharge through the different types of orifices and mouthpieces are discussed in details by Modi and Seth (2005).

5.3.14 Meter Gate

A meter gate is basically a modified submerged orifice, so arranged that the orifice is adjustable in area. They are manufactured commercially and are used to control water flowing from one channel to another. They may serve the rate of flow, if the head and area of opening can be determined and the gate has been calibrated the head is same as that under submerged orifice.

Normally the gate and the opening in to outlet are circular and their area can be determined easily if the gate is fully open. However, under most conditions the gate is only partially open, leaving a crescent shaped area which is difficult to measure. Consequently, most gates are calibrated and tables are supplied, giving the rate of flow as a function of the head and the degree of gate opening as measured by the displacement of the gate stem.

5.3.15 Water Meters

Water meters utilizes a multi-blade propeller made of metal, plastic or rubber, rotating in a vertical or horizontal plane and geared to a totaliser in such a manner that a numerical counter can the flow in any volumetric units. Water meters are available for a range of sizes suiting the pipe size commonly used on the farm.

There are two basic requirements for accurate operation of water meter:

1. The pipe must flow full at all the times, and

2 The rate of flow must exceed the minimum for the rated range.

Meters are calibrated in the factory and field adjustments are usually not required. When water meters are installed in open channels, the flow must be brought through a pipe of known cross-sectional area. Care must be taken that no debris or other foreign materials obstruct the propeller.

5.3.16 Parshall Flume

Parshall flumes consists of (i) a converging inlet section (ii) a throat section and (iii) a diverging outlet section. The throat section lies inbetween the inlet and outlet sections. Fig. 5.22 shows the geometry of a Parshall flume. Parshall flumes may be built of wood, concrete, galvanized sheet metal or other materials. Large flumes are usually constructed on site, while smaller ones can be purchased as prefabricated units that are installed in the channel. Dimensions and ranges of flow for the various standard Parshall flumes as well as the

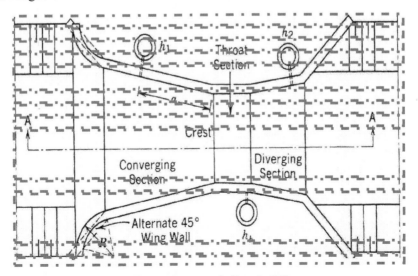

Fig. 5.22: Geometry of a Parshall flume

design procedures can be found in Water Measurement Manual of the U.S. Bureau of Reclamation (1975).

The primary advantage of Parshall flumes is their ability to provide accurate flow measurements over a wide range of flows with minimum of head loss (U.S. Bureau of Reclamation, 1975). This allows their uses in relatively shallow channels with flat grades. The main disadvantages of Parshall flumes are their relatively large size (Walker and Skogerboe, 1987).

The rate of flow through a Parshall flume may or may not be affected by the tail water depth in the downstream of the throat. Free flow condition exists when the tail water depth is not high enough to affect the flow. A flume is considered to be submerged when the tail water depth is sufficient to affect the flow and the submergence ratio, S defined as the ratio of head in the diverging section (H_b) to that in the converging section (H_a) exceeds the transition ratio S_t. Values of S_t for different sizes of Parshall flumes as decided by the throat widths are presented in Table 5.1 (James, 1988).

Table 5.1: Free flow and submerged flow coefficients and exponents for Parshall flume

Throat width, W, cm	C_f for Q in (lit/s)	C_s for Q in (lit/s)	n_f	n_s	S_t
2.54	9.57	8.47	1.55	1.00	0.56
5.08	19.14	17.33	1.55	1.00	0.61
7.62	28.09	25.91	1.55	1.00	0.64
15.24	58.34	47.01	1.58	1.00	0.55
22.86	86.94	71.08	1.53	1.08	0.63
30.48	113.28	88.08	1.52	1.06	0.62
45.72	169.92	125.17	1.54	1.08	0.64
60.96	226.56	168.22	1.55	1.115	0.66
76.20	283.2	204.47	1.555	1.14	0.67
91.44	339.84	243.55	1.56	1.15	0.68
121.92	453.12	314.35	1.57	1.16	0.70
152.40	566.40	383.74	1.58	1.185	0.72
182.88	679.68	448.87	1.59	1.205	0.74
213.36	792.96	514.01	1.60	1.230	0.76
243.84	906.24	577.73	1.60	1.250	0.78
304.8	1136.48	702.05	1.59	1.260	0.80

The following head discharge equation is used for free flow Parshall flumes (James, 1988).

$$Q = C_f \left(K H_a \right)^{n_f} \tag{5.55}$$

where, Q = discharge, lit/sec; C_f = free flow coefficient depending on throat width (Table 5.1); n_f = free flow exponent (Table 5.1); H_a = head measured in the converging section of the flume, m and K = unit constant taken as 3.28 when H_a is in m.

In submerged case, the ratio of head in the diverging section (H_b) and that in the converging section (H_a) i.e $S = H_b/H_a$ > transition submergence, S_t. Table 5.1 can be referred for deciding submergence case considering the value of S_t. In submergence case, discharge is given (James, 1988) as:

$$Q = \frac{C_s \left[K \left(H_a - H_b \right) \right]^{n_f}}{\left[-\left(\log S + C \right) \right]^{n_s}} \tag{5.56}$$

where, Q = discharge, lit/sec; C_s = submerged flow coefficient depending on throat width (Table 5.1); n_f and n_s = free and submerged flow exponent (Table 5.1); H_a and H_b = heads measured in the converging and diverging sections of the flume, m, respectively and K = unit constant taken as 3.28 when H_a is in m, S = ratio H_b to H_a, and C = 0.0044.

Sample Calculation 5.9

A Parshall flume of throat width 7.5 cm is used to measure discharge. If the head measured in the converging section of the flume is 30 cm and the flow is under free flow, find out the discharge measured by it.

Solution

Given throat width, W = 7.5 cm; H_a = 30 cm = 0.3 m and free flow condition.

Referring Table 5.1 for W = 7.5 cm, C_f = 28.09; n_f = 1.55.

Using H_a = 0.30 m and K = 3.28 (when H_a is in m),

Discharge is obtained as

Q = 28.09 x (3.28 x 0.3) $^{1.55}$ = 27.4 lit/sec.

Sample Calculation 5.10

The discharge is measured by a Parshall flume having the following data.

Throat width, W = 7.5 cm; head measured in the converging and diverging section of the flume are 45 and 30 cm, respectively. Find out the discharge measured by the flume.

Solution

Given, throat width, W = 7.5 cm; H_a = 45 cm = 0.45 m, H_b = 0.35 m.

Value of $S = H_b/H_a = 0.30/0.45 = 0.67$.

Referring Table (5.1) for W = 7.5 cm, we find S_t = 0.64

Since $S > St$, it is submerged case.

Now refereeing Table (5.1), for W = 7.5 cm and under submerged case, C_s = 25.91; n_f = 1.55, n_s = 1.00; C = 0.0044; S = 0.67; K = 3.28 (H_a is taken in m).

Using the above data and Eq. (5.56), discharge is obtained as = 50.90 lit/sec.

Sample Calculation 5.11

A parshall flume is used to irrigate a field with tomato crop .The throat width of the flume is 15 cm.The head measured in the converging and diverging sections are 50 cm and 35 cm, respectively. Whether the flow is free or submerged? Find out the discharge measured by flume?

Solution

Given data

Throat width, W=15 cm; H_a=30 cm =0.3 m and H_b= 50 cm = 0.5 m

Value of $S = H_a/H_b = 0.3/0.5 = 0.7$

Referring Table (5.1) for W=15 cm, we find S_t = 0.55 (by interpolating)

Since $S > S_t$, it is submerged case.

Now referring Table (5.1) and by interpolating, for W = 15 cm and under submerged condition, C_s = 46.35

$n_f = 1.58$, $n_s = 1$, $C = 0.0044$, $S = 0.7$, $K = 3.28$ (where H_a and H_b are taken in meter)

Discharged can be computed from the following formula

$Q = Cs [K(Ha - Hb)]^{n_f}/[- (log S+C)ns$

$= 46.34[3.28(0.5–0.35)]^{1.58}/[-(log 0.7+0.0044)]^1$

$= 100.00$ lit/sec.

The discharge of the parshall flume is obtained as = 100.00 lit/sec.

5.3.17 Cut-throat Flume

A cut-throat flume consists of (i) level floor (ii) a uniformly converging inlet section (iii) a uniformly diverging outlet section and (iv) a throat occurring at the intersection of the inlet and outlet sections. A plane view of a cut-throat flume and construction details are shown in Fig. 5.23.

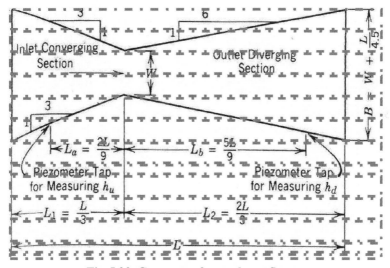

Fig. 5.23: Geometry of a cut-throat flume

Cut-throat flumes are easier and more economical to build and install than the Parshall flume. The same equations as used for discharge determination for free and submerged flow conditions in Parshall flume (Eqns. 5.55 and 5.56, respectively) are also used for discharge computation in cut-throat flume for free and submerged flow conditions, respectively. However, the parameters of these equations like C_f, C_s, n_f, n_s, S_f, K_f and K_s are different and are given in Fig.5.24 (James, 1988). Value of C is considered as zero. The following equations are used to calculate C_f and C_s.

$$C_f = K \, K_f \, W^{1.025} \tag{5.57}$$

$$C_s = K \, K_s \, W^{1.025} \tag{5.58}$$

where, W = throat width, ft; K_f and K_s are constants for free and submerged flow conditions, respectively (taken from Fig. 5.24); K = unit constant (1 for Q and H in Eqns. 5.55 and 5.56 in ft³/sec and ft, respectively and also for H_a and H_b in Eq. 5.56 in ft).

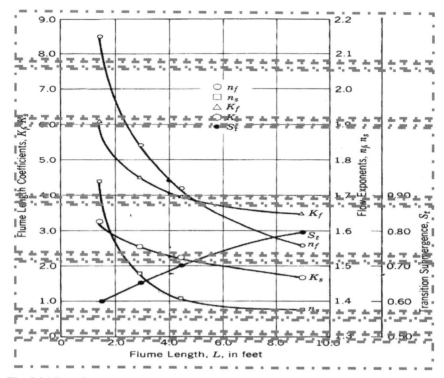

Fig. 5.24 Free flow and submerged flow coefficients (K_f and K_s), exponents (n_f and n_s) and transition submergence (S_t) for cut-throat flumes

Sample Calculation 5.12

A cut-throat flume is used to measure discharge under free flow condition for irrigating ground nut field. The flume has a length 4 ft. The throat width of the flume is 3 ft. The head measured in the converging inlet section of the flume is 2 ft.

Solution

Given length of flume, $L = 4$ ft.

Width at throat, $W = 3$ ft.

Head at the converging inlet section of the flume, $H_a = 2$ ft.

From Fig. 5.24, when length, $L = 4$ ft, $n_f = 1.74$ and $K_f = 4.1$.

From Eq. (5.57),

Putting $K = 1$, $K_f = 4.1$ and $W = 3$, C_f is found as $= 12.64$

Under free flow case, discharge

Putting $C_f = 12.64$, $K = 1$, $H_a = 2$ and $n_f = 1.74$,

Discharge is obtained as 42.23 ft³/sec which is equivalent to 1195.54 lit/sec.

Sample Calculation 5.13

A cut-throat flume is installed in an irrigation channel to measure the discharge. Length of flume is 6 ft. Width of flume at throat is 3.5 ft. Head measured at upstream and downstream end of the flume are 2.2 and 1.8 ft respectively. Decide the flow conditions and find out the discharge.

Solution

Given data:

Throat width $(W) = 3.5$ ft, $H_a = 2.2$ ft, $H_b = 1.8$ ft, Length of flume $(L) = 6$ ft

From graph (Fig. 5.24), for flume length of 6 ft, the transition ratio $(S_t) = 0.75$

Again submergence ratio $(S) = H_b/H_a = 1.8/2.2 = 0.82$

As the submergence ratio is > transition ratio, so, the flow through the flume is submerged flow.

The discharge can be found out from the equation as:

$Q = Cs [K (Ha - Hb)]n_f / (- log S)n_s$

In above equation H_a and H_b are in ft and Q is in cusec and so $K = 1$

Again, $Cs = K\ Ks\ W^{1.025}$

From graph (Fig. 5.24) for the throat length of 6 ft

$Ks = 2$, $n_f = 1.64$, $n_s = 1.38$, and so, $Cs = 1 \times 2 \times 3.5^{1.025} = 7.22$

So, discharge $= 7.22 [1(2.2-1.8)]^{1.64} / (-log\ 0.82)^{1.38} = 47.32$ cusec which is equivalent to 1339.64 lit/sec.

5.3.18 Chemical Dilution Method

In this method of flow measurement, a relatively large quantity of dye, called tracer is dissolved in a small quantity of water and placed in a bottle so that the tracer solution can be discharged at a known rate into the water flowing in the stream. By measuring the concentration of tracer at upstream and downstream side of the stream, the discharge is computed as (Michael, 1987):

$$Q = q_1 \frac{(C_2 - C_1)}{(C_0 - C_2)} \tag{5.59}$$

where, Q = Discharge flowing in the stream, q_1 = quantity of tracer added, C_0 = original tracer concentration, C_1 = concentration of the tracer in the bottle, C_2 = concentration of the tracer at the downstream side of the stream.

The tracer used to mix with water should not react with the fluid. It should completely mix with water also. The technique in which discharge flowing the stream Q is measured by knowing the values of q_1, C_0, C_1 and C_2 is known as *constant rate injection method or plateau gauging.* There is another method called as *sudden injection or gulp or integration method* to measure the stream flow also. The chemical dilution method used to measure stream flow is based on the principle that the stream flow is steady. For unsteady case, there will be a change in the storage volume in the reach. Systematic errors may arise in such cases.

The tracers used should have the following properties:

- It should be completely mixable with water.

- It should not react with water and should be non-toxic.

- It should not be absorbed by the sediment, vegetations etc.

- It should not be lost by evaporation.

- It should be capable of being detected distinctibly in small concentrations.

- It should not be very costly.

The tracers used are of mainly three types. They are:

- Chemicals like common salt and sodium dichromate

- Fluorescent dyes like Rhodamine –WT and Sulpho-Rhodamine B Extra and

- Radioactive materials such as Bromine-82, Sodium-24 etc.

Sample Calculation 5.14

A 30 gm/l solution of a fluorescent tracer was discharged in to a stream at a constant arte of 12 cm³/sec. The background concentration of the tracer in the stream water was found zero. At a downstream section sufficiently far away, the tracer was found to reach an equilibrium concentration of 6 parts per billion. Compute the stream flow.

Solution

By Eqn. (5.59), we have

$$Q = q_1 \frac{(C_2 - C_1)}{(C_0 - C_2)}$$

Given, q_1 = 12 cm³/sec = 12 x 10⁻⁶ m³/sec.

C_1 = 30 gm/l = 0.030, C_2 = 6.0 x 10⁻⁹ and C_0 = 0

Hence, $Q = \dfrac{12 \, x \, 10^{-6}}{6 \, x \, 10^{-9}} \, x \left(0.030 - 6 \, x \, 10^{-9}\right) = 60.0$ m³/sec.

5.3.19 Radioisotope Method to Measure Discharge

It may be used in place of chemical or dye tracer and the degree of dilution is determined by counting the gamma ray emissions from the diluted isotope solution using Geiger counters. In the total count method, a known amount of radioisotope is introduced into the flow in a relatively small time. At the downstream side measuring station where the isotope is thoroughly mixed, concentration of radioisotope tracer is determined from the gamma ray emissions detected and counted by the counter. The discharge is given as (Michael, 1987):

$$Q = \frac{F \, A}{N} \tag{5.60}$$

where, Q = discharge flowing in the stream/channel, F = counts per unit of radioactivity per unit volume of water per unit of time, A = total units of radioactivity to be introduced for each discharge measurement and N = total counts.

Radioactive materials must be handled very carefully. Emissions of Gamma radiations are highly injurious. Only licensed personnel are allowed to handle it.

5.3.20 Electromagnetic Method to Measure Discharge

Faraday's principle can be employed in electromagnetic method. As par Faraday's principle, an emf is induced in a conductor when it cuts a normal magnetic field. Large coils buried at the bottom of the stream or channel carry a current, I to produce a vertical magnetic field. Electrodes provided at the sides of the channel section measure the voltage produced due to flow of water in the channel (Fig. 5.25). The signal output which is the generated voltage produced due to flow of water is related to discharge, Q as:

$$Q = K_1 \left(\frac{E\,d}{i} + K_2 \right)^n \tag{5.61}$$

where, d is depth of flow, i is current in the coil, and n, K_1 and K_2 are system constants. This method is suitable for stream having width up to 70 m. It is also suitable for those streams in which considerable changes takes place in the flow of the stream due to excessive weed growth, sedimentation as well as there is variations in flow stages. This method uses sophisticated and costly instruments. It is suitable for tidal channels where the flow undergoes rapid changes both in magnitude and in directions. Presently available electromagnetic flowmeters can measure discharge up to an accuracy of ± 3%.

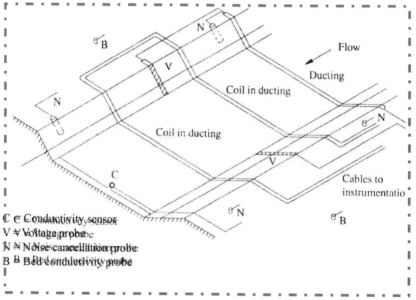

Fig. 5.25: Electromagnetic method to measure streamflow (Subramanya, 2003)

5.3.21 Ultrasonic Method to Measure Discharge

This method was introduced by Swingel in 1955. This method follows the principle of area-velocity method in which average velocity is measured by ultrasonic signals. In this method, two transducers A and B are installed at same level from the stream bed on either side of the stream (Fig. 5.26). These transducers can receive as well as send ultrasonic signals. The elapsed time between emitting and receiving of the signals by both transducers are recorded as observation for calculating the flow velocity. Let A send an ultrasonic signal which is received by B after an elapse time t_1. Similarly let B sends an ultrasonic signal which is

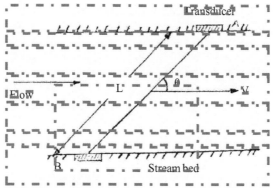

Fig. 5.26: Streamflow measurement by ultrasonic method

received by A after an elapse time t_2. Let The velocity of sound in water be C. If L is the length of path from A to B, then we can write:

$$t_1 = \frac{L}{C + V_p} \tag{5.62}$$

where, V_p is the velocity component in the sound path ($V_p = V \cos \theta$) and other terms are as defined above. Similarly, we can write:

$$t_2 = \frac{L}{C - V_p} \tag{5.63}$$

From Eqns. (5.62) and (5.63), we get

$$\frac{1}{t_1} - \frac{1}{t_2} = \frac{2V_p}{L} = \frac{2V \cos \theta}{L}$$

Solving above equation, we get

$$V = \frac{L}{2 \cos \theta} \left(\frac{1}{t_1} - \frac{1}{t_2} \right) \tag{5.64}$$

Thus, for a given L and , the average velocity, V along the path AB can be determined if we know t_1 and t_2.CThe above obtained value V is the average velocity at some height above the stream bed. To get the average velocity of the whole cross-section (V_a), a calibration curve between V/V_a and depth of flow, h is derived for the given stream cross-section.

After knowing the average velocity of the whole cross-section (V_a), the stream discharge can be computed by multiplying the velocity V_a with the cross-sectional area of flow. Estimating the discharge by using one signal path as discussed above is called as *single-path gauging*. However, for a given depth of flow, multiple single paths can be used to obtain V for different h values. Mean velocity of flow through the cross-section is obtained by taking the average of these V values. This technique is called as *multiple-path gauging.*

Ultrasonic flowmeters using the above principle have frequencies of the order of 500 kHz. Sophisticated and costly instruments are needed to transmit, detect and evaluate the mean velocity of flow along the flow path. Currently available systems can measure the discharge up to an accuracy of 2% in *single-path gauging* and 1% in *multiple-path gauging.* The system is now available for stream up to 500 m width.

Advantages of Ultrasonic Method: Followings are the advantages

- It is rapid and gives high accuracy.
- It is suitable for automatic recording of data.
- Installation cost is independent of stream size.
- It easily incorporates the effect of quick change of flow rate and its direction as in tidal channels.

The accuracy of this method is affected by

- Instability of stream cross-section.
- Entertainment of air.
- Presence of suspended solids in flowing water.
- Change in temperature and salinity level of the flowing water.
- Fluctuating weed growth.

5.4 Hydrometry Stations

Hydrometry stations are meant to measure the stream flow data. The WMO recommendations for the minimum number of hydrometry stations for various geographical regions are given in Table 5.2.

Table 5.2: WMO recommendations for deciding the density of hydrometry stations

Sl. No.	Geographical region	Minimum density (km²/station)	Acceptable density under difficult situations (km²/station)
1.	Flat region of temperate, mediterranean and tropical zones	1000-25000	3000-10000
2.	Mountainous regions of temperate, mediterranean and tropical zones	300-1000	1000-5000
3.	Arid and polar regions	5000-20000	

As per WMO recommendations India needs 1700 hydrometry stations to accurately assess the water resources potential. But till 1984, there were 542 stations which are under the control of Central Water Commission (CWC). Amongst 542 stations, 224 are used for measurement of flow stage and discharge and rest 318 are used for multiple uses like measurement of sediment, stage, discharge and water quality data. In addition to the above 542 CWC operated and managed stations, about 800 hydrometry stations have been installed by various state governments. There are also a number of stations where only gauge readings are taken.

Question Banks

Q1. Define streamflow. What is its importance in study of hydrology?

Q2. What do you mean by *hydrometry?* Discuss its importance in the study of surface water hydrology.

Q3. What is the definition of discharge? What are the different units used to find out discharge?

Q4. What are the objectives of streamflow measurement?

Q5. Write the different methods used to estimate streamflow.

Q6. How is the volume method of discharge measurement done?

Q7. What are the two methods used in measurement of streamflow by velocity-area method? Briefly discuss them.

Q8. Describe in details the velocity-area method used to estimate the discharge.

Q9. What are the approaches followed for discharge computation by velocity-area method?

Q10. What are the differences between the mid section and mean section method for discharge measurement?

Q11. Explain with figures how the mid section and mean section methods are used to measure discharge in a channel.

Q12. Narrate how the slope-area method is used to estimate the discharge?

Q13. Write down the Manning's and Chezy's formula for estimation of velocity of flow.

Q14. Explain how the water surface slope and energy slope of a channel are determined.

Q15. Write down the various steps adopted to compute streamflow by slope-area method?

Q16. What are different methods used to find out the depth of flow in a channel?

Q17. How are the sounding rods used to measure depth of flow?

Q18. How does the echo sounder help to measure the depth of flow in a channel?

Q19. Briefly describe how the float method is used to measure velocity of water.

Q20. How is sub-surface float helpful in measurement of velocity of water?

Q21. Explain in details how the double float and twin float are used in measurement of velocity of water in a stream.

Q22. A rectangular channel section is 15 m wide. The depth of flood water at two sections in this channel at 200 m apart is 3.0 and 2.9 m, respectively. The two sections have elevations of 15.3 and 15.0 m, respectively above datum. Assume Manning's coefficient at both the sections as same with a value of 0.025 and considering the flow to be uniform, compute the flood discharge through the channel section.

Q23. What is the use of pitot tube? Describe how it is used to measure the stream velocity.

Q24. Enumerate the working principles of the following instruments in measurement of depth of water in a stream.

(i) suspension rod (ii) sounding rod (iii) lead lines (iv) reel line and cranes and (v) echo sounder

Q25. A pitot tube is used to measure the flow velocity in a channel having depth of flow 2.2 m. The position of the pitot tube above the channel bed and the height of rising water in the tube are as follows. Compute the flow velocities at different positions of the tube.

Position of the pitot tube above channel bed, m	0.2	0.5	0.8	1.0	1.4	1.9
Depth of water in the pitot tube, m	0.75	0.80	0.92	1.0	1.1	1.2

Q26. What is a current meter? What are the various types of current meters?

Q27. With neat sketch describe the working and construction of vertical axis current meter.

Q28. What is the difference between vertical axis and horizontal axis current meter? Draw the diagram of a horizontal axis current meter and describe its working principles.

Q29. What is a Pigmi current meter? Describe its advantages in measurement of velocity.

Q30. Describe how the current meter is calibrated to measure the velocity.

Q31. Following velocities are recorded in an irrigation channel by a current meter.

Depth above the channel bed, m	0	1.0	2.0	3.0	4.0
Velocity, m/sec	0	0.4	0.5	0.7	0.8

Find out the discharge per unit width of the channel near the point of measurement if the total depth at the point of measurement is 5 m.

Q32. Following readings are taken by a current meter in a main canal section.

Distance from one bank to the other, m	0	0.5	1.0	1.5	2.0	2.5	3.0	3.5	4.0
Velocity at 0.2 depth, m/s	0	0.4	0.6	0.7	0.9	1.2	0.9	0.6	0
Velocity at 0.8 depth, m/s	0	0.3	0.5	0.7	0.8	1.0	0.8	0.5	0
Depth, m	0	0.3	0.8	1.0	1.5	1.2	1.0	0.5	0

Compute the total discharge flowing in the canal by both mid section and mean section method.

Q33. Calculate the discharge in a channel section from the following observations taken by a current meter whose rating curve is $V = 0.04 + 0.3 N$

Distance from bank, m	2	4	6	8	10	12	14
Depth, m	0.4	0.8	1.0	1.2	1.0	0.8	0.5
Revolution of current meter/sec	0.9	1.0	1.2	2.2	2.0	2.0	1.0

Q34. Water is flowing in a river cross section. The velocity of a surface float of the river cross section as observed is given below. Estimate the total discharge of the river cross section passing at the gauging site. Assume the velocity factor to convert the surface float to mean velocity of water as 0.89.

Width of section (b), m	Mean depth of each section (d), m	Surface velocity observed (V_s), m/sec
25	1.6	0.9
30	1.6	1.0
30	2.2	1.1
28	2.5	0.8
25	2.0	0.8
25	2.2	1.2
30	1.8	1.1
30	1.8	1.2
30	2.0	0.9
30	2.0	0.9

Q35. How a gauge-discharge curve is prepared? Explain how this curve helps in estimation of discharge.

Q36. Write notes on the followings.
(i) Permanent control and (ii) Shifting control

Q37. Explain the followings methods used for extrapolation of rating curves.

(i) Logarithm method (ii) Steven's method and (iii) Conveyance method

Q38. What are the essential requirements of a gauge site?

Q39. Write short notes on the followings.

(i) Staff gauge (ii) vertical staff gauge (iii) inclined gauge (iv) hook gauge (v) self recording gauge

Q40. Describe the construction and function of a stage recorder.

Q41. Write down the essential requirements of a discharge measuring site.

Q42. Write down the essential requirements of a discharge measuring site in case of a hilly area.

Q43. Draw a typical stage-discharge curve and explain how does this curve help in computation of discharge.

Q44. Following gauge and discharge data of a river site is collected by stream gauging operation. Develop a gauge-discharge relationship for the stream at this section. Assume the value of gauge at zero discharge as 7.2 m. Using the developed relation, estimate the value of discharge at gauge height of 11 m at this stream section.

Gauge reading, m	Discharge, m³/sec	Gauge reading, m	Discharge, m³/sec
7.5	15.3	8.4	170.0
7.7	28.0	8.5	260.8
7.8	55.5	8.6	312.8
7.9	60.8	8.7	355.8
8.0	75.9	8.8	432.0
8.1	100.5	9.2	518.5
8.2	140.5	9.4	550.0

Q45. Describe the Running's method used for finding out stage at zero discharge.

Q46. Explain the effects of unsteady flow and backwater on stage discharge relationship.

Q47. What are the differences between the weir and flume?

Q48. What is a weir? What are the different types of weirs?

Q49. With a neat sketch, explain the working of a rectangular weir for measurement of discharge.

Q50. Write short notes on the followings.

(i) Francis' formula (ii) Bazin's formula (iii) Rehbock's formula

Q51. Write down the formula to compute the discharge through trapezoidal and triangular weir.

Q52. What is a cipoletti weir? Draw a diagram of the cipoletti weir and describe how the discharge is measured by it.

Q53. Find the discharge over a suppressed rectangular weir having 4 m length and the height of water above the crest level is 60 cm under (i) no end contraction and (ii) with one side end contraction.

Q54. Find the discharge over a rectangular weir using the following data.

The length of the weir crest = 5 m. The crest level height above the channel bed = 1.2 m and the height of water flowing over the crest is 2.0 m above the channel bed.

Q55. What are the differences between ogee spillway and siphon spillway?

Q56. With a neat sketch describe the construction and function of an ogee spillway and a siphon spillway.

Q57. What is a proportional weir? Draw a diagram of proportional weir and describe the working of it.

Q58. How can you measure the discharge flowing in a pipe line? Discuss the use of orifices and mouthpieces for measurement of discharge in a pipe line.

Q59. Differentiate between a parshall flume and cutthroat flume.

Q60. With a neat label sketch, explain the construction and working a parshall flume.

Q61. How can you know the flow through a parshall flume and cutthroat flume are under free or submerged condition?

Q62. Draw a label sketch of a cutthroat flume and describe its principle to use to measure the discharge.

Q63. Write down the formula to measure the discharge through a cutthroat flume under submerged condition.

Q64. Compute the discharge passing through a parshall flume under the following condition.

(i)   Throat width = 15 cm and (ii) Head measured in the inlet and out let sections of the flume are 50 and 40 cm, respectively.

Q65. A cutthroat flume is installed in a channel to measure the flow of water passing through it. Length of the flume = 6 ft, width of the flume at throat = 3.5 ft, Head of water measured at the upstream and down stream end of the flume are 2.0 and 1.7 ft, respectively. Find out the flow condition and accordingly find out the discharge.

Q66. Describe the chemical dilution method and radioisotope method for measurement of streamflow.

Q67. Describe the construction and functions of the following streamflow measuring instruments.

(i) Ultrasonic method (ii) Electromagnetic method

Q68. What are the properties of the tracers used in streamflow measurement by chemical dilution method?

Q69. A 28 gm/l solution of a fluorescent tracer was discharged in to a stream at a constant rate of 11 cm^3/sec. The background concentration of the tracer in the stream water was found zero. At a downstream section sufficiently far away, the tracer was found to reach an equilibrium concentration of 5 parts per billion. Compute the stream flow.

Q70. Briefly enumerate the advantages of ultrasonic method for measurement of velocity of water.

Q71. Write notes on WMO recommendations for the minimum number of hydrometry stations for various geographical regions.

References

Garg, S.K. 1978. Irrigation Engineering and Hydraulic Structures. Khanna Publishers, Delhi.

James, L.G. 1988. Principles of Farm Irrigation System Design. John Wiley & Sons, New York, pp. 543.

Lenka, D. 2001. Irrigation and Drainage. Kalyani Publishers, New delh.

Michael, A.M. 1978. Irrigation Theory and Practices. Vikash Publishing House Pvt. Ltd., New Delhi.

Modi, P.N. and Seth, S.M. 2005. Hydraulics and Fluid Mechanics Including Hydraulic Machines. Standard Book house, Delhi.

Murty, V.V.N. 1998. Land and Water Management Engineering. 2nd Edition, Kalyani Publishers, New Delhi, pp. 586.

Sharma, R.K. and Sharma, T.K. 2002. Irrigation Engineering including Hydrology, S. Chand and Company Ltd. New Delhi.

Sharma, R.K., 1987. A Text book of Hydrology and Water Resources. Dhanpat Rai and Sons, Delhi.

Subramanya, K. 2003. Engineering Hydrology, Tata McGraw-Hill Publishing Co. Ltd., New Delhi, 392 pp.

Suresh, R. 2008. Watershed Hydrology. Standard Publishers Distributors, Delhi, pp. 692.

U.S. Bureau of Reclamation 1975. Water Measurement Manual, U.S. Government Printing Office, Washington, D.C. pp. 327.

Walker, W.R and Skogerboe, G.V. 1987. The Theory and Practices of Surface Irrigation. Prentice-Hall, Inc., Englewood Cliffs, N.J.

CHAPTER 6

Hydrograph

6.1 Introduction

When rainfall strikes a ground surface, some losses like initial and infiltration will occur. After these losses are fulfilled, runoff will begin. The runoff will ultimately reach a stream. The main routes through which runoff flows are (i) overland flow, (ii) interflow and (iii) groundwater flow. Overland flow which is also called as surface flow flows over the land mass and reaches a stream. Overland flow soon reaches a stream and if it occurs in sufficient quantity, it forms an important element in the formation of flood peaks. In small and moderate storms, there may have predominant initial and infiltration losses resulting minimum amount of surface runoff and this is especially true in case of permeable soils. If the soil is impermeable, then there may have reasonable runoff. However, surface runoff is an important factor in streamflow only for heavy or high intensity rainfall.

Some of the infiltrated water moves laterally within the upper soil layers until it reaches a stream. This lateral movement of infiltrated water is called as *interflow or subsurface flow*. *Interflow* moves more slowly than the surface flow and reaches the stream somewhat later. The amount of *interflow* contributing the total runoff to a stream depends on the geology of the basin. A thin soil cover overlying a hard rock layer favours formation of more *interflow* whereas uniformly permeable soil enhances percolation resulting reduced interflow. In some cases, *interflow* forms major component than the overland flow especially when there is slow or moderate storms falling over permeable surface layer.

Some precipitation moves vertically downwards as deep percolation and contributes to groundwater. This groundwater component eventually discharges into the stream as *groundwater flow or base flow or dry weather flow* when the water table intersects the stream channels of the basin. In this event, the streams are called as *effluent streams*. There is another stream called *as influent stream* where water moves away from the stream to outsides. It is to be remembered that groundwater flow moves extremely slowly and takes very long time to reach a stream. Sometimes, it may take two years also to reach a stream. Of course, the rate of movement of groundwater flow depends on the voids and interstices available in the groundwater layers. Groundwater flow is not an important component that contributes to flood peaks. Basins having permeable surface soils and large effluent groundwater bodies show sustained high flow throughout the year with a relatively small ratio of flood flow to mean flow.

Now consider an isolated storm which produces a fairly uniform rainfall of duration, T_r over a basin. After the initial and infiltration losses are met, rainfall excess reaches the stream through overland and channel flow. During translation, a certain amount of storage is built-up in the overland and channel flow phases. This storage depletes slowly after the cessation of rainfall. Thus, there is a time lag between the occurrence of rainfall in the basin and the time when that water passes the stream-gauging station at the basin outlet. The runoff measured at the stream-gauging station at the basin outlet will produce a typical *hydrograph* as shown in Fig. 6.1. Duration of rainfall is also marked in this figure to indicate the time lag

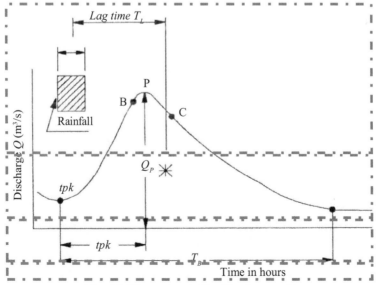

Fig. 6.1: Elements of a flood hydrograph

in the rainfall and runoff. If there is an isolated rainfall, then the hydrograph produced due to this rainfall is single peaked and has skew distribution of discharge. This hydrograph is called as *storm hydrograph, flood hydrograph or simply hydrograph*. This hydrograph has three characteristic regions: (i) The rising limb AB which joins point A (starting point of the rising curve) and B (point of inflection), (ii) the crest segment BC between the two points of inflection with a peak P in-between and (iii) the falling limb/recession limb CD starting from the second point of inflection C. The time to peak represents the duration from the point A to occurrence of peak discharge, Q_p (point P) and is represented by term t_{pk}. Time interval from the centre of mass of rainfall to the centre of mass of hydrograph is called as lag time, T_L. Time base of hydrograph is the duration from the point A to the point D, T_B.

It is to be noted here that a hydrograph is a response of a given catchment (runoff) to a rainfall input. It comprises runoff in all three stages i.e. surface runoff, interflow and base flow and embodies in itself the integrated effects of a wide range of catchment and rainfall parameters and this integrated effect has complex interactions. The hydrographs of two identical and isolated storms produce same kind of hydrograph in the same catchment or basin. But for two different storms of varying nature, the nature of hydrograph is different even though the catchment is same. Even when the two storms are same in nature (identical), they may produce different hydrographs when they apply to different catchment/basins. It is, indeed, difficult in nature to have two same types of hydrograph because of as interaction effects of catchment and storm characteristics. When one examines a number of flood hydrographs, of a stream, it is observed that they have multiple peaks, kinks etc. These features make the resulting hydrograph much different than the one shown in Fig. 6.1 which a single peak.

6.2 Factors Affecting Shape of Hydrograph

The factors that affect the shape of the hydrograph are broadly classified into two categories i.e. (i) Climatic factor and (ii) Physiographic factors.

6.2.1 Physiographic Factors

The various physiographic factors are:

- Basin characteristics

- Infiltration characteristics and

- Channel characteristics

Basin characteristics are:

- Shape
- Size
- Slope
- Nature of valley
- Elevation and
- Drainage density

Infiltration characteristics are:

- Land use and land cover
- Soil type and geological condition and
- Depression storages like lakes and swamps

The various climatic factors that affect the shape of the hydrograph are:

- Storm characteristics like precipitation and its intensity, duration, magnitude and direction of movement
- Initial loss and
- Losses like evaporation and transpiration

Of the various factors, the climatic factors affect the rising whereas the recession limb of a hydrograph is affected by the physiographic factor. The various factors mentioned as above are interrelated to themselves and so let us consider the salient factors in qualitative terms.

6.2.1.1 Shape of the Basin

The shape of the catchment or basin affects the time of concentration of the runoff of the catchment. Time of concentration is the time to travel water from the remotest point to the outlet point in the catchment where runoff is measured. If the shape of the catchment is fan shaped which is nearly semi-circular, then time of concentration is less. As a result, the hydrograph gives high peak and narrow hydrograph with less time base of the hydrograph. On the other hand, a fern shaped catchment is elongated in nature. The time of concentration for it is more. Hence, runoff takes much time to reach at the catchment outlet resulting low runoff and low peak discharge and wide hydrograph. The time base of such a hydrograph is more. Fig. 6.2 represents systematically the hydrographs from three catchments having identical rainfall and infiltration characteristics over the catchment. In catchment A (fan shaped), hydrograph is skewed to left

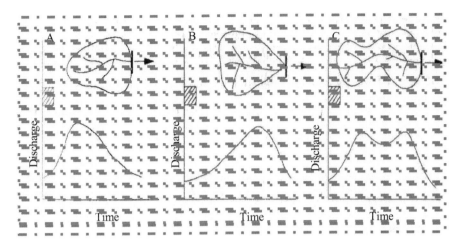

Fig. 6.2: Shape of catchment affecting hydrograph

with high peak discharge and narrow time base whereas in catchment B (fern shaped), hydrograph is skewed to right with low peak discharge and wide time base. In catchment C which has composite shape, the hydrograph is complex.

6.2.1.2 Size of the Basin

Size of the basin affects the shape of the hydrograph by changing the discharge rate with time. Overland flow generally dominates a small basin and so in case of small basin, peak discharge occurs soon resulting high peak and narrow hydrograph. In large basins, channel flow is more predominant than the overland flow. The peak discharge is observed to vary with A^n where A is the catchment area and n is an exponent having value less than 1.0 (generally the value of n is about 0.5). In large basins, the duration of surface runoff from time to peak is observed to vary with A^m where A is the catchment area and m is an exponent having value less than 1.0 (generally the value of n is about 0.2). For large basins, the peak discharge is less with wider time base of the hydrograph.

6.2.1.3 Slope

The slope of the stream controls the velocity of flow in the channel. It affects the shape of the recession limb by affecting the withdrawal of water from the storage made in the basin. If the stream slope is high, then the slope of the recession limb is more and as a result, the time base of the hydrograph is less. On the other hand, a low stream slope results in flatter recessing limb causing wider time base of the hydrograph.

6.2.1.4 Nature of Valley

Nature of valley refers to the slope of the bed and side of valley. Its effect on shape of hydrograph is same as slope of the stream i.e. greater valley slope increases the slope of the recession limb and decreases the time element of the hydrograph.

6.2.1.5 Drainage Density

Ratio of the total channel lengths in the basin to the total area of the basin is called as drainage density. A large drainage density helps in quick disposal of runoff from a basin resulting pronounced peak discharge. In a basin having small value of drainage density, overland flow is predominant which produced hydrograph squat type with a slowly rising limb (Fig. 6.3).

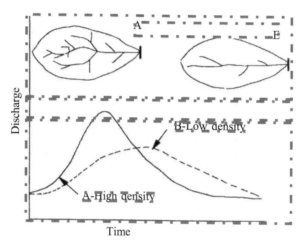

Fig. 6.3: Effect of drainage density on hydrograph

6.2.1.6 Land Use

A basin which has more vegetation and has more forest cover affects the movement of runoff more. It increases the interception loss, enhances infiltration opportunity time and also enhances the evapotranspiration loss. All these effects the runoff of the basin. The effect is predominant in case of small basins having size less than 150 km². Further, the effect of vegetation on runoff is predominant in case of a small storm. Comparing vegetation density, the catchment which has higher vegetation density will give rise to lower peak discharge than the other one having lesser vegetation density provided other catchment characteristics are same. This affects the shape of the hydrograph.

6.2.1.7 Depression Storage

Presence of depression storages like ponds, lakes, swamps etc. in the catchment delays and modify the flow pattern of the runoff which affects the occurrence of peak flow. In general, a catchment with more depression storages produces reduced peak discharge and wider time base of the hydrograph.

6.2.2 Climatic Factors

Among the different climatic factors, intensity, duration and direction of storm movements are important ones affecting the shape of a hydrograph. For a given duration, if the intensity of rainfall is more than the peak as well as volume of surface runoff will be more Intensity of rainfall influences the shape of hydrograph more in a small watershed. For a given rainfall intensity, the amount of runoff is directly proportional to the duration of rainfall. But in practice a storm with higher intensity continues for a small duration and so the generated runoff is less. The peak runoff is resulted after a longer period and the resulting hydrograph is flatter with wider time base. The orientation of the storm i.e. movement of the storm affects shape of hydrograph. If the storm moves from upstream of the catchment to the downstream end, there will be a quicker concentration of flow at the basin outlet. This will give rise to a high peaked hydrograph. On the other hand, if the storm moves in reverse direction, then the peak of the hydrograph will be flat with more time base. This effect becomes more prominent when the shape of the catchment is long and narrow and the storm moves in opposite direction to the outlet of the catchment.

6.3 Components of a Hydrograph

The essential components of a hydrograph are:

- Rising Limb
- Crest Segment and
- Recession Limb

6.3.1 Rising Limb

Rising limb of a hydrograph is also called as *concentration curve.* It represents the rising of discharge due to gradual formation of storage in the channels existing in the area and also over the land masses in the catchment. The slope of the rising limb depends on both climatic and physiographic factors. The initial losses and infiltration losses during the early period of a storm causes runoff to be less and so the discharge rises rather slowly in the early stage. This makes the rising limb to be flatter at the beginning. However, as the storm continues, the losses are met and then the runoff increases. More and more runoff passes through the outlet causing higher peaked hydrograph. The general shape of the rising limb is concave. As shown in Fig. 6.1, the portion AB of the hydrograph represents the rising limb.

6.3.2 Crest Segment

It is the most important part of hydrograph. It contains the peak flow. The peak flow occurs when all the parts of the watershed start to yield the runoff simultaneously to the outlet. In large watersheds, it occurs after the end of rainfall. The time interval between the centre of mass of the rainfall to the peak of the hydrograph is affected by both the basin and storm characteristics. When there are two or more storms that occur one after the other in a given catchment, then there will be multiple peaked complex hydrograph. The crest segment of the hydrograph is shown as BC in Fig. 6.1.

6.3.3 Recession Limb

Recession limb is also called as falling limb or depletion curve of the hydrograph. It extends from the point of inflection at the end of crest segment to the commencement of groundwater flow or base flow in the catchment. Starting point of the recession limb i.e. inflection point represents the condition of maximum storage. Recession limb represents the withdrawal of water from the storage built up in the basin during the early phase of the hydrograph. The shape of the recession limb of the hydrograph is independent of storm characteristics but it depends on the basin characteristics. The nature of the falling limb is generally convex in shape due to continuous decrease of storage and runoff rate. The falling limb of the hydrograph is the portion CD of the hydrograph (Fig. 6.1).

The storage of water in the basin exists as (i) surface storage which includes both surface detention and channel storage, (ii) interflow storage and (iii) groundwater storage. Barnes (1940) reported that the recession of storage can be expressed as:

$$Q_t = Q_0 \, K_r^{\,t} \tag{6.1}$$

where, Q_0 is the initial storage, Q_t is the storage at time interval of t i.e. storage at day t, K_r is recession constant having a value less than 1.0. The recession constant comprises three components i.e. recession constant for surface storage (K_{rs}), recession constant for interflow (K_{ri}), and recession constant for base flow (K_{rb}). Thus,

$$K_r = K_{rs} \cdot K_{ri} \cdot K_{rb} \tag{6.2}$$

Values of K_{rs} ranges from 0.05 to 0.20, K_{ri} ranges from 0.50 to 0.85 and K_{rb} ranges from 0.85 to 0.99. Value of K_{ri} can be assumed as 1.0 if the interflow is insignificant.

6.3.3.1 Determination of Recession Constant

If the data of Q_t vs. t are plotted in a semi-logarithmic graph paper with Q_t taken in logarithmic side and time, t in arithmetic side then a straight line is obtained. Then K_r is determined by selecting the point in such a way that there is little or no direct runoff contribution. Since, the value of K_r is not constant, it is essential to predict its value at different points of recession curve within a given flow range. Towards this purpose, recession curve is divided into a series of time segments in such a way that the segment should not be exactly straight line. Rather it should be in the form of a curve having gradual decreasing slope.

Following steps have been proposed by Barnes (1940) for determining various components of recession constants.

Step I: Plot data of Q_t vs. t in a semi-logarithmic graph paper with Q_t taken in logarithmic side as ordinate and time, t in arithmetic side as abscissa and draw the hydrograph.

Step II:. Extend the tail end of the hydrograph backward up to the point below the point of inflection on recession limb as shown in Fig. 6.4. This is shown by straight line DE. The line DE is also called as groundwater recession curve.

Step III: Calculate the slope of the straight line DE at the tail end of the hydrograph. This gives the value of recession constant for base flow i.e. K_{rb}.

Step IV: Deduct the base flow from the total runoff hydrograph and plot the new ordinates vs. time data on the same graph paper as discussed in Step I.

Step V: Extend the tail end of the new hydrograph backward up to the point below the point of inflection on recession limb for separating the interflow from the new hydrograph. In Fig. 6.4, it is shown by line GL and LF.

Step VI: Find out the value of recession constant for interflow i.e. K_{ri} which is the slope of the straight line at tail end of the new hydrograph.

Step VII: Calculate the ordinates of surface runoff by deducting the interflow from the new hydrograph (in Step VI) and plot the ordinates against time and draw the hydrograph. The resulting hydrograph is called as surface flow hydrograph.

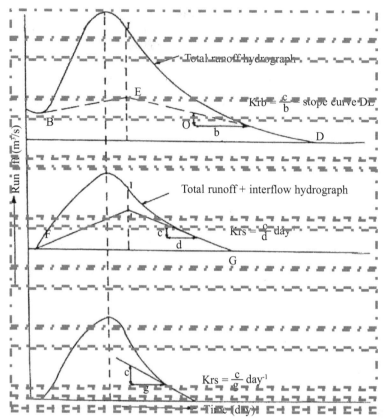

Fig. 6.4: Computation of recession constant of hydrograph (Suresh, 2008)

Step VIII: Extend the tail end of surface flow hydrograph backward and compute the value of surface flow recession constant (K_{rs}).

6.4 Base Flow Separation

The contribution of base flow to the total runoff is generally small. Because of this reason, errors made in the separation of base flow are small and not so important as long as one method is used consistently. In the case of simple hydrographs unaffected by rainfall prior or subsequent to the period under study, any of the following separation technique may apply.

6.4.1 Straight Line Method

Method I

It is a simple procedure of drawing a straight line from a point in the beginning of surface runoff to a point in the recession limb representing the end of direct runoff. In Fig. 6.5, it is shown by straight line AB , in which A is the starting point and B is the end point of surface runoff. The point A is easy to identify in

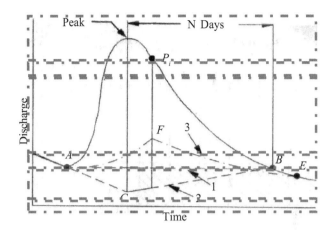

Fig. 6.5: Base flow separation technique

view of the sharp change in the runoff rate at that point. The point B is located at a time $N = 0.83 A^{0.2}$ days after the peak of hydrograph where A is the area of the watershed in km^2 (Linsley *et al.,* 1940). The area of the hydrograph below the line AB represents base flow whereas the area above line AB is surface runoff or called as direct runoff. It is to be noted that the value of N obtained as above is approximate and the position of B should be decided by considering a number of isolated hydrographs, keeping in mind that the total time base should not be excessively long and the rise of the groundwater should not be too great. According to Linsley et al. (1940), these values of N should be slightly reduced for mountainous regions while for long narrow basins or for basins of flat slopes, value of N should be increased by as much as 50 percent.

6.4.2 Method II

The most widely used separation method consists of extending the recession existing prior to the storm to a point under the peak of the hydrograph (point C in Fig. 6.5). From this point a straight line is drawn to the hydrograph at a point N days after the peak (point B). Segment AC and CB demarcates the bae flow and surface runoff. The reason behind this hypothesis is that as the stream rises, there is flow from the stream in to the banks. Hence the base flow should decrease until the stages in the streams begin to drop and the bank storage returns to the channel. Some opinions are that this method of base flow separation is also arbitrary and not better than the straight line method as discussed above.

6.4.3 Method III

In this method, the base flow recession curve after the depletion of flood water is extended backwards till it intersects the ordinate at the point of inflection in the recession limb. This is represented by line EF in Fig. 6.5. The tail end of the hydrograph is generally found in the form of a straight line that is extended as recession curve. The points A and F are then joined by an arbitrary smooth curve. This method of base flow separation may have some advantages where groundwater is relatively large in quantity and reaches the stream fairly rapidly.

6.4.4 Chow Method

In this method, the following steps are followed (Suresh, 2008):

- Extend the base flow curve before the rising limb of hydrograph up to the point M (Fig. 6.6).The point M is arbitrarily located at the distance of one-tenth of the base width of the hydrograph from the peak.

- Extend the recession curve inward starting from point G.

- Locate a point N at the middle of M and G on the recession limb arbitrarily.

- Joins the points M and N by smooth curve in convex nature shooting upward.

- The curve BMNG represents the base flow line.

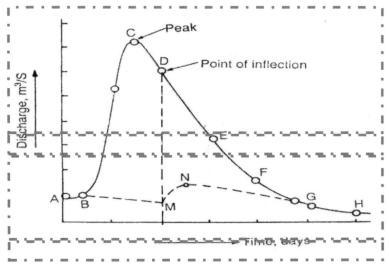

Fig. 6.6: Base flow separation by Chow method

It is seen that all the above methods of base flow separation are arbitrary. The selection of any one of them depends on the local practice. The surface runoff hydrograph obtained after the base flow separation is also called as *direct runoff hydrograph*.

6.4.5 Base Flow Separation from Complex Hydrograph

Complex hydrographs are produced when there are several storms of varying intensities that occur successively in a basin. Separation of base flow from the complex hydrographs requires the knowledge of effect of rainfall event on base flow instead of separating the base flow from the total runoff.

Let us consider a two peaked complex hydrograph (Fig. 6.7) which needs base flow separation. The following steps are required to do it.

- Let points 1 and 2 represent the peak of hydrograph 1 and 2, respectively. Locate points B and D after N days from the peaks 1 and 2 of these two hydrographs. N (days) is given by equation $N = 0.83\,A^{0.2}$ where A is the area of the watershed in km^2.

- Extend the initial curve of hydrograph forward to cut the perpendicular line drawn from peak (1) and let it be line EF (Fig. 6.7).

- Join the point A on recession limb of peak (1) to the point B by using composite recession curve.

- If point B falls before the peak of the second hydrograph, then base flow is separated by joining the line BC and CD by straight line. But if B falls after the peak of second hydrograph, then point D is directly joined to the recession curve AB by straight line.

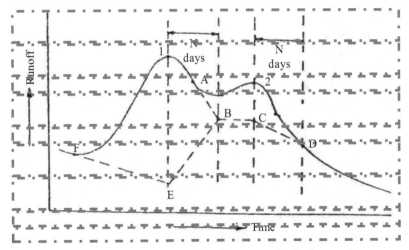

Fig. 6.7: Separation of base flow from complex hydrograph

6.5 Effective Rainfall and Effective Rainfall Hyetograph

The part of rainfall that contributes to the formation of direct runoff is called as effective rainfall. For the purpose of correlating *direct runoff hydrograph*

(DRH) with the rainfall which produces the flow, the hyetograph of rainfall is also pruned by deducting the losses. The hyetograph of a typical storm is shown in Fig. 6.8. When the initial losses and infiltration losses are subtracted from it, the resulting hyetograph is called as *effective rainfall hyetograph (ERH)*. *ERH* is also called as *hyetograph of rainfall excess or supra rainfall*. Initial losses together with the infiltration losses are called as ϕ–index.

Area under the *ERH* gives the total depth of effective rainfall whereas area under the rainfall hyetograph or simply called as hyetograph is the total depth of rainfall. Total depth of effective rainfall is expressed as:

$$ER = \sum_{i=1}^{i=n} I_i \cdot \Delta t \tag{6.3}$$

where, *ER* is total depth of effective rainfall or excess rainfall, I_i is intensity of effective rainfall at any instant *i* and Δt is time interval of effective rainfall hyetograph. Multiplying the total depth of effective rainfall with the basin area that produces the runoff, we get volume of direct runoff or volume of rainfall excess. The area under direct runoff hydrograph (DRH) also gives volume of direct runoff or volume of rainfall excess. It is to be remembered that the initial loss is given as the difference between the area under hyetograph and that under effective rainfall hyetograph.

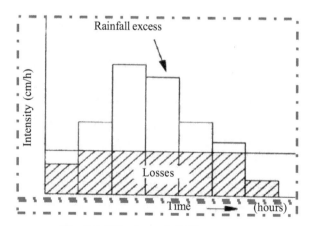

Fig. 6.8: Effective rainfall hyetograph

6.5.1 Relation between DRH and ERH

The DRH and ERH represent the same total quantity of direct runoff but in different units. The hyetograph minus the abstractions when replotted gives the ERH. The area under the ERH gives runoff in terms of depth. When this area is multiplied with the area of the watershed, then we get volume of direct runoff. It is also to be noted here that the area under the DRH gives and DRH can be written as:

Area of ERH x Area of the watershed = Area under DRH

In other words, $\dfrac{\text{Area under DRH}}{\text{Area under ERH}}$ = *Area of the watershed*

Sample Calculation 6.1

There are two consecutive storms of magnitude 3.5 cm and 2.5 cm occurring in two consecutive 4-hr duration in a catchment of 30 km² which produces a hydrograph with the observed flow as recorded below. Estimate the effective rainfall and losses.

Time since rainfall occurs, hr	-6	0	6	12	18	24	30	36	42	48	54	60
Observed flow, m³/s		6.1	4.4	11	24	20	16	11	8	6.5	5	4.4	4.4

Solution

The hydrograph is plotted to scale (Fig. 6.9). It is seen that the storm hydrograph has a base flow component. Base flow separation may be done by simple straight line method.

Now $N = 0.83\ A^{0.2} = 0.83$ x $30^{0.2} = 1.64$ days = 39.4 hr.

But by inspection, direct runoff hydrograph (DRH) starts at t = 0, has a peak at t = 12 hr and ends at t = 54 hr. This gives value of N as 54 – 12= 42 hr. It is revealed that N = 42 appears to be more realistic than N = 39.4 hr. In the present case the DRH is assumed to exist from t = 0 to 54 hr. A straight line base flow separation method gives a constant value of base flow as 4.4 m³/sec. The ordinates of DRH are given in table below.

Time since rainfall occurs, hr	-6	0	6	12	18	24	30	36	42	48	54	60
Observed flow, m³/s		6.1	4.4	11	24	20	16	11	8	6.5	5	4.4	4.4
Ordinates of DRH, m³/s		-	0	6.6	19.6	15.6	11.6	6.6	3.6	2.1	0.6	-	-

Area of DRH = 6 x 60 x 60 [0.5 x 6.6 + 0.5 x (6.6 + 19.6) + 0.5 x (19.6 + 15.6) + 0.5 x (15.6 + 11.6) + 0.5 x (11.6 + 6.6) + 0.5 x (6.6 + 3.6) + 0.5 x (3.6 + 2.1) + 0.5 x (2.1 + 0.6) + 0.5 x 0.6] = 1432080 m³ which is the total volume of direct runoff from the catchment of area 30 km².

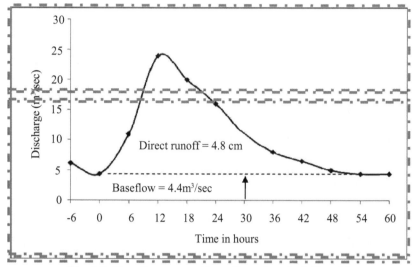

Fig. 6.9: Base flow separation – Sample calculation 6.1

Hence the depth of direct runoff = Effective rainfall = 1432080 / (30 x 10^6) = 0.048 m = 4.8 cm

Total rainfall = 3.5 + 2.5 = 6.0 cm and duration = 4 + 4 = 8 hr

Hence, the losses = 6.0 – 4.8 = 1.2 cm and the loss rate = ϕ- index = 1.2 /8 = 0.15 cm/hr.

Sample Calculation 6.2

A storm over a catchment of 6.2 km² had a duration of 16 hours.The mass curve of the rainfall of the storm over the catchemnt is as follows.

Time from start of storm, hr	0	2	4	6	8	10	12	14	16
Auumulated rainfall, cm	0	0.5	2.6	5.0	6.5	7.2	9.0	9.5	9.8

If the valueof ϕ- index is 0.40 cm/hr, estimate the effective rainfall hyetograph and the volume of direct runoff from the catchment due to the storm.

Solution

The solution is explained by computation of data like depth of rainfall and losses in each 2 hours duration, effective rainfall and intensity of effective rainfall as provided in Table below. In a given time interval, the effective rainfall is the difference of depth of rainfall and losses. If the difference is positive then it is taken as effective rainfall. If the difference is zero or negative, then, effective rainfall is considered as zero.

Time from start of storm, hr	Time interval, Δt, hr	Accumulated rainfall in Δt, cm	Depth of rainfall in Δt, cm	Loss ($\phi \Delta t$), cm	Effective rainfall, ER, cm	Intensity of ER, cm/hr
0	2	0	-	-	-	-
2	2	0.5	0.5	0.8	0	0
4	2	2.6	2.1	0.8	1.3	0.65
6	2	5.0	2.4	0.8	1.6	0.8
8	2	6.5	1.5	0.8	0.7	0.35
10	2	7.2	0.7	0.8	0	0
12	2	9.0	1.8	0.8	1.0	0.5
14	2	9.5	0.5	0.8	0	0
16	2	9.8	0.3	0.8	0	0

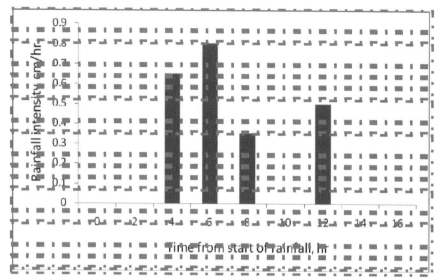

Fig. 6.10: Hyetograph of effective rainfall – Sample calculation 6.2

The ordinates of the effective rainfall hyetograph are the values shown in last column of above table. The effective rainfall hyetograph is drawn by plotting the data of last column of the above mentioned table against the time from the start of rainfall (first column) of above mentioned table and is shown in Fig. 6.10.

Total effective rainfall = Direct runoff due to storm = Area under ERH

$= (0.65 + 0.8 + 0.35 + 0.5) \times 2 = 4.6$ cm

Volume of direct runoff $= (4.6 / 100) \times 6.2 \times 10^6 = 28.52 \times 10^4 \, m^3$

6.6 Unit Hydrograph

Unit hydrograph was first originated by Sherman (1932) and has undergone many refinements since then. It is widely used for flood estimation and prediction. Unit hydrograph is a hydrograph of direct runoff resulting from one cm of effective rainfall occurring uniformly over the basin area at a uniform rate for a specified duration (D-hours). We can also assume 1 mm or 1 inch as effective rainfall instead of 1 cm though 1 cm is generally used. Here the effective rain refers to that rainfall that appears as runoff in a stream and is often referred as rainfall excess. The effective rainfall is assumed to be uniformly distributed over the entire basin so that the direct runoff originates at the beginning of rainfall excess. The unit rainfall duration refers to the smallest period of time applicable to and commensurate with the basin in which the rainfall can be assumed to be uniform. It should not exceed the period during which the design storm rainfall is considered to be approximately uniform in intensity in various portion of the basin under study. A rough guide for the choice of the duration of unit hydrograph is that it should not exceed the least of (i) time of rise, (ii) the basin lag and (iii) time of concentration. The best unit hydrograph is about one-fourth of the basin lag time which is the time from middle of the block of rainfall excess to the peak of observed hydrograph. This duration may vary from 12 hours in case of large basin of area more than 2600 km^2 to 6, 8 or 10 hours for small basin of area 250 to 2600 km^2. For areas of 50 km^2, 2-hr unit hydrograph is used (Varshney, 1979). For convenience, it is taken in terms of 6, 12 or 24 hours. For example, a 6-hour unit hydrograph refers to 6-hour duration of rainfall excess. The duration is used as a prefix to a specific unit hydrograph.

The definition of unit hydrograph refers to:

(i)   The unit hydrograph represents the lumped response of the catchment to a unit rainfall excess of D-hr duration to produce a direct runoff hydrograph. It relates rainfall excess to the direct runoff only.

(ii) Area under unit hydrograph refers to the volume of direct runoff resulted by unit depth of rainfall excess over the catchment.

(iii) The rainfall is considered to have a uniform intensity of rainfall excess of 1/D cm/hr for the duration D-hr of the storm.

(iv) The distribution of the storm is considered to be uniform all over the catchment.

6.6.1 Assumptions of Unit Hydrograph

The two basic assumptions of unit hydrograph are:

(i)   Time invariance

(ii) Linear response

6.6.1.1 Time Invariance

It states that the direct runoff due to the rainfall excess over the watershed is time invariant. In other words, the direct runoff hydrograph of given effective rainfall in the watershed is always same irrespective of when it occurs.

6.6.1.2 Linear Response

The direct runoff response to the rainfall excess is assumed to be linear. This is the most important assumptions of the unit hydrograph theory. This assumption is also called as principle of superposition or principle of proportionality. The basic premise of unit hydrograph principle is that R cm of effective rainfall over a basin in the same unit duration will produce a direct runoff hydrograph of the same time base as the unit hydrograph but having ordinates R times those of unit hydrograph. Since the area of the resulting DRH should increase by the ratio R, the base of the DRH will remain same as that of the unit hydrograph. The ordinates of the direct runoffs resulting from effective rainfalls of equal durations but with different intensities are proportional to the total amount of direct runoff represented by each hydrograph. L.K.Sherman indicated that if two identical storms could occur over a given basin with identical conditions prior to the occurrence of the storms, the hydrographs of runoff from the two storms would be expected to be same. This principle may be extended to a series of storms of varying intensities. These series of the storms will produce a series of overlapping hydrographs resulting from separate periods of uniform effective rain and having ordinates proportional to the unit hydrograph. The hydrograph of total direct runoff is obtained by summing the ordinates of the individual hydrographs and this principle of obtaining the DRH is known as principle of superposition.

Sample Calculation 6.3

The ordinates of a 6-hr unit hydrograph (UH) of catchment of area 300 km^2 are given below. Calculate the ordinates of the DRH due to a rainfall excess of 3 cm occurring in 6 hr.

Time, hr	0	3	6	9	12	15	18	24	30	36	42	48	54	60
UH ordinates, m^3/sec	0	22	48	82	120	150	180	150	100	60	30	20	10	0

Solution

By using linear response theory of unit hydrograph, the ordinates of a DRH due to 3 cm of rainfall excess are obtained by multiplying the ordinates of unit hydrograph as given in the problem by 3. The resulting DRH is shown in Fig. 6.11 below along-with unit hydrograph.

A sample calculation of derivation of DRH from unit hydrograph due to the occurrence of multiple storms in succession by superposition theorem is explained as below.

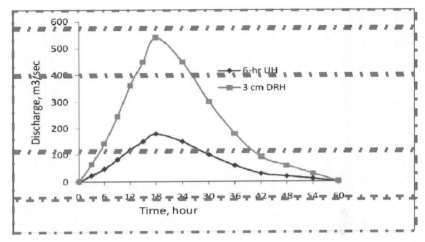

Fig. 6.11: 3 cm DRH derived from 6-hr unit hydrograph (Sample calculation 6.3)

Sample Calculation 6.4

Two storms each of 6-hr duration have rainfall excess values of 2 and 3 cm each. The 2 cm excess rainfall is followed by the 3 cm excess rainfall. The 6-hr unit hydrograph for the catchment has the following ordinates. Calculate the DRH. If there is a constant base flow of 5 m³/sec, compute the total flood hydrograph.

Time, hr	0	3	6	9	12	15	18	24	30	36	42	48	54	60	69
UH ordinates, m³/sec	0	22	48	82	120	150	180	150	100	60	30	20	10	4	0

Solution

First we have to compute the DRH of 2 cm of excess rainfall from the given unit hydrograph and then we have to compute the DRH of 3 cm of excess rainfall from the given unit hydrograph as explain by the above sample calculation. Then we have to lag the ordinates of 3 cm excess rainfall by 6-hour since the 3 cm excess rainfall occurs after 2 cm excess rainfall. These lagged values are shown in column (4) of Table 6.1 for the sample calculation. Values of ordinates of 2 and 3 cm DRH are shown in proper sequences in columns (3) and (4) of Table 6.1. Using the method of superposition, the ordinates of resulting DRH are obtained by combining the ordinates of 2 and 3 cm DRH at any instant (column (5) of Table 6.1. Adding constant base flow of 5 m³/sec, we get the ordinates of total flood hydrograph (col. 6) of Table 6.1. Fig. 6.12 shows the 2 cm, 3 cm and composite 5 cm (2 + 3 = 5 cm) excess rainfall DRHs obtained by method of superposition.

Table 6.1: Calculation of DRH by method of superposition (Sample calculation 6.4)

Time, hr	Ordinates of 6-hr UH, m³/sec	Ordinates of 2 cm DRH, m³/sec	Ordinates of 3 cm DRH, m³/sec (lagged by 6 hr)	Ordinates of 5 cm DRH, m³/sec 5: (col.5 = col.3 +col. 4)00	Ordinates of flood hydrograph, m³/sec	Remarks
1	2	3	4	5	6	
0	0	0	0	0	5	
3	22	44	0	44	49	
6	48	96	0	96	101	
9	82	164	66	230	235	
12	120	240	144	384	389	
15	150	300	246	546	551	
18	180	360	360	720	725	
(21)	(165)	(330)	450	(780)	(785)	Interpolated value
24	150	300	540	840	845	
30	100	200	450	650	655	
36	60	120	300	420	425	
42	30	60	180	240	245	
48	20	40	90	130	135	
54	10	20	60	80	85	
60	4	8	30	38	43	
(66)	(1.33)	(2.66)	(12)	(14.66)	(19.66)	Interpolated value
69	0	0	(8)	(8)	(13)	Interpolated value
75	0	0	0	0	5	

N.B: (i) The entries in col. (4) are shifted by 6 hr in time relative to col. (2).

(ii) Due to unequal time interval of ordinates, a few entries have been interpolated to complete the table and these interpolated values are shown in parenthesis.

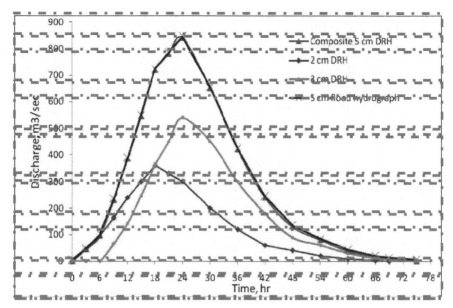

Fig. 6.12: Principle of superposition- (Sample calculation 6.4)

6.6.2 Applications of Unit Hydrograph

It is possible to develop a DRH in a catchment due to a given storm or succession of storms provided the basic principles of unit hydrograph and ordinates of a unit hydrograph for the catchment are known. For instance, we want to develop a DRH of D-hr duration and we know the D-hr unit hydrograph of the catchment and storm hyetographs of the catchment are available. If the initial and infiltration losses are known, then we can calculate the effective rainfall hyetographs (ERH) from the available hyetographs as discussed earlier. The ERH is then divided into N blocks of D-hr duration each. The rainfall excess in each D-hr duration is then operated upon the unit hydrograph successively to get a number of DRH curves. The ordinates of these DRH curves are lagged upon to obtain the proper time sequence and are then summed up at each time element to obtain the needed net DRH due to the storm.

Let us consider a sequence of N blocks of excess rainfall values $R_1, R_2, R_3....R_N$ each of D-hr duration (Fig. 6.13). Let the ordinate of D-hr unit hydrograph (UH) at any instance of time (t) from the beginning be $u(t)$. The direct runoff due to excess rainfall R_1 at any time t is:

$Q_1 = R_1. u (t)$

The direct runoff due to excess rainfall R_2 at time $(t-D)$ i.e. D- hr lagging is:

$Q_2 = R_2. u (t - D)$

Thus, $Q_3 = R_3$. u $(t - 2D)$

$Q_i = R_i$. u $\{t - (i - 1)D\}$

and proceeding in the similar way

$Q_N = R_N$. u $\{t - (N - 1)D\}$

At any instance of time, t, the total direct runoff is:

$$Q_t = \sum_{i=1}^{N} Q_i = \sum_{i=1}^{N} R_i . u\{t - (i - 1) D\} \qquad (6.4)$$

After obtaining the DRH, the estimated base flow is then added to the ordinates of the DRH to get the total flood hydrograph. The following sample calculation illustrates the development of DRH and total flood hydrograph using the application of unit hydrograph theory.

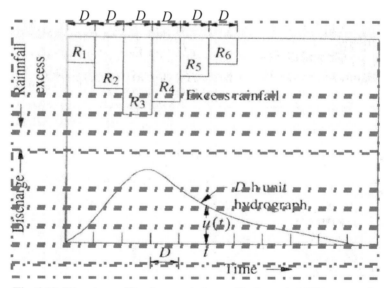

Fig. 6.13: Direct runoff hydrograph from effective rainfall hyetograph

Sample Calculation 6.5

The ordinates of a 3-hr unit hydrograph of a catchment are given below.

Time, hr	0	3	6	9	12	15	18	21	24	27	30	33	36	39	42	45
Ordinates of 6-hr UH	0	28	68	102	148	176	200	167	100	82	58	50	35	25	10	0

The base flow is constant and occurs at 20 m³/sec rate. The average rate of the storm loss is 0.25 cm/hr. Derive the flood hydrograph in the catchment due to the following storms.

Time since start of storm (hr)	0	3	6	9
Cumulative rainfall, cm	0	3.0	9.0	14.0

Solution

First we calculate the ERH from the given storm data and losses and show it in table as given below.

Time interval, Δt, hr	0 - 3	3 - 6	6 - 9
Storm loss @0.25 cm/hr for 3 hr	0.75	0.75	0.75
Depth of rainfall, cm	3.0	6.0	5.0
Excess rainfall (ER), cm	2.25	5.25	4.25

Now we compute the ordinates of DRH due to the ER of 2.25, 5.25 and 4.25 cm by superposition theorem by lagging 3 hr for each ER of 5.25 and 4.25 cm he ER of 2.25 cm (columns 4 and 5 of table as given below). Then we compute the ordinates of composite DRH by adding the ordinates of 2.25, 5.25 and 4.25 cm DRH during a given time interval (column 6 of table below). Finally add the base flow to the ordinates of composite DRH to get the flood hydrograph (column 8 of table below).

The figure of flood hydrograph along-with composite DRH is shown below (Fig. 6.14).

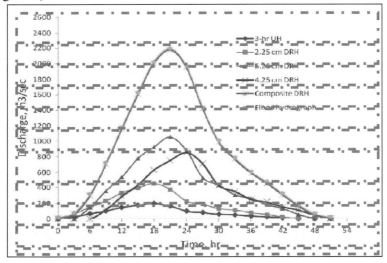

Fig. 6.14 Flood hydrograph (Sample calculation 6.5)

Time, hr	Ordinates of . . . ³/sec	Ordinates of 2.25 cm DRH, m³/sec	Ordinates of 5.25 cm DRH, m³/sec (lag 3 hr)	Ordinates of 4.25 cm DRH, m³/sec (lag by 6.0 hr)	Ordinates of composite DRH, m³/sec	Base flow, m³/sec	Ordinates of flood hydrograph, m³/sec
(1)	(2)	(3)*	(4)**	(5)***	(6)****	(7)	(8)*****
0	0	0	0	0	0	20	20
3	28	63	0	0	63	20	83
6	68	153	147	0	300	20	320
9	102	229.5	357	119	705.5	20	725.5
12	148	333	535.5	289	1157.5	20	1177.5
15	176	396	777	433.5	1606.5	20	1626.5
18	200	450	924	629	2003	20	2023
21	167	375.8	1050	748	2173.8	20	2193.8
24	100	225	876.8	850	1951.8	20	1971.8
27	82	184.5	525	709.8	1419.3	20	1439.3
30	58	130.5	430.5	425	986	20	1006
33	50	112.5	304.5	348.5	765.5	20	785.5
36	35	78.75	262.5	246.5	587.8	20	607.8
39	25	56.25	183.8	212.5	452.5	20	472.5
42	10	22.5	131.3	148.8	302.5	20	322.5
45	0	0	52.5	106.3	158.8	20	178.8
48			0	42.5	42.5	20	62.5
51				0	0	20	20

N.B. * Col. 3 = Col. 2 x 2.25, ** Col. 4 = Col. 2 x 5.25, *** Col. 5 = Col. 2 x 4.25, **** Col.6 = Col. 3 + Col. 4 + Col. 5, ***** Col. 8 = Col. 6 + Col. 7

6.6.3 Derivation of Unit Hydrograph

A number of isolated storm hydrographs resulted from short spells of rainfall excess are selected from a study of continuously gauged runoff of the stream. These isolated storm hydrographs are so selected that they durations remain almost same ranging from 0.90 to 1.1 D-hr. For each of these hydrographs, base flow is separated by adopting a suitable method as discussed earlier. The area under each DRH is computed which gives volume of runoff. The volume of runoff so computed is divided by the catchment area to give depth of ER. The ordinates of the respective DRHs are divided by the respective ER values as calculated above which gives ordinates of unit hydrograph. If flood hydrographs re used in the analysis following points must be considered.

(i)   Isolated storms occurring individually must be selected.

(ii)   The rainfall should be fairly uniform over the whole catchment and also should be fairly uniform for the entire duration.

(iii) Duration of the rainfall should be 1/5 to 1/3 of the basin lag of the catchment.

(iv)   The rainfall excess of the selected storm should be high. Preferably depth of ER of 1 to 4 cm is normally preferred.

A number of unit hydrographs of a given duration are derived by the above method and then are plotted on a common pair of axis as shown in Fig.6.15. These unit hydrographs so developed will not be same because rainfall is spatio-temporally variable. Hence, as suggested by Subramanya (2003), a mean value of all the developed curves of unit hydrographs is adopted as the unit hydrographs of a given duration for the catchment. It should be noted that the average of the peak and time to occur the peak of all the developed unit hydrographs are first considered while deriving the mean unit hydrograph curve. Next a best fit mean curve (judged by eye) is drawn through the averaged peak to close on an average base length. The volume of the DRH is estimated and any departure from the unity is corrected by adjusting the value of the ERH of unit depth is drawn in the plot of the unit hydrograph to show the type and duration of rainfall causing the unit hydrograph. It is to be remembered that the duration of unit hydrograph should not exceed 1/5 to 1/3 basin lag. For catchment of sizes more than 250 km^2, a 6-hr duration unit hydrograph is generally satisfactory.

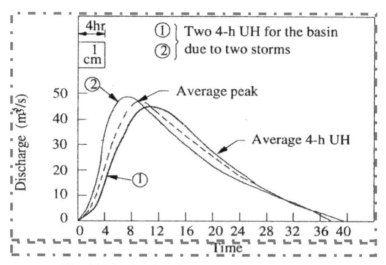

Fig. 6.15: Derivation of an average unit hydrograph

Sample Calculation 6.6

Derive the ordinates of a 3-hr unit hydrograph for a catchment of size 240 km². The discharge measured at the catchment outlet due to 3-hr isolated storm is given below.

Time, hr	-3	0	3	6	9	12	15	18	21	24	27
Discharge, m³/sec	9	9	10	35	50	74	112	110	100	76	65
Time, hr	30	33	36	39	42	45	48	51	54	57	60
Discharge, m³/sec	60	55	50	45	40	30	28	25	15	9	9

Solution

The storm hydrograph is plotted to scale as shown in Fig. 6.16. By inspection of the storm hydrograph (Fig. 6.16), we find that the peak of DRH occurs at time, t = 15 hr. The base flow is constant and its value is 9 m³/sec. The beginning of DRH occurs at time, t = 0 and ends at time, t = 57 hr.

Now, $N = 57 - 0 = 57$ hr = 2.375 days

Again, $N = 0.83\ A^{0.2} = 0.83 \times (240)^{0.2} = 2.48$ days

In this sample calculation, let us consider $N = 57$ hours for convenience. This means we assume the direct runoff ends at time $t = 57$ hr. A straight line is drawn joining the beginning and end point of DRH as shown in Fig. 6.16 which shows the base line separation. The ordinates of DRH are calculated by subtracting base flow from the ordinates of storm hydrograph. The calculations are shown in tabular form as given below.

Time, hr	Ordinates of hydrograph, m³/sec	Base flow, m³/sec	Ordinates of DRH, m³/sec	Ordinates of 3-hr UH, m³/sec
(1)	(2)	(3)	(4)	(5)
-3	9	9	0	0
0	9	9	0	0
3	10	9	1	0.27
6	35	9	26	7.03
9	50	9	41	11.08
12	74	9	65	17.56
15	112	9	103	27.84
18	110	9	101	27.30
21	100	9	91	24.60
24	76	9	67	18.11
27	65	9	56	15.13
30	60	9	51	13.78
33	55	9	46	12.43
36	50	9	41	11.08
39	45	9	36	9.73
42	40	9	31	8.38
45	30	9	21	5.68
48	28	9	19	5.14
51	25	9	16	4.32
54	15	9	6	1.62
57	9	9	0	0
60	9	9	0	0

Volume of DRH is then calculated by the formula:

Volume of DRH = 60 x 60 x 3 x (sum of the ordinates of DRH)

$$= 60 \times 60 \times 3 \times (818) = 8834400 \ m^3$$

Area of the catchment = 240 km² = 240000000 m²

Depth of ER = 8834400/240000000 = 0.037 m = 3.7 cm

Ordinates of 3-hr unit hydrograph is calculated by dividing the ordinates of DRH (column 4 in table below) by depth of ER i.e. 3.7. Column 5 of table gives the ordinates of 3-hr unit hydrograph. Fig. 6.16 shows the ordinates of 3-hr UH.

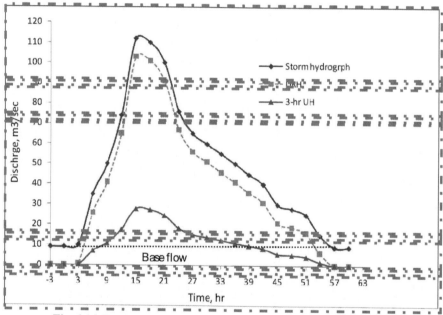

Fig. 6.16: Derivation of unit hydrograph (Sample calculation 6.6)

Sample Calculation 6.7

The peak of flood hydrograph due to a 3-hr duration isolated storm in a catchment is 250 m³/sec. Total depth of rainfall is 6 cm. Assume average loss of infiltration as 0.3 cm/hr. The base flow is assumed to be constant at the rate of 15 m³/sec. Compute the peak of the 3-hr unit hydrograph of the catchment. If the catchment has an area of 500 km², estimate the base width of the 3hr unit hydrograph by assuming it to have triangular shape.

Solution

Duration of excess rainfall (ER) = 3 hr

Total depth of rainfall = 6 cm

Average loss of infiltration = 0.3 cm/hr and in 3 hr, the loss = 0.9 cm

Hence ER = 6 – 0.9 = 5.1 cm

Peak flow of flood hydrograph = 250 m^3/sec

Base flow = 15 m^3/sec

Peak of DRH = 250 – 15 = 235 m^3/sec

Peak of 3-hr unit hydrograph = Peak of DRH/ ER = 235/5.1 = 46.08 m^3/sec

Now area under unit hydrograph = volume of 1 cm depth of ER over the whole catchment area

If we assume the base width of the unit hydrograph as B hours and we have computed earlier the peak of unit hydrograph as 46.08 m^3/sec, then for a triangular hydrograph, area under it is 0.5 x B x 60 x 60 x 46.08 = 82944 B m^3.

Again volume of 1 cm ER over a catchment area of 500 km^2 = 0.01 x 500, 000000 = 5000000 m^3

Thus, 82944 B = 5000000 and hence B = 60.3 hr H" 60 hr.

The base width of the unit hydrograph is 60 hr.

6.6.4 Development of Unit Hydrograph from Complex Storm

Where simple hydrograph from isolated unit storm are not available for derivation of unit hydrograph, a complex hydrograph resulting from a storm of various intensities is studied. In case of two peaked complex hydrographs resulting from sufficiently separated storms, the two storms are separated by the use of a direct runoff depletion curve and base flow separated. The unit hydrographs are then developed from each storm. However, if the period of rainfall is extended for a long period such that the hydrographs cannot be separated into portions contributed by each rain storm, the complex hydrographs are analyzed by the method of successive approximation by Collin's method (Sharma, 1987).

Let us consider a rainfall excess made up of three consecutive duration of D-hr and ER values of R_1, R_2 and R_3. Fig. 6.17 shows the ERH. By base flow separation of the resulting composite flood hydrograph, a composite RH is obtained. Let the ordinates of the composite DRH be drawn at a time interval of D hr. At various time intervals *1D, 2D, 3D* ….from the start of ERH, let the

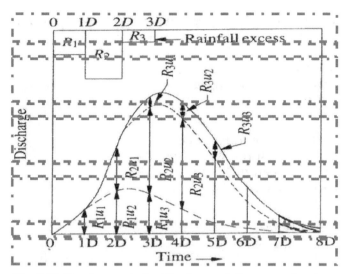

Fig. 6.17: Development of unit hydrograph from complex storm

ordinates of the unit hydrograph be u_1, u_2, u_3... and let the ordinates of composite DRH be Q_1, Q_2, Q_3....

Then $Q_1 = R_1 . u_1$

$Q_2 = R_1 . u_2 + R_2 . u_1$

$Q_3 = R_1 . u_3 + R_2 . u_2 + R_3 . u_1$

$Q_4 = R_1 . u_4 + R_2 . u_3 + R_3 . u_2$ (6.5)

and so on.....

With known values of Q_1, Q_2, Q_3.... and R_1, R_2 and R_3........ and using Eqn. (6.5), we can find out the values of u_1, u_2, u_3......There is a limitation of this method. The derived unit hydrograph can show erratic variation and even may give negative values of ordinates. This kind of variation is due to non-linearity in the effective rainfall-direct runoff relationship in the catchment. It may be time consuming to solve Eqn. (6.5), to get the ordinates of unit hydrograph. However, matrix method with optimization techniques are now available to solve Eqn. (6.5) using a high speed digital computer.

The matrix formation of Eqn. (6.5) of a unit hydrograph can be done as:

$$
\begin{bmatrix}
R_1 & 0 & 0 & \cdots\cdots 0 & 0 & \cdots\cdots 0 & 0 \\
R_2 & R_1 & 0 & \cdots\cdot 0 & 0 & \cdots\cdots 0 & 0 \\
R_3 & R_2 & R_1 & \cdots\cdot 0 & 0 & \cdots\cdots 0 & 0 \\
\cdot & & & & & & \\
\cdot & & & & & & \\
\cdot & & & & & & \\
R_m & R_{m-1} & \cdots\cdots R_1 & 0 & \cdots\cdot 0 & 0 & \\
0 & R_m & R_{m-1} & \cdots R_2 & R_1 & \cdots\cdot 0 & 0 \\
\cdot & & & & & & \\
\cdot & & & & & & \\
0 & 0 & 0 & \cdots\cdot 0 & 0 & \cdots R_m & R_{m-1} \\
0 & 0 & 0 & \cdots\cdot 0 & 0 & \cdots\cdot 0 & R_m
\end{bmatrix}
\begin{bmatrix}
u_1 \\
u_2 \\
u_3 \\
u_4 \\
\cdot \\
\cdot \\
\cdot \\
u_{n-m+1}
\end{bmatrix}
=
\begin{bmatrix}
Q_1 \\
Q_2 \\
Q_3 \\
Q_4 \\
\cdot \\
Q_m \\
Q_{m+1} \\
\cdot \\
\cdot \\
Q_{n-1} \\
Q_n
\end{bmatrix}
$$

or

$$[R] \cdot [u] = [Q] \tag{6.6}$$

6.6.5 Derivation of Unit Hydrograph for Different Durations

Generally unit hydrographs are derived from simple isolated storms and if the durations of different storms are close to each other with a deviation of 20% D, they are grouped together under one average duration of D hr. But in practical applications, if unit hydrographs of various durations are required, they are derived from field data. If there is no adequate data available, then it becomes difficult to develop unit hydrographs covering a wide range of durations for a given catchment. Under such cases, a D-hr unit hydrograph is used to develop unit hydrographs of differing durations nD where n is the integer multiple i.e. $n = 1, 2, 3, 4.....$ Following two methods are available for this purpose.

(i) S-Hydrograph (S-Curve) method

(ii) Superposition method

6.6.5.1 Superposition Method

This method is used when a unit hydrograph for the duration as multiple of the duration of given unit hydrograph is required to develop. In other words, if a D-

hr unit hydrograph is available, it is possible to derive a nD-hr unit hydrograph by the method of superposition. Method of superposition mainly involves the superposition of n unit hydrographs with each graph separated from the previous ones by D-hr. Following steps are needed for this method:

(i)   Consider a D-hr (say 4-hr) unit hydrograph of a catchment.

(ii) Suppose we are interested to derive a 12 hr unit hydrograph. Here $n = 3$.

(iii) Take 3 4-hr unit hydrographs (say A, B and C) such that they differ from each other by 4 hours i.e. curve B begins 4 hours after A and curve C begins 4 hours after B or 8 hours after A (Fig. 6.18).

(iv) Add the ordinates of all these three unit hydrographs (A, B and C).The obtained ordinates are the ordinates of DRH for 3 cm ER of 12-hr duration.

(v)   Divide the ordinates of this composite DRH by 3 which give the ordinates of 12-hr unit hydrograph.

Following sample calculation helps in understanding how to derive a nD unit hydrograph from a D-hr unit hydrograph.

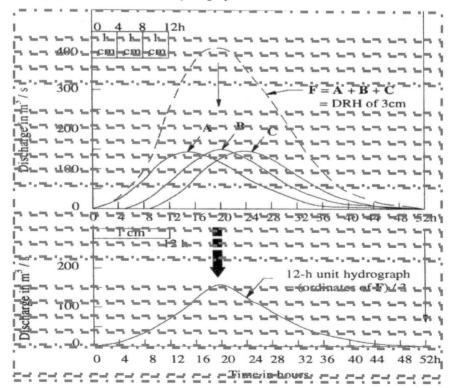

Fig. 6.18: Construction of 12-hr unit hydrograph from 4-hr unit hydrograph

Sample Calculation 6.8

Derive a 9-hr unit hydrograph for a catchment whose 3 hr unit hydrograph ordinates are as follows:

Time, hr	0	3	6	9	12	15	18	21	24	27	30	33	36	39	42
Ordinates of 3-hr UH	0	30	72	106	150	180	210	172	90	76	50	35	21	10	0

Solution

The given ordinates of 3- hr UH, are lagged by 3 hours each for 2 times as shown in columns (3) and (4) of table as given below. Then we add the ordinates of UHs i.e. col. (2) + col. (3) + col. (4) to get col.(5) which gives the ordinates of 3 cm ER DRH. Finally, the values of col. (5) are divided by 3 to get the ordinates of 9-hr UH. Solution of above sample calculation is explained in the tabular form as mentioned below.

Fig. 6.19 shows the derivation of 9-hr UH from 3-hr UH by method of superposition.

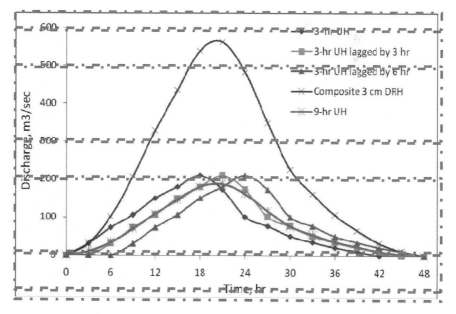

Fig. 6.19: Derivation of 9-hr UH from 3-hr UH (Sample calculation 6.8)

Time, hr	Ordinates of 3-hr UH, m³/sec	Ordinates of 3-hr UH lagged by 3 hr, m³/sec	Ordinates of 3-hr UH lagged by 6 hr, m³/sec	Ordinates of 3 cm DRH, m³/sec	Ordinates of 9-hr UH, m³/sec
(1)	(2)	(3)	(4)	(5)	(6)
0	0	0	0	0	0
3	30	0	0	30	10
6	72	30	0	102	34
9	106	72	30	208	69.33
12	150	106	72	328	109.33
15	180	150	106	436	145.33
18	210	180	150	540	180
21	172	210	180	562	187.33
24	100	172	210	482	160.67
27	76	100	172	348	116
30	50	76	100	226	75.33
33	35	50	76	161	53.67
36	21	35	50	106	35.33
39	10	21	35	66	22
42	0	10	21	31	10.33
45		0	10	10	3.33
48			0	0	0

6.6.5.2 S-Curve Method

S-hydrograph is a graph showing the summation of the ordinates of a series of unit hydrographs spaced at unit rainfall duration intervals. It represents the hydrograph of average rate of effective rainfall of the unit duration continued indefinitely. The method is very much suitable when uniform spells of rainfall have been recorded at greater time interval than the unit duration applicable to the catchment. However, it is not so suitable for smaller catchments where rainfall is reordered at shorter intervals not exceeding the unit duration.

It has advantage over the superposition method is that it can compute the unit hydrograph for the fraction of duration of rainfall excess of a given unit hydrograph. For example if we want to develop a unit hydrograph of duration mD, where m is a fraction having rational value, superposition method is not suitable but S-hydrograph method is suitable.

The S-hydrograph also called as S-curve or summation curve is, essentially, a hydrograph produced by a continuous effective rainfall at a constant rate for an indefinite period. It is a curve obtained by summation of an infinite series of D-hr unit hydrographs each separated from preceding curve by D-hr (Fig. 6.20).

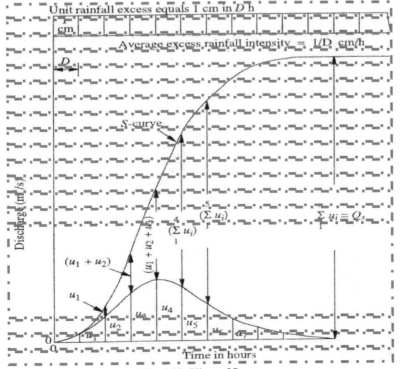

Fig. 6.20: View of S-curve

The ordinate of S-curve at any time is the sum of all the ordinates of different D-hr unit hydrographs occurring at that time. The smooth plotting of the ordinates on a linear graph paper gives a S shaped curve called as S-curve. Fig. 6.20 shows the view of a S-curve.

The S-curve has an initial steep portion which attains maximum equilibrium discharge at a time equal to the time base of the first unit hydrograph. The average intensity of effective rainfall of S-curve is given as:

$$I_{aer} = 1/D \text{ cm/hr} \tag{6.7}$$

The equilibrium discharge (which represents maximum discharge rate at the rainfall intensity of $1/D$ cm/hr) of S-curve is given as:

$$Q_{eq} = \left(\frac{A}{D} \times 10^4 \right) m^3 / hr \tag{6.8}$$

where, Q_{eq} is the equilibrium discharge, m³/hr; A is the catchment area, km² and D is the duration of effective rainfall of the given unit hydrograph, hr.

Eqn. (6.8) can also be written as:

$$Q_{eq} = 2.778 \frac{A}{D} m^3 / sec \tag{6.9}$$

where, A is the catchment area, km² and D is the duration of effective rainfall of the given unit hydrograph, hr.

In actual practical condition, S-curve is found to oscillate at the top around the equilibrium value. This is mainly due to magnification and accumulation of errors in the hydrograph. To rectify this nature of S-curve, an average smooth curve is drawn in such a way that it attains equilibrium discharge at the time base of the unit hydrograph. If 2 D-hr S-curves i.e. A and B are lagged by T-hr, then difference of the ordinates of both the S-curves results DRH of the excess rainfall for T-hr duration. The magnitude of ER is equal to $(1/D) \times T$ cm where $1/D$ is the rainfall intensity, cm/hr. The difference of ordinates of 2 S-curves (i.e. $S_A - S_B$) divided by rainfall excess (i.e. T/D) computes the ordinates of T-hr unit hydrograph.

Following sample calculation illustrates the use of S-curve technique to derive unit hydrograph.

Sample Calculation 6.9

Solve Sample calculation 6.8 by S-curve method.

Solution: Ordinates of 3-hr UH are shown in column 2 of table below. Col. 3 gives the s-curve addition and Col. 4 gives the s-curve ordinates At $t = 3$ hr, the

Time, hr	Ordinates of 3-hr UH, m³/sec	S-curve addition, m³/sec	S-curve ordinate, m³/sec	S-curve lagged by 9 hr, m³/sec	Col 4 - Col 5	$\frac{\text{Col }6}{3} \times 9 =$ 9-hr UH ordinates, m³/sec
(1)	(2)	(3)	(4) = (2) + (3)	(5)	(6)	(7)
0	0	-	0	-	0	0
3	30	0	30	-	30	10
6	72	30	102	-	102	34
9	106	102	208	0	208	69.33
12	150	208	358	30	328	109.33
15	180	358	538	102	436	145.33
18	210	538	748	208	540	180
21	172	748	920	358	562	187.33
24	100	920	1020	538	482	160.67
27	76	1020	1096	748	348	116
30	50	1096	1146	920	226	75.33
33	35	1146	1181	1020	161	53.67
36	21	1181	1202	1096	106	35.33
39	10	1202	1212	1146	66	22
42	0	1212	1212	1181	31	10.33
45		1212	1212	1202	10	3.33
48			1212	1212	0	0

ordinates of 3-hr UH = ordinates of the S-curve. This value becomes the S-curve addition at $t = 2 \times 3 = 6$ hr. At this $t = 6$ hr, the ordinate of UH (72) + S-curve addition (30) = S-curve ordinate (102). The S-curve addition at $t = 3 \times 3$ = 9 hr is 102, and so on. Col. 5 shows the S-curve lagged by 9 hr. Col. 6 gives the subtraction of lagged S-curve (col. 5) from the S-curve (col. 4). Ordinates given in col. 6 are divided by $T/D = 9/3 = 3$ to obtain the ordinates of the 9-hr UH which is shown in col. 7 of the table.

Sample Calculation 6.10

Ordinates of 4-hr UH are given as below. Using this, derive the ordinates of 2-hr UH by S-curve method.

Time, hr	0	4	8	12	16	20	24	28	32	36	40	44
Ordinates of 4-hr UH, m3/sec	0	20	80	120	150	130	100	50	25	15	6	0

Solution

The ordinates of 4-hr UH are given and we have to find out the ordinates of 2-hr UH. So in this case, the time interval of the ordinates of the given unit hydrograph should be 2 hr. But the ordinates are given at 4-hr interval. So we have to plot the 4 hr UH and find out the ordinates at 2 hr interval. The ordinates at 2 hr interval are shown in column 2 of the following table (plot of 4-hr UH is not shown in this sample calculation). The S-curve addition and the S-curve ordinates are shown in columns 3 and 4, respectively. First the S-curve ordinates corresponding to the time interval equal to successive durations of the given unit hydrograph (in this case at 0, 4, 8…hr) are determined by following the method of superposition. Next, the ordinates at intermediate intervals (i.e. at t= 2,6, 10…hr) are determined by having another series of S-curve addition. To obtain a 2-hr UH, the S-curve is lagged by 2 hr (column 5) and then this is subtracted from column 4. The results are mentioned in column 6. The ordinates of column 6 are finally divided by $T/D = 2/4 = 0.5$ to get the ordinates of 2-hr UH (column 7). The solution of the above sample calculation is given in tabular form as below.

Time, hr	Ordinates of 4-hr UH, m³/sec	S-curve addition, m³/sec	S-curve ordinate, m³/sec	S-curve lagged by 2 hr, m³/sec	Col 4 - Col. 5	= 2-hr UH ordinates, m³/sec
(1)	(2)	(3)	(4) = (2) + (3)	(5)	(6)	(7)
0	0	-	0	-	0	0
2	8	-	8	0	8	16
4	20	0	20	8	12	24
6	48	8	56	20	36	72
8	80	20	100	56	44	88
10	100	56	156	100	56	112
12	120	100	220	156	64	128
14	138	156	294	220	74	148
16	150	220	370	294	76	152
18	142	294	436	370	66	132
20	130	370	500	436	64	128
22	115	436	551	500	51	102
24	100	500	600	551	49	98
26	75	551	626	600	26	52
28	50	600	650	626	24	48
30	35	626	661	650	11	(22) 30
32	25	650	675	661	14	28
34	19	661	680	675	5	10
36	15	675	690	680	10	(20) 8
38	11	680	691	690	1	(2) 4
40	6	690	696	691	5	(10) 2
42	4	691	695	696	-1	(-2) 0
44	0	696	696	695	1	(2) 0

N.B. Final adjusted values are given in column 7 and unadjusted values are given in parenthesis in column 7.

Sometimes it is observed that there are errors in interpolation of unit hydrograph ordinates which often results in oscillation of S-curve at the equilibrium value. This results in the derived T-hr UH having abnormal sequences of discharges and sometimes we may get negative values at the tail end. In the above examples we get -2 as the value of discharge at 42 hr. This is adjusted by fairing the S-curve and also the resulting T-hr UH by smooth curves. The abnormal values which are shown in parenthesis in column 7 are adjusted.

6.6.6 Use and Limitation of Unit Hydrograph

6.6.6.1 Use of Unit Hydrograph

Unit hydrographs are used for the following purposes:

(i)   Development of flood hydrograph for extreme rainfall magnitudes which are used in the design of hydraulic structures such as culverts, piers and bridges.

(ii)   Extension of flood-flow records based on rainfall records and

(iii) Development of flood forecasting and warning systems in a catchment based on rainfall.

6.6.6.2 Limitations of Unit Hydrograph

In the study of unit hydrograph, it is assumed that there is uniform distribution of rainfall over the whole catchment. The intensity of rainfall is also assumed to be constant for the duration of rainfall excess. However, in practice, the above two assumptions do not remain valid. In a catchment, there is variation of rainfall in distribution aerially and also the intensity of rainfall within the storm varies. However, if the areal distribution is constant between different storms, the unit hydrograph can still be used. But one should remember that the upper limit of catchment size plays a crucial role in such case. The upper limit is considered as 5000 km² for the application of unit hydrograph. For larger size basins having area more than 5000 km², it can be divided into a number of sub-catchments and for each sub-catchment, DRHs are to be developed by the UH method. These DRHs may be routed through their respective channels to obtain the composite DRH at the basin outlet. The unit hydrographs have also some limitations in application with respect to smaller sizes of catchments. The lower limit is 2 km² or 200 ha. As the size is less than this value, the overland flow dominates and the relationship between rainfall and runoff is affected by a number of factors. The UH developed from the records of rainfall –runoff is not accurate enough for the prediction of DRHs.

In addition to the above limitations, other limitations are:

(i) The precipitation should always be in the form of rainfall. Snow-melt runoff cannot satisfactorily be used in representation of UH.

(ii) The catchment should not have large size storages such as tanks, ponds etc. which affect the relationship between the storage and discharge.

(iii) A decidedly non-uniform precipitation results in-accurate UH.

It is to be noted that in the use of UH, very accurate results are not expected. Variations in peak discharge by ± 10% and in time base by ± 20% are accepted.

6.7 Distribution Graph

Bernard (1935) first introduced the concept of distribution graph. It is a special type of unit hydrograph. Distribution graph is basically a D-hr UH with ordinates showing the percentage of surface runoff occurring in successive periods of equal time intervals of D hr. The duration of rainfall excess i.e. D hr is considered as the unit interval and distribution-graph ordinates are indicated at successive such unit intervals. Fig. 6.21 shows a typical 4-hr distribution graph. It is to be noted that the total area under the distribution graph adds up to 100%. Distribution graphs are used to compare the characteristics of runoff of various catchments.

Let P_1, P_2, P_3 ..... are the volume of rainfall excess in successive unit storm periods and u_1, u_2, u_3... are the percentage of the distribution graph. Then discharge volumes Q_1, Q_2, Q_3... of each unit storm are given as:

$$Q_1 = u_1 P_1$$

$$Q_2 = u_2 P_1 + u_1 P_2$$

$$Q_3 = u_3 P_1 + u_2 P_2 + u_1 P_3$$

$$Q_4 = u_4 P_1 + u_3 P_2 + u_2 P_3 + u_1 P_4$$

$\cdot \cdot \cdot \cdot \cdot \cdot \cdot \cdot$

$\cdot \cdot \cdot \cdot \cdot \cdot \cdot \cdot$

$\cdot \cdot \cdot \cdot \cdot \cdot \cdot \cdot$

The general expression is written as:

$$Q_n = \sum_{i=1}^{i=n} u_i P_{n-(i-1)} \tag{6.10}$$

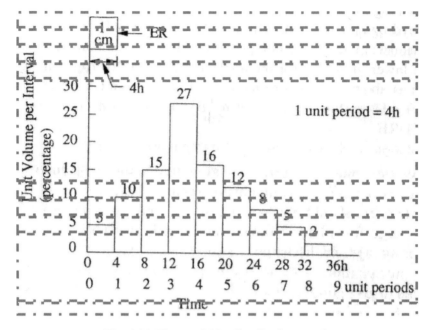

Fig. 6.21: Figure of 4-hr distribution graph

Use of distribution graph to generate a DRH is explained in sample calculation as given below.

Sample Calculation 6.11

A catchment of area 300 ha receives rainfall of 7.0, 2.0 and 5.0 cm in three consecutive days. The average losses in all these three days can be assumed as 2.0 cm/day. Distribution graph percentages of the surface runoff that extended over 6 days for every rainfall of 1 day duration are 5, 15, 40, 25, 10 and 5. Determine the ordinates of the discharge hydrograph. Neglect base flow.

Solution

Solution of above sample calculation 6.11 is given in tabular form as follows.

Time interval, days	Rainfall, cm	Losses,— cm/day	Effective rainfall, cm	Av. Distribution ratio (%)	Distributed runoff (cm) for effective rainfall of		Runoff cm m³/sec (x 10⁻²)
					5 cm	3 cm	
0-1	7.0	2.0	5.0	5	0.250	0	0.250 8.68
1-2	2.0	2.0	0	15	0.750	0	0.750 26.04
2-3	5.0	2.0	3.0	40	2.000	0.150	2.150 74.65
3-4				25	1.250	0.450	1.700 59.03
4-5				10	0.500	1.200	1.700 59.03
5-6				5	0.250	0.750	1.000 34.72
6-7				0	0	0.300	0.300 10.42
7-8						0.150	0.150 5.21
8-9						0	0 0

6.8 Synthetic Unit Hydrograph

In order to develop a unit hydrograph, we require data of rainfall and data of the resulting flood hydrograph. But these data are not available in some catchments. Catchments located at the remote places may not have this facility. Development of unit hydrographs for such areas including ungauged catchments requires empirical equations of regional validity that relate salient hydrograph characteristics to the basin characteristics or morphological characteristics. Unit hydrographs developed by such relationship is called as synthetic unit hydrograph. A number of methods are available for development of synthetic unit hydrograph. All these methods are based on empirical correlation and are meant for regional basis in which they were developed. They cannot be considered as general use for all regions.

6.8.1 Snyder's Method

The synthetic unit hydrograph (SUHG) named after Snyder (1938) is called as Snyder's synthetic unit hydrograph method. It is applicable for watersheds ranging from 30 to 30,000 km² in area of the watersheds. Snyder made experiment on a large number of watersheds having area range as mentioned above in Appalachian highlands of USA and developed a set of empirical equations for synthetic unit hydrographs in those areas. These equations were meant for use for USA condition and with some modifications; they are now being used in other countries and constitute the methodology called as Snyder's synthetic unit hydrograph. The Snyder's method consists of the following sets of relationships for computing different components of SUGH based on morphological characteristics of the watersheds

1. Basin Lag (t_p): It is the most important characteristic that affect a hydrograph. Basin lag refers to the time interval between the mid-point of unit effective rainfall and peak of the unit hydrograph. Basin lag is sometimes referred to lag time. Physically, it refers to the mean time of travel of water from all parts of a catchment to the outlet of the catchment during a given storm. It is determined based on the morphological characteristics of the watershed like size of the watershed, length of the watershed, stream grade, stream frequency and drainage density of the watershed etc. However, for its determination, a few of the important watershed characteristics are required. It is computed as:

$$t_p = C_t \left(L \cdot L_{ca} \right)^{0.30} \tag{6.11}$$

where, t_p = basin lag (hr); C_t = regional constant which is a function of storage and slope of the watershed; L = basin length measured along the water course from the basin divide to the gauging station, km and L_{ca} = distance along the main water course from the gauging station to a point opposite to the watershed centroid, km. Value of C_t in Snyder's study ranged from 1.35 to 1.65. However, studies by many investigators have shown that C_t depends on the region under study and wide variations with value of C_t ranging from 0.30 to 6.0 have been reported (Sokolov, 1976). Linsley (1958) found that t_p is better correlated with the catchment parameter $\left(\dfrac{L \cdot L_{ca}}{\sqrt{S}}\right)$ where S is the basin slope. Hence, a modified form of Eqn. (6.11) to compute t_p is given as:

$$t_p = C_{tL}\left(\frac{L \cdot L_{ca}}{\sqrt{S}}\right)^n \tag{6.12}$$

where, C_{tL} and n are basin constants. For USA condition, value of n was found by them as 0.38 and that of C_{tL} was 1.715, 1.03 and 0.50 for mountainous drainage areas, foot hill drainage areas and valley drainage areas, respectively.

2. Standard Duration of Effective Rainfall (t_r): It is required to determine the peak discharge for derivation of synthetic unit hydrograph. It is computed as:

$$t_r = \frac{t_p}{5.5} \tag{6.13}$$

where both t_p and t_r are in hr.

3. Peak Discharge, Q_{ps}: Peak discharge, Q_{ps} for standard duration t_r is given as:

$$Q_{ps} = \frac{2.78\, C_p \cdot A}{t_p} \tag{6.14}$$

where, A is catchment area in km²; C_p is regional constant which depends on the retention and storage capacities of the watershed. The value of C_p ranges from 0.56 to 0.69 for Snyder's study areas and is considered as an indication of the retention and storage capacity of the watershed. Like C_t, value of C_p also varies quite considerably depending on the characteristics of the area. Values of C_p have been reported to range from 0.31 to 0.93.

4. Non-standard Duration of Effective Rainfall: It is the duration of effective rainfall other than standard duration and is given by symbol, t_R. It is basically the modified formula of basin lag. It is computed as:

$$t_{pr} = t_p + \left(\frac{t_R - t_r}{4}\right) \tag{6.15}$$

Using Eqns. (6.13) and (6.15) we get

$$t_{pr} = \frac{21}{22} t_p + \frac{t_R}{4} \tag{6.16}$$

t_{pr} is basin lag in hours for an effective duration of t_R hr and t_p is given by Eqn. (6.11) or (6.12). In computation of peak discharge by Eqn. (6.14), one has to use t_{pr} instead of t_p. Thus, the peak discharge for a non-standard duration of effective rainfall of duration t_R is given as:

$$Q_{ps} = \frac{2.78 \, C_p \cdot A}{t_{pr}} \tag{6.17}$$

When $t_R = t_p$ we have $Q_p = Q_{ps}$.

5. Time Base of Unit Hydrograph (t_b): It refers to the duration of direct surface runoff. It is given by Snyder as:

$t_b = 3 + t_p/8$ (days)

$\quad = 72 + 3 \, t_p$ (hour) for non-standard duration $\tag{6.18}$

$\quad = 3 + t_{pr}/8$ (days)

$\quad = 72 + 3 \, t_{pr}$ (hour) for standard duration $\tag{6.19}$

Eqns. (6.18) and (6.19) give reasonable estimate time base for large catchemnts. But for small catchments, they yield very high values. Butler (1957) has recommended

$$t_b = 5\left(t_{pr} + \frac{t_R}{2} \right) \text{hour} \tag{6.20}$$

Value of t_b is assumed three to five times the time to peak (Mutreja, 1986).

6. Shape of Unit Hydrograph: After determining the values of basin lag (t_p), standard or non-standard duration of effective rainfall (t_r or t_{pr}), peak discharge for standard or non-standard duration of effective rainfall (Q_p or Q_{ps}), and time base (t_b), sketching of synthetic unit hydrograph can be done. To assist in sketching of the unit hydrograph, width of unit hydrographs at 50 and 75% of the peak discharge rate are considered by the formulae as:

$$W_{50} = \frac{5.87}{q^{1.08}}$$ (6.21)

and $W_{75} = \frac{W_{50}}{1.75}$ (6.22)

where, W_{50} is width of unit hydrograph in hr at peak discharge of 50%, W_{75} is width of unit hydrograph in hr at peak discharge of 75% and q is peak discharge per unit catchment area (m³/sec/km²) = Q_p/A.

Snyder's synthetic unit hydrograph is shown in Fig. 6.22.

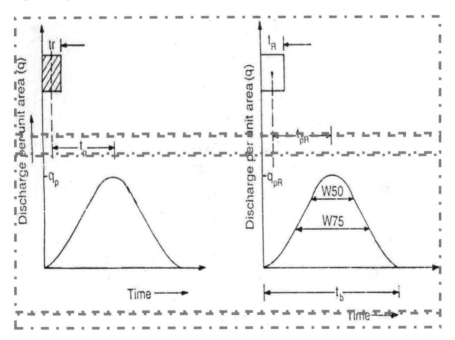

Fig. 6.22: Snyder's synthetic unit hydrograph

Sample Calculation 6.12

Two catchments A and B which are hydro-meteorologically similar have the following characteristics.

Catchment A	Catchment B
L = basin length measured along the water course from the basin divide to the gauging station = 32 km L_{ca} = distance along the main water course from the gauging station to a point opposite to the watershed centroid = 15 km A = Area of catchment = 240 km²	L = basin length measured along the water course from the basin divide to the gauging station = 44 km L_{ca} = distance along the main water course from the gauging station to a point opposite to the watershed centroid = 24 km A = Area of catchment = 400 km²

A 2 hr UH has been developed for catchment A which has a peak discharge of 45 m³/sec and time to peak from the beginning of the rainfall excess is 8 hr. Develop a UH for the catchment B using Snyder's method.

Solution

Catchment A

$t_R = 2$ hr

Time to peak from beginning of ER = 8 hr = $t_R/2 + t_{pr}$

$t_{pr} = 7$ hr

From Eqn. (6.16), $t_{pr} = \dfrac{21}{22} t_p + \dfrac{t_R}{4}$

$$7 = \dfrac{21}{22} t_p + 0.5$$

$$t_p = 6.81 \text{ hr}$$

From Eqn. (6.11), we have

$$t_p = C_t \left(L \cdot L_{ca} \right)^{0.30}$$

$6.81 = C_t . (32 \times 15)^{0.3}$

$C_t = 1.07$

From Eqn. (6.17) we have

$$Q_{ps} = \dfrac{2.78 \, C_p . A}{t_{pr}}$$

$$45 = \frac{2.78\, C_p \,.\, 240}{7}$$

$$C_p = 0.472$$

Catchment B

Using the values of $C_t = 1.07$ and $C_p = 0.472$ for catchment B, the parameters of the synthetic UH of the catchment B are computed as:

From Eqn. (6.11), we have $t_p = 1.07\,(44 \times 24)^{0.3} = 8.64$ hr

By Eqn. (6.13) we have

$$t_r = \frac{t_p}{5.5} = \frac{8.63}{5.5} = 1.57 \text{ hr}$$

For the 2 hr UH (as given in the problem), we have $t_R = 2$ hr.

Using Eqn. (6.16) we have

$$t_{pr} = \frac{21}{22} t_p + \frac{t_R}{4} = \frac{21}{22} \times 8.63 + \frac{2}{4} = 8.74 \text{ hr}$$

From Eqn. (6.17) we have

$$Q_{ps} = \frac{2.78\, C_p \,.\, A}{t_{pr}}$$

$$= \frac{2.78 \times 0.472 \times 400}{8.74} = 60.0 \text{ m}^3/\text{sec}$$

From Eqn. (6.21)

$$W_{50} = \frac{5.87}{q^{1.08}} = \frac{5.87}{(60/400)^{1.08}} = 45.54 \text{ hr} \approx 46 \text{ hr}$$

From Eqn. (6.22)

$$W_{75} = \frac{W_{50}}{1.75} = 46/1.75 = 26.3 \text{ hr H" } 26 \text{ hr}$$

Time base of the hydrograph from Eqn. (6.19) is

$$t_b = 72 + 3\, t_{pr} \text{ (hour)} = 72 + 3 \times 8.74 = 98.22 \text{ hr} \approx 99 \text{ hr}$$

By Eqn. (6.20),

$$t_b = 5\left(t_{pr} + \frac{t_R}{2}\right) \text{ hour} = 5 \times (8.74 + 0.5) = 46.2 \text{ hr} = 46 \text{ hr}$$

Considering the values of W_{50} and W_{75} and noting that the area of the catchment B is 400 km² which is less, we can assume time base of the hydrograph (t_b) as 46 hr which is more reasonable.

After obtaining the parameters, we have to sketch the UH in such a way that the area under the UH should represent unit depth of effective rainfall.

6.8.2 Finalizing of Synthetic Unit Hydrograph

After obtaining the values of Q_{ps}, t_R, W_{50}, W_{75}, t_b and t_{pr} by Snyder's method we have to sketch a tentative unit hydrograph. After that an S-curve is developed and plotted. The obtained S-curve is tentative and has kinks since the ordinates of the UH are tentative. They are, therefore, smoothened and a logical pattern of the S-curve is sketched. Using this S-curve, the t_R hour UH is then derived back. Now the area under the synthetic UH is checked which should represent unit depth of effective rainfall. If it is not equal to unit depth of effective rainfall, then repeat the procedure of adjustment through the S-curve till satisfactory results are obtained. It may be noted that out of the different parameters of synthetic UH, time base of the UH will be least accurate. However, this can be changed to meet other parameters.

6.8.3 Dimensionless Unit Hydrograph

In order to compare the unit hydrograph from basins of different sizes and shapes or those resulting from different storm patterns, they are reduced to dimensionless form. It can be derived from (i) an observed storm flow hydrograph of a flood; (ii) a unit hydrograph or (iii) a summation graph. This is a type of synthetic unit hydrograph which is developed by U.S. Soil Conservation Service in the year 1972. So it is also called as SCS dimensionless unit hydrograph. A typical dimensionless unit hydrograph developed by U.S. Soil Conservation Service is shown in Fig. 6.23 below. The ordinate of this unit hydrograph is expressed as ratio of discharge to peak discharge (Q/Q_p) and abscissa represents the ratio of time to time of peak (t/t_{pk}).

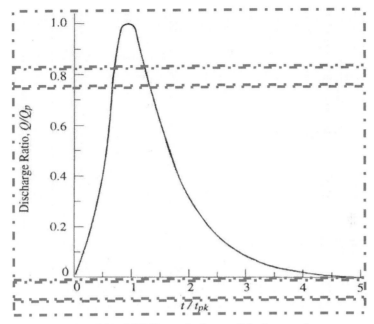

Fig. 6.23: SCS dimensionless unit hydrograph

A dimensionless unit hydrograph has the following characteristics:

- For a given peak discharge and lag time, a unit hydrograph can be developed from the synthetic dimensionless unit hydrograph for any watershed.

- For the development of unit hydrograph by using dimensionless unit hydrograph, the equations followed are (Suresh, 2008):

$$t_r = 1.33\ t_c \tag{6.23}$$

$$\text{and } Q_p = \frac{5.36\ A}{t_{pk}} \tag{6.24}$$

$$\text{where, } t_{pk} = t_{pr} + \frac{t_R}{2} \tag{6.25}$$

- When $Q/Q_p = 1$, $t/t_{pk} = 1$

The coordinates of the SCS dimensionless unit hydrograph are given in Table 6.2 which can be used in developing a synthetic unit hydrograph in place of Snyder's equations.

Table 6.2: Coordinates of SCS dimensionless unit hydrograph (Gray, 1970)

t/t_{pk}	Q/Q_p	t/t_{pk}	Q/Q_p	t/t_{pk}	Q/Q_p
0	0	1.0	1.000	2.4	0.180
0.1	0.015	1.1	0.980	2.6	0.130
0.2	0.075	1.2	0.920	2.8	0.098
0.3	0.160	1.3	0.840	3.0	0.075
0.4	0.280	1.4	0.750	3.5	0.036
0.5	0.430	1.5	0.660	4.0	0.018
0.6	0.600	1.6	0.560	4.5	0.009
0.7	0.770	1.8	0.420	5.0	0.004
0.8	0.890	2.0	0.320		
0.9	0.970	2.2	0.240		

6.8.4 The Indian Practice

After studying a large number of catchments of different sizes varying from 25 to 500 km² in India, Central Water Commission (CWC) of India has recommended the following equations for developing synthetic unit hydrograph:

The peak discharge for a D-hr UH (Q_{pd}) in m³/sec is given as:

$$Q_{pd} = 4.44 \, A^{3/4} \text{ for weighted mean slope } (S_m) > 0.0028 \tag{6.26}$$

and $Q_{pd} = 222 \, A^{3/4} \, S_m^{\,2/3}$ for weighted mean slope $(S_m) < 0.0028$ (6.27)

where, A = catchment area, km², and S_m = weighted mean slope expressed as:

$$S_m = \left[\frac{L_{ca}}{\dfrac{L_1}{\sqrt{S_1}} + \dfrac{L_2}{\sqrt{S_2}} + \dfrac{L_3}{\sqrt{S_3}} + \cdots + \dfrac{L_n}{\sqrt{S_n}}} \right]^2 \tag{6.28}$$

where, L_{ca} = distance along the main water course from the gauging station to a point opposite to the watershed centroid; $L_1, L_2, L_3 \ldots \ldots L_n$ = length of the channels 1 to n, respectively and $S_1, S_2, S_3 \ldots \ldots S_n$ = slope of the channels 1 to n, respectively.

The channel lengths and their slopes are computed from the topographic maps of the watershed.

The lag time in hours (lag time is the time interval from the mid point of rainfall excess to the peak of the hydrograph) for a 1-hr UH is given as:

$$t_{pl} = \frac{3.95}{\left(\dfrac{Q_{pd}}{A}\right)^{0.9}} \qquad (6.29)$$

where, t_{pl} = lag time in hours for a 1-hr UH, A = area of the watershed in km^2 and Q_{pd} = peak discharge for a D-hr UH in m^3/sec.

For design purpose, the duration of rainfall excess in hours is taken as:

$$D = 1.1\, t_{pl} \qquad (6.30)$$

The time to peak is determined by using Snyder's formula of basin lag (Eqns. 6.11 or Eqns. 6.12).

Eqn. (6.26) to Eqn. (6.30) help to determine the peak discharge and duration of a design unit hydrograph.

For development of synthetic UH of un-gauged basins in Indian condition, Mutreja (1986) has suggested the following set of formulae:

$$\text{Peak discharge, } Q_{pd} = 0.315\ A^{0.93}\ S^{0.53} \qquad (6.31)$$

$$\text{Basin lag, } t_p = 1.13 \cdot \frac{(L \cdot L_{ca})^{0.277}}{S^{0.5}} \qquad (6.32)$$

$$\text{Time base, } t_b = 4.3\ \frac{(L \cdot L_{ca})^{0.28}}{S^{0.5}} \qquad (6.33)$$

$$\text{Width of synthetic UH at 50\% peak } (W_{50}) = 2.18 \left(\frac{Q_{pd}}{A}\right)^{-1.12} \qquad (6.34)$$

$$\text{Width of synthetic UH at 75\% peak } (W_{75}) = 0.81\ W_{50}{}^{0.72} \qquad (6.35)$$

6.9 Instantaneous Unit Hydrograph (IUH)

For a given catchment, there may be several D-hr unit hydrographs. Depending on the values of D, the shape of these different unit hydrographs will be different. As the value of D decreases, intensity of rainfall excess (intensity of rainfall excess = $1/D$) increases which makes the shape of the unit hydrograph to become more skewed and vice versa. When D tends to zero, resulting unit hydrograph is finite which is a limiting case of unit hydrograph called as instantaneous unit hydrograph (IUH). Thus IUH is a fictitious and conceptual unit hydrograph that represents the surface runoff from the catchment due to instantaneous precipitation of the rainfall excess volume of 1 cm. IUH is designated as $u(t)$ or at times as $u(0,t)$ in which '0' refers to the duration of excess rainfall and 't' to any time. It is a single peaked hydrograph with a finite base width. The properties of IUH are:

- $0 \leq u(t) \leq a$ *positive value*   for $t > 0$;

- $u(t) = 0$ for $t \leq 0$;

- $u(t) \to 0$ *for* $t \to \infty$

- $\sum_{0}^{\infty} u(t)\, dt$ = unit depth over the catchment

- Time to peak < time to the centroid of the curve

Consider an effective rainfall of duration t_0 which is applied to a catchment as shown in Fig. 6.24. Each infinitesimal element of this ERH will operate on the IUH to produce DRH whose discharge at time t is expressed as:

$$Q(t) = \int_{0}^{t} u(t-\tau)\, I(\tau)\, d\tau \qquad\qquad (6.36)$$

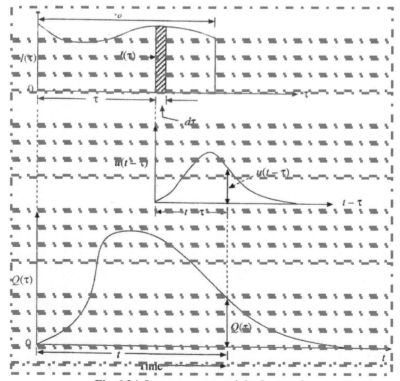

Fig. 6.24: Instantaneous unit hydrograph

where, $t^l = t$ when $t < t_0$

and $t^l = t_0$ when $t \geq t_0$

Eqn. (6.36) is called as *convolution integral or Duhamel integral* in which u $(t - \tau)$ is the corneal function, I (τ) is input with boundary condition $t' = t$ when $t < t_0$ and $t^l = t_0$ when $t \geq t_0$.

The main advantages of IUH are (i) it is independent of the duration of the effective rainfall and

thus it has one parameter less than the D-hr UH, (ii) it is suitable for theoretical analysis of excess rainfall-runoff relationship for watersheds and (iii) it is useful for analyzing the storage characteristics of the watershed since IUH is independent of rainfall characteristics.

6.9.1 Development of IUH

Following methods are used to develop IUH based on the excess rainfall and its response to direct runoff.

- Chow method
- Laplace transformation
- Harmonic analysis and
- Conceptual models

6.9.1.1 Chow Method

It was developed by V.T. Chow in the year 1962. In this method, the slope of the S-curve at any time, t gives the ordinate of IUH at that time. The available data base of the rainfall and runoff of a watershed helps in development of S-curve. This method follows the basis that the ordinate of S-curve at any time t is equal to the integration of area enclosed by IUH with the limit 0 to t.

Let us assume that for a D-hr UH S_1 represents the S-hydrograph and S_2 is another S-hydrograph with intensity of excess rainfall equals to i cm/hr. Let the time lag between these two S-hydrographs be *"dt"*. The ordinate of DRH due to excess rainfall of duration dt is the difference of the ordinates of the two S-hydrographs. The unit hydrograph of dt hours is determined by dividing the difference of S-curve or S-hydrograph ordinates with $i.dt$ in which the intensity of rainfall excess of the S-hydrograph is $1/D$ cm/hr. The ordinate of dt-hour UH is expressed as:

Ordinate of dt-hour UH $= \dfrac{S_2 - S_1}{i \cdot dt}$ (6.37)

In eqn. (6.37), if the time dt approaches zero, then the resulting ordinate gives the ordinate of IUH. At any time, t the ordinate of IUH is expressed as:

$$u(t) = \lim_{dt \to 0} \left(\frac{S_2 - S}{i \cdot dt} \right) = \frac{1}{i} \cdot \frac{dS}{dt}$$ (6.38)

If we assume the intensity of rainfall excess (i) as equals to 1 cm/hr, then Eqn. (6.38) reduces to

$$u(t) = \frac{dS'}{dt}$$ (6.39)

where, is the S-hydrograph with intensity of rainfall excess as 1 cm/hr. Thus, the ordinate of an IUH at any time t is the slope of the S-hydrograph of intensity 1 cm/hr at the corresponding time. Eqn. (6.38) can be used to derive an approximate IUH.

6.9.1.2 Laplace Transformation Method

This method is a theoretical procedure for deriving IUH. In this method, the ordinate of IUH is expressed as:

$$\bar{u}(p) = \int_0^{\infty} e^{-pt} \cdot u(t)\, dt$$ (6.40)

where, p is the variable of laplace transform in the unit of time^{-1}. In the above mentioned equation, since $u(t)$ is a bounded function of exponential order, $\bar{u}(p)$ is finite for all the positive values of p. Value of is positive because u(t) and e^{-pt} are positive functions. The term approaches zero as p increases. When $p = 0$, $= 1$ and when $p = 1$, $\dfrac{d\bar{u}(p)}{dt} = -t_L$. The term $\dfrac{d\bar{u}(p)}{dt}$ is always negative and tends to zero as p tends to ∞ The laplace transform of convolution integral as given by equation is equal to the product of laplace transforms of two functions involved in the integral i.e.

$$\bar{Q}(p) = \bar{u}(p) \cdot \bar{I}(p)$$ (6.41)

where, and are the functions of laplace transform of DRH $Q(t)$ and *ERH I(t),* respectively.

6.9.1.3 Harmonic Analysis Method

This method was first proposed by Donnel (1960). In this method, ERH, DRH and the resulting IUH are presented by three harmonic series. Harmonic coefficients for IUH are determined from ERH and DRH based on which the ordinates of IUH are determined. Details of the method are available in the reference of Donnel (1960).

6.9.1.4 Conceptual Models for development of IUH

The conceptual models are discussed in a separate chapter under the title Flood Routing in this book (Chapter 9).

6.9.2 Derivation of D-hour Unit Hydrograph from IUH

Consider Eqn. (6.39) i.e.

$$u(t) = \frac{dS'}{dt} = u(t) \cdot dt$$

$$dS' = u(t) \cdot dt \tag{6.42}$$

Integrating Eqn. (6.42) between two points

$$S_2' - S_1' = \int_{t_1}^{t_2} u(t) \, dt \tag{6.43}$$

If $u(t)$ is linear within the limit t_1 and t_2, then for small value of , the ordinate $u(t)$ can be written as:

$$u(t) = \bar{u}(t) = \frac{1}{2}\left[u(t_1) + u(t_2)\right] \tag{6.44}$$

From Eqns. (6.43) and (6.44), we have,

$$S_2' - S_1' = \frac{1}{2}\left[u(t_1) + u(t_2)\right] = dt$$

$$= \frac{1}{2}\left[u(t_1) + u(t_2)\right] [t_2 - t_1]$$

$$\frac{S_2' - S_1'}{t_2 - t_1} = \frac{1}{2}\left[u(t_1) + u(t_2)\right] \tag{6.45}$$

But the term $\dfrac{S_2' - S_1'}{t_2 - t_1}$ is the ordinate of the UH of duration D_I (i.e. $D_I = t_2 - t_1$). Thus, the general relationship between IUH and D-hr UH can be written as:

$$\left(D_1 - hr\ UH \right)_t = \frac{1}{2}\left[\left(IUH \right)_t + \left(IUH \right)_{t - D_1} \right] \tag{6.46}$$

Eqn. (6.46) reveals that if two IUHs are lagged by D_I hr duration then sum of their corresponding ordinates of IUH divided by 2 results the ordinate of D_I –hr UH. The calculated ordinates of D_I –hr UH are then converted for any D –hr UH ordinate by using S-curve method.

Following sample calculation will help to understand the method.

Sample Calculation 6.13

The ordinates of an IUH of a catchment at different time intervals are given below. If a storm of duration 4 hours having a rainfall excess of 5 cm occurs in the catchment, then find out the direct runoff hydrograph (DRH) for the catchment.

Time, hr	0	1	2	3	4	5	6	7	8	9	10	11	12
Ordinate of IUH, m³/sec	0	7	30	50	45	40	30	20	15	10	5	2	0

Solution

The calculations are performed and shown in Table 6.3

1. Using Eqn. (6.46), determine the ordinates of 1-hr UH.

   In Table 6.3, Col. 2 = ordinate of given IUH = u(t)

   Col. 3 = ordinate of IUH lagged by 1-hr

   Col. 4 = average of the values of Col. 2 and Col. 3 which is the ordinates

   of 1-hr UH by Eqn. (6.46).

2. Using 1-hr UH, the S-curve is obtained and lagging it by 4 hours, ordinates of 4-hr UH are obtained.

   In Table 6.3, Col. 5 = S-curve additions

   Col. 6 = Col. 4 + Col. 5 = S-curve ordinates

Table 6.3: Calculation of direct runoff hydrograph (DRH) from IUH of Sample calculation 6.13

Time, hr	$u(t)$, m³/sec	$u(t)$ lagged by 1-hr	Ordinate UH, m³/sec of 1-hr	S-curve addition, m³/sec	S-curve ordinate, m³/sec	S-curve lagged by 4-hr	DRH of 4 cm in 4-hrs, m³/sec	Ordinate of 4-hr UH, m³/sec	DRH of 5 cm ER in 4-hr, m³/sec
0	0	-	0	-	0	0	0	0	0
1	7	0	3.5	0	3.5	0	3.5	0.88	4.40
2	30	7	18.5	3.5	22	0	22	5.5	27.50
3	50	30	40	22	62	0	62	15.5	77.50
4	45	50	47.5	62	109.5	0	109.5	27.38	136.90
5	40	45	42.5	109.5	152	3.5	148.5	37.13	185.65
6	30	40	35	152	187	22	165	41.25	206.25
7	20	30	25	187	212	62	150	37.5	187.50
8	15	20	17.5	212	229.5	109.5	120	30	150.00
9	10	15	12.5	229.5	242	152	90	22.5	112.50
10	5	10	7.5	242	249.5	187	62.5	15.63	78.15
11	2	5	3.5	249.5	253	212	41	10.25	51.25
12	0	2	1	253	254	229.5	24.5	6.13	30.65
13		0	0	254	254	242	12	3	15.00
14				254	254	249.5	4.5	1.13	5.65
15				254	254	253	1	0.25	1.25
16				254	254	254	0	0	0

Col. 7 = Col. 6 lagged by 4 hours = S-curve ordinates lagged by 4 hours

Col. 8 = Col. 6 – Col. 7 = Ordinates of a DRH due to 4 cm of ER in 4 hr

Col. 9 = (Col. 8)/4 = Ordinates of a 4-hr UH

3. The required DRH ordinates for 5 cm ER in 4 hours are obtained by multiplying the ordinates of 4-hr UH by 5.

In Table 6.3, Col. 10 = (Col. 9) x 5 = Ordinates of required DRH.

6.9.3 Geomorphological Instantaneous Unit Hydrograph

Instantaneous unit hydrograph derived on the basis of geomorphological parameters of a watershed is called as geomorphological instantaneous unit hydrograph (GIUH). The philosophy of GIUH is based on the distribution of arrival time of unit instantaneous impulse injected throughout the channel network which is affected by the morphological and hydrological characteristics of the channel.

6.9.3.1 Development of GIUH

For development of GIUH, a contour map of the watershed is required. The methodology for development of GIUH consists of the following steps:

(i) Computation of geomorphological parameters of watershed

The stream order, stream length, drainage area of each order stream that conveys the discharge to the stream, stream length ration, bifurcation ratio, etc. are the different geomorphological parameters of the watershed which should be computed from the contour map of he watershed. The procedure suggested by Strahler (1964) can be used to compute the above mentioned parameters.

(ii) Computation of flow velocity

The dynamic property of the watershed gives the flow velocity which is computed as (Suresh, 2008):

$$V_{peak} = \alpha_{\Omega}^{0.6} \cdot A_{\Omega} \cdot (I_c)^{0.4} \qquad \text{for } t_e \geq t_c \tag{6.47}$$

$$V_{peak} = \alpha_{\Omega} \cdot (t_c \cdot i_e)^{2/3} \left(\frac{A_{\Omega}}{L_{\Omega}} \right)^{2/3} \qquad \text{for } t_e < t_c \tag{6.48}$$

$$\text{where, } \alpha_{\Omega} = \frac{S_{\Omega}^{0.5}}{n \cdot b_{\Omega}^{2/3}} \tag{6.49}$$

where, t_e = duration of excess rainfall, sec; t_c = time of concentration, sec; α_Ω = kinematic wave parameter, ($m^{-1/2} \cdot \sec^{1/2}$); A_Ω = area of watershed, m²; i_e = intensity of rainfall, m/sec; L_Ω = length of main stream, km; S_Ω = slope of main stream, m/m; b_Ω = width of main stream, m and n = Manning's roughness coefficient.

(iii) Computation of Shape and Scale Parameters

Shape and scale parameters are determined on the basis of geomorphological parameters of the watershed. Shape parameter is defined as:

$$n = 3.29 \left(\frac{R_B}{R_A} \right)^{0.78} \cdot R_L^{0.07} \tag{6.50}$$

Scale parameter is defined as:

$$K = 0.70 \frac{R_A^{0.48}}{R_B \cdot R_L} \cdot \frac{L_\Omega}{V_{peak}} \tag{6.51}$$

where, R_B and R_A are bifurcation and stream area ratio, respectively; R_L is stream length ratio; V_{peak} is flow velocity and other terms are defined as earlier.

(iv) Computation of peak runoff and time to peak

Peak runoff (q_p) and time to peak (t_p) are the most important parameters of GIUH. Peak runoff is expressed as:

$$q_p = 1.31 \frac{R_L^{0.43}}{L_\Omega} \tag{6.52}$$

Time to peak is expressed as:

$$t_p = \frac{0.44 L_\Omega}{V_{peak}} \left(\frac{R_B}{R_A} \right)^{0.55} R_L^{-0.38} \tag{6.53}$$

(v) Computation of ordinates of GIUH

It is computed as:

$$u(t) = \frac{1}{K \Gamma(n)} \left(\frac{t}{K} \right)^{n-1} e^{-\left(\frac{t}{K} \right)} \tag{6.54}$$

where, K is scale parameter, Γ is gamma function and other terms are defined earlier. Values of gamma functions for K are given in Table 6.4 below.

In above Eqn. (6.54), $u(t)$ is given in cm/hr. To convert it in m³/sec, we use formula:

$$u(t)_{m3/sec} = 2.78 \ x \ A_\Omega \ x \ u(t)_{cm/hr} \tag{6.55}$$

where, is area of watershed in km².Following sample calculation helps in development of GIUH.

Sample Calculation 6.14

Following geomorphological parameters were derived from toposheet of a watershed. Bifurcation ratio $(R_B) = 4.5$, stream length ratio $(R_L) = 1.25$, stream area ratio $(R_A) = 3.70$, length of main stream $(L_\Omega) = 1.5$ km, width of main stream $(b_\Omega) = 20$ m, slope of the main stream $(S_\Omega) = 0.0175$ percent, area of the watershed $(A_\Omega) = 29.93$ km² and time of concentration $(t_c) = 2.5$ hour. Compute the ordinates of GIUH if effective rainfall intensity $(i_e) = 1.4 \times 10^{-6}$ m/sec, duration of effective rainfall $(t_e) = 1$ hr, rainfall excess $= 1.35$ mm, kinematic wave parameter $(\alpha_\Omega) = 0.0455$ and Manning's roughness coefficient $(n) = 0.045$. Also derive a 1-hr UH from 1-hr GIUH.

Solution

Since , $t_e < t_c$, the flow velocity is determined by Eqn. (6.48)

$$V_{peak} = \alpha_\Omega \cdot \left(t_c \cdot i_e \right)^{2/3} \left(\frac{A_\Omega}{L_\Omega} \right)^{2/3}$$

$$= 0.0455 \ x \ (3600 \ x \ 1.4 \ x \ 10^{-6})^{2/3} x \left(\frac{29.93 \ x \ 10^6}{1500} \right)^{2/3}$$

$$= 1.03 \ \text{m/sec.}$$

Shape parameter is computed by Eqn. (6.50):

$$n = 3.29 \left(\frac{R_B}{R_A} \right)^{0.78} \cdot R_L^{\ 0.07}$$

$$= 3.29 \ x \left(\frac{4.5}{3.7} \right)^{0.78} x \ (1.25)^{0.07}$$

$$= 3.89 = 3.9$$

Table 6.4: Values of K and its gamma function

K	$\lfloor\overline{K}$	ϕK	$\lfloor\overline{K}$	ϕK	$\lfloor\overline{K}$	ϕK	$\lfloor\overline{K}$	ϕK	$\lfloor\overline{K}$
1.00	1.000000	1.205	0.916857	1.405	0.8870028	1.605	0.894088	1.805	0.932720
1.01	0.994325	1.21	0.915576	1.41	0.886764	1.61	0.894680	1.81	0.934076
1.02	0.988844	1.22	0.913106	1.42	0.886356	1.62	0.895924	1.82	0.936845
1.03	0.983549	1.23	0.910754	1.43	0.886036	1.63	0.897244	1.83	0.939690
1.04	0.978438	1.24	0.908521	1.44	0.885805	1.64	0.898642	1.84	0.942612
1.05	0.973504	1.25	0.906402	1.45	0.885661	1.65	0.900116	1.85	0.945611
1.06	0.968744	1.26	0.904397	1.46	0.885694	1.66	0.901668	1.86	0.948687
1.07	0.964152	1.27	0.902503	1.47	0.885633	1.67	0.903296	1.87	0.951840
1.08	0.959725	1.28	0.900718	1.48	0.885747	1.68	0.905001	1.88	0.955071
1.09	0.955459	1.29	0.899041	1.49	0.885945	1.69	0.906781	1.89	0.958379
1.10	0.951351	1.30	0.897471	1.50	0.886227	1.70	0.908639	1.90	0.961765
1.11	0.947395	1.31	0.896004	1.51	0.886591	1.71	0.910571	1.00	0.961766
1.12	0.943590	1.32	0.894640	1.52	0.887039	1.72	0.912581	1.92	0.968774
1.13	0.939931	1.33	0.893378	1.53	0.887567	1.73	0.914665	1.93	0.972376
1.14	0.936416	1.34	0.892216	1.54	0.888178	1.74	0.916826	1.94	0.976099
1.15	0.933040	1.35	0.891151	1.55	0.888868	1.75	0.919062	1.95	0.979880
1.16	0.929803	1.36	0.890185	1.56	0.889639	1.76	0.912375	1.96	0.983743
1.17	0.926699	1.37	0.889313	1.57	0.890489	1.77	0.923763	1.97	0.987684
1.18	0.923728	1.38	0.888537	1.58	0.891420	1.78	0.926227	1.98	0.991708
1.19	0.920885	1.39	0.887854	1.59	0.892428	1.79	0.928767	1.99	0.995813
1.20	0.918169	1.40	0.887264	1.60	0.893515	1.80	0.931384	2.00	1.000000

Computation of scale parameter is given by Eqn. (6.51):

$$K = 0.70 \frac{R_A^{0.48}}{R_B \cdot R_L} \cdot \frac{L_\Omega}{V_{peak}}$$

$$= 0.70 \, x \frac{3.7^{0.48}}{4.5 \, x \, 1.25} x \frac{1500}{1.03}$$

$$= 0.34$$

Ordinates of GIUH are given by Eqn. (6.54) as:

$$u(t) = \frac{1}{K \Gamma(n)} \left(\frac{t}{K}\right)^{n-1} e^{-\left(\frac{t}{K}\right)}$$

$$= \frac{1}{0.34 \, \Gamma(3.9)} \left(\frac{t}{0.34}\right)^{2.90} e^{-\left(\frac{t}{0.34}\right)}$$

Value of $\Gamma(3.9) = 2.9 \, x \, 1.9 \, x = 2.9 \, x \, 1.9 \, x \, 0.961766 = 5.2933$ (value of is taken from Table 6.4.

$$\text{Hence, } u(t) = \frac{1}{0.34 \, x \, 5.2933} \left(\frac{t}{0.34}\right)^{2.90} e^{-\left(\frac{t}{0.34}\right)}$$

$$= 0.556 \, x \left(\frac{t}{0.34}\right)^{2.90} e^{-\left(\frac{t}{0.34}\right)} \tag{6.56}$$

Computations of ordinates of GIUH (cm/hr) of 1-hr for various values of time, t is shown in Table 6.5 below.

Now the value of $u(t)$ in m³/sec is determined using Eqn. (6.55) as:

$$u(t)_{m3/sec} = 2.78 \, x \, A_\Omega \, x \, u(t)_{cm/hr}$$

$$= 2.78 \, x \, 29.93 \, x \, u(t)_{cm/hr} \qquad = 83.20 \, x \, u(t)_{cm/hr} \tag{6.57}$$

The ordinates of GIUH are computed in m³/sec by Eqn. (6.57). Computation of 1-hr UH from 1-hr GIUH are then computed by the methods as described earlier and shown in table below.

Table 6.5 Computations of ordinates of GIUH of 1-hr

Time, t (hr)	$u\ (t),$ cm/hr (computed by Eqn. 6.56)	$u\ (t), m^3/sec$ (computed by Eqn. 6.57)	Ordinates of 1-hr UH	
			$u(t)$ lagged by 1-hr	Unit Hydrograph, m^3/sec (m^3/sec)
(1)	(2)	(3)	(4)	= (Col. 3 + Col. 4)/2
0	0	0	0	0
1	0.670578	55.79208	0	27.89604
2	0.264301	21.98988	55.79208	38.89098
3	0.04523	3.763143	21.98988	12.87651
4	0.005501	0.457653	3.763143	2.110398
5	0.000555	0.046157	0.457653	0.251905
6	4.97E-05	0.004135	0.046157	0.025146
7	4.1E-06	0.000341	0.004135	0.002238
8	3.19E-07	2.66E-05	0.000341	0.000184
9	2.37E-08	1.97E-06	2.66E-05	1.43E-05
10	1.7E-09	1.41E-07	1.97E-06	1.06E-06

Question Banks

Q1. What are the different routes through which runoff flows?

Q2. Differentiate between interflow and overland flow.

Q3. What is an interflow? What are the factors on which it depends?

Q4. Differentiate between effluent and influent streams.

Q5. What do you mean by base flow? Which factors decide the velocity of base flow?

Q6. Define hydrograph. Describe its various components.

Q7. Explain the various factors that affect the shape of a hydrograph.

Q8. Explain how a hydrograph represents total runoff.

Q9. The hydrographs of two identical and isolated storms produce same kind of hydrograph in the same catchment. But when the catchments are different, the shapes of the hydrographs are different. Explain why?

Q10. Describe how the climatic factors affect the shape of hydrograph.

Q11. How the drainage density influences the shape of a hydrograph?

Q12. Define recession constants and describe its various types.

Q13. Write the different steps used to estimate the recession constants.

Q14. With diagram enumerate the different methods employed for separation of base flow.

Q15. Describe Chow's method for separation of base flow.

Q16. Differentiate between surface runoff hydrograph and direct runoff hydrograph.

Q17. Write the process of base flow separation for a complex hydrograph.

Q18. Define effective rainfall and describe its uses.

Q19. What is an effective rainfall hyetograph? With a sketch, describe it.

Q20. Derive the relationship between direct runoff hydrograph and effective rainfall hyetograph.

Q21. There are two consecutive storms of magnitude 3.8 cm and 2.8 cm occurring in two consecutive 4-hr duration in a catchment of 32 km^2 which produces a hydrograph with the observed flow as recorded below. Estimate the effective rainfall and losses.

Time since rainfall occurs, hr	-6	0	6	12	18	24	30	36	42	48	54	60
Observed flow, m³/s		6.4	4.4	11.7	24	20	16	11	8	6.5	5.5	4.0	4.0

Q22. A storm over a catchment of 6.0 km² had a duration of 16 hours. The mass curve of the rainfall of the storm over the catchemnt is as follows.

Time from start of storm, hr	0	2	4	6	8	10	12	14	16
Auumulated rainfall, cm	0	0.8	2.5	5.0	6.9	7.5	9.2	9.5	10.0

If the value of ϕ- index is 0.40 cm/hr, estimate the effective rainfall hyetograph and the volume of direct runoff from the catchment due to the storm.

Q23. Define unit hydrograph.

Q24. Describe the procedure for development of unit hydrograph.

Q25. What are the parameters to which the definition of unit hydrograph refers to?

Q26. What are the two assumptions for development of unit hydrograph? Describe them in short.

Q27. The ordinates of a 6-hr unit hydrograph (UH) of catchment of area 320 km² are given below. Calculate the ordinates of the DRH due to a rainfall excess of 3 cm occurring in 6 hr.

Time, hr	0	3	6	9	12	15	18	24	30	36	42	48	54	60
UH ordinates, m³/sec	0	25	50	80	120	160	185	150	100	60	30	22	11	0

Q28. Define direct runoff and explain the methods to compute it.

Q29. Describe various uses and limitations of unit hydrograph.

Q30. Write a procedure to derive unit hydrogarph from a complex storm.

Q31. Two storms each of 6-hr duration have rainfall excess values of 2.5 and 3 cm each. The 2.5 cm excess rainfall is followed by the 3 cm excess rainfall. The 6-hr unit hydrograph for the catchment has the following ordinates. Calculate the DRH. If there is a constant base flow of 5.5 m³/sec, compute the total flood hydrograph.

Time, hr	0	3	6	9	12	15	18	24	30	36	42	48	54	60	69
UH ordinates, m³/sec	0	25	50	80	120	160	185	150	100	60	30	20	12	5	0

Q32. The ordinates of a 3-hr unit hydrograph of a catchment are given below.

Time, hr	0	3	6	9	12	15	18	21	24	27	30	33	36	39	42	45
Ordinates of 6-hr UH	0	30	70	100	158	179	210	167	100	80	58	50	30	20	10	0

The base flow is constant and occurs at 18 m³/sec rate. The average rate of the storm loss is 0.25 cm/hr. Derive the flood hydrograph in the catchment due to the following storms.

Time since start of storm (hr)	0	3	6	9
Cumulative rainfall, cm	0	3.5	9.0	15.0

Q33. The peak of flood hydrograph due to a 3-hr duration isolated storm in a catchment is 260 m³/sec. Total depth of rainfall is 6.5 cm. Assume average loss of infiltration as 0.4 cm/hr. The base flow is assumed to be constant at the rate of 16 m³/sec. Compute the peak of the 3-hr unit hydrograph of the catchment. If the catchment has an area of 500 km², estimate the base width of the 3- hr unit hydrograph by assuming it to have triangular shape.

Q34. Describe the S-curve technique and Superposition methods used for derivation of unit hydrograph for different durations.

Q35. Using method of superposition as well as S-curve technique, derive a 9-hr unit hydrograph for a catchment whose 3 hr unit hydrograph ordinates are as follows: Draw the 9-hr and 3-hr unit hydrograph.

Time, hr	0	3	6	9	12	15	18	21	24	27	30	33	36	39	42
Ordinates of 3-hr UH	0	32	70	100	150	190	200	172	90	76	50	35	20	12	0

Q36. Ordinates of 4-hr UH are given as below. Using this, derive the ordinates of 2-hr UH by S-curve method.

Time, hr	0	4	8	12	16	20	24	28	32	36	40	44
Ordinates of 4-hr UH, m3/sec	0	24	86	120	158	130	110	50	25	15	5	0

Q37. How is the duration of unit hydrograph decided? Write how the area of the watersheds decided the duration of unit hydrograph.

Q38. What is a distribution graph. How is it used in development of unit hydrograph?

Q39. A catchment of area 400 ha receives rainfall of 7.0, 2.5 and 5.0 cm in three consecutive days. The average losses in all these three days can be assumed as 2.0 cm/day. Distribution graph percentages of the surface runoff that extended over 6 days for every rainfall of 1 day duration are 5, 15, 40, 20, 15 and 5. Determine the ordinates of the discharge hydrograph. Neglect base flow.

Q40. What is a synthetic unit hydrograph? What are its uses?

Q41. Describe Snyder's method to derive synthetic unit hydrograph.

Q42. Two catchments A and B which are hydro-meteorologically similar have the following characteristics.

Catchment A	Catchment B
L = basin length measured along the water course from the basin divide to the gauging station = 30 km	L = basin length measured along the water course from the basin divide to the gauging station = 40 km
L_{ca} = distance along the main water course from the gauging station to a point opposite to the watershed centroid = 15 km	L_{ca} = distance along the main water course from the gauging station to a point opposite to the watershed centroid = 25 km
A = Area of catchment = 250 km²	A = Area of catchment = 400 km²

A 2 hr UH has been developed for catchment A which has a peak discharge of 48 m³/sec and time to peak from the beginning of the rainfall excess is 8 hr. Develop a UH for the catchment B using Snyder's method.

Q43. Enumerate the procedure to finalise the shape of a Synthetic unit hydrograph.

Q44. What is a dimensionless unit hydrograph? What are its advantages?

Q45. How is a dimensionless unit hydrograph is prepared?

Q46. Mention the relations used to develop a synthetic unit hydrograph for Indian condition as proposed by CWC.

Q47. Define instantaneous unit hydrograph andexplain its various properties.

Q48. Describe the procedures for development of an instantaneous unit hydrograph.

Q49. Describe the steps followed for derivation of a D-hr unit hydrograph from instantaneous unit hydrograph.

Q50. The ordinates of an IUH of a catchment at different time intervals are given below. If a storm of duration 4 hours having a rainfall excess of 4 cm occurs in the catchment, then find out the direct runoff hydrograph (DRH) for the catchment.

Time, hr	0	1	2	3	4	5	6	7	8	9	10	11	12
Ordinate of IUH, m³/sec	0	10	36	57	49	38	30	22	15	10	5	3	0

Q51. Define a geomorphological instantaneous unit hydrograph? Describe the various steps required to develop it.

Q52. Following geomorphological parameters were derived from contour map of a watershed. Bifurcation ratio (R_B) = 4.6, stream length ratio (R_L)= 1.27, stream area ratio (R_A) = 3.75, length of main stream (L_Ω) = 1.4 km, width of main stream (b_Ω) = 20 m, slope of the main stream (S_Ω) = 0.0185 percent, area of the watershed (A_Ω) = 30.0 km² and time of concentration (t_c) = 2.2 hour. Compute the ordinates of GIUH and 1-hr UH if effective rainfall intensity (i_e) = 1.5 x 10⁻⁶ m/sec, duration of effective rainfall (t_e) = 1 hr, rainfall excess = 1.37 mm, kinematic wave parameter (α_Ω) = 0.0555 and Manning's roughness coefficient (n) = 0.042.

References

Barnes, B.S. 1940. Discussion on analysis of runoff characteristics by O.H. Meyer, Transactions of Am. Soc. Civil Engrs, Vol. 105: 104-106.

Bernard, M.M. 1935. An approach to determine stream flow. Transactions of Am. Soc. Civil Engrs, Vol. 100: 347-395.

Butler, S.C. 1957. Engineering Hydrology. Prentice Hall Inc., USA.

Gray, D.M. 1970. Principles of Hydrology. Water Information Centre, New York.

Linsley, R.K. 1958.Hydrology for Engineers. Mc-Graw Hill Book Co. Inc. NewYork.

Linsley, R.K., Kohler, M.A. and Paulhus, J.L.H. 1940. Applied Hydrology. Mc-Graw Hill Book Co. Inc. NewYork.

Mutreja, K.N. 1986. Applied Hydrology. Tata Mc-Graw Hill Book Co. New Delhi.

O' Donnel, T. 1960. Instantaneous unit hydrograph derivation by harmonic analysis. Intern, Assoc. Sci. Hydrology, Pub. 51: 546-557.

Sharma, R.K. 1987. A Text book of Hydrology and Water Resources. Dhanpat Rai and Sons, Delhi.

Sherman, L.K. 1932. Stream flow from rainfall by the unit graph method. Eng-News-Rec, Vol. 108: 501-505.

Snyder, F.F. 1938. Synthetic unit graphs. Trans. Am.Geophys Union, Vol. 19: 447-454.

Sokolov, A.A. 1976. Flood Flow Computation. The UNESCO Press, Paris.

Strahler, A.N. 1964. Quantitative geomorphology of drainage basins and channel networks. Handbook of Applied Hydrology: Ed. V.T. Chow, Mc-Graw Hill Book Co. Inc. NewYork.

Subramanya, K. 2003. Engineering Hydrology, Tata McGraw-Hill Publishing Co. Ltd., New Delhi, pp. 392.

Suresh, R. 2008. Watershed Hydrology. Standard Publishers Distributors, Delhi, pp. 692.

Varshney, R.S. 1979. Engineering Hydrology. Nem Chand and Bros, Roorkee, pp. 915.

CHAPTER 7

Flood

7.1 Introduction

Flood is an unusually high stage in a river resulted due to high rainfall or snow melt which causes large flow of runoff in it. During flood, the river overflows the banks and inundates the adjoining areas. Flood causes heavy damages to buildings and hydraulic structures besides causing catastrophic loss to human life and property. It causes great economic loss due to disruption of many structures and damage to agricultural fields. Many forest and aquatic lives are affected by the flood. Hundreds of crores of rupees are spent every year to control, forecast and manage the flood.

Almost all the rivers of India carry heavy discharge during the rainy season when their catchments receive intense and heavy rainfall. In the upper reaches, where the rivers flow through mountainous regions or undulating terrains, there is generally no overtopping of banks during periods of high discharge. However, in lower reaches especially where the land is flat and where the rivers fall and flow, the high discharges in the rivers overtop the bank and creates flood which submerges crop fields, smashes the buildings and different hydraulic structures and disrupt communication. From the point of view of flood problem, rivers in India can be grouped into the following four regions (Sharma, 1987):

- Brahmaputra region
- Ganga region
- Northwest region and
- Central India and Deccan region

Brahmaputra Region: The main rivers in this region are Brahmaputra and Barak and their tributaries and cover the states of Assam, Meghalaya, Manipur, Tripura, Nagaland, northern parts of West Bengal and the union territories of Arunachala Pradesh and Mizoram. This region receives heavy rainfall to a tune of 600 cm annually. The main problems in the Brahmaputra region are landslides, bank erosion, drainage congestion and overspills.

Ganga Region: The main rivers in this region are the Ganga and its numerous tributaries. It covers the states of Uttar Pradesh, Bihar, south and central portion of West Bengal, parts of Haryana, Himachal Pradesh, Rajasthan, Madhya Pradesh and union territory of Delhi. The region receives annual rainfall which varies from about 60 cm in the western part to about 175 cm in the eastern. The main flood problems in the region are mostly confined to the northern tributaries which bring a lot of silts and deposit them in the river bed which consequently causes the rivers to change their courses.

Northwest Region: The main rivers in this region are the Indus and its tributaries Jhelum, Chenab, Ravi, Beas and Sutlej. It includes the river Ghaggar also. The region covers the states of Jammu and Kashmir, Punjab, parts of . Himachal Pradesh, Haryana and Rajasthan. The annual rainfall varies from 75 cm in Jammu and Kashmir to 175 cm in hilly areas of Himachal Pradesh. The flood problems in the region are less severe as compared to the Brahmaputra region. This is mainly due to creation of large storage structures in the rivers of Beas and Sutlej.

Central India and Deccan Region: The main rivers in this region are the west flowing rivers Narmada and Tapi and the east flowing rivers Mahanadi, Subarnrekha, Brahmani, Baitarani, Godavari, Krishna and Cauveri. These rivers generally have some well defined and stable courses. The regions cover all the southern parts of the country including Tamil Nadu, Andhra Pradesh, Karnataka, Kerala and states of Odisha, Maharashtra, Gujarat and parts of Madhya Pradesh. This region gets reach rainfall due to south-west monsoon to a tune varying from 75 to 125 cm annually. The region, in general, has no serious flood problems. However, there may have some occasional severe flood problems in some of the rivers due to heavy intense storms. In the deltaic region, there is deposition of silts which cause raising of flood levels and drainage congestion.

7.2 Classification of Flood

- Design flood
- Maximum probable flood
- Maximum observed flood

- Standard project flood
- Peak flood
- Maximum known flood
- Annual flood and
- Ordinary flood and

Design Flood: It is also known as maximum flood. It is considered for design of hydraulic structures like bridge, piers, drop structures etc. Design floods may be Standard Project Flood (SPF), Maximum Probable Flood (MPF) or flood of any desired recurrence interval depending on the degree of flood protection measures employed against possibilities of failure of structures. Basing on cost-benefit-ratio, selection of design flood is done.

Maximum Probable Flood: This type of flood is caused due to the severe most combinations of critical hydrological as well as meteorological characteristics of the watershed and happens rarely but becomes catastrophic. It is also otherwise named as probable maximum flood (PMF). The PMF is used in situations where the failure of the structure would result in loss of life and catastrophic damage. Standard project flood (SPF) is often used where the failure of the structure would result in less loss of life. Typically SPF is about 40 to 60% of the PMF for the catchment.

Maximum Observed Flood: It is the highest flood during a specified length of record which may range from a few weeks to a year or so.

Standard Project Flood: It is the flood that is likely to occur from the most severe combinations of hydrological as well as meteorological characteristics that are applicable for the region. Extremely rare combinations of factors are excluded. This flood is approximately 80% of the value of maximum probable flood.

Maximum Known Flood: It is the flood which has occurred with the highest value during past years and is ascertained by gathering information from the inhabitants of the area where the flood has occurred.

Peak Flood: Peak flood is also known as maximum intensity flood and is the maximum instantaneous flow during the occurrence of the flood.

Annual Flood: The highest flood that occurs in a water year is called as annual flood. The annual flood is either equal to or greater than the magnitude of specified flood, generally once in a year.

Ordinary Flood: Ordinary flood is the flood that occurs once or more times during the project period.

7.3 Guidelines for Selection of Design Flood

As defined earlier, design flood is the flood which is considered for design of hydraulic structures. The small hydraulic structures like culverts, storm drainage lines etc. are designed relatively for less severe floods. This is done since the consequences due to failure of the structures to handle the flood is less. However, there are some massive structures like storage reservoirs including dams which need careful attention in their design. Failure of these structures to handle the flood cause catastrophic incidences. In such case, we have to consider flood which cause maximum runoff to be handled for design purpose. The storage capacity and hydraulic head of dam decides size of dam. For example, a storage capacity of 0.5 to 10.0 Mm^3 and hydraulic head of 7.5 to 12.0 m indicates the dam as small type. On the other hand, a storage capacity of 10.0 to 60.0 Mm^3 and hydraulic head of 12.0 to 30.0 m indicates the dam as medium and storage capacity more than 60 Mm^3 and hydraulic head more than 30.0 m indicate the dam as large.

Regarding safety point of view and for other requirements, selection of flood for design and construction of hydraulic structures is very important. The guidelines for selection of design floods for various purposes as laid down by CWC (1973) are as follows:

1. For spillways of major and medium projects whose capacity is more than 60 Mm^3:

   (i) Probable maximum flood or probable maximum precipitation can be used as the design flood.

   (ii) If the above design flood is not possible, then a flood of 100 years recurrence interval should be used as design flood.

2. For permanent barrage and minor dams whose capacity is less than 60 Mm^3:

   (i) Standard project flood computed by unit hydrograph and standard project storm which is the largest storm of the area

   (ii) Flood of 100 years recurrence interval should be selected.

   Out of the above mentioned two, any one which is the highest value flood should be considered as design flood.

3. For pickup weirs:

   Flood with recurrence interval of 50 or 100 years depending on the importance of the project is used as design flood.

4. For aqueducts:

(a) Waterways: Flood with 50 years recurrence interval is taken as the design flood.

(b) Foundation and freeboard: Flood with 100 years recurrence interval is taken as the design flood.

5. For project with inadequate data:

For this purpose, the design flood can be evaluated based on available empirical formulae for the region.

The generalized design criteria as proposed by the National Academy of Science, USA (1983) for different hydraulic structures are mentioned in Table 7.1 below.

Table 7.1: Generalised design criteria for different hydraulic structures

Sl.No.	Structure	Recurrence interval
1.	Highway	
	(a) Low traffic	5- 10
	(b) Intermediate traffic	10 – 25
	(c) High traffic	50 - 100
2.	Highway bridge	
	(a) Secondary system	10 - 50
	(b) Primary system	50 - 100
3.	Farm Drainage	
	(a) Culverts	5 - 50
	(b) Ditches	5 - 50
4.	Urban Drainage	
	(a) Storm sewers (small cities)	2 - 25
	(b) Storm sewers (big cities)	25 - 50
5.	Levees	
	(a) On farms	2 - 50
	(b) Around cities	50 - 200
6.	Dams (low hazard)	
	(a) Small	50 - 100
	(b) Medium	>100
	(c) Big	>100
7.	Dams (High hazard)	
	(a) Small	>100
	(b) Medium	>100
	(c) Big	>100

7.4 Causes of Flood

A river system acts like a control volume in which there is some inflow, some out flow and the difference of the two is the change in storage. If the inflow to the system exceeds the outflow then there is some storage in the river which results in raising the water level in the river. As the inflow continues to exceed the outflow, the water level will go on increasing and when the rising water level continued for a longer period, there is overtopping of banks of the river. This causes flood which inundates the adjoining areas and creates great loss to both animal and plant lives. Flood is affected by two factors i.e. precipitation characteristics and watershed characteristics. Among the different precipitation characteristics, forms of precipitation, magnitude and intensity of precipitation, duration and frequency of rainfall, its orientation and distribution etc. are the important parameters. Similarly the watershed characteristics like size and shape of the watershed, area of the watershed, soil type, slope, structure and texture of the soil etc. affect the runoff and hence flood. Amongst the different characters as mentioned above, following three factors are important.

- Occurrence of high intensity rainfall over a small hilly catchment
- Intense rainfall over large catchment and
- Occurrence of rainfall over accumulated snow

If a high intensity rainfall occurs over a small catchment and if the slope of the catchment is high such as a hilly catchment then there will be high accumulation of runoff in a river. As the accumulation exceeds the storage capacity of the river, the excess amount overtops the bank causing flood. Further if the duration of this high intense storm is more then there will be more rush of runoff to the river causing high flood. A greater land slope results in accelerating the velocity of flow which enhances the runoff and thus favouring occurrence of a flood. Land use affects the runoff producing characteristics of a watershed and thereby influences flood. A vegetative land produces less runoff than a bare land and thus reduces the magnitude of flood. Antecedent moisture also affects runoff and flood. A wet soil mass will generate more runoff and more flood than a dry soil. Size and shape of the catchment has pronounced affect on generating flood in a catchment. If the shape of the catchment is fab shaped then it will produce more runoff and more flood. A large size catchment, generally, produces more runoff and hence more flood. Simultaneous occurrence of rainfall and melting of snow causes high flood in rivers.

7.5 Estimation of Flood Peak

Estimation of flood volume and rate is important for planning and design of flood control structures and other hydraulic structures. It helps in planning and design of flood regulating structures, irrigation and hydropower projects. Flood peak values are also used in estimation of scour at a hydraulic structure, design of bridges and culverts, design of spillways and cross drainage structures. To estimate the peak flood, following methods are employed (Mutreja, 1986):

- Rational method
- Empirical formulae
- SCS curve number method
- Envelope curve method
- Unit hydrograph technique
- Flood frequency studies and

The use of a particular method depends on desired objective, availability of data and importance of the project. It is to be remembered that flood frequency analysis and unit hydrograph approach are used for flood estimation in gauged catchment whereas other methods as listed above are used for un-gauged catchment.

7.5.1 Flood Peak Estimation for Un-gauged Catchment

7.5.1.1 Rational Method

Rational method is widely used in computation of the magnitude of flood peak in small watersheds upto 50 km^2. In this method, the peak rate of runoff or peak flood is given by the equation as:

$$Q = \frac{CIA}{360} \tag{7.1}$$

where, Q is peak runoff rate, m^3/sec; I is rainfall intensity, mm/hr; C is runoff coefficient, and A is catchment area generating runoff, ha.

Values of runoff coefficient (C) vary from 0 to 1. Details of the Rational method used for computation of flood peak are described in Chapter 3.

7.5.1.2 Empirical formulae

A number of empirical methods have been developed by different scientists to estimate the flood peak values. But these methods or formulae are meant for use in specified regions. They are regional formulae and cannot be used universally for all the regions to estimate the peak flood values. They are developed on statistical correlation between observed peak discharge and catchment properties. These formulae do not consider all the catchment characters for estimating runoff rate. Hence, they give approximate value of peak runoff rate. It is observed that area of the catchment is taken as unique character in most of the empirical formulae to estimate the peak rate of runoff. Many of the empirical formulae neglect flood frequency as a parameter to estimate the peak runoff rate. Some of the important empirical formulae used to estimate the flood peak are as follows:

Inglis Formula (1940): This formula was developed to estimate the flood flow from catchments of Maharashtra and Gujarat and is expressed as:

$$Q = \frac{124\,A}{\sqrt{A + 10.4}} \tag{7.2}$$

where, Q is discharge, m³/sec and A is area of catchment in km².

Nawab Jang Bahadur Formula: This formula was developed for the catchments of old Hyderabad states. It is given as:

$$Q = C\,A^{(0.993 - 1/14\,\log A)} \tag{7.3}$$

where, Q is discharge, m³/sec, C = constant which may vary from 48 to 86 and A is area of catchment in km².

Creager's Formula: This formula was developed for the American catchments and is:

$$Q = 0.386\,C\,A^{0.894\,(0.86\,A)^{-0.043}} \tag{7.4}$$

where, Q is discharge, m³/sec, C is a constant and A is area of catchment in km².

Modified Myer's Formula: It is given as:

$$Q = 175\,P\,\sqrt{A} \tag{7.5}$$

where, Q is discharge, m³/sec, C is a constant varying from 0.002 to 1 (usually taken as 1) and A is area of catchment in km².

Fuller's Formula: This formula is used for catchments of USA and is:

$$Q_{Tp} = C_f\,A^{0.8}\,(1 + 0.8\,\log T) \tag{7.6}$$

where, Q_{Tp} is peak discharge with a frequency of T years in m³/sec, A is area of catchment in km² and C_f is a constant varying from 0.18 to 1.88.

Dredge and Burge Formula: It is given as:

$$Q = 19.5 \frac{A}{L^{2/3}} \qquad (7.7)$$

where, Q is discharge, m³/sec, L is length of the basin, km and A is area of catchment in km².

Fanning Formula: This formula is meant for USA. It is expressed as:

$$Q = 2.64 \left(A^{0.8} \right) \qquad (7.8)$$

where, Q is discharge, m³/sec and A is area of catchment in km².

Boston Society of Civil Engineers Formula: The peak discharge by this formula is given as:

$$Q = C \left(A^{0.5} \right) \qquad (7.9)$$

where, Q is discharge, m³/sec, C is a constant having a value 3.5 for rainfall < 500 mm, 8.4 for rainfall between 500 to 750 mm and 35 for rainfall > 750 mm and A is area of catchment in km².

Horton's Formula: The peak discharge is computed as (Panigrahi, 2013):

$$q_p = 71.2 \ T^{1/4} \ A^{-1/2} \qquad (7.10)$$

where, q_p is discharge, m³/sec/km², A is area of catchment in km² and T is return period in years.

U.S. Geological Survey (1955) Formula: This formula helps in estimation of mean annual average flood ($Q_{2.33}$ m³/sec) with a return period of 2.33 years and is expressed as:

$$Q_{2.33} = 0.0147 \ C \ A^{0.7} \qquad (7.11)$$

where, Q is discharge, m³/sec, C is a constant varying from 1 to 100 and A is area of catchment in km².

Pettis Formula: By this formula

$$Q_p = C \left(P \cdot B \right)^{5/4} \qquad (7.12)$$

where, Q_p is discharge, m³/sec at 100 years return period, C is a constant (0.195 for desert and 1.51 for humid regions), P is one day rainfall at 100 years return period, cm and B 5 is average width of basin.

Other formulae like Dicken's formula, Coutagne formula and Ryve's formula are discussed in Chapter 3.

Sample Calculation 7.1

A catchment has an area of 20 km². Estimate the peak flood value by Inglis formula and Nawab Jang Bahadur formula. Also calculate the flood peaks for a return period of 100 years by Fuller's formula as well as Horton's formula. Assume the value of constant, C in Nawab Jang Bahadur formula and Fuller's formula as 50 and 1.5, respectively.

Solution

Area of catchment, A = 20 km²

Return period, T = 100 year

Value of constant, C in Fuller's formula = 1.5 and in Nawab Jang Bahadur formula = 50

Flood peak by Inglis formula is computed by Eqn. (7.2) as:

$$Q = \frac{124\,A}{\sqrt{A+10.4}}$$

$$= \frac{124 \times 20}{\sqrt{20+10.4}} = 449.8 \; m^3 / \sec$$

Flood peak by Nawab Jang Bahadur formula is computed by Eqn. (7.3) as:

$$Q = C\,A^{(0.993-1/14\log A)}$$

$$= 50 \times 20^{(0.993-1/14\log 20)} = 741.3 \; m^3 / \sec$$

Flood peak by Fuller's formula is computed by Eqn. (7.6) as:

$$Q_{Tp} = C_f\,A^{0.8}\left(1+0.8\log T\right)$$

$$= 1.5 \times 20^{0.8} \times \left(1+0.8\log 100\right) = 42.84 \; m^3 / \sec$$

Flood peak by Horton's formula is computed by Eqn. (7.10) as:

$$q_p = 71.2 \ T^{1/4} \ A^{-1/2}$$

$$= \ 71.2 \ x \ 20 \ x \ 100^{1/4} \ x \ 20^{-1/2} = 1006.9 \ m^3 \ / \sec$$

7.5.1.3 SCS Curve Number Method

SCS curve number method also called as hydrologic soil cover complex number method is used to estimate peak rate of runoff from small watersheds. A runoff curve *(CN)* number is used in computation of runoff. Details of the method are presented in Chapter 3.

7.5.1.4 Envelope Curve Method

In the regions having similar climatological characteristics, if the availability of hydrological data are inadequate, then enveloping curve technique can be adopted to develop a relationship between the peak floods and drainage areas. In this method, flood peak data from a large number of catchments which are hydro-meteorologically and topographically similar are collected. These data are plotted in a log-log graph paper where the drainage areas are plotted in abscissa and the flood peaks are plotted as ordinates. This results in a plot in which the data are in scattered form. If an enveloping curve that would encompass all the plotted data points is drawn, it can be used to get maximum peak discharge for any given area. Envelop curves thus obtained are useful in obtaining quick rough estimates of peak discharges. The empirical flood formulae developed using the data points of discharge and area are in the form of $Q = f(A)$.

Kanwar Sain and Karpov (1967) have presented envelope curves representing the relationship between peak flood values and catchment areas for Indian conditions. They obtained two curves; one for south Indian rivers and the other for north Indian and central Indian rivers (Fig. 7.1). These curves are based on data covering catchment areas ranging from 10^3 to 10^6 km².

For all other regions of the world, J.M. Baird in 1951 proposed the following formula to compute the peak flood values:

$$Q_{mp} = \frac{3025 \ A}{\left(278 + A\right)^{0.78}} \tag{7.13}$$

where, Q_{mp} = peak flood value in m³/sec and A = area of catchment, km². Eqn. (7.13) is also called as world enveloping flood equation.

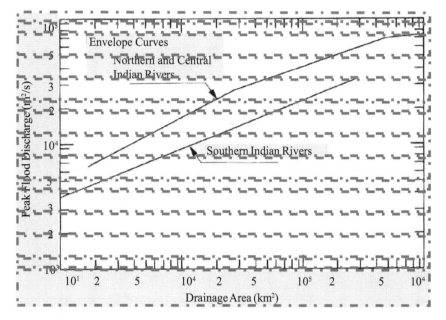

Fig. 7.1: Enveloping curves for rivers in India (Subramanya, 2003)

Based on the peak flood data of a number of river basins having identical drainage features, Creager developed the following envelope curve equation:

$$Q = 0.3389 \cdot C \cdot A^{0.864\left(\frac{A}{2.95}\right)^{-0.048}} \tag{7.14}$$

where, Q = peak flood value in m³/sec, C = constant taken as 130 and A = area of catchment, km².

Meyer proposed the following modified envelope curve to estimate flood peak values which is expressed as:

$$Q = 5682.7 \, P \, A^{1/2} \tag{7.15}$$

where, Q = peak flood value in m³/sec, P = numerical percentage depending on the basin characteristics and A = area of catchment, km².

Sample Calculation 7.2

Estimate the maximum flood peak value for a catchment having area of 50 km² by world experience method. Also compute the flood peak value for the same catchment by Creager's envelope curve equation.

Solution

Area of catchment, A = 50 km²

Flood peak value by world experience method is given by Eqn. (7.13)

$$Q_{mp} = \frac{3025\,A}{(278 + A)^{0.78}}$$

$$= \frac{3025\,x\,50}{(278 + 50)^{0.78}} = 1649.35 \text{ m}^3/\text{sec}$$

Flood peak value by Creager's envelope curve equation is computed by Eqn. (7.14)

$$Q = 0.3389 \cdot C \cdot A^{0.864\left(\frac{A}{2.95}\right)^{-0.048}}$$

Value of constant, C in Creager's equation = 130

$$Q = 0.3389\,x\,130\,x\,50^{0.864\left(\frac{50}{2.95}\right)^{-0.048}} = 842.29 \text{ m}^3/\text{sec}$$

7.5.2 Flood Peak Estimation for Gauged Catchment

7.5.2.1 Unit Hydrograph Technique

Unit hydrograph technique to compute flood peak can be employed if we know the value of rainfall that produces flood, infiltration characteristics of the soil of the catchment and the appropriate unit hydrograph is available. For design purpose, extreme rainfall situations are used to obtain the design storm. The known or derived unit hydrograph of the catchment is then operated upon by the design storm to generate the desired flood hydrograph. It is to be noted that unit hydrograph technique can be employed for catchments having area less than 5000 km². If catchment of larger than this value is present, then it has to be divided into sub-catchments and then for each sub-catchment, unit hydrograph is developed. Total flood is obtained by adding all together. Unit hydrograph technique used for estimation of peak flood values has been discussed in Chapter 6.

7.5.2.2 Flood Frequency Studies

Hydrological events like floods are complex natural events. They are difficult to be modeled analytically since they depend on a number of components which are stochastic. For example the flood in a catchment depends on catchment characteristics as well as climatic factors like rainfall, its intensity, duration, frequency etc. Each of these factors again depends on a host of factors. This makes the prediction of flood in a catchment extremely uncertain. Empirical formulae, unit hydrograph method and some other methods as discussed above are used to predict the flood peak. Flood frequency study is yet another method used to predict the flood peak. It is to be noted the flood frequency is used to predict other hydrological events like rainfall, drought etc.

In flood frequency study we mostly use the annual maximum flood data of a given catchment for large number of years in succession that constitute an *annual series*. These data are then arranged in descending order by their magnitude and the probability P of each event being equaled to or greater (plotting position) is computed by the plotting position formula as:

$$P = \frac{m}{N+1} \tag{7.16}$$

where, m = rank or order number of the event and N = total number of events in the data series. In above equation, P is also called as probability of exceedance and is given in decimal. When Eqn. (7.16) is multiplied by 100 then probability is expressed in percent. Eqn. (7.16) is known as Weibul's plotting position formula or simply called as Weibul's formula. It is to be noted that probability of non exceedance of an event is equal to 1 minus probability of exceedance.

The recurrence interval otherwise called as return period or frequency *(T)* is given as:

$$T = 1/P = \frac{N+1}{m} \tag{7.17}$$

where, T is in years and P is in decimal. If the probability of exceedances, P is 40 percent, then return period, T is $T = 1/0.40 = 2.5$ years. Return period is the average time interval of an event of a given magnitude being equaled or exceeded but not the actual time interval.

Following steps are used to compute the flood peaks by Weibul's formula.

- Arrange the data in descending order by their magnitude.
- Give rank number (m) to each data with rank number one for the highest data and least value for the lowest data. Value of m will be 1, 2, 3......If there are some data having same value for a number of years then they are given rank numbers chronologically.

- Find out the value of N i.e. total number of data.

- Compute plotting position, P (probability of exceedance) by Eqn. (7.16) against each data. Since, $P = 1/T$, we can also calculate the flood peak data at each return period.

- Plot the flood peak data and their probability of exceedance which yields the probability distribution or flood frequency curve. For small return periods (i.e. large probability of exceedance) or where limited extrapolation is required, a simple best fitting curve through the plotted points can be used as the probability distribution. A logarithmic scale for P or T is advantageous. It is to be remembered that probability distribution curves can be used to predict the flood data at any return period or probability of exceedances by extrapolation for small values of extrapolation. When large extrapolation of T or P is required, theoretical probability distribution is used.

Sample Calculation 7.3

Following measurements of flood peak data for 24 years of a basin are recorded. Using Weibul's formula, draw the probability distribution curve and compute the flood peak value at return period of 2 and 20 years. Also compute the flood peak value at 20 and 40 percent of probability levels.

Year	1990	1991	1992	1993	1994	1995	1996	1997	1998	1999	2000	2001
Flood data, m³/sec	180	187	234	210	162	172	148	154	150	159	140	144

Year	2002	2003	2004	2005	2006	2007	2008	2009	2010	2011	2012	2013
Flood data, m³/sec	116	122	128	119	147	146	130	134	106	109	104	113

Solution

Calculation of plotting position for the above sample calculation based on the steps described as above is given in table below.

Flood data, m³/sec	Rank number, m	Probability, P (%)P = (m/N+1) x 100	Return period, T = (N + 1) /m, year
234	1	4	25.0
210	2	8	12.5
187	3	12	8.33
180	4	16	6.25
172	5	20	5.0
162	6	24	4.17
159	7	28	3.57
154	8	32	3.13
150	9	36	2.78
148	10	40	2.5
147	11	44	2.27
146	12	48	2.08
144	13	52	1.92
140	14	56	1.78
134	15	60	1.67
130	16	64	1.56
128	17	68	1.47
122	18	72	1.39
119	19	76	1.32
116	20	80	1.25
113	21	84	1.19
109	22	88	1.14
106	23	92	1.09
104	24	96	1.04

Probability distribution curve for the above calculated values of probability in percent (column 3) vrs. flood data (column 1) are plotted in a semi-log graph paper and probability distribution curve is drawn (Fig. 7.2). Using this curve, flood data at 20 and 40 percent probability levels are computed as 172 and 148 m³/sec, respectively.

Now we have to compute the flood data at 2 and 20 years return period. 2 and 20 years of return period are equal to 50 and 5 percent of probability levels, respectively. Using Fig. 7.2, flood data at 50 and 5 percent probability levels are estimated as 145 and 228 m³/sec, respectively.

There are a number of other formulae available for computing plotting position. However, Weibull's plotting position is the most widely used formula to compute the plotting position. Table 7.2 lists different used formulae to compute the plotting position. Each method has some advantages and disadvantages also. California method has a disadvantage that it gives maximum probability value of 100% when rank number equals to the total number of data in the series i.e. $m = N$ which is difficult to plot in a probability paper.

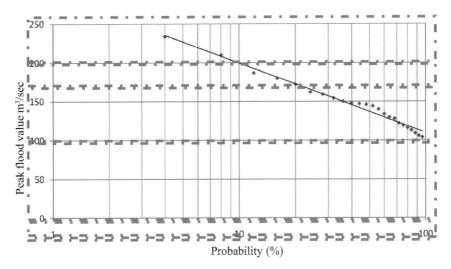

Fig. 7.2: Probability distribution curve for Sample calculation 7.3

Table 7.2: Various plotting position formulae (Suresh, 2008)

Sl. No.	Formula name	Plotting position or probability, P, %
1.	California (1923)	m/N
2.	Hazen (1914)	$(m - 0.5)/N$
3.	Weibull (1939)	$m/(N + 1)$
4.	Beard (1943)	$(m - 0.31)/(N + 0.38)$
5.	Chegodayev (1955)	$(m - 0.30)/(N + 0.40)$
6.	Blom (1958)	$(m - 3/8)/(N + 1/4)$
7.	Tukey (1962)	$(3m - 1)/(3N + 1)$
8.	Gringorten (1963)	$(m - 0.44)/(N + 0.12)$
9.	Cunnane (1978)	$(m - 0.4)/(N + 0.20)$
10.	Adamowski (1981)	$(m - 1/4)/(N + 1/2)$

7.6 Hydrologic Frequency Analysis

Chow (1951) has shown that most of the frequency distribution functions applicable in hydrologic studies can be expressed by the *general equation of hydrologic frequency analysis* which is expressed as:

$$X_T = \overline{x} + K\sigma \qquad (7.18)$$

where, X_T = value of the variate X of a random hydrologic series with a return period of T years, $\overline{x}$ = mean of the variates, σ = standard deviation of the variates and K = frequency factor which depends on return period T and the used frequency distribution. Standard books on statistics give the relationship of K-T (frequency factor vrs. return periods) for various frequency distributions in the form of tables. A number of frequency distributions are used to compute the hydrologic variates including flood data. In this chapter a few important and frequently used frequency distributions are discussed. Details of various theoretical probability distributions including discrete and continuous distributions and frequency distributions are presented and discussed in Chapter 8 of this book.

7.6.1 Foster Method

Eqn. (7.18) is used to compute the value of the variate at any return period, T. The frequency factor, K is computed by the procedures as outlined below.

- For the given data series, find out the coefficient of skewness (C_s) of the data by the formula:

$$C_s = \frac{\sum (x_i - \bar{x})^3}{(N-1)\sigma^3}$$

(7.19)

where, x_i = value of variate, $\bar{x}$ = mean of the variates, σ = standard deviation of the variates and N = number of data i.e. variates. The summation is carried out from $i = 1$ to $i = n$.

If the length of record is shorter, then C_s has to be adjusted. For *Foster Type I distribution*, adjust C_s by multiplying it with factor ($1 + 6/N$). For *Foster Type III distribution*, adjust C_s by multiplying it with factor ($1 + 8.5/N$) (Patra, 2001).

- Find out the values of frequency factor, K from Table 7.3 for *Foster Type I distribution* and Table 7.4 for *Foster Type III distribution*. Use the adjusted values of coefficient of skewness for each type of distribution at a given value of probability level in percent (frequency in percent). If return period is given then convert it to probability or frequency in percent as discussed earlier. For example return period of 100 years is equal to 1/ 100 = 0.01 = 1%. Then take this value of frequency in percent and use Table 7.3 or Table 7.4 for Type I or Type III distribution for computed values of adjusted C_s. Sample calculation 7.4 helps in computation of frequency factor, K by Foster distribution.

- Find out the mean and standard deviation of the given data series. Mean and standard deviation are computed by Eqn. (7.20) and (7.21), respectively as:

$$\bar{x} = \sum_{1}^{n} x_i / N$$

(7.20)

and $\sigma = \sqrt{\dfrac{\sum_1^n (x_i - \bar{x})^2}{(N-1)}}$ (7.21)

where, the terms are defined earlier.

- Using the values of mean, standard deviation and frequency factor, compute the value of the variate at any return period by Eqn. (7.18).

- For a set of values of return periods, find out the values of variates. Now plot the data of variates and their return periods in a probability paper or semi-log graph paper taking variates in ordinate and probability in abscissa and draw curve.

- From the developed curve, determine the value of the variate at any return period.

Sample Calculation 7.4

The peak flood data of a gauging station from 1990 to 2013 are given below. Determine the flood peak values for 100 and 1000 years by *Foster Type I* and *Foster Type III* distribution method.

Year	1990	1991	1992	1993	1994	1995	1996	1997	1998	1999	2000	2001
Peak flood, m³/sec	180	187	234	210	162	172	148	154	150	159	140	144

Year	2002	2003	2004	2005	2006	2007	2008	2009	2010	2011	2012	2013
Peak flood, m³/sec	116	122	128	119	147	146	130	134	106	109	104	113

Solution

Given total number of data, $N = 24$.

The mean of the variates ($\bar{x}$) is computed by Eqn. (7.20) and is obtained as $3514/24 = 146.4$ m³/sec. Using the mean and variates data for different years, the parameters $(x - \bar{x})$, $(x - \bar{x})^2$ and $(x - \bar{x})^3$ are computed and presented in table below.

Year	Flood data, m³/sec (x)	$(x - \overline{x})$	$(x - \overline{x})^2$	$(x - \overline{x})^3$
1990	180	33.6n	1128.96	37933.06
1991	187	40.6	1648.36	66923.42
1992	234	87.6	7673.76	672221.4
1993	210	63.6	4044.96	257259.5
1994	162	15.6	243.36	3796.42
1995	172	25.6	655.36	16777.22
1996	148	1.6	2.56	4.09
1997	154	7.6	57.76	438.98
1998	150	3.6	12.96	46.66
1999	159	12.6	158.76	2000.38
2000	140	-6.4	40.96	-262.14
2001	144	-2.4	5.76	-13.82
2002	116	-30.4	924.16	-28094.50
2003	122	-24.4	595.36	-14526.80
2004	128	-18.4	338.56	-6229.50
2005	119	-27.4	750.76	-20570.80
2006	147	0.6	0.36	0.22
2007	146	-0.4	0.16	-0.06
2008	130	-16.4	268.96	-4410.94
2009	134	-12.4	153.76	-1906.62
2010	106	-40.4	1632.16	-65939.30
2011	109	-37.4	1398.76	-52313.60
2012	104	-42.4	1797.76	-76225.00
2013	113	-33.4	1115.56	-37259.70
Total		0.40	24649.84	749648.5

Table 7.3: Frequency factor, K for Foster Type I distribution

C_s	Frequency (%)										
	0.0001	0.001	0.01	0.1	1	5	20	50	80	95	99
0	2.62	2.59	2.53	2.39	2.08	1.64	0.92	0	-0.92	-1.64	-2.08
0.2	3.00	2.94	2.83	2.66	2.25	1.72	0.89	-0.05	-0.93	-1.56	-1.91
0.4	3.44	3.35	3.18	2.95	2.42	1.79	0.87	-0.09	-0.93	-1.47	-1.75
0.6	3.92	3.80	3.59	3.24	2.58	1.85	0.85	-0.13	-0.92	-1.38	-1.59
0.8	4.43	4.27	4.00	3.55	2.75	1.90	0.83	-0.17	-0.91	-1.30	-1.44
1.0	4.95	4.75	4.42	3.85	2.92	1.95	0.80	-0.21	-0.89	-1.21	-1.30
1.2	5.50	5.25	4.83	4.15	3.09	1.99	0.77	-0.25	-0.86	-1.12	-1.17
1.4	6.05	5.75	5.25	4.45	3.25	2.03	0.73	-0.29	-0.83	-1.03	-1.06
1.6	6.65	6.25	5.57	4.75	3.40	2.07	0.69	-0.32	-0.80	-0.95	-0.96
1.8	7.20	6.75	6.08	5.05	3.54	2.10	0.64	-0.35	-0.76	-0.87	-0.87
2.0	7.80	7.25	6.50	5.35	3.67	2.13	0.58	-0.37	-0.71	-0.79	-0.80

Table 7.4: Frequency factor, K for Foster Type III distribution

C_s	Frequency (%)										
	0.0001	0.001	0.01	0.1	1	5	20	50	80	95	99
0	4.76	4.27	3.73	3.09	2.33	1.64	0.84	0	-0.84	-1.64	-2.33
0.2	5.48	4.84	4.16	3.08	2.48	1.69	0.83	-0.03	-0.85	-1.58	-2.18
0.4	6.24	5.42	4.60	3.67	2.62	1.74	0.82	-0.06	-0.85	-1.51	-2.03
0.6	7.02	6.01	5.04	3.96	2.77	1.79	0.80	-0.09	-0.86	-1.45	-1.88
0.8	7.82	6.61	5.48	4.25	2.90	1.83	0.78	-0.13	-0.86	-1.38	-1.74
1.0	8.63	7.22	5.92	4.54	3.03	1.87	0.76	-0.16	-0.86	-1.31	-1.59
1.2	9.45	7.85	6.37	4.82	3.15	1.90	0.74	-0.19	-0.85	-1.25	-1.45
1.4	10.28	8.50	6.82	5.11	3.28	1.93	0.71	-0.22	-0.84	-1.18	-1.32
1.6	11.12	9.17	7.28	5.39	3.40	1.96	0.68	-0.25	-0.82	-1.11	-1.19
1.8	11.96	9.84	7.75	5.66	3.50	1.98	0.64	-0.28	-0.80	-1.03	-1.08
2.0	12.81	10.51	8.21	5.91	3.60	2.00	0.61	-0.31	-0.78	-0.95	-0.99
2.2				6.20	3.70	2.01	0.58	-0.33	-0.75	-0.89	-0.90
2.4				6.47	3.78	2.01	0.54	-0.35	-0.71	-0.82	-0.83
2.6				6.73	3.87	2.01	0.51	-0.37	-0.68	-0.76	-0.77
2.8				6.99	3.95	2.02	0.47	-0.38	-0.65	-0.71	-0.71
3.0				7.25	4.02	2.02	0.42	-0.40	-0.62	-0.66	-0.67

Standard deviation is computed by the Eqn. (7.21) as:

$$\sigma = \sqrt{\frac{\sum_1^n (x_i - \bar{x})^2}{(N-1)}} = \sqrt{\frac{24649.84}{(24-1)}} = 32.73 \text{ m}^3/\text{sec}$$

Coefficient of skewness is given by Eqn. (7.19)

$$C_s = \frac{\sum (x_i - \bar{x})^3}{(N-1)\sigma^3} = \frac{749648.5}{(24-1) \times 32.73^3} = 0.93$$

Adjusted C_s for *Foster Type I* distribution $= C_s x \ (1 + K/N) = 0.93$ x $(1 + 6/24)$ $= 1.16$

We have to estimate the variate (flood peak data) at 100 and 1000 years of return periods by *Foster Type I* distribution.

$T = 100$ years is equal to 1 percent frequency and $T = 1000$ years is equal to 0.1 percent frequency. Using Table 7.3, frequency factor, K for adjusted C_s value of 1.16 at 1 and 0.01 percent frequency is estimated as 2.216 and 2.606.

Using Eqn. (7.18), flood peaks are computed as:

$$X_{100} = \bar{x} + K \sigma$$
$$= 146.4 + 2.216 \text{ x } 32.73 = 218.93 \text{ m}^3/\text{sec}$$
$$X_{1000} = \bar{x} + K \sigma$$
$$= 146.4 + 2.606 \text{ x } 32.73 = 231.69 \text{ m}^3/\text{sec}$$

Adjusted C_s for *Foster Type III* distribution $= C_s x \ (1 + K/N) = 0.93$ x $(1 + 8.5/24) = 1.26$

Using Table 7.4, frequency factor, K for adjusted C_s value of 1.26 at 1 and 0.1 percent frequency is estimated as 3.189 and 4.907.

Using Eqn. (7.18), flood peaks are computed as:

$$X_{100} = \bar{x} + K \sigma$$
$$= 146.4 + 3.189 \text{ x } 32.73 = 250.78 \text{ m}^3/\text{sec}$$
$$X_{1000} = \bar{x} + K \sigma$$
$$= 146.4 + 4.907 \text{ x } 32.73 = 307.01 \text{ m}^3/\text{sec}$$

7.6.2 Chow's Regression Method

Chow proposed the following linear equation to predict the flood value at a return period, T year:

$$Q_T = a + b X_T \tag{7.22}$$

where, Q_T is the flood value at a return period of T years, a and b are constants of regression computed by method of least square and X_T is given as:

$$X_T = \log \log \left(\frac{T}{T-1} \right) \tag{7.23}$$

Values of constants a and b are computed by method of least square as:

$$\sum Q_T = a N + b \sum X_T \tag{7.24}$$

$$\text{and} \sum Q_T X_T = a \sum X_T + b \sum X_T^2 \tag{7.25}$$

The return period or frequency (T) is given as:

$$T = 1/P = \frac{N+1}{m} \tag{7.26}$$

Using Eqns.(7.23) and (7.26) we get:

$$X_T = \log \log \left(\frac{N+1}{N+1-m} \right) \tag{7.27}$$

where, N is total number of data and m is rank number when the data are arranged in descending order. Following sample calculation helps in using the Eqns. (7.22) to (7.27) to predict the flood data at a return period of T years by Chow's regression method.

Sample Calculation 7.5

Solve sample calculation 7.4 by Chow's approach.

Solution

Calculations are shown in table below.

$$\sum Q_T = 3514 \qquad\qquad \sum X_T = -14.210$$

$$\sum Q_T X_T = -2441.18 \qquad\qquad \sum X_T^2 = 13.760$$

Using Eqn. (7.24), $\sum Q_T = a N + b \sum X_T$

$3514 = 24\ a\ -14.210\ b$

Using Eqn. (7.25), $\sum Q_T X_T = a \sum X_T + b \sum X_T^2$

$-2441.18 = -14.210\ a + 13.760\ b$

Solving the above two equations, we get $a = 106.48$ and $b = -67.45$

Hence Eqn. (7.22) now becomes $Q_T = 106.48 - 67.45\ X_T$

We have to compute the flood data at 100 and 1000 years return period.

Put $T = 100$ and 1000, respectively in Eqn. (7.23) i.e.

$$X_T = \log\log\left(\frac{T}{T-1}\right)$$

$$X_{100} = \log\log\left(\frac{100}{100-1}\right) = -2.360$$

$$X_{1000} = \log\log\left(\frac{1000}{1000-1}\right) = -3.362$$

(1) Year	(2) Flood data, Q_T, m³/sec	(3) Flood data in descending order, Q_T, m³/sec	(4) Rank number, m	(5) $\log\log[(N+1)/(N+1-m)] = X_T$	(6) $X_T \cdot Q_T$ (col. 5 x col. 3)	(7) X_T^2 (col. 5 x col. 5)
1990	180	234	1	-1.75	-409.809	3.067127
1991	187	210	2	-1.44	-302.641	2.0769
1992	234	187	3	-1.26	-234.792	1.57646
1993	210	180	4	-1.12	-201.741	1.25616
1994	162	172	5	-1.01	-174.345	1.027449
1995	172	162	6	-0.92	-149.651	0.853357
1996	148	159	7	-0.85	-134.462	0.715166
1997	154	154	8	-0.78	-119.505	0.602189
1998	150	150	9	-0.71	-106.890	0.507801
1999	159	148	10	-0.65	-96.783	0.427641
2000	140	147	11	-0.60	-88.041	0.35871
2001	144	146	12	-0.55	-79.810	0.298866
2002	116	144	13	-0.50	-71.510	0.24655
2003	122	140	14	-0.45	-62.703	0.200599
2004	128	134	15	-0.40	-53.624	0.160146
2005	119	130	16	-0.35	-45.878	0.124548
2006	147	128	17	-0.31	-39.107	0.093346
2007	146	122	18	-0.26	-31.403	0.066254
2008	130	119	19	-0.21	-24.723	0.043163
2009	134	116	20	-0.16	-18.043	0.024193
2010	106	113	21	-0.10	-11.204	0.009831
2011	109	109	22	-0.035	-3.905	0.001283
2012	104	106	23	0.040	4.258	0.001614
2013	113	104	24	0.145	15.131	0.021167
Total	3514	3514		-14.210	-2441.18	13.760

Q_{100} = *106.48- 67.45 x -2.360* = 265.662 m³/sec

and Q_{1000} = *106.48- 67.45 x -3.362* = 333.25 m³/sec

7.6.3 Stochastic Method

For the computation of annual flood peak values based on probability law, a well known equation (Patra, 2001) used is:

$$Q_T = Q_{min} + 2.303 \left(Q_{av} - Q_{min} \right) \log \left(\frac{nT}{N} \right) \tag{7.28}$$

where, T (return period) is given by California formula as $T = N/m$, n is number of recorded floods which assigns only one value to the same event, N is total number of data, Q_{min} is minimum value of the data in the series, Q_{av} is average of the data series and Q_T is flood value at a return period of T years. If a particular data, say 200 m³/sec occurs three times in a data series containing 10 numbers of data, then $n = 8$ and $N = 10$. Following sample calculation helps in using the stochastic method.

Sample Calculation 7.6

Flood data of a basin recorded for 24 years are same as that mentioned in sample calculation 7.4 except for the years 2005 and 2006 when data recorded are 129 m³/sec each for both the years. Estimate the flood data at 100 and 1000 years return periods by stochastic method.

Solution

Using the data of sample calculation 7.4, we get $Q_{min} = 104$ m³/sec, $Q_{av} = 146.4$ m³/sec and $N = 24$. Flood data of 2005 and 2006 are same and so $n = 23$.

Flood peak at 100 years return period is computed by Eqn. (7.28) as:

$$Q_T = Q_{min} + 2.303 \left(Q_{av} - Q_{min} \right) \log \left(\frac{nT}{N} \right)$$

= 104 + 2.303 (146.4 − 104.0) log [(23 x 100)/24] = 297.49 m³/sec

Flood peak at 1000 years return period is computed as:

= 104 + 2.303 (146.4 – 104.0) log [(23 x 1000)/24] = 395.14 m³/sec

7.7 Regional Flood Frequency Analysis

Regional analysis is adopted in a catchment where the availability of data is too short to conduct the frequency analysis. In such a case we can consider the data of neighbouring catchments which are hydro-meteorologically similar. However, the available long term data of the neighbouring stations are tested for homogeneity before the data of these catchments are considered. These group of stations are constitute a region and all the station data of this region are pooled and analysed as a group to find the frequency characteristics of the region. The mean annual flood (Q_{ma}) which corresponds to a return period of 2.33 years is used for non-dimensionalising the results. The variations of Q_{ma} with catchment area and the variation of Q_T/Q_{ma} with return period, T where Q_T is the flood discharge for ant T are the basic plots prepared in this analysis.

7.8 Partial Duration Series

In the study of annual hydrologic data series, only the maximum value of flood in a year is considered. It may, sometimes, happen in some of the catchments that there is more than one independent flood in a year and many of these may be appreciably high. We may have large number of data and in some years, we may have also two or more than two large data. In such case, we have to screen out the data by assigning some threshold value that floods of magnitude larger than an arbitrarily chosen value (threshold value) are included in the analysis. Such a data series is called as *partial duration series*. In partial duration series, it is necessary to see that all the events are independent and as such it is suitable to use in rainfall analysis where conditions of independency of events are easy to establish. The return period of annual series (T_A) and that of the partial duration series (T_P) are related as (Subramanya, 2003):

$$T_P = \frac{1}{\ln T_A - \ln (T_A - 1)} \tag{7.29}$$

Eqn. (7.29) indicates that for $T_A < 10$ years, the difference between T_A and T_P is significant and when $T_A > 20$, difference is very small.

7.9 Risk, Reliability and Safety Factor

7.9.1 Risk and Reliability

The design of any hydraulic structure involves some risk of failure. It may not be 100% perfect. The reason is due to several hydrological factors including the reliability of getting true and representative data of the basin for which the structure is to be designed. If there is any instrumental or human error in measurement of discharge data or stage data, then the design of the structure involving these data may lead to some error in design f the hydraulic structures. There are many such errors which influence the design of the structure. We can consider one more example say a weir is designed with an expected life of 50 years and is designed with flood discharge at 100 years return period. The weir may fail if a flood magnitude larger than the designed flood occurs within the life period of 50 years.

The probability of occurrence of an event $(x \geq x_T)$ at least once over a period of n successive years is known as *risk*. The risk ($\overline{R}$) is defined as:

$\overline{R}$ = 1 – (probability of non-occurrence of an event $x \geq x_T$ in n years.

$= 1 - (1 - P)^n$

$= 1 - (1 - 1/T)^n$ \hfill (7.30)

where, P is the probability of exceedence $(x \geq x_T) = 1/T$ where T is return period, years.

The reliability (R_e) is defined as:

$R_e = (1-P)^n = (1 - 1/T)^n$ \hfill (7.31)

7.9.2 Safety Factor

There are a number of causes including structural, constructional, operational and environmental causes which influences the design of a hydraulic structure. In addition, technological considerations such as economic, sociological and political factors may also influence the structures. So, any hydraulic structure should have some safety factor in its design. The safety factor for a given hydrological parameter M is defined as under (Subramanya, 2003).

Safety factor for parameter $M = (SF)_m$

$$= \frac{Actual\ value\ of\ parameter\ M\ adopted\ in\ design\ of\ the\ project}{Value\ of\ the\ parameter\ M\ obtained\ from\ hydro\log ical\ consideraion\ only}$$

$$= \frac{C_{am}}{C_{hm}} \hfill (7.32)$$

The parameter, M includes items like flood discharge values, maximum stage of rivers, reservoir capacity and free board. In hydraulic design we sometime use a term *safety margin* which is $(C_{am} - C_{hm})$.

Sample Calculation 7.7

A hydraulic structure is designed with a life span of 50 years. It is designed considering peak flood discharge with return period of 100 years. (i) What is the risk of the hydrologic design? (ii) Find out the reliability of design too. (iii) Considering 20% risk, what return period will have to be adopted?

Solution

Given life span, $n = 50$ years

Return period, $T = 100$ years

(i)   Risk (Eqn. 7.30) is given as:

$$\overline{R} = 1 - (1 - 1/T)^n$$
$$= 1 - (1 - 1/100)^{50} = 0.395 = 39.5\%.$$

(ii)   Reliability, $R_e = 1 - \overline{R} = 1 - 0.395 = 0.605 = 60.5\%$.

(iii)   Given $\overline{R} = 20\% = 0.20$

$$0.20 = 1 - (1 - 1/T)^{50}$$

Hence, $T = 225$ years.

Sample Calculation 7.8

Annual flood data of river Baitarani at Anandapur covering a period of 50 years yielded a mean and standard deviation of 25,000 m³/sec and 10,000 m³/sec, respectively. It is proposed to construct a bridge across the river near this site with a risk value of 15%. The proposed life span of the structure is 100 years. Estimate the flood discharge by Gumbel's distribution which is used for design of the structure. If the actual flood value adopted in the design is 80,000 m³/sec, compute the safety factor and safety margin relating to maximum flood discharge.

Solution

$$0.15 = 1 - (1 - 1/T)^{100}$$

$$T = 616 \text{ years}$$

Gumbel's method is used for design. Given number of years of data of flood discharge used in design of the structure $N = 50$ years

Using Table 8.5 (Chapter 8), reduced mean, for $N = 50$ years is 0.5485. Using Table 8.6 (Chapter 8), reduced standard deviation, for $N = 50$ years is 1.1607.

Now using Eqn. (8.53) of Chapter 8 we have

$$y_T = - \ln\left\{ \ln\left(\frac{T}{T-1} \right) \right\}$$

$$= - \ln\left\{ \ln\left(\frac{616}{616-1} \right) \right\} =$$

$$= 6.422$$

Now using Eqn. (8.58) of Chapter 8 we have

$$K = \frac{y_T - \overline{y_n}}{S_n}$$

$$= (6.422 - 0.5485) / 1.1607 = 5.06$$

Value of the variate i.e. floods data at 616 years return period is then computed by Eqn. (8.57) Chapter 8 is:

$$x_{616} = \overline{x} + K\, \sigma_{n-1}$$

$= 25000 + 5.06 \times 10000 = 75600\ \text{m}^3/\text{sec}$ (hydrological design flood magnitude).

Actual flood value adopted in the design $= 80,0000\ \text{m}^3/\text{sec}$.

Hence, safety factor (Eqn. 7.32) $= 80,000 / 75,600 = 1.06$

Safety margin for flood magnitude $= 80,000 - 75,600 = 4,400\ \text{m}^3/\text{sec}$.

7.10 Flood Forecasting

Floods cause a lot of damages and sometimes create catastrophes causing losses to animal lives and crops in vast areas. It disrupts roads, bridges etc causing communication problems. It also damages many buildings and hydraulic structures and wash away crop fields. Thus, it affects the gross domestic income of a country. If a number of cascade reservoirs are available, then these reservoirs can modulate the flood peak and sometimes may store all the flood water. But in developing countries, it is costly and hence cannot be afforded. Sometimes, the siltation of reservoirs reduces the water storage capacity which aggravates the flood problems. Occupation by habitats in the flood plains, construction of buildings and industries in the flood plain also creates havocs due to flood in the flood plain and deltaic regions. In addition, man made factors like deforestation, shifting cultivation, improper watershed management and after all climatic change affects the flood situation in the catchments. Under such circumstances, it is better that along with flood control measures, flood forecasting measures need to be emphasized so that people can be aware of the situation and can take effective steps to avert the alarming situation. They can be evacuated from the flood would hit areas beforehand. This saves enormous lives and national economy. Thus, flood forecasting plays an important role for preventing the consequences of floods by giving an advance ultimatum to the people dwelling in those areas. The flood forecasting should give accurate and meaningful information. An inaccurate forecasting may lead to loss of faith of the people with the system which later on may create a lot of problem in evacuation program.

Sometimes, we use the term river forecasting. River forecasting means determination of flood stages at a section in downstream location at definite time in advance from the knowledge of the information of the upstream location. Such forecasting may be short term or long term type. Hydrologic forecasting includes both short term and long term forecasting. Hydrologic forecasting includes information like rainfall, infiltration, antecedent precipitation index (API), river stages and other data as input and runoff hydrograph or the peak flows as output. A short term forecast is used to range for a few hours whereas a long term forecast ranges for a few days. Followings are the different flood forecasting methods.

7.10.1 Method I

1. Computations of flood hydrograph at an upstream location is made by either of the following ways:

    (i) Using antecedent precipitation index (API) and a coaxial graph method, the storm runoff depth is estimated. From the knowledge of unit hydrograph for the basin, runoff can be distributed over the time span.

    (ii) Average storm rainfall over the basin can be used to construct a rainfall histogram. A loss rate when separated on the basis of the past experience of the floods gives an effective rainfall hyetograph (ERH). The ERH is convoluted with the unit hydrograph for the basin to give necessary DRH. When the base flow is added to the ordinates of DRH, we get a storm or flood hydrograph. This method has been discussed in the chapter Unit Hydrograph.

(iii) Flood magnitudes at the outlet of a basin can also be estimated from various available rainfall and runoff models.

2. Muskingum method (discussed in Chapter 9) can be employed to route the flood hydrograph at any desired location at the downstream end of the river so that flood hydrograph at that location can be found out.

7.10.1.1 Antecedent Precipitation Index and Coaxial Graph Method to Compute Runoff

Antecedent precipitation index (API) is expressed as:

$$API = a_1 P_1 + a_2 P_2 + a_3 P_3 + \ldots\ldots\ldots + a_t P_t \tag{7.33}$$

where, $a_1, a_2, a_3 \ldots a_t$ are coefficients with values less than 1.0 and $p_1, p_2, p_3 \ldots p_t$ are the precipitations at 1-day, 2-day, 3-days .... t- days before the present day. It is assumed that a_t decreases with time. It is found that $a_t = K^t$ where, t is time in days and K is a constant ranging from 0.30 to 0.95. Replacing the value of $a_t = K^t$ in Eqn. (7.33) we get:

$$API = K^1 P_1 + K^2 P_2 + K^3 P_3 + \ldots\ldots\ldots + K^t P_t \tag{7.34}$$

Following sample calculation helps in understanding how to compute API.

Sample Calculation 7.9

Compute API from the given rainfall data for the day of 11th July, 2014, 12th July, 2014 and 16th July, 2014. Assume value of K =0.85.

Date	10/7/ 2014	11/7/ 2014	12/7/ 2014	13/7/ 2014	14/7/ 2014	15/7/ 2014	16/7/ 2014	17/7/ 2014
Rainfall, mm	10.0	0	0	5	0	0	20	0

Solution

API on 11th July, 2014 due to rainfall of 10 mm on 10th July, 2014 is $= 0.85^1$ x 10 $= 8.5$ mm.

API on 12th July, 2014 $= 0.85^2$ x 10 $+ 0.85^1$ x 0 $= 7.225$ mm.

Similarly API on 16th July, 2014 $= 20 + 0$ x $0.85^1 + 0$ x $0.85^2 + 5$ x $0.85^3 + 0$ x $0.85^4 + 0$ x $0.85^5 + 10$ x $0.85^6 = 20 + 3.07 + 3.77 = 26.84$ mm

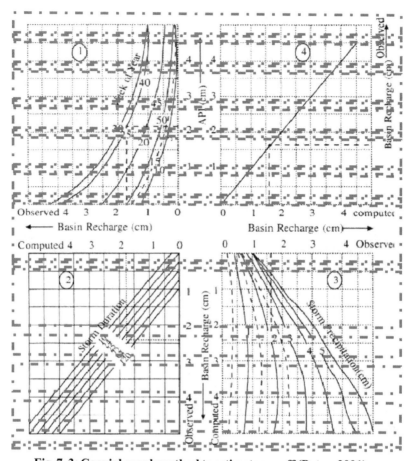

Fig. 7. 3: Coaxial graph method to estimate runoff (Patra, 2001)

To estimate runoff from a known API value, a typical coaxial graph (Fig. 7. 3) is used (Patra, 2001). In the coaxial graph, block 1 gives a relation between the time period (week of the year) versus API. Knowing API and week of the year, a point in block 1 is located. Then we move from block 1 to block 2 vertically down from the point located in the block 1 and with known duration of the storm in hours, a point in block 2 is located. From block 2 we move horizontally to block 3 which contain storm precipitation depth. This horizontal line from known storm duration of block 2 touches the storm precipitation depth line of block 3. Finally we move from block 3 to block 4 which give storm runoff in cm. Following data are needed to proceed in this procedure:

- Value of API
- Storm duration

- Storm precipitation depth
- Coaxial chart of the basin

Once we determine the runoff depth from the storm precipitation depth, next we distribute it over time in the basin. This is done by knowledge of data of time distribution of storm rainfall from a self recording raingauge. Time distribution of storm rainfall is convoluted over the unit hydrograph of the basin to get the flood hydrograph. The obtained flood hydrograph is routed by Muskingum method (discussed in Chapter Flood Routing) to obtain the peak as well as flood hydrograph at downstream location in the river.

7.10.2 Method II

In this method, a multiple correlation between the stages of upstream end of a river with that of the forecasting station at the downstream end of the same river is established. This established correlation between the stages helps in forecasting the flood at the downstream end in the river by knowing the stages at the upstream end. Multiple stage correlation is now discussed here.

A simple gauge to gauge relation between the two stations at up and down stream end in a river basin is quite helpful to predict flood at downstream end provided we can develop such a relation for all stages. Such relationships are generally used in India to forecast flood. However inclusion of other parameters like rainfall and API for the basin between the base station and the forecasting station gives a more realistic forecast of flood. All these parameters are used to establish a multiple stage correlation which is discussed as below.

1. Let us consider a uniform stream section (river section) having uniform slope and cross section. Let there be no addition or loss of water between the two sections/stations of the river. Let us consider two stations i.e. (i) base station at upstream end of river basin and (ii) forecasting station at downstream end of the river basin. Let stage at Nth hour of the base station is known. Let the flood water takes T hour (lag time) to reach a particular forecasting station in the downstream end of the river. Let the stage at $(N + T)$th hour at the downstream end (forecasting station) of the river is known. For different known values of stages at Nth and $(N + T)$th hour between the two stations, we can develop a coaxial relationship such as a one shown in Fig 7.4. Once we know the stage at the downstream end and the stage-discharge relation at the downstream end is known, then we can forecast the flood/discharge value at the downstream end for the known value of the stage.

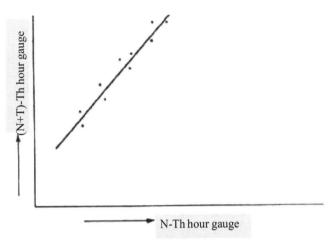

Fig. 7.4: Correlation between N and (N + T)ᵗʰ hour stations

2. A correlation between the change in stage/gauge of the upstream end and of the base station and change in stage of the downstream forecasting station in *T* hour is plotted (Fig. 7.5).

   For a large river with permanent control, such a relation fives good result. When a change in the stage at the base station takes place, the corresponding change at the forecasting station after the lag time can easily be predicted from such relation.

3. There is another way to correlate the stages in which we plot a relation between *N*th and (*N* + *T*)th hour stages of the base and forecasting stations with change in the stage of the upstream base station during *T* hour as shown in Fig. 7.6 (Patra, 2001).

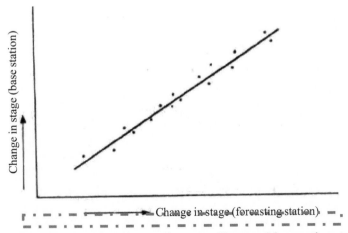

Fig. 7.5: Correlation between changes in stage of base and forecasting station

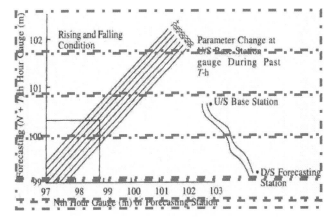

Fig. 7.6: Correlation graph between N^th and (N + T)^th hour stages at forecasting stations (Patra, 2001)

4. The principle of multiple correlations can be extended to forecast flood at the downstream end of a river when one or more than one tributary join the main river at its upstream end. In such case a graph paper is divided into quadrants and in each quadrant, change in the stage levels of different tributaries is plotted forming a multiple correlation curve. To incorporate time of travel of water in different tributaries which varies according to distance of the base station from the forecasting station and the value of the flood in the tributaries, the parameter is considered as the third variable in each quadrant. For example if tree tributaries join a main river, the plot covers three quadrants of the graph paper. This type of flood forecasting is very common in India.

Sample Calculation 7.10

A tributary joins a main river at its upstream end of a flood forecasting station. Data of the flood stages of previous storms for the main river and its tributary at the upstream end and that at the downstream end of the forecasting station of the main river are given as below. Develop a flood forecasting model for the system. Find out the coefficient of correlation between the stage data of all these three stations.

Stage, U/S main river, m	165.0	165.5	166.1	166.5	170.0	171.2	171.6	172.6	173.0
Stage, U/S tributary, m	112.0	112.8	113.2	113.7	114.0	114.4	114.8	115.1	115.3
Stage, D/S forecasting station, m	80.0	80.6	82.1	83.4	85.0	86.1	86.8	87.2	88.0

Solution

Let the forecasting model relating stages upstream of main and tributary river and that of the downstream forecasting station on the main river be in the form of

$$X_1 = a + b X_2 + c X_3$$

where, X_1 is the stage of U/S in main river, X_2 is that of tributary, X_3 is that of main river at D/S where forecasting station is located and a, b and c are the coefficients. We can find out the values of a, b and c by method of least square. As discussed earlier, the above equation can be written as:

$$\sum X_1 = a N + b \sum X_2 + c \sum X_3$$

$$\sum X_1 X_2 = a \sum X_2 + b \sum X_2^2 + c \sum X_2 X_3$$

$$\sum X_1 X_3 = a \sum X_3 + b \sum X_2 X_3 + c \sum X_3^2$$

The calculations are entered in table below.

X_1	X_2	X_3	$X_1.X_2$	X_2^2	$X_2.X_3$	$X_1.X_3$	X_3^2
(1)	(2)	(3)	(4)	(5)	(6)	(7)	(8)
165	112	80	18480	12544	8960	13200	6400
165.5	112.8	80.6	18668.4	12723.84	9091.68	13339.3	6496.3
166.1	113.2	82.1	18802.5	12814.24	9293.72	13636.8	6740.4
166.5	113.7	83.4	18931.0	12927.69	9482.58	13886.1	6955.5
170	114	85	19380	12996	9690	14450	7225
171.2	114.4	86.1	19585.2	13087.36	9849.84	14740.3	7413.2
171.6	114.8	86.8	19699.6	13179.04	9964.64	14894.9	7534.2
172.6	115.1	87.2	19866.2	13248.01	10036.7	15050.7	7603.8
173	115.3	88	19946.9	13294.09	10146.4	15224	7744
(1521)	(1025)	(759)	(173360)	(116814.3)	(86515.5)	(128422)	(64112)

N.B. Values in parenthesis in last row of the above table represents sum of values of individual columns.

Using the tabular values we get three equations as:

$1521 = 9 a + 1025 b + 759 c$

$173360 = 1025 a + 116814.3 b + 86515.5 c$

$128422 = 759 a + 86515.5 b + 64112 c$

Solutions of the above three equations gives $a = $ -11.232, $b = $ 1.059 and $c = 0.707$

Hence, the forecasting model relating stages upstream of main and tributary river and that of the downstream forecasting station on the main river is expressed as:

$$X_1 = -11.232 + 1.059 \, X_2 + 0.707 \, X_3$$

Now the coefficient of correlation between X_1 and X_2 is obtained as 0.952 (see Eqn. 8.77, Chapter 8).

Similarly, coefficient of correlation between X_1 and X_3 is obtained as 0.978 and that between X_2 and X_3 is 0.986.

7.11 Flood Control Measures

As discussed, flood creates a lot of damage in the river basins causing loss of lives and property. It is of utmost importance to take appropriate control measures along with forecasting measures to save the damage. The flood control measures are site specific. There are a number of factors that decide what protection and control measures are to be taken at a particular site for flood control. Selection of any particular measure depends on the experience, intelligence, design consideration, and suitability of the structure to fit the locality. Various flood control measures are described as under (Varshney, 1979):

- By constructing storage reservoirs and retarding basins
- By constructing dykes/levees and floodwalls
- By constructing diversion or by-pass channels
- Improvement of channel capacity
- By constructing spurs or groynes
- Adoption of soil conservation measures
- Flood plain management for flood control

7.11.1 Storage Reservoir

Reservoirs are formed when an obstruction is created across a river by constructing a dam across it so it creates a pull of water upstream the dam. It is built for multiple purposes including flood control, irrigation, navigation, inland fisheries, power generation etc. Thus a reservoir constructed across a river is used for multiple purposes. Depending on the storage capacity, a reservoir may be small like a tank, medium or large. In India, thousands of small reservoirs are constructed and are now used for irrigation and domestic uses. They help in modulating the flood peak during rainy season and thus avert the severe problems of floods. It is less effective in flood control works in a large river basin since operation of such a large number of small reservoirs during the flood peak is not synchronous and hence is rather difficult to regulate in a definite pattern. On the other hand, a large reservoir is easy to regulate. These reservoirs increase the economy of the region due to both direct and indirect benefits. From the reservoir, excess water than its safe storage capacity is discharged into the downstream side of the reservoir (area back to the dam) through spillways. The stored water from the upstream side of the reservoir is released through head regulator to the canals which serve as irrigation to the command areas. Irrigation water is supplied to the command through canal distribution systems constituting main canal, branch canal, distributaries, minors, sub-minors, water courses and finally to the crop fields. The storage capacity of a reservoir is discussed elsewhere in this book.

7.11.2 Dykes/Levees

Dykes or levees are simple earth dam sections. They are simple and economical structures constructed parallel to river flow section in the flood plain of the river at suitable distance away from the main channel. The structure controls flood effectively in the reach where it is constructed. It may be constructed at one or both sides of a river depending on the area to be protected. When a masonry wall or R.C.C. structure is constructed instead of an earthen section which is also parallel to the river flow, then the structure is called as flood wall. Flood wall is a type of retaining wall. Dykes are properly designed to withstand water pressure including sliding pressure likely to be developed on the wall. Sometimes, a second dyke is constructed on the same side of the river parallel to the first one which guards important places or where the flood wall is designed as retaining wall. Dykes are less costly and simple structures and locally available materials are generally used in its construction. However they are likely to breach requiring high maintenance cost. Failure due to wave action, seepage and flood impact are high. The structure also hampers in smooth drainage of water from the flood protected area. A large area is lost due to the construction of the levee near the river side. It deteriorates the river curse by high silt charge. It creates the problem of rising ground water table in the adjacent areas. Another disadvantage of levee is that there is chance of failure of the structure due to piping. Design of levee must consider the following parameters.

Cross Section: Cross section of levee should be such that it suits the site condition and materials available at the site. Height of the levee should be adequate so that there is no overtopping of flood water. It should be large enough to withstand the seepage pressure.

Top Width: If the levee is to be used as a road on which trucks/buses will fly or farm machineries are allowed to move on it, then the top width may be kept within 2.5 to 3.5 m.

Side Slope: Side slope depends on angle of repose of the soil materials. It may vary from 3:1 to 5:1 depending on the type of soil materials used for construction. The slope is more in downstream side than the upstream side i.e. flatter at the downstream side as compared to the upstream side. Height of the levees may be kept within 9 m.

Key Trench: It is provided at the base of the levee to control the flow of water from the base. It acts like an anchorage giving strength to the section and protects the structure from sliding. It is also called as muck ditch. Fig.7.7 shows the typical section of a levee with key trench.

Berm: Berm is used to protect the levee from failure due to overtopping. Berm width may vary from 3 to 4.5 m depending on the height of levee. If the height is more, then the berm width is also more. For levees of height less than 2.5 m, berm width may not be required. Side slope of the berm is taken as 4:1 (H:V).

Free Board: To protect the structure from overtopping due to large flood, a free board of height of 20 percent of the height of levee is provided at the top at high flood level (H.F.L.).

Stone pitching in the upstream slope side of the levee is done to protect the structure from wave action and flood impact. Sometimes, turfing, sodding tree or shrub planting at the upstream slope side are done to protect it from wave action.

A levee may fail because of (i) overtopping due to wav action or high level of flood water (ii) seepage flow through the cross section (iii) caving and piping action (iv) breaching due to saturation of levee's section and (v) direct flow of water through the cracks developed at the levee's section.

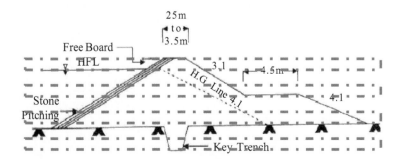

Fig. 7.7: Typical cross section of a levee (Sharma and Sharma, 2002)

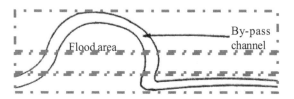

Fig. 7.8: View of a by-pass channel

7.11.3 Diversion or By-Pass Channels

Large cities are situated at the bank of the rivers. To protect these cities from flood problem, especially during rainy season, by pass channels are constructed to divert or by pass the floodwater of the river from the adjoining areas. Fig.7.8 shows a typical by pass channel. A by-pass channel acts like a branch channel which is taken out from the upstream end of a river section. It protects the area from peak flood impact downstream of the section by diverting a part of flood water. In India, a diversion channel taken off at 5 km upstream of Srinagar city from the river Jhelum protects the city from flood.

7.11.4 Improvement of Channel Capacity

Lining of channels increases velocity of water and hence improves the channel carrying capacity. At times, cutoff is provided in the meandering channel to increase the flow velocity and hence improves the channel carrying capacity (Fig. 7. 9). Straightening the channel, widening and deepening the channel section also enhances the carrying capacity of the channel. When the channel carrying capacity increases, ultimately the stage of the river from which the channels originate decreases. This reduces the flood problem in the river. Desilting the channel and river sections, cleaning the channel sections periodically such that there is no weed growth and flood water can easily and quickly move in the channels, are some other ways to improve the channel capacity and reduce the flood hazards. The above mentioned measures are undertaken in the areas where levees or dykes cannot be constructed due to non-availability of space and storage reservoir cannot be constructed upstream of the section.

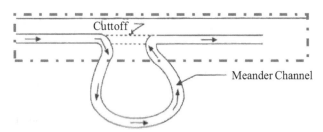

Fig. 7.9: Cutoff in meandering channel to improve channel capacity

7.11.5 Spurs or Groynes

Spurs or groynes are constructed projecting from the banks into the rivers which protect the sides of the rivers from flood impact and wave action. There are many types of groynes available on construction. When a solid groyne is constructed by earth material, the section is designed like an earth dam section. One of the main problems with groynes is to protect its nose which projects into the river. Large stones are normally dumped at the nose of the groynes to protect it from high water velocities and the section is also pitched by large stones. There are three types of groynes. They are (i) repelling groynes which project upstream into the river from the point of its origin from the bank (Fig. 7.10 a), (ii) attracting groynes which project downstream into the river from the point of its origin from the bank (Fig. 7.10 b), and (iii) perpendicular groynes which project perpendicular to the river from the point of its origin from the bank (Fig. 7.10 c).

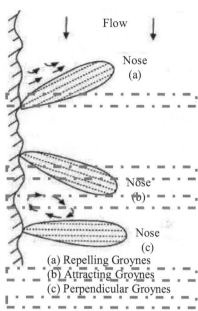

(a) Repelling Groynes
(b) Attracting Groynes
(c) Perpendicular Groynes

Fig. 7.10: View of different types of groynes used in flood protection

7.11.6 Soil Conservation Measures

Water enters into a reservoir from its catchment area. Proper care must be taken at the upstream side of the reservoir catchment area so that it does not get silted. Deposit of sediment or silt that is carried by the runoff from the upstream side of the catchment gets deposited in the reservoir and reduces its storage capacity. As a result, the reservoir cannot store maximum runoff during the flood time and hence flood water creates havoc in the downstream side of the reservoir. Various soil conservation measures must be taken at the upstream side of the catchment so that less runoff enters into the reservoir and silt free water enters into the reservoir. Various soil conservation measures are (i) biological or agronomic measures and (ii) engineering measures. The various biological measures are crop cultivation, buffer strip cropping, contour strip cropping, crop rotation, mixed cropping, intercropping, prohibiting shifting cultivation, prohibiting jhoom cultivation etc. The various engineering measures are construction of engineering structures like V-ditch, contour bund, terrace, bench terrace, contour trench, miniature bund, construction of water harvesting structures like farm pond, percolation tank etc. It is reported that a thick grass cover may reduce the runoff by 80 to 90 percent whereas row crops may reduce the same by about 50 to 65 percent, close growing crops reduce by 30 to 35 percent as compared to a barren land. Deep tillage, summer ploughing, addition of organic manures etc. also improves the infiltration capacity of the soil and reduce runoff and hence flood. Details of soil and water conservation measures are available in various standard text books of Soil And Water Conservation Engineering.

7.11.7 Flood Plain Management

The areas adjacent to the river where there are chances of frequent occurrence of floods should not be used for dwelling purposes. Rather these areas may be used for recreational or parks purposes so that the losses would be less. Further inundation of these areas would not hinder human population. The total area where flood occurs may be taken for flood plain management. Depending on the degree of hazard, flood plains are divided into different zones. The different zones are divided into the following three types:

- Warning zone
- Restrictive zone and
- Prohibitive zone

Effect of flood hazard is more in the warning zone. This zone extends from the river bank up to the boundary of the flood affected area. Adoption of flood protection measures is very much essential in these zones since risk of flood is more in these areas. The area ahead of the warning zone is restrictive zone which is also called as flood freezing zone. Effect of flood hazard is not that severe as compared to the warning zone. Construction of houses may be undertaken in this zone. Prohibitive zone extends from the middle of the river course to some areas adjoining both sides of the rivers. Effect flood hazard is severe in these zones and so these zones are completely prohibited for construction of houses.

7.12 Economics of Flood Control

Flood causes a lot of losses in terms of economics. On the other hand to tackle the flood situation and minimize the flood losses, flood control measures are taken up which requires huge sum for investment. Hence, it is essential to work out economic analysis of effective flood control management. The economic analysis may be taken up in terms of cost-benefit ratio (CBR) which consists of computing the total cost invested for constructing different flood control measures and benefits realized from them by protecting the areas against flood. Various cots involved include as follows:

- Capital cost which includes cost of investigation, planning, purchase of materials, payment of labours etc.

- Interest cost which is calculated as per the prevailing bank interest rate.

- Depreciation cost which is calculated on the basis of capital cost which may be taken as 5 to 10% of capital cost.

- Operation cost which varies as per the local situation.

- Repair and maintenance cost which is also taken as 5 to 10% of capital cost.

The benefits from the project of flood control are of two types i.e. (i) direct benefit and (ii) indirect benefit. Direct benefits are by preventing loss of land, human lives etc. The area saved due to drainage and waterlogging by flood control measures and consequently crop saved and its economic returns are included in the benefits. In addition, buildings, hydraulic structures etc. are saved and are included in benefit sides. The soil and water conserved by flood control measures are further included in benefit sides. Indirect benefit includes money saved under the insurance, benefits due to increase in crop production due to the protection of cultivable lands, benefits due to elimination of environmental pollution, diseases etc.

7.12.1 Benefit-Cost-Ratio Analysis

Benefit-cost-ratio (BCR) analysis forms main part of economic analysis of flood control works. In the analysis, total benefits including all direct and indirect benefits are calculated. Then total cost of flood control measures including capital, depreciation, operational cost, interest and repair and maintenance costs are also computed. Then reduction in the flood peaks due to flood control measures is calculated. In the next step, curves relating total benefits, total costs and reduction in flood peak values are drawn. Reduction in flood peak values are drawn as abscissa in the graph whereas costs and benefits are located as ordinate in the graph (Fig. 7.11). As shown in the Fig. 7.11, value of CBR is 1.0 at point A which indicates cost is same as benefit and hence the flood control project is viable. If value of BCR < 1, then the project is economically unviable. At point A, where value of CBR = 1, reduction in flood peak is maximum (Suresh, 2008).

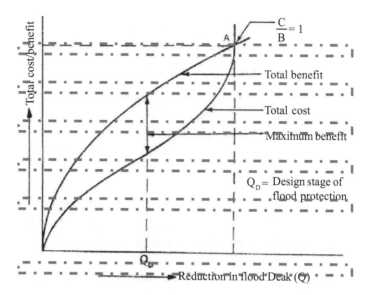

Fig. 7.11: Benefit-cost analysis of flood control (Suresh, 2008)

Question Banks

Q1. What is the literal meaning of flood? How does it occur? What are its consequences?

Q2. Briefly describe the flowing flood regions of India.

    (i)   Brahmaputra region

    (ii) Ganga region

    (iii) North-west region

    (iv) Central India and Deccan region

Q3. Write short notes on followings:

    (i)   Design flood

    (ii)   Maximum probable flood

    (iii) Maximum observed flood

    (iv) Standard project flood

    (v)   Peak flood

    (vi) Maximum known flood

    (vii) Annual flood and

    (viii) Ordinary flood and

Q4. Why is it essential to consider flood of maximum magnitude for design of storage structures?

Q5. Classify dams based on the storage capacity and hydraulic head of reservoirs.

Q6. Mention the guidelines for selection of design floods for various purposes as laid down by Central Water Commission.

Q7. Discuss the generalized design criteria as proposed by the National Academy of Science, USA for different hydraulic structures.

Q8. How do the following factors affect flood. Explain them in brief.

(i) Occurrence of high intensity rainfall over a small hilly catchment

(ii) Intense rainfall over large catchment

(iii) Occurrence of rainfall over accumulated snow

Q9. Describe the various methods to estimate the peak floods in un-gauged catchments and gauged catchments.

Q10. Write down the rational formula to estimate the peak floods. What are its limitations?

Q11. Write down the formulae of the followings empirical methods to estimate the peak flood.

(i) Inglis Formula (ii) Nawab Jang Bahadur Formula, (iii) Creager's Formula, (iv) Modified Myer's Formula, (v) Fuller's Formula, (vi) Dredge and Burge Formula, (vii) Fanning Formula, (viii) Boston Society of Civil Engineers Formula, (ix) Horton's Formula, (x) U.S. Geological Survey (1955) Formula, (xi) Pettis Formula

Q12. A catchment has an area of 50 km^2. Estimate the peak flood value by Inglis formula and Nawab Jang Bahadur formula. Also calculate the flood peaks for a return period of 200 years by Fuller's formula as well as Horton's formula. Assume the value of constant, C in Nawab Jang Bahadur formula and Fuller's formula as 50 and 1.5, respectively.

Q13. Narrate the envelope curve method to estimate the peak discharge of a catchment.

Q14. Estimate the maximum flood peak value for a catchment having area of 100 km^2 by world experience method. Also compute the flood peak value for the same catchment by Creager's envelope curve equation.

Q15. What is probability of exceedance? How is it related to recurrence interval?

Q16. Write the formula to compute the probability of exceedance of a hydrologic event by Weibul's plotting position method. Write down the various steps used to compute the flood peaks by Weibul's formula.

Q17. Following measurements of flood peak data for 24 years of a basin are recorded. Using Weibul's formula, draw the probability distribution curve and compute the flood peak value at return period of 2 and 20 years. Also compute the flood peak value at 20 and 40 percent of probability levels.

Q18. Write down the different plotting position formulae including Weibul's formula.

Q19. What is the *general equation of hydrologic frequency analysis* as given by Chow? Describe how it is useful in estimation of flood values at a particular return period.

Q20. Describe the following method for computation of flood peak values.

- Foster method

- Chow's Regression Method

- Stochastic method

Q21. The peak flood data of a gauging station from 1990 to 2013 are given below. Determine the flood peak values for 500 and 1000 years by *Foster Type I* and *Foster Type III* distribution method.

Year	1990	1991	1992	1993	1994	1995	1996	1997	1998	1999	2000	2001
Peak flood, m^3/sec	185	181	244	250	162	172	148	154	150	159	140	144

Year	2002	2003	2004	2005	2006	2007	2008	2009	2010	2011	2012	2013
Peak flood, m^3/sec	116	122	128	119	157	146	130	134	106	109	114	118

Q22. Solve Q. 21 by Chow's regression method.

Q23. Flood data of a basin recorded for 24 years are given in Q. 21 as above except for the years 2008 and 2009 when data recorded are 139 m^3/sec each for both the years. Estimate the flood data at 500 and 1000 years return periods by stochastic method.

Q24. What do you mean by regional flood frequency analysis? How is it carried out?

Q25. What is a partial duration series? How is it different from annual series?

Q26. Summarise the terms used in design of hydraulic structures as given below:

- Risk,
- Reliability
- Safety Factor

Q27. A hydraulic structure is designed with a life span of 100 years. It is designed considering peak flood discharge with return period of 200 years. (i) What is the risk of the hydrologic design? (ii) Find out the reliability of design too. (iii) Considering 25% risk, what return period will have to be adopted?

Q28. Annual flood data of river Brahmani covering a period of 50 years yielded a mean and standard deviation of 27,000 m³/sec and 10,500 m³/sec, respectively. It is proposed to construct a bridge across the river with a risk value of 20%. The proposed life span of the structure is 100 years. Estimate the flood discharge by Gumbel's distribution which is used for design of the structure. If the actual flood value adopted in the design is 90,000 m³/sec, compute the safety factor and safety margin relating to maximum flood discharge.

Q29. Describe the various methods adopted in short and long term flood forecasting.

Q30. Write down the antecedent precipitation index and coaxial graph method to compute runoff in a basin.

Q31. Compute value of API from the given rainfall data for the day of 11th July, 2015, 12th July, 2015 and 16th July, 2015. Assume value of $K=0.80$.

Date	10/7/ 2015	11/7/ 2015	12/7/ 2015	13/7/ 2015	14/7/ 2015	15/7/ 2015	16/7/ 2015	17/7/ 2015
Rainfall, mm	10.0	0	0	5	0	0	20	0

Q32. How does a multiple stage correlation help in flood forecasting? Enumerate the procedure to work it out.

Q33. A tributary joins a main river at its upstream end of a flood forecasting station. Data of the flood stages of previous storms for the main river and its tributary at the upstream end and that at the downstream end of the forecasting station of the main river are given as below. Develop a flood forecasting model for the system. Find out the coefficient of correlation between the stage data of all these three stations.

Stage, U/S main river, m	165.0	165.5	166.1	166.5	170.0	171.2	171.6	172.6	173.0
Stage, U/S tributary, m	112.0	112.8	113.2	113.7	114.0	114.4	114.8	115.1	115.3
Stage, D/S forecasting station, m	80.0	80.6	82.1	83.4	85.0	86.1	86.8	87.2	88.0

Q34. With sketches narrate the different control measures used to control flood in a river basin.

Q35. How do the various soil conservation measures help in flood control work?

Q36. Explain how the flood plain management help in reducing flood in a river basin.

Q37. How the economics of flood control work is is carries out? With diagram, explain the benefit-cost analysis of flood control measures.

References

Central Water Commission, India. 1973. Estimation of flood peak. Flood Estimation Directorate, Report No. 1/73, New Delhi.

Mutreja, K.N. 1986. Applied Hydrology. Tata McGraw-Hill, New Delhi.

Panigrahi, B. 2013. A Handbook of Irrigation and Drainage. New India Publishing Agncy, New Delhi, pp. 602.

Patra, K.C. 2001. Hydrology and Water Resources Engineering. Narosa Publishing House. New Delhi, pp. 561.

Sharma, R.K. 1987. A Text Book of Hydrology and Water Resources. Dhanpat Rai and Sons, Delhi.

Sharma, R.K. and Sharma, T.K. 2002. Irrigation Engineering including Hydrology. S. Chand and Company Ltd. , New Delhi.

Subramanya, K. 2003. Enginering Hydrology, Tata McGraw-Hill Publishing Company Limited,New Delhi, pp. 392.

Suresh, R. 208. Watershed Hydrology. Standard Publishers Distributors, Nai Sarak, Delhi, pp. 692.

Varshney, R.S. 1979. Engineering Hydrology. Nem Chand & Bros., Roorkee, U.P. India, pp. 917.

CHAPTER 8

Statistics and Probability in Hydrology

8.1 Introduction

The knowledge of statistics and probability is very helpful for correct prediction of hydrological data and correct prediction of hydrological data is useful in giving good performance in all water resources projects. Since hydrological data are highly stochastic varying both in space and time, it is essential to predict them as correct as possible. The hydrologic evens do not depend on physical and chemical laws but depend entirely on nature. They cannot be predicted exactly by using any relationship but can be predicted at different probability levels by fitting suitable probability distribution functions. Application of statistics and probability in the field of hydrology can help to predict the hydrologic events in a better way. In hydrology, objectives of statistics are as follows (Suresh, 2008):

- Presentation of hydrologic data such as rainfall, stream flow etc. in different forms.

- Interpretation of hydrologic data.

- Investigation of hydrologic probabilistic regularities and

- Derivations of various relationships concerned with agriculture on basis of available flood.

In hydrology, the use of physical models to predict future sequences is less attempted than abstract models which describe the system in mathematical terms. Depending on the approach, an abstract model may be deterministic, stochastic or probabilistic with considerations of features given in Table 8.1 (Patra, 2001).

Table 8.1: Various forms of abstract models

Deterministic	(i)	Sequencing of occurrence of variables is considered.
	(ii)	Model is considered to follow a definite law of certainty but not the law of probability.
	(iii)	Chance of occurrence of variables is considered.
Probabilistic	(i)	Sequence of occurrence of the variables is not considered.
	(ii)	Chance of occurrence of variables follows a definite probability distribution.
	(iii)	Variables are considered as pure random.
Stochastic	(i)	Sequencing of occurrence of variables is considered.
	(ii)	Probability distribution of variables may or may not be time dependent.
	(iii)	Variables may or may not be pure random.

It is to be remembered that most of the hydrologic models are either deterministic or stochastic. The difference between the two can be understood from the following example. A deterministic model can forecast with reasonable accuracy the daily evaporation from a lake but the prediction of rainfall for a particular day at a particular place is purely stochastic as the random component is very large.

Sequencing of occurrences of any particular hydrologic data is called a time series. The time series may be annual duration series, partial duration series or extreme value series. In annual duration series, we consider the highest value in a year. In extreme value series, we consider all yearly or monthly largest or smallest values. Sometimes, we consider partial duration series. In a partial duration series, we consider a certain base value above which the data are considered. For example, in some river basins, there may be more than one independent flood exceeding certain base value say (x) in a year and many of them may be very high in magnitude and there may have some years having no flood more than this base value. All these flood values which are more than the base value are considered in partial duration series. It is important to remember that in partial duration series, all events are considered to be independent. Partial duration series is adopted mostly for rainfall analysis where conditions of independency of events are easy to establish. Its use in flood studies is rare where the flood data are complex in nature where independency of events is difficult to establish.

The recurrence interval of an event obtained by annual series (T_A) and partial duration series (T_P) are related (Subramanya, 2003) as:

$$T_P = \frac{1}{\ln T_A - \ln \left(T_A - 1 \right)}$$

When $T_A < 10$ years, difference between T_A and T_P is significant. When $T_A > 20$ years, difference between T_A and T_P is negligibly small.

Time series may be stationary if the distribution characteristics like mean and standard deviation remain constant for the series. For example we have records of 100 years rainfall. If we divide them into 10 equal groups having 10 data in each group and all these 10 groups have same mean and standard deviation, then the series containing 100 years data is stationary. In some hydrologic events, the annual values, say annual runoff in a basin, are considered to be stationary than the monthly values or seasonal values.

Sometimes we have continuous and discrete data series. Hydrologic phenomenon like stream flow is continuous. Sometimes we break down the series into intervals and group them to a discrete series. Contrarily, we can make discrete variables continuous by fitting continuous functions to the series.

8.2 Elements of Statistics

Followings are the main elements of statistics:

- Measures central tendency
- Measures of dispersion
- Measures of symmetry and
- Measures of peakedness

8.2.1 Measures of Central Tendency

Mean, median and mode are the parameters of central tendency. Central tendency represents the concentration of distribution about the central values.

8.2.1.1 Mean

Mean represents the value which is centrally located. In other words, it is the average of all the records about which the distribution is equally weighed. The distance of the mean from the origin is obtained by taking first moment from the origin. It represents a location from the origin the first moment abut which all positive and negative points balance. Sum of all the departures from the mean is taken zero. There are four types of mean. There are:

- Arithmetic mean
- Geometric mean
- Harmonic mean and
- Weighted mean

All these means are described as under.

Arithmetic mean

For a grouped series with N number of data, mean is computed by taking the first moment of the distribution of the sample from origin. It is expressed as:

$$\mu = \frac{\sum X_i f(X_i)}{\sum f(X_i)} = \frac{\sum X_i f(X_i)}{N} \qquad (8.1)$$

where, μ is population mean $f(X_i)$ is number of occurrence of each event called as frequency, X_i is the variate or hydrologic data and $\sum f(X_i) = N$. For discrete variables, the above equation can be reduced to as follows:

$$\bar{X} = \frac{\sum X_i}{N} \qquad (8.2)$$

Summation in the above equation (Eqn. 8.2) continues from $i = 1$ to N. In Eqn. (8.2), arithmetic mean of ungrouped data or discrete variables. Arithmetic mean is simply called as average.

Geometric mean

The sample estimate of geometric mean (X_{gm}) is expressed as:

$$X_{gm} = (X_1 \cdot X_2 \cdot X_3 \ldots \ldots X_n)^{1/N} \qquad (8.3)$$

Harmonic mean

The sample estimate of harmonic mean (X_{hm}) is expressed as:

$$X_{hm} = \frac{N}{\sum_{i=1}^{n} \frac{1}{X_i}} = \frac{N}{\left(\dfrac{1}{X_1} + \dfrac{1}{X_2} + \dfrac{1}{X_3} + \cdots + \dfrac{1}{X_n} \right)} \qquad (8.4)$$

Weighted mean

In arithmetic mean, all the occurring events are assumed to have equal importance. But in actual cases, all the events are not given same importance. In that case, weighted mean is computed by assigning different weights to different variables. The equation is:

$$X_w = \frac{\sum_{i=1}^{n} w_i X_i}{\sum_{i=1}^{n} w_i} = \frac{w_1 X_1 + w_2 X_2 + w_3 X_3 + \cdots\cdots + w_n X_n}{w_1 + w_2 + w_3 + \cdots\cdots + w_n} \qquad (8.5)$$

where, w_1, w_2, w_3.... are the weights assigned to variables X_1, X_2, X_3..... Weighted mean is assigned to compute average curve number (CN) to a catchment which has a number of sub-catchments having different areas and curve numbers. Similarly if a catchment has different sub-catchments each having different areas and each sub-catchment has some recorded rainfall, then we can use weighted mean to compute average rainfall of the catchment.

Sample Calculation 8.1

A catchment has six stations named A, B, C, D, E and F. The area enclosed by each station and the rainfall recorded by each station is given as below. Compute the mean rainfall of the catchment by (i) arithmetic mean, (ii) geometric mean, (iii) harmonic mean and (iv) weighted mean method.

Station	A	B	C	D	E	F
Rainfall, cm	30.0	50.5	67.2	45.8	36.9	40.0
Area, km²	41.8	35.7	45.0	55.2	48.0	50.1

Solution

Arithmetic mean: It is given by Eqn. (8.2) as:

$$\bar{x} = \frac{\sum X_i}{N} = \frac{30 + 50.5 + 67.2 + 45.8 + 36.9 + 40.0}{6} = 45.07 \text{ cm}$$

Geometric mean: It is given by Eqn. (8.3) as:

$$X_{gm} = (X_1 . X_2 . X_3 ....X_n)^{1/N} = (30 \times 50.5 \times 67.2 \times 45.8 \times 36.9 \times 40.0)^{1/6} = 43.61 \text{ cm}$$

Harmonic mean: It is given by Eqn. (8.4) as:

$$X_{hm} = \cfrac{N}{\left(\cfrac{1}{X_1} + \cfrac{1}{X_2} + \cfrac{1}{X_3} + \cdots\cdots + \cfrac{1}{X_n}\right)} = \cfrac{6}{\left(\cfrac{1}{30.0} + \cfrac{1}{50.5} + \cfrac{1}{67.2} + \cdot\cfrac{1}{45.8} + \cfrac{1}{36.9} + \cfrac{1}{40.0}\right)}$$

$= 42.27$ cm.

Weighted mean: It is given by Eqn. (8.5) as:

$$X_w = \frac{w_1\,X_1 + w_2\,X_2 + w_3\,X_3 + \cdots\cdots + w_n\,X_n}{w_1 + w_2 + w_3 + \cdots\cdots + w_n}$$

$$= \frac{41.8 \, x \, 30.0 + 35.7 \, x \, 50.5 + 45.0 \, x \, 67.2 + 55.2 \, x \, 45.8 + 48.0 \, x \, 36.9 + 50.1 \, x \, 40.0}{41.8 + 35.7 + 45.0 + 55.2 + 48.0 + 50.1} = 44.90 \text{ cm}$$

We find that arithmetic mean > geometric mean > harmonic mean.

8.2.1.2 Median

Median is the middlemost data of the series containing odd number of samples or arithmetic average of the two middlemost values of the series containing even number of samples when the data are arranged in ascending order. Median represents a better estimate of the central tendency parameters as fifty percent of the data are distributed equally on either side. When the data is in the form of groups, median, M_d is computed by interpolation techniques as:

$$\text{Median} = M_d = L_1 + h\,\frac{N/2 - S_{bm}}{f_{median}} \tag{8.6}$$

where, L_1 is lower boundary/lower limit of class containing median, N is total number of data in the sample i.e. cumulative frequency, S_{bm} is sum of all frequencies below median class, f_{median} is frequency of median class and h is size of median class interval.

Sample Calculation 8.2

Find the median value of the following rainfall data recorded at 7 stations.

Station	A	B	C	D	E	F	G
Rainfall, cm	23.7	30.0	33.6	20.5	25.0	33.8	24.8

Solution

Since there are odd number of stations (7 stations), the median will be the middlemost data when the data are arranged in ascending order. The above mentioned data series when arranged in the ascending order, the middlemost value is the data containing 4th value and that is 25.0 cm.

Hence median is 25 cm.

If we have even number of stations then there will be two middle values. For example we have rainfall data of 6 stations (stations A, B, C, D, E and F) and rainfall recorded in each of these stations is same as mentioned in above sample calculation. In this case we have two middle values i.e. 3rd and 4th value. When the rainfall data are arranged in ascending order, the 3rd and 4th data give values of rainfall 25 and 30 cm, respectively. Hence the median value is average of 3rd and 4th data which is average of 25 and 30 cm i.e. 27.5 cm.

Sample Calculation 8.3

Calculate the median of the following data.

Rainfall, cm	8.5	3.5	7.5	10.0	9.5	7.0	8.6
Frequency	3	4	2	1	1	1	2

Solution

Computation is shown in the following table.

Rainfall, cm	Frequency after arranging in ascending order	Cumulative frequency
3.5	4	4
7.0	1	5
7.5	2	7
8.5	3	10
8.6	2	12
9.5	1	13
10.0	1	14

Here total number of data $N = 14$ which is even. Hence the median will be average of the two middle values. The two middle values are 7th and 8th value. The 7th value of the above table gives rainfall 7.5 cm. The 8th value is 8.5 cm. It is to be noted that for the above mentioned data 8th, 9th and 10th value gives rainfall 8.5 cm. In other words 8.5 cm rainfall has been recorded three times i.e. frequency = 3.

Hence the median is the average of 7th and 8th value i.e. average of 7.5 and 8.5 cm which is 8.0 cm.

8.2.1.3 Mode

Mode is the most frequently occurring values in a series. For continuous variables, mode is the maximum value in the frequency distribution. Suppose in a series the data 23 occurs 2 times, 35 occur 4 times, 50 occur 1 time and 12 occur 2 times. Then we see that the data 35 occurs maximum times i.e. 4 times. Hence, the mode is 35.

Calculation of mode is usually carried out for the grouped data and is expressed as:

$$Mode = L + \left[\frac{(f_1 - f_0)}{(f_1 - f_0) + (f_1 - f_2)} \right] h \tag{8.7}$$

where, L is lower limit of the modal f_1 is frequency of modal class, f_0 is frequency before the modal class, f_2 is frequency after the modal class, and h is the length of interval of modal class. Modal class is the class with maximum frequency. For a symmetrical distribution, mean, mode and median are same. The relationship between mean, mode and median is given as:

Mode = 3 median − 2 mean $\qquad\qquad$ (8.8)

Eqn. (8.8) holds good for large number of data in group having small class interval. For all practical calculations, a hydrologist generally prefers to use arithmetic mean than mode or median since it is considered that mean represents the central tendency better.

Sample Calculation 8.4

The annual rainfall of a station in Odisha for 45 years are recorded as follows. Group the data into interval of 10 and estimate the mean, mode and median of the sample.

Rainfall data (cm): 55, 65, 70, 45, 64, 86, 64, 68, 45, 69, 39, 40, 53, 33, 80, 74, 47, 42, 33, 44, 66, 78, 87, 96, 91, 82, 84, 100, 39, 110, 112, 105, 69, 96, 119, 102, 100, 97, 78, 30, 73, 59, 101, 106, 99.

Solution

Computations are carried in tabular form as follows:

The lowest data of the above series is 30 and the highest data is 119. We have to take class interval 10. We group the data considering the lowest and the

Sl.No.	Rainfall class	x_i	Frequency, f_i	Cumulative frequency	$f_i x_i$	Remarks
(1)	(2)	(3)	(4)	(5)	(6)	(7)
1.	30-40	35	5	5	175	Median class is 45/2 = 22.5
2.	40-50	45	6	11	270	lying in the 5th group.
3.	50-60	55	3	14	165	Modal class has highest
4.	60-70	65	7	21	455	frequency 7 lying in 4th group
5.	70-80	75	5	26*	375	
6.	80-90	85	5	31	425	
7.	90-100	95	5	36	475	
8.	100-110	105	6	42	630	
9.	110-120	115	3	45	345	
	Total		45		3315	

highest data and the class interval into 9 groups as mentioned in the 2nd column of the above table. The first class is 30-40 and the last class is 110-120. We counted the number of data falling in each group and mention it as frequency (column 3). Then we take the cumulative frequency (column 4). If a data is the value in upper limit of a particular class mark/group then we do not count it in frequency of that class interval. For example data 70 in the above sample calculation is not counted towards frequency in the class mark/group of 60-70 but counted as a frequency in the group 70-80. After computing the frequency and cumulative frequency, we find out the modal class and median class as mentioned in remarks column of the table.

The arithmetic mean of the grouped data is computed by Eqn. (8.1) as:

$$\frac{\sum X_i \, f(X_i)}{N} = 3315/45 = 73.67 \text{cm}.$$

Median of the grouped data, M_d is computed by Eqn. (8.6) as:

$$M_d = L_1 + h \frac{N/2 - S_{bm}}{f_{median}}$$

$$= 70 + 10 \, x \, \frac{45/2 - 19}{5} = 77.00 \; cm$$

Mode of the grouped data is given by Eqn. (8.7) as:

$$Mode = L + \left[\frac{(f_1 - f_0)}{(f_1 - f_0) + (f_1 - f_2)} \right] h$$

$$= 60 + \left[\frac{(7-3)}{(7-3) + (7-5)} \right] x \, 10 = 66.67 \; cm$$

8.2.2 Measures of Dispersion

Variability or dispersion of data is measured by the following parameters:

8.2.2.1 Range

Range is the difference between the lowest and the highest data in a data series. For example the data set as mentioned in Sample Calculation 8.4 has lowest data 30 and the highest data is 119. Hence, the range is 89. The range of a population is many times the interval from $-\infty$ to $+\infty$ or from 0 to ∞. Sometimes we use relative range which is range divided by mean.

8.2.2.2 Mean Deviation (MD)

It is the mean of absolute deviation from the mean of all data. It is expressed as:

$$MD = \frac{\sum(x_i - \bar{X})}{N} \tag{8.9}$$

where, N is total number of data, x is the variate or data and $\bar{X}$ is the mean of all variates.

8.2.2.3 Variance

Variance is the most common measure of dispersion. Variance is the second moment of all the points of the sample about the mean. It is expressed as:

$$\sigma_n^2 = \frac{1}{N}\sum(X_i - \bar{X})^2 \tag{8.10}$$

where, σ_n^2 is the population variance. For grouped data, Eqn. 8.10 becomes

$$\sigma_n^2 = \frac{1}{N}\sum f_i(X_i - \bar{X})^2 \tag{8.11}$$

Thus variance is the average squared deviation from mean. When the sample length is less than or equal to 30, the variance is termed as sample variance, σ_{n-1}^2 which is in unbiased form and is expressed as:

$$\sigma_{n-1}^2 = \frac{1}{N-1}\sum(X_i - \bar{X})^2 \tag{8.12}$$

For grouped data, the sample variance is expressed as:

$$\sigma_{n-1}^2 = \frac{1}{N-1}\sum f_i(X_i - \bar{X})^2 \tag{8.13}$$

8.2.2.4 Standard Deviation

Variability is better expressed by the parameter standard deviation which is the positive square root of the variance. It is given as:

$$\sigma_n = \sqrt{\left\{\frac{\sum(X_i - \bar{X})^2}{N}\right\}} \tag{8.14}$$

For unbiased estimate, the sample standard deviation σ_{n-1} is:

$$\sigma_{n-1} = \sqrt{\left\{ \frac{\sum(X_i - \bar{X})^2}{N-1} \right\}} \tag{8.15}$$

Standard deviation represents how far the data are spread from the mean. Larger is the value, larger is the spread of the data from their mean. Unit of standard deviation is the unit of the variate. Unit of variance is the square of the unit of the variate. For grouped data, standard deviation is:

$$\sigma_{n-1} = \sqrt{\left\{ \frac{\sum f(X_i - \bar{X})^2}{N-1} \right\}} \tag{8.16}$$

8.2.2.5 Coefficient of Variation (C_v)

It is a dimensionless term which is used to measure the variability of data. It is computed as the ratio of standard deviation to mean and is:

$$C_v \frac{\sigma_{n-1}}{\bar{X}} \tag{8.17}$$

Eqn. (8.17) is multiplied by 100 when C_v is expressed in percent.

8.2.2.6 Standard Error of Estimate, S_e

Standard error of standard deviation, S_{es} is given as:

$$S_{es} = \frac{\sigma_{n-1}}{\sqrt{2N}} \tag{8.18}$$

Standard error of mean, S_m is given as:

$$S_m = \frac{\sigma_{n-1}}{\sqrt{N}} \tag{8.19}$$

Units of the above mentioned two parameters depend on the unit of standard deviation.

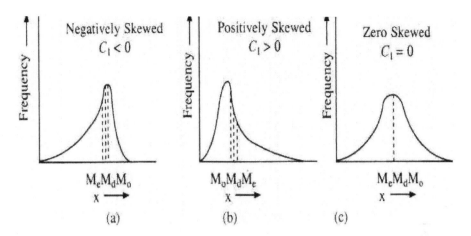

Fig. 8.1: Various types of skewness

8.2.3 Measures of Symmetry

Skewness is referred to as a measure of symmetry. It is defined as the third moment about the mean. It represents the symmetricity of the data about the mean. The distribution may be negatively skewed for which $C_s < 0$, positively skewed for which $C_s > 0$ or no skew (zero skewed) for which $C_s = 0$. For a negatively skewed/right skewed distribution, the data are distributed to the right with a long tail to the left. For a positively skewed/left skewed distribution, the data are distributed to the left with a long tail to the right. For a symmetrical distribution, $C_s = 0$. It is neither skewed to left nor to right. Fig. 8.1 shows the various types of skewness. For a symmetric distribution, mean, mode and median coincide.

Skewness, α or μ_3 is given as:

$$\alpha \text{ or } \mu_3 = \frac{1}{N} \sum \left(X_i - \bar{X} \right)^3 \tag{8.20}$$

For grouped data skewness, μ_3 is given as:

$$\mu_3 = \frac{1}{N} \sum f \left(X_i - \bar{X} \right)^3 \tag{8.21}$$

When data length is less than or equal to 30, the unbiased estimate of the sample skewness (μ_3) is:

$$\mu_3 = \frac{N}{(N-1)(N-2)} \sum \left(X_i - \bar{X} \right)^3 \tag{8.22}$$

Coefficient of skewness is defined as the ratio of third moment about the mean to the cube of the standard deviation and is:

$$C_s = \frac{\mu_3}{\sigma^3}$$
(8.23)

Pearson's First Coefficient of Skewness: It is given as:

$$S_{k1} = \frac{3(\text{Mean - Mode})}{\text{Standard Deviation}}$$
(8.24)

Pearson's Second Coefficient of Skewness: It is given as:

$$S_{k2} = \frac{3(\text{Mean - Median})}{\text{Standard Deviation}}$$
(8.25)

8.2.4 Measures of Peakedness

A fourth property of random variables based on moments is kurtosis, μ_4 which is the fourth moment of the data about the mean and is given for grouped data as under:

$$\mu_4 = \frac{1}{N} \sum f_i (X_i - \bar{X})^4$$
(8.26)

For unbiased estimate kurtosis is given as:

$$\mu_4 = \frac{N^2}{(N-1)(N-2)(N-3)} \sum f_i (X_i - \bar{X})^4$$
(8.27)

Kurtosis is a measure of peakedness or flatness because this value tends to become zero faster as it represents the fourth power deviation from the mean. When the deviation from mean increases, kurtosis approaches infinity. Kurtosis is the ratio of fourth power about the mean to the square of variance ad is given as:

$$\beta = \frac{\mu_4}{\sigma^4}$$
(8.28)

The value of β_2 and $\gamma_2 = (\beta_2 - 3)$ represents grouping of data at the central place. β_2 is known as coefficient of excess. For flat curve, $\beta_2 < 3$ and γ_2 is negative. For peaked curve, $\beta_2 > 3$ and γ_2 is positive. For symmetrical distribution like normal distribution, $\beta_2 = 3$ and $\gamma_2 = 0$. Fig. 8.2 represents the different types of kurtosis.

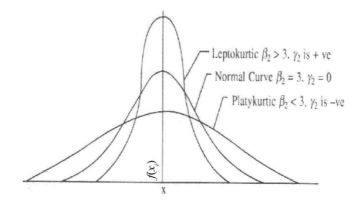

Fig. 8.2: Different types of kurtosis

8.3 Graphical Presentation of Data

Hydrologists often collect large quantity of data which require proper analysis. These data are to be collected in tabular form or to be represented in some graphical form. A useful first step in data analysis is to plot the data as frequency histogram. This is done by grouping the data into classes and then plotting a bar graph with number or relative frequency of observations in a class versus the mid point of the class interval. A graph can also be drawn between the mid classes of the class interval versus cumulative relative frequency. A graph between the mid classes of the class interval versus cumulative frequency is called as Ogive curve. Mid point of a class is called as class mark. The class interval is the difference between the upper and lower class boundary. In assigning the data to classes, we find sometimes some data fall in the class boundary. In such a case, we can (i) define the boundary to more significant figures than the actual data, (ii) assign a data falling on a boundary to the next higher (lower) class or (iii) to alternatively assign the data to the next higher and then the next lower class. In sample calculation 8.4 and 8.5, we have used the second criteria i.e. assign the data falling on a boundary to the next higher (lower) class.

It is to be noted that the selection of class interval and location of first class mark can appreciably affect the appearance of frequency histogram. The appropriate width of a class interval depends on the range of data, number of observations and behavior of data. Spiegel (1961) suggested that there should be 5 to 20 classes. Steel and Torrie (1960) state that class interval should not exceed one-fourth to one-half of the standard deviation of the data. Sturges (1926) recommends that the number of classes be determined from

$m = 1 + 3.3 \log n$ 　　　　　　　　　　　　　　　　　　　(8.29)

where m is the number of classes, n is the number of data and logarithm to base 10 is used. If too many or too less classes are used, then sensitivity of the analysis is lost.

Sample Calculation 8.5

Considering the annual rainfall of 45 years as given in Sample calculation 8.4, plot a curve of frequency histogram and cumulative frequency. Calculate the standard deviation, variance, coefficient of variation, skewness and kurtosis of the data.

Solution

Let us group the data taking 10 as class interval. The procedures for finding out rainfall class, frequency and cumulative frequency as entered in the table below are already explained in solution for sample calculation 8.4.

Fig. 8.3 shows the frequency histogram of the data. Fig. 8.4 shows the cumulative frequency diagram. Cumulative frequency diagram shows the relative cumulative frequency of an event which is equal to or less than the given data. We can also draw a cumulative frequency diagram showing the relative cumulative frequency of an event which is equal to or greater than the given data. This is done by calculating the cumulative relative frequency an event which is equal to or greater than the given data by subtracting the relative cumulative frequency of an event which is equal to or less than the given data from one.

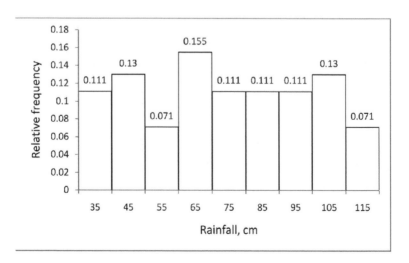

Fig. 8.3: Frequency histogram of rainfall (sample calculation 8.5)

Solution of sample calculation 8.5 in tabular form:

Sl. No.	Rainfall class	Mean of class/mid class, x_i	Frequency, f_i	Relative frequency $\cdot f_i/N$	Cumulative frequency	Cumulative relative frequency
(1)	(2)	(3)	(4)	(5)	(6)	(7)
1.	30–40	35	5	0.111	5	0.111
2.	40–50	45	6	0.130	11	0.244
3.	50–60	55	3	0.071	14	0.311
4.	60–70	65	7	0.155	21	0.467
5.	70–80	75	5	0.111	26*	0.578
6.	80–90	85	5	0.111	31	0.689
7.	90–100	95	5	0.111	36	0.800
8.	100–110	105	6	0.130	42	0.933
9.	110–120	115	3	0.071	45	1.00.
	Total		45	1.000		

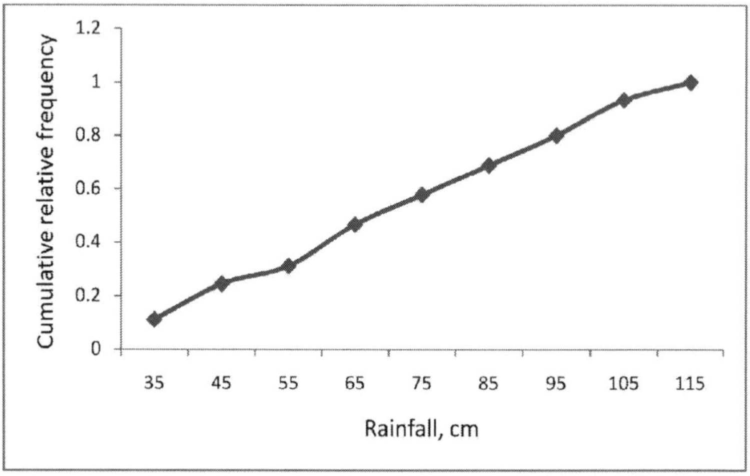

Fig. 8.4 Cumulative frequency of rainfall (sample calculation 8.5)

The mean of the grouped data has been computed in sample calculation 8.4 and it is 73.67 cm.

Using this mean value, values of x_i (col.3 of above table) and values of f_i (col.4 of above table), we calculate the values of $(x_i - x_{av})^2$, $(x_i - x_{av})^3$, $(x_i - x_{av})^4$, f_i $(x_i - x_{av})^2$, f_i $(x_i - x_{av})^3$ and f_i $(x_i - x_{av})^4$ and show them in the table below.

Using these tabular values, we get variance (Eqn. 8.11):

$$\sigma_n^2 = \frac{1}{N} \sum f_i \left(X_i - \bar{X} \right)^2$$

$= 27920/45 = 620.44$

Standard deviation $= (620.44)^{1/2} = 24.91$

Coefficient of variation, C_v = Standard deviation/mean = 24.91/73.67 = 0.3381 = 33.81%.

Skewness is given as;

$$\mu_3 = \frac{1}{N} \sum f \left(X_i - \bar{X} \right)^3$$

$= -2492.53/45 = -55.39$

Solution of sample calculation 8.5 in tabluar form

Mid class, x_i	$(x_i - x_{av})^2$	$(x_i - x_{av})^3$	$(x_i - x_{av})^4$	Frequ-ency, f_i	$f_i(x_i - x_{av})^2$	$f_i(x_i - x_{av})^3$	$f_i(x_i - x_{av})^4$
(1)	(2)	(3)	(4)	(5)	(6)	(7)	(8)
35	1495.37	-57825.9	2236128.0	5	7476.845	-289130.0	11180641.0
45	821.969	-23565.8	675632.9	6	4931.813	-141395.0	4053797.0
55	348.569	-6507.78	121500.3	3	1045.707	-19523.3	364500.8
65	75.1689	-651.714	5650.364	7	526.1823	-4562.0	39552.54
75	1.7689	2.352637	3.129007	5	8.8445	11.76319	15.64504.0
85	128.369	1454.42	16478.57	5	641.8445	7272.098	82392.87
95	454.969	9704.487	206996.7	5	2274.845	48522.43	1034983.0
105	981.569	30752.55	963477.5	6	5889.413	184515.3	5780865.0
115	1708.17	70598.62	2917841.0	3	5124.507	211795.9	8753523.0
Total	6015.92	23961.17	7143709.0		27920.0	-2492.53	31290271.0

Coefficient of skewness is:

$$C_s = \frac{\mu_3}{\sigma^3}$$

$$= -55.39/(24.91)^3 = -0.00358$$

For calculating kurtosis, we have to calculate first fourth power about the mean.

$$\mu_4 = \frac{1}{N}\sum f_i\left(X_i - \bar{X}\right)^4$$

$$= 31290271/45 = 695339.36$$

Coefficient of kurtosis is

$$\beta = \frac{\mu_4}{\sigma^4}$$

$$= 1.81$$

Now $\gamma_2 = (\beta_2 - 3) = 1.81 - 3.0 = -1.19$ i.e. the curve is flat (platykurtic).

8.4 Theoretical Probability Distribution

A hydrologist predicts hydrological events with their frequency of occurrence. This helps to assess a flood of a particular magnitude that can be expected to occur in the life time of a project. Probability of occurrence of the random variable like flood can be determined by fitting a frequency distribution to the hydrologic data set. Fitting of the frequency distribution can be carried out either by (i) method of moment or (ii) by method of maximum likelihood.

In this chapter some of the theoretical distributions normally used for hydrologic analysis are discussed.

8.4.1 Discrete Distribution

8.4.1.1 Binomial Distribution

The probability density function of binomial distribution is:

$$p(x) = c_x^n\, p^x\, q^{n-x} = \left\{ \frac{n!}{\left[x!(n-x)!\right]} \right\} p^x\, q^{n-x} \tag{8.30}$$

In the above expression, p is probability of success whereas q is probability of failure and n is total number of trials. Mean of the probability is np and variance of distribution is $n.p.q$. This type of distribution best suits to rainy and non-rainy days; the events being mutually exclusive. In the above trial, x can be number of rainy days out of n days under consideration. If $p = q$, then distribution is symmetrical. If $q > p$, distribution skews to the right and if $q < p$, then it skews to left.

Sample Calculation 8.6

A hydraulic structure needs to be cast which needs no rainfall during 5 days of its casting. The month is July in which there is chance of 40% rainfall on any day. Compute the probability of having no rainfall during these casting periods of 5 days and one rainy day during the casting period.

Solution

Let us use Binomial distribution. Here $n = 5$, $p = 0.4$ and $q = 1- 0.4 = 0.6$.

First we have to compute the probability of having no rainfall during casting periods i.e. $x = 0$.

Substituting the above values in Eqn. (8.30) we get:

$$p\left(x = 0\right) = c_x^n \, p^x \, q^{n-x} = \left\{ \frac{n!}{\left[x!\left(n-x\right)!\right]} \right\} p^x \, q^{n-x}$$

$$= \left\{ \frac{5!}{\left[0!\left(5-0\right)!\right]} \right\} 0.4^0 \, 0.6^{5-0}$$

$$= 0.078$$

In the second case, we have one rainy day during casting period of 5 days i.e. $x = 1$.

Hence, probability of having one rainy day during casting period of 5 days is:

$$p\left(x = 1\right) = c_x^n \, p^x \, q^{n-x} = \left\{ \frac{n!}{\left[x!\left(n-x\right)!\right]} \right\} p^x \, q^{n-x}$$

$$= \left\{ \frac{5!}{\left[1!\left(5-1\right)!\right]} \right\} 0.4^1 \, 0.6^{5-1}$$

$$= \left\{ \frac{5!}{\left[1! \left(5 - 1 \right)! \right]} \right\} 0.4^{1} \, 0.6^{5-1}$$

$$= 0.259$$

8.4.1.2 Poisson Distribution

Poisson distribution is a special case of Binomial distribution where (i) n is large, (ii) probability, p is small and (iii) the product of p and n is finite (say $p.n = \lambda$).

The probability distribution function is expressed as:

$$p(x) = \frac{\lambda^{x} \, e^{-\lambda}}{x!} \qquad \lambda > 0; \, x = 0, 1, 2\ldots\ldots\ldots \qquad (8.31)$$

This distribution has both mean and variance equal to ë. Skewness of the distribution is $1/(\lambda)^{0.5}$.

Sample Calculation 8.7

Rainfall data of 100 years of a station were scanned. It was observed that the probability of occurrence of rainfall of 200 mm in a day is 5% $i.e.$ 0.05. Find out the probability of three rainfall of one day magnitude exceeding 200 mm in the next 10 days.

Solution

Let us use Poisson distribution.

Given $p = 0.05$ and $n = 10$.

$n.p = 0.05 \times 10 = 0.5 = \lambda$

Using Eqn. (7.89):

$$p(x = 3) = \frac{0.5^{3} \, e^{-0.5}}{3!}$$

$$= 0.0126$$

Hence, probability of three rainfall of one day magnitude exceeding 200 mm in the next 10 days is 0.0126 = 1.26 percent.

8.4.2 Continuous Distribution

In hydrology, many events are considered to be a part of continuous events. For such events, continuous distributions like Normal, Log-Normal, Gamma, Pearson Type III, Log- Pearson Type III etc are applied to the observed hydrologic variables. The theoretical concepts along-with the application of these continuous distributions are discussed as under.

8.4.2.1 Normal Distribution

A random variable x is said to follow normal distribution with parameter μ (mean) and variance σ^2 when its probability distribution function *(PDF)* is given as (Mutreja, 1986):

$$f(x) = \frac{1}{\sigma\sqrt{2\pi}} \exp\left[\frac{-1}{2}\left\{\frac{x-\mu}{\sigma}\right\}^2\right] \qquad -\infty < x < \infty \qquad (8.32)$$

It is where, and are population mean and standard deviation and *f(x)* is *PDF* of variable x varying between $-\infty$ to ∞. Frequency distribution curve for normal distribution has the following properties:

- A symmetrical and bell shaped curve.

- Mean of the sample is at the top of the bell and

- For normal distribution, mean, mode and median coincide.

Let us replace the term $\frac{(x-\mu)}{\sigma}$ in Eqn. (8.32) by a standard normal variate t and integrated between $-\infty$ to ∞, the equation represents a standard normal distribution. As the variate varies between $-\infty$ to ∞, the area under the curve between any two of its limits gives the probability of occurrence of the event. The variate t is normally distributed with zero mean and unit standard deviation.

The area of the standard normal distribution curve for the ranges of t can be obtained from any standard text book of Statistics. The area of the curve for x lying between - to + (i.e. t lying between -1 to 1), - 2 to + 2 (i.e. t lying between -2 to 2) and - 3 to + 3 (i.e. t lying between -3 to 3) is given in Fig. 8.5. Table 8.2 gives area under the standard normal curve for all standard normal variates, t. For any cumulative probability, the value of t can be found out from the tables of area from which the variate x is calculated from the equation:

$$t = \frac{(x-\mu)}{\sigma} \qquad (8.33)$$

Eqn. (8.33) gives

$$x = \mu + \sigma t \tag{8.34}$$

For sample data Eqn. (8.34) is taken as:

$$x = x_{av} + \sigma_s t \tag{8.35}$$

where, x_{av} is the sample mean or simply arithmetic mean and is σ the sample standard deviation.

Normal distribution is generally fitted to annual series data and not to daily, weekly or monthly data. It is to be remembered that normal distribution is a special case of Pearson Type III distribution in which $C_s = 0$. Therefore Table 8.4 can be used to find out the frequency factor (K) for normal distribution by taking $C_s = 0$. In Table 8.4, the row that contains $C_s = 0$ is taken to read the K values. The procedure to compute the flood value at any return period is same as that described by Pearson Type III distribution. When C_s is not equal to zero but the distribution assumes to follow the normal distribution, then the procedure to solve such kind of problem using area under the standard normal curve is illustrated in sample calculation 8.8. The area under the standard normal curve is given in Table 8.2. Using Table 8.2, procedure to find value of frequency factor, K for normal distribution and the event magnitude of the desired return period, T can be seen from this sample calculation 8.8.

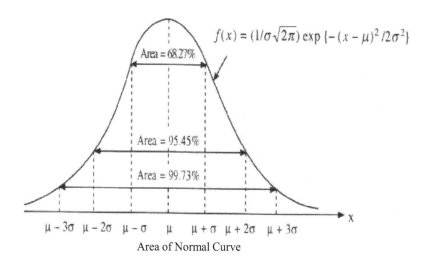

Area of Normal Curve

Fig. 8.5: Area under the Normal probability distribution curve

Table 8.2 :Area under standard normal curve $t=$

t	0.00	0.01	0.02	0.03	0.04	0.05	0.06	0.07	0.08	0.09
0.0	0.0000	0.0040	0.0080	0.0120	0.0159	0.0199	0.0239	0.0279	0.0319	0.0359
0.1	0.0398	0.0438	0.0478	0.0517	0.0557	0.0596	0.0636	0.0675	0.0714	0.0735
0.2	0.0793	0.0832	0.0871	0.0910	0.0948	0.0987	0.1026	0.1064	0.1103	0.1141
0.3	0.1179	0.1271	0.1255	0.1293	0.1331	0.1368	0.1406	0.1443	0.1480	0.1517
0.4	0.1554	0.1591	0.1628	0.1664	0.1700	0.1736	0.1772	0.1808	0.1884	0.1879
0.5	0.1915	0.1950	0.1985	0.2019	0.2054	0.2088	0.2123	0.2157	0.2190	0.2224
0.6	0.2257	0.2291	0.2324	2357	0.2389	0.2422	0.2454	0.2486	0.2518	0.2549
0.7	0.2580	0.2611	0.2642	0.2673	0.2704	0.2734	0.2464	0.2794	0.2823	0.2852
0.8	0.2881	0.2910	0.2939	0.2967	0.2995	0.3023	0.3051	0.3078	0.3106	0.3133
0.9	0.3159	0.3186	0.3212	0.3238	0.3264	0.3289	0.3315	0.3340	0.3365	0.3389
1.0	0.3413	0.3438	0.3461	0.3485	0.3508	0.3531	0.3554	0.3577	0.3599	0.3621
1.1	0.3643	0.3665	0.3686	0.3708	0.3729	0.3749	0.3770	0.3790	0.3810	0.3830
1.2	0.3849	0.3869	0.3888	0.3907	0.3925	0.3944	0.3962	0.3980	0.3997	0.4015
1.3	0.4032	0.4049	0.4066	0.4082	0.4099	0.4115	0.4131	0.4147	0.4162	0.4177
1.4	0.4192	0.4207	0.4222	0.4236	0.4251	0.4265	0.4279	0.4292	0.4306	0.4319
1.5	0.4332	0.4345	0.4657	0.4370	0.4382	0.4394	0.4406	0.4418	0.4430	0.4441
1.6	0.4452	0.4463	0.4474	0.4485	0.4495	0.4505	0.4515	0.4525	0.4535	0.4545
1.7	0.4554	0.4564	0.4573	0.4582	0.4591	0.4599	0.4608	0.4616	0.4625	0.4633
1.8	0.4641	0.4649	0.4656	0.4664	0.4671	0.4678	0.4686	0.4693	0.4699	0.4706

t	0.00	0.01	0.02	0.03	0.04	0.05	0.06	0.07	0.08	0.09
1.9	0.4713	0.4719	0.4726	0.4732	0.4738	0.4744	0.4750	0.4756	0.4762	0.4767
2.0	0.4772	0.4778	0.4783	0.4788	0.4793	0.4798	0.4803	0.4808	0.4812	0.4817
2.1	0.4821	0.4826	0.4830	0.4835	0.4838	0.4842	0.4846	0.4850	0.4854	0.4857
2.2	0.4861	0.4865	0.4868	0.4871	0.4875	0.4878	0.4881	0.4884	0.4887	0.4890
2.3	0.4893	0.4896	0.4898	0.4901	0.4904	0.4906	0.4909	0.4911	0.4913	0.4916
2.4	0.4918	0.4990	0.4922	0.4925	0.4927	0.4929	0.4931	0.4932	0.4934	0.4936
2.5	0.4938	0.4940	0.4941	0.4943	0.4945	0.4946	0.4948	0.4949	0.4951	0.4952
2.6	0.4953	0.4955	0.4956	0.4957	0.4959	0.4960	0.4961	0.4962	0.4963	0.4964
2.7	0.4965	0.4966	0.4967	0.4968	0.4969	0.4970	0.4971	0.4972	0.4973	0.4974
2.8	0.4974	0.4975	0.4976	0.4977	0.4977	0.4978	0.4979	0.4980	0.4980	0.4981
2.9	0.4981	0.4982	0.4983	0.4983	0.4984	0.4984	0.4985	0.4985	0.4986	0.4986
3.0	0.4986	0.4987	0.4987	0.4988	0.4988	0.4989	0.4989	0.4989	0.4980	0.4990

8.4.2.2 Log-Normal Distribution

It is a special case of normal distribution in which the variates are replaced by their logarithmic transformed values with base e. After logarithmic transformation, the data are assumed to follow normal distribution. The data are then anaysed in the same way as the normal distribution is done. If x_i is a variate then its logarithmic transformation i.e. $\ln x_i$ is taken as z_i. Mean and standard deviation of the transformed data i.e. z_i data are written as $\mu_{z\,and}\,\sigma_z$, respectively. Following Chow, the statistical parameters for x series can be obtained as:

Mean $\mu_x = e^{\mu_z + \sigma_z^2/2}$ $\qquad\qquad\qquad\qquad\qquad\qquad\qquad$ (8.36)

Variance $\sigma_x^2 = \mu_x^2 \left(e^{\sigma_z^2 - 1}\right)$ $\qquad\qquad\qquad\qquad\qquad\qquad$ (8.37)

Probability density function $f(x)$ is expressed as:

$$f(x) = \frac{1}{x\,\sigma\sqrt{2\pi}}\, e^{\left[-1/2\left\{\frac{\ln x - \mu_z}{\sigma_z}\right\}^2\right]} \quad x > 0,\ \mu_z = z_{av} \qquad (8.38)$$

Using tables of standard normal distribution, the log normal distribution may be fitted to the series.

It is to be remembered that log-normal distribution is same as log-Pearson Type III distribution with coefficient of skewness (C_s) taken as zero. The procedure outlined for log-Pearson Type III distribution is followed in this case and Table 8.4 is used to take values of frequency factor, K at $Cs = 0$ and for various assumed vales of T.

Chow (1964) has reported frequency factor as given Table 8.3 for use in log-normal distribution which is more appropriate to use in hydrology since the log transformed data in some cases are found to have some value of C_s instead of zero.

Table 8.3: Chow's frequency factor used for log-normal distribution

C_s	Probability (%) equal to or greater than the given variate									C_v
	99	95	80	50	20	5	1	0.1	0.01	
	−	−	−	−	+	+	+	+	+	
0.0	2.33	1.65	0.84	0	0.84	1.64	2.33	3.09	3.72	0.000
0.1	2.25	1.62	0.85	0.02	0.84	1.67	2.40	3.22	3.95	0.033
0.2	2.18	1.59	0.85	0.04	0.83	1.70	2.47	3.39	4.18	0.067
0.3	2.11	1.56	0.85	0.06	0.82	1.72	2.55	3.56	4.42	0.100
0.4	2.04	1.53	0.85	0.07	0.81	1.75	2.62	3.72	4.70	0.136
0.5	1.98	1.49	0.86	0.09	0.82	1.77	2.70	3.88	4.96	0.166
0.6	1.91	1.46	0.85	0.10	0.79	1.79	2.77	4.05	5.24	0.197
0.7	1.85	1.43	0.85	0.11	0.78	1.81	2.84	4.21	5.52	0.230
0.8	1.79	1.40	0.84	0.13	0.77	1.82	2.90	4.37	5.81	0.262
0.9	1.74	1.37	0.84	0.14	0.76	1.84	2.97	4.55	6.11	0.292
1.0	1.68	1.34	0.84	0.15	0.75	1.85	3.03	4.72	6.40	0.234
1.1	1.63	1.31	0.83	0.16	0.73	1.86	3.09	4.87	6.71	0.351
1.2	1.58	1.29	0.82	0.17	0.72	1.87	3.15	5.04	7.02	0.381
1.3	1.54	1.26	0.82	0.18	0.71	1.88	3.21	5.19	7.31	0.409
1.4	1.49	1.23	0.81	0.19	0.69	1.88	3.26	5.35	7.62	0.436
1.5	1.45	1.21	0.81	0.20	0.68	1.89	3.31	5.51	7.92	0.462
1.6	1.41	1.18	0.80	0.21	0.67	1.89	3.36	5.66	8.26	0.490
1.7	1.38	1.16	0.79	0.22	0.65	1.89	3.40	5.80	8.58	0.517
1.8	1.34	1.14	0.78	0.22	0.64	1.89	3.44	5.96	8.88	0.544
1.9	1.31	1.12	0.78	0.23	0.63	1.89	3.48	6.10	9.20	0.570
2.0	1.28	1.10	0.77	0.24	0.61	1.89	3.52	6.25	9.51	0.596

Contd.

C_s	Probability (%) equal to or greater than the given variate									C_v
	99	95	80	50	20	5	1	0.1	0.01	
	−	−	−	−	+	+	+	+	+	
2.1	1.25	1.08	0.76	0.24	0.60	1.89	3.55	6.39	9.79	0.620
2.2	1.22	1.06	0.76	0.25	0.59	1.89	3.59	6.51	10.12	0.643
2.3	1.20	1.04	0.75	0.25	0.58	1.88	3.62	6.65	10.43	0.667
2.4	1.17	1.02	0.74	0.26	0.57	1.88	3.65	6.77	10.72	0.691
2.5	1.15	1.00	0.74	0.26	0.56	1.88	3.67	6.90	10.95	0.713
2.6	1.12	0.99	0.73	0.26	0.55	1.87	3.70	7.02	11.25	0.734
2.7	1.10	0.97	0.72	0.27	0.54	1.87	3.72	7.13	11.55	0.755
2.8	1.08	0.96	0.72	0.27	0.53	1.86	3.74	7.25	11.80	0.776
2.9	1.06	0.95	0.71	0.27	0.52	1.86	3.76	7.36	12.10	0.796
3.0	1.04	0.93	0.71	0.28	0.51	1.85	3.78	7.47	12.36	0.818
3.2	1.01	0.90	0.69	0.28	0.49	1.84	3.81	7.65	12.85	0.857
3.4	0.98	0.88	0.68	0.29	0.47	1.83	3.84	7.84	13.36	0.895
3.6	0.95	0.86	0.67	0.29	0.46	1.81	3.87	8.00	13.83	0.930
3.8	0.92	0.84	0.66	0.29	0.44	1.80	3.89	8.16	14.23	0.966
4.0	0.90	0.82	0.65	0.29	0.42	1.78	3.91	8.30	14.70	1.000
4.5	0.84	0.78	0.63	0.30	0.39	1.75	3.93	8.60	15.62	1.081
5.0	0.80	0.74	0.62	0.30	0.37	1.71	3.91	8.86	16.45	1.155

In log-normal distribution, C_s and C_v are related as:

$$C_s = 3\,C_v + C_v^3 \qquad\qquad (8.39)$$

Sample Calculation 8.8

Following measurements of flood peak data for 24 years of a basin are recorded.

Year	1990	1991	1992	1993	1994	1995	1996	1997	1998	1999	2000	2001
Flood data, m³/sec	180	187	234	210	162	172	148	154	150	159	140	144

Year	2002	2003	2004	2005	2006	2007	2008	2009	2010	2011	2012	2013
Flood data, m³/sec	116	122	128	119	147	146	130	134	106	109	104	113

Estimate the flood data at 100 and 1000 years return periods by norml and log normal distribution.

Solution

Normal Distribution:

Here $N = 24$

For 100 years return period, $T = 100$, P $(X > x)$ is the area of the normal curve bounded between 0 and $1 - 1/T = 1 - 1/100 = 0.99 = 99\%$ on either side. From the standard normal curve for area up to $+ 49\%$ (99% - 50%) i.e. 0.49, we get $t = 2.325$. The value of $t = 2.325$ is obtained from Table 8.2 by interpolation as discussed below.

Area of 0.49 (49%) lies between 0.4898 and 0.4901. In Table 8.2 for area 0.4898, we *get* $t = 2.32$. Similarly for area 0.4901, we get $t = 2.33$. Now by interpolation, we get $t = 2.325$ when area is 0.49.

The mean of the variates ($\overline{x}$) is computed as 3514/24 = 146.4 m³/sec. Using the mean and variates data for different years, the parameters $(x - \overline{x})^2$ and $(x - \overline{x})^3$ are computed and presented in table below.

Year	Flood data, m³/sec (x)	$(x - \overline{x})^2$	$(x - \overline{x})^3$	$y = \log x$
1990	180	1128.96	37933.06	2.255273
1991	187	1648.36	66923.42	2.271842
1992	234	7673.76	672221.4	2.369216
1993	210	4044.96	257259.5	2.322219
1994	162	243.36	3796.42	2.209515
1995	172	655.36	16777.22	2.235528
1996	148	2.56	4.09	2.170262
1997	154	57.76	438.98	2.187521
1998	150	12.96	46.66	2.176091
1999	159	158.76	2000.38	2.201397
2000	140	40.96	-262.14	2.146128
2001	144	5.76	-13.82	2.158362
2002	116	924.16	-28094.50	2.064458
2003	122	595.36	-14526.80	2.08636
2004	128	338.56	-6229.50	2.10721
2005	119	750.76	-20570.80	2.075547
2006	147	0.36	0.22	2.167317
2007	146	0.16	-0.06	2.164353
2008	130	268.96	-4410.94	2.113943
2009	134	153.76	-1906.62	2.127105
2010	106	1632.16	-65939.30	2.025306
2011	109	1398.76	-52313.60	2.037426
2012	104	1797.76	-76225.00	2.017033
2013	113	1115.56	-37259.70	2.053078
Total		24649.84	749648.5	51.7425

Standard deviation is computed by the Eqn. (8.15) as:

$$\sigma = \sqrt{\frac{\sum_{1}^{n}\left(x_i - \overline{x}\right)^2}{\left(N-1\right)}} = \sqrt{\frac{24649.84}{\left(24-1\right)}} = 32.73 \text{ m}^3/\text{sec}$$

Now flood value at 100 years return period is given as:

$X_{100} = \overline{x} + \sigma t$

$= 146.4 + 32.73 \times 2.325 = 222.50 \text{ m}^3/\text{sec}$

Similarly for 1000 years return period, $T = 1000$, $P\ (X > x)$ is the area of the normal curve bounded between 0 and $1 - 1/T = 1 - 1/1000 = 0.99 = 99.9\%$ on either side. From the standard normal curve for area up to + 49.5% (99.9% - 50%) i.e. 0.495, we get $t = 2.575$.

Flood value at 1000 year return period is:

$X_{1000} = \bar{x} + \sigma\, t$

$= 146.4 + 32.73 \times 2.575 = 230.68$ m³/sec

Log-Normal Distribution:

Here $N = 24$

The flood data are transformed logarithmically which are shown in column 5 of above table. Applying Eqn. (8.2) and (8.15) to the log transformed data, we get mean, y and standard deviation, σ_y as 2.1559 and 0.0921, respectively.

Coefficient of variation, C_v = standard deviation/mean = 0.0921/2.1559 = 0.04272

Using Eqn. (8.39) for log-normal distribution,

$C_s = 3\, C_v + C_v^3$

$= 3 \times 0.04272 + (0.04272)^3 = 0.128$

Using Table 8.3 for $C_s = 0.128$ and $T = 100$ years, we get frequency factor, $K_y = 2.4196$

Similarly for $C_s = 0.128$ and $T = 1000$ years, we get frequency factor, $K_y = 3.2676$.

Hence flood value at T year return period is:

$y_T = \bar{y} + K_y \sigma_y$

$y_{100} = 2.1559 + 2.4196 \times 0.0921 = 2.378745$

Thus, $y_{1000} = 2.1559 + 3.2676 \times 0.0921 = 2.456846$

Taking anti log, we get flood value at 100 years return period, $x_{100} = 239.18$ m³/sec.

Thus, taking anti log, we get flood value at 1000 years return period, $x_{1000} = 286.32$ m³/sec.

We can also compute the value of K_y from Table 8.4 (frequency factor for Pearson Type III distribution) considering $C_s = 0$ for log normal distribution which is read as 2.326 and 3.090, respectively for return periods of 100 and 1000 years, respectively.

Now, $y_{100} = 2.1559 + 2.326 \times 0.0921 = 2.37012$

and $y_{1000} = 2.1559 + 3.090 \times 0.0921 = 2.44049$

Taking anti log, we get flood value at 100 years return period, $x_{100} = 234.48$ m³/sec.

Thus, taking anti log, we get flood value at 1000 years return period, $x_{1000} = 275.69$ m³/sec.

we find out that both the approaches give comparable value of flood estimates. However, the first method is preferred since coefficient of skewness of the data series is high.

Final results are tabulated as:

Distribution	Flood value at 100 years return period, m³/sec (x_{100})	Flood value at 1000 years return period, m³/sec (x_{1000})
Normal	222.50	230.68
Log-normal	239.18	286.32

8.4.2.3 Gamma Distribution

A continuous random hydrologic varaite of series x having probability density function is said to have a gamma distribution when

$$f(x) = \left\{ \frac{\lambda^\gamma \, x^{\gamma-1}}{\Gamma(\gamma)!} \right\} \left\{ e^{-\lambda(x)} \right\} \quad x > 0, \lambda > 0, \lambda = \frac{x_{av}}{\sigma^2}, \ \gamma = (\frac{1}{C_v})^2 \tag{8.40}$$

This distribution helps in formulating flood forecasting techniques when the tributaries join the main river.

8.4.2.4 Pearson Type III Distribution

The basic equation defining probability density of Pearson Type III distribution is:

$$f(x) = \left\{ \frac{1}{\Gamma(b)} \right\} \left\{ \lambda^b (x-c)^{b-1} e^{-\lambda(x-c)} \right\} \quad x \geq c, x = \frac{\sigma}{\sqrt{b}}, \ b = (\frac{2}{C_s})^2 \tag{8.41}$$

The above distribution reduces to Gamma distribution by substituting $c = 0$. It is a skewed distribution having long tail to the right that can be reduced to a normal distribution as a special case. Foster applied this distribution to describe the probability distribution of annual maximum floods.

This distribution is normally used for skewed data. The frequency factor, K is related to coefficient of skewness and return period. Table 8.4 gives the values of frequency factor for various values of coefficient of skewness and return periods for Pearson Type III distribution. Following steps are employed in computation of flood peak values by Pearson Type III distribution.

- Given the data series, find out the total number of data (N) of the series.

- Compute mean, standard deviation and coefficient of skewness of the data series by the formulae discussed earlier.

- Compute the adjusted coefficient of skewness if the length of data is short (normally less than 100) by multiplying the computed coefficient of skewness (comuted by step II as above) with a factor ($1+ 8.5/N$) as suggested by Foster.

- To predict the flood peak value at any return period, T, read the frequency factor, K from Table 8.4.

- Finally compute the flood peak value of desired return period by using Chow's general equation.

Table 8.4: Frequency factor for Pearson Type III distribution

Coefficient of skew, C_s	Recurrence interval T in years						
	2	10	25	50	100	200	1000
3.0	-0.396	1.180	2.278	3.152	4.051	4.970	7.250
2.5	-0.360	1.250	2.262	3.048	3.845	4.652	6.600
2.2	-0.330	1.284	2.240	2.970	3.705	4.444	6.200
2.0	-0.307	1.302	2.219	2.912	3.605	4.298	5.910
1.8	-0.282	1.318	2.193	2.848	3.499	4.147	5.660
1.6	-0.254	1.329	2.163	2.780	3.388	3.990	5.390
1.4	-0.225	1.337	2.128	2.706	3.271	3.828	5.110
1.2	-0.195	1.340	2.087	2.626	3.149	3.661	4.820
1.0	-0.164	1.340	2.043	2.542	3.022	3.489	4.540
0.9	-0.148	1.339	2.018	2.498	2.957	3.401	4.395
0.8	-0.132	1.336	1.998	2.453	2.891	3.312	4.250
0.7	-0.116	1.333	1.967	2.407	2.824	3.223	4.105
0.6	-0.099	1.328	1.939	2.359	2.755	3.132	3.960
0.5	-0.083	1.323	1.910	2.311	2.686	3.041	3.815
0.4	-0.066	1.317	1.880	2.261	2.615	2.949	3.670
0.3	-0.050	1.309	1.849	2.211	2.544	2.856	3.525
0.2	-0.033	1.301	1.818	2.159	2.472	2.763	3.380

Contd.

Coefficient of skew, C_s			Recurrence interval T in years				
0.1	-0.017	1.292	1.785	2.107	2.400	2.670	3.235
0.0	0.000	1.282	1.751	2.054	2.326	2.576	3.090
-0.1	0.017	1.270	1.716	2.000	2.252	2.482	2.950
-0.2	0.033	1.258	1.680	1.945	2.178	2.388	2.810
-0.3	0.050	1.245	1.643	1.890	2.104	2.294	2.675
-0.4	0.066	1.231	1.606	1.834	2.029	2.201	2.540
-0.5	0.083	1.216	1.567	1.777	1.955	2.108	2.400
-0.6	0.099	1.200	1.528	1.720	1.880	2.016	2.275
-0.7	0.116	1.183	1.488	1.663	1.806	1.926	2.150
-0.8	0.132	1.166	1.448	1.606	1.733	1.837	2.035
-0.9	0.148	1.147	1.407	1.549	1.660	1.749	1.910
-1.0	0.164	1.128	1.366	1.492	1.588	1.664	1.880
-1.4	0.225	1.041	1.198	1.270	1.318	1.351	1.465
-1.8	0.282	0.945	1.035	1.069	1.087	1.097	1.130
-2.2	0.330	0.844	0.888	0.900	0.905	0.907	0.910
-3.0	0.396	0.660	0.666	0.666	0.667	0.667	0.668

N.B. $C_s = 0$ corresponds to log-normal distribution.

8.4.2.5 Log Pearson Type III Distribution

This distribution is widely used in United States, India and other countries as the standard distribution for flood frequency analysis of annual maximum floods. The distribution has the added advantage of providing skew adjustment. The distribution can be reduced to log normal distribution if skew adjusted is zero.

The variates of the hydrologic series is represented by x. If x follows the Pearson Type III, then log (x) has the log-Pearson type III distribution.

The PDF is given as

$$f(x) = \left(\frac{1}{a x \Gamma(b)}\right)\left\{\frac{y-c}{a}\right\}^{b-1} e^{\frac{-(y-c)}{a}}$$

(8.42)

where $y = \log x$, mean $\mu_y = c + ab$, variance $= \sigma_y^2 = a^2 b$

It follows that if the log transferred series of the sample is fitted with Pearson type-III distribution then the x_i (untransferred) sample should follow log-Pearson type –III distribution.

For skewed data, log-Pearson type III gives a better fit and is widely used. Zero skew reduces log Pearson –III distribution to log normal distribution.

If is the variate of a random hydrologic series, then the series of variates are computed as:

$$y = \log x \tag{8.43}$$

Compute mean , standard deviation and coefficient of skewness, C_s for the y variates/y series.

For this series, for any recurrence interval T compute

$$y_T = \overline{y} + K_y \sigma_y \tag{8.44}$$

where $K_y = a$ frequency factor which is a function of recurrence interval T and the coefficient of skew .

σ_y = standard deviation of the y variate sample is given as:

$$\sigma_y = \sqrt{\Sigma(y-\overline{y})^2 / (N-1)} \tag{8.45}$$

C_s = coefficient of skew of variate y is given as:

$$C_s = \frac{N\Sigma(y-\overline{y})^3}{(N-1)} \frac{}{(N-2)(\sigma_y)^3} \tag{8.46}$$

where, N = sample size = number of years of record and other terms defined earlier.

The variation of is same as that of frequency factor for Pearson Type III distribution as given in Table 8.4.

For a set of given y variates, we use Eqn. (8.44) to find out the value of y_T. After finding, the corresponding value of x_T is estimated by taking anti log of y_T.

When data series is not very large i.e. for finite data series, the coefficient of skew is adjusted to account for the size of the sample by using the following relation as proposed by Hazen (1930):

$$C_s = C_s \left(\frac{1+8.5}{N} \right) \tag{8.47}$$

where, C_s in left hand side of Eq. (8.47) is adjusted coefficient of skew.

However, the standard procedure for use of log-Pearson Type III distribution adopted by U.S.Water Resources Council does not include this adjustment for skew.

Stepwise Procedure to fit Log-Pearson type-III distribution to the observed series are outlined below:

(i)   Transfer the observed data series to the logarithmic values. Generate series as $y_i = \log \chi_i$

(ii)   Find mean, standard deviation and coefficient of skewness for the log transferred series.

(iii) Multiply the coefficient of skewness with a factor (1+8.5/N) as suggested by Hazen to overcome the short length of data. For records exceeding 100, the factor need not be multiplied. This is done because it is assumed that by multiplying the factor, the sample statistics of the skewness coefficient is converted to be the representative of the population. However, as suggested by U.S.Water Resources Council, the C_s need not be adjusted.

(iv)   To compute flood of required return period T, frequency factor K_y can be calculated from K-T relation of Table 8.4 corresponding to the skewness coefficient of the log transferred series.

(v)   Knowing mean, standard deviation and K_y and using Eqn. (8.44) compute the event (y_T) of desired return period T in logarithmic scale.

(vi)   By taking antilog of y_T, find which is the desired value of the event for the return period T.

8.4.2.6 Extreme Value Distribution

There are three types of extreme value distribution used in hydrologic studies. Extreme value type I distribution also called as Gumbel's distribution is widely used in hydrologic and meteorological studies including flood study. The Extreme Value Type III distribution is called as Weibull's distribution mainly used for study of drought. The Type II extreme Value Distribution is known as Frechet Distribution and is hardly used in study of hydrology.

8.4.2.7 Gumbel's Distribution

This extreme value named as extreme value type I distribution was introduced by Gumbel (1941) and is commonly known as Gumbel's distribution. It is one of the most widely used probability distribution functions for extreme values in hydrologic and meteorologic studies for prediction of flood peaks, maximum rainfalls, maximum wind speed etc.

Gumbel defined a flood as the largest of the 365 daily flows and the annual series of flood flows constitute a series of largest values of flows. The cumulative probability of occurrence of an event is

$$F(x) = e^{-e^{-\left(\frac{a+x}{c}\right)}} = e^{-e^{-y}} \qquad (8.48)$$

where $y = (a + x)/c$, a and c are constants.

Therefore probability of occurrences of an event equal to or larger than a value x_0 is

$$P(x \geq x_0) = 1 - e^{-e^{-y}} \qquad (8.49)$$

The value of c and a in equation (8.48) is usually taken as

$$c = \left(\frac{\sqrt{6}}{\pi}\right) \sigma \text{ and } a = \gamma c - x_{av}$$

where, γ is 0.57721

Putting the values of a, c and γ in the equation $y = (a + x)/c$ and rearranging we get,

$$y = \frac{1.2825 (x - x_{av})}{\sigma_x} + 0.577 \qquad (8.50)$$

where, x_{av} is mean, σ_x is standard deviation of variates x.
Rearranging equation (8.48) we get

$$y = \frac{(a+x)}{c} = -\ln\left[-\ln F(x)\right] \qquad (8.51)$$

$$F(x) = 1 - P(x \geq x_0) = 1 - P = \frac{T-1}{T} \qquad (8.52)$$

Using Eqn. (8.51), value of y at any return period, T from Eqn. (8.50) comes as:

$$y_T = -\ln\left\{\ln\left(\frac{T}{T-1}\right)\right\} \qquad (8.53)$$

$$= -\left[0.834 + 2.303 \log \log \frac{T}{T-1} \right] \tag{8.54}$$

Now rearranging Eqn. (8.50), the value of the variate x with a return period T is

$$x_T = \overline{x} + K\,\sigma_x \tag{8.55}$$

where $K = \dfrac{\left(y_T - 0.577\right)}{1.2825}$ \hfill (8.56)

In Eqn. (8.56), y_T is reduced variate which is a function of T.

Eqns. (8.55) and (8.56) constitutes basic Gumbel's equations and are applicable to infinite sample size i.e. N tends to infinite. But since in practice, the hydrological data like rainfall, flood are finite, Eqn. (8.56) is modified for finite N values and are discussed below.

Gumbel's Equation for Practical Use:

For practical conditions, Eqn. (8.55) is taken as:

$$x_T = \overline{x} + K\,\sigma_{n-1} \tag{8.57}$$

where σ_{n-1} = standard deviation of the sample of size $N = \sqrt{\dfrac{\Sigma\left(x - \overline{x}\right)^2}{N-1}}$

K = frequency factor expressed as $K = \dfrac{y_T - \overline{y}_n}{S_n}$ \hfill (8.58)

In which y_T = reduced variate, a function of T (computed by Eqn. 8.53 or 8.54)

$\overline{y}_n$ = reduced mean, a function of sample size N and is given in the Table 8.5.

For , $N \to \infty$ $\overline{y}_n \to 0.577$

S_n = reduced standard deviation, a function of sample size N and is given in Table 8.6.

For $N \to \infty, S_n \to 1.2825$

These equations are used under the following procedure to estimate the flood magnitude corresponding to a given return based on an annual flood series.

Table 8.5 Reduced mean in Gumbel's Extreme value distribution (Subramanya, 2003)

N = sample size

N	0	1	2	3	4	5	6	7	8	9
10	0.4952	0.4996	0.5035	0.5070	0.5100	0.5128	0.5157	0.5181	0.5202	0.5220
20	0.5236	0.5252	0.5268	0.5283	0.5296	0.5309	0.5320	0.5332	0.5343	0.5353
30	0.5362	0.5371	0.5380	0.5388	0.5396	0.5402	0.5410	0.5418	0.5424	0.5430
40	0.5436	0.5442	0.5448	0.5453	0.5458	0.5463	0.5468	0.5473	0.5477	0.5481
50	0.5485	0.5489	0.5493	0.5497	0.5501	0.5504	0.5508	0.5511	0.5515	0.5518
60	0.5521	0.5524	0.5527	0.5530	0.5533	0.5535	0.5538	0.5540	0.5543	0.5545
70	0.5548	0.5550	0.5552	0.5555	0.5557	0.5559	0.5561	0.5563	0.5565	0.5567
80	0.5569	0.5570	0.5572	0.5574	0.5576	0.5578	0.5580	0.5581	0.5583	0.5585
90	0.5586	0.5587	0.5589	0.5591	0.5592	0.5593	0.5595	0.5596	0.5598	0.5599
100	0.5600									

Table 8.6 Reduced Standard deviation S_n in Gumbel's Extreme value distribution (Subramanya, 2003)

N = sample size

N	0	1	2	3	4	5	6	7	8	9
10	0.9496	0.9676	0.9833	0.9971	1.0095	1.0206	1.0316	1.0411	1.0493	1.0565
20	1.0628	1.0696	1.0754	1.0811	1.0864	1.0915	1.0961	1.1004	1.1047	1.1086
30	1.1124	1.1159	1.1193	1.1226	1.1255	1.1285	1.1313	1.1339	1.1363	1.1388
40	1.1413	1.1436	1.1458	1.1480	1.1499	1.1519	1.1538	1.1557	1.1574	1.1590
50	1.1607	1.1623	1.1638	1.1658	1.1667	1.1681	1.1696	1.1708	1.1721	1.1734
60	1.1747	1.1759	1.1770	1.1782	1.1793	1.1803	1.1814	1.1824	1.1834	1.1844
70	1.1854	1.1863	1.1873	1.1881	1.1890	1.1898	1.1906	1.1915	1.1923	1.1930
80	1.1938	1.1945	1.1953	1.1959	1.1967	1.1973	1.1980	1.1987	1.1994	1.2001
90	1.2007	1.2013	1.2020	1.2026	1.2032	1.2038	1.2044	1.2049	1.2055	1.2060
100	1.2065									

1. Assemble the discharge data and note the sample size N. Here the annual flood value is the variate x. Find $\overline{x}$ and σ_{n-1} for the given data.

2. Using Tables 8.5 and 8.6 determine $\overline{y}_n$ and S_n appropriate to given N.

3. Find y_T for a given T by Eqn. (8.53 or 8.54).

4. Find K (Eqn. 8.58).

5. Determine the required x_T (Eqn. 8.57).

Sample Calculation 8.9

Using the flood data of sample calculation 8.8, estimate the flood data at 100 and 1000 years return periods by Gumbel's, Pearson Type III and Log Pearson Type III distribution.

Solution

Gumbel's Distribution

Here, $N = 24$. mean, $\overline{x}$ and standard deviation, σ_{n-1} of the data are computed as 146.4 and 32.73 m³/sec, respectively. Using $N = 24$, from Tables 8.5 and 8.6, we get $\overline{y}_n$ and S_n as 0.5296 and 1.0864, respectively.

Next we calculate the value of reduced variate, y_T by Eqn. (8.53) as:

$$y_T = -ln\left\{ ln\left(\frac{T}{T-1} \right) \right\}$$

For $T = 100$ and $T = 1000$, we get $y_T = 4.600$ and 6.907, respectively. Values of frequency factor, K are then computed by Eqn. (8.58) as:

$$K = \frac{y_T - \overline{y}_n}{S_n}$$

$= (4.600 - 0.5296) / 1.0864 = 3.747$ when $T = 100$ years.

When $T = 1000$ years, we get $K = (6.907 - 0.5296) / 1.0864 = 5.8702$.

The value of the variate i.e. floods data at 100 years return period is then computed by Eqn. (8.57) as:

$$x_{100} = \bar{x} + K \sigma_{n-1}$$

$$= 146.4 + 3.747 \times 32.73 = 269.04 \text{ m}^3/\text{sec.}$$

Similarly at $T = 1000$ years, we get flood value $= x_{1000} = 338.53$ m³/sec.

Pearson Type III Distribution

$N = 24$. Mean, and standard deviation, of the data are computed as 146.4 and 32.73 m³/sec, respectively. Coefficient of skewness, C_s is computed as 0.93. Since data length is only 24 years, it is adjusted as suggested by Foster by multiplying the computed C_s with a factor $(1 + 8.5/N)$.

The adjusted Cs $= 0.93 \times (1 + 8.5/24) = 1.26$.

Using Table 8.4, for value of $C_s = 1.26$ and $T = 100$ years, we get frequency factor, $K = 3.183$

Similarly when $C_s = 1.26$ and $T = 1000$ years, we get frequency factor, $K = 4.898$.

Now the flood value at 100 year return period is:

$$x_{100} = \bar{x} + K \sigma_{n-1}$$

$$= 146.4 + 3.183 \times 32.73 = 250.58 \text{ m}^3/\text{sec.}$$

Similarly at $T = 1000$ years, we get flood value $= x_{1000} = 146.4 + 4.898 \times 32.73 = 306.71$ m³/sec.

Log-Pearson Type III Distribution

The flood discharge values (x) are transformed to logarithmic form (log to base 10) and they are entered in tabular form as below.

Year	Flood, m³/sec = x	$y = \log x$
1990	180	2.255273
1991	187	2.271842
1992	234	2.369216
1993	210	2.322219
1994	162	2.209515
1995	172	2.235528
1996	148	2.170262
1997	154	2.187521

Contd.

Year	Flood, m³/sec = x	$y = \log x$
1998	150	2.176091
1999	159	2.201397
2000	140	2.146128
2001	144	2.158362
2002	116	2.064458
2003	122	2.08636
2004	128	2.10721
2005	119	2.075547
2006	147	2.167317
2007	146	2.164353
2008	130	2.113943
2009	134	2.127105
2010	106	2.025306
2011	109	2.037426
2012	104	2.017033
2013	113	2.053078
Total	3514	51.7425

The statistical parameters like mean $\bar{y}$, standard deviation σ_y and coefficient of skewness, C_s for the "y" variates/ "y" series are calculated using the log transformed data (col. 3 of above table) and the values are:

$\bar{y} = 2.1559$, $C_s = 0.587$ and $\sigma_y = 0.0921$

For $C_s = 0.587$, and at $T = 100$ years = 1% frequency, the value of K_y from Table 8.4 is read as 2.705. Similarly for $C_s = 0.587$, and at $T = 1000$ years = 0.1% frequency, the value of K_y from Table 8.4 is read as 3.835.

Flood value at T years return period is given by Eqn. (8.44) as:

$$y_T = \bar{y} + K_y \sigma_y$$

$$y_{100} = 2.1559 + 2.705 \times 0.0921 = 2.4050$$

Thus, $y_{1000} = 2.1559 + 3.835 \times 0.0921 = 2.5091$

Taking anti log, we get flood value at 100 years return period, $x_{100} = 254.51$ m³/sec.

Thus, taking anti log, we get flood value at 1000 years return period, $x_{1000} = 323.10$ m³/sec.

Final results are tabulated as:

Distribution	Flood value at 100 years return period, m³/sec (x_{100})	Flood value at 1000 years return period, m³/sec (x_{1000})
Gumbel	269.04	338.53
Pearson Type III	250.58	306.71
Log Pearson Type III	254.51	323.10

Sample Calculation 8.10

The mean annual flood peak of a river basin measured at a site is 300 m³/sec with standard deviation of 30 m³/sec. Compute the probability that a flood of 390 m³/sec will occur in the site in the next 10 years.

Solution

Given mean, $\bar{x} = 300$ m³/sec and standard deviation, $\sigma = 30$ m³/sec.

Assume Gumbel's distribution fits the data and using Chow's general frequency equation,

$$X_T = \bar{x} + K\sigma$$

$$390 = 300 + 30K$$

Which gives $K = 3.0$. Using Eqn. (8.56) we get:

$$K = \frac{(y_T - 0.577)}{1.2825}$$

$$3 = \frac{(y_T - 0.577)}{1.2825}$$

$$y_T = 4.4245$$

Using Eqn. (8.54) we have

$$y_T = -\left[0.834 + 2.303 \log \log \frac{T}{T-1}\right]$$

$$4.4245 = -\left[0.834 + 2.303 \log \log \frac{T}{T-1}\right]$$

Which gives $T = 84$ years.

The probability of the event occurring in the next 10 years is:

$$= 1 - (1 - 1/T)^{10} = 1 - (1 - 1/84)^{10} = 0.1128 = 11.3\%.$$

Sample Calculation 8.11

The flood data of a particular site in a basin has the following particulars.

Length of record for which data are recorded is 30.The observed flood at 50 and 100 years return periods are 42 and 48 m³/sec, respectively. Assuming Gumbel's distribution holds god, compute the flood data at 200 years return period.

Solution

Given $N = 30$

$X_{50} = 42$ m³/sec and $X_{100} = 48$ m³/sec.

From Eqn. (8.54)

$$y_T = -\left[0.834 + 2.303 \log \log \frac{T}{T-1}\right]$$

$$y_{50} = -\left[0.834 + 2.303 \log \log \frac{50}{50-1}\right]$$

$y_{50} = 3.902$

Similarly for $T = 100$ years, we have

$y_{100} = 4.601$

From Eqn. (8.56), we have

$$K = \frac{\left(y_T - 0.577\right)}{1.2825}$$

For $T = 50$ years, $K_{50} = \dfrac{\left(3.902 - 0.577\right)}{1.2825} = 2.593$

Similarly for $T = 100$ years, $K_{100} = \dfrac{\left(4.601 - 0.577\right)}{1.2825} = 3.138$

Chow's general frequency equation is:

$$X_T = \bar{x} + K\sigma$$

$$42 = \bar{x} + 2.593\,\sigma$$

and $48 = \bar{x} + 3.138\,\sigma$

Solving the above two equations, we get mean and standard deviations as 13.45 and 11.01, respectively.

Now for $T = 200$ years,

$$Y_{200} = -\left[0.834 + 2.303 \log \log \frac{200}{200-1}\right] = 5.297$$

From Eqn. (8.56), we have

$$K_{200} = \frac{(5.297 - 0.577)}{1.2825} = 3.680$$

Hence flood at $T = 200$ years is:

$$X_{200} = 13.45 + 3.680 \times 11.01 = 53.97 \; m^3 / \sec.$$

8.4.2.8 Exponential Distribution

It is given as:

$$f(x) = \lambda\, e^{-\lambda x} \tag{8.60}$$

and the cumulative distribution function is given as:

$$F(x) = 1 - e^{-\lambda x} \tag{8.61}$$

This distribution has parameter as the mean rate of occurrence of the event. It has mean $(1/\lambda)$ and variance $(1/\lambda^2)$. It is used to obtain time interval between the occurrence of hydrologic events like flood, cyclone etc.

8.5 Construction of Probability Paper

The cumulative probability of distribution can be presented graphically on probability paper. The Y-axis (ordinate) represents the value of the variables in certain scale and X-axis (abscissa) is probability scale. On the probability scale, return period is also represented. The two axes of the pair are so adjusted that the distribution plot is obtained in straight line. The main objective of probability graph paper is to linearise the distribution for easy extrapolation and comparison. The linearization of probability plotting was first proposed by Hazen in 1914 and since then several advances have been made by a number researchers.

All types of probability papers are not available in the markets. However, a probability paper can be constructed for any type of distribution knowing their frequency-return period (k-T) relationship. For a particular distribution, values of k for various corresponding values of T are computed. Then in a rectangular plot, abscissa consisting of k values varying from 0 to +7 is marked and the corresponding T values are noted. The ordinate scale plots either arithmetic or logarithmic values of the variates. This gives the probability paper for the selected distribution.

8.6 Selection of Type of Distribution

A number of probability distributions are now used to fit the hydrologic data series. However, one should have some common idea which type of distribution should be fitted to which hydrologic data series. Following points should be followed to select the appropriate distribution that fits to a particular type of hydrologic data series:

- Extreme value distribution (Gumbel) is used for analysis of peak flood discharge, maximum rainfall depth etc.

- Extreme value type III distribution is used for drought study.

- Exponential distribution is more suitable for extrapolation of partial duration series.

8.6.1 Selection of Best Fit Distribution

For a particular hydrologic data series, a number of probability distribution functions can be fitted. For a given frequency or return period, each of the fitted probability distribution function will predict the hydrologic variable and all these values may not be same. Hence, it becomes difficult to decide which the appropriate value of hydrologic variable at a given return period. In this context, it is very important to know how to select the best fit distribution out of a number of distributions. Following criteria are used to select the best fit distribution:

8.6.1.1 Chi-Square Test of Goodness of Fit

In this method, the following formula is used to test the goodness of fit:

$$\chi^2 = \sum_{i=1}^{v} \frac{(O_i - P_{ei})^2}{P_{ei}} \qquad (8.62)$$

where, χ^2 is chi-square with v degrees of freedom, O_i and P_{ei} are the observed and predicted number of occurrences. The value of v is equal to $(N - h - 1)$ where, N is total number of data, h is number of parameters used in fitting the proposed distribution and 1 is subtracted since chi-square parameter is also computed. Values of χ^2 for various degrees of freedom and cumulative distribution (percentage level of significance) are presented in Table 8.7. In hydrology, 95% level of confidence is considered as the typical value. A null hypothesis is proposed in which we compare the values of χ^2 obtained from Eqn. (8.62) with the limiting value given in Table 8.7. The fitted probability distribution is accepted if the value of χ^2 obtained from Eqn. (8.62) is less than its corresponding value read from Table 8.7 for the degree of freedom of $(N - h - 1)$. The values of χ^2 are computed for different types of distributions fitted to the data series to predict the hydrologic variables and the distribution that gives the lowest value of χ^2 is considered as the best fit distribution.

Sample Calculation 8.12

The annual rainfall data of a station in Jharsuguda of Odisha grouped in class interval of 10 along with the frequency are given as below. Use chi-square test to check whether the normal distribution can be fitted to the data sets or not.

Rainfall class, cm	30-40	40-50	50-60	60-70	70-80	80-90	90-100	100-110	110-120
Frequency, f	5	6	3	7	5	5	5	6	3

Solution

The calculations are shown in tabular form

The data of this sample calculation is same as that mentioned for sample calculations 8.4 and 8.5. The mean (μ) calculated earlier in sample calculation 8.4 is 73.67 and standard deviation (σ) is calculated in sample calculation 8.5 is 24.91.

Using these values of mean (μ) and standard deviation (σ) and using Eqn. (8.33) which is:

Table 8.7: Chi-Square distribution percentages

Degree of freedom	0.005	0.010	0.025	0.05	0.10	0.20	0.30	0.40	0.50	0.60	0.70	0.80	0.90	0.95	0.99	0.995
1	0.04393	0.3175	0.02982	0.02393	0.0158	0.642	0.148	0.275	0.455	0.708	1.07	1.64	2.71	3.84	6.63	7.88
2	0.0100	0.0201	0.0506	0.103	0.211	0.446	0.713	1.02	1.39	1.83	2.41	3.22	4.61	5.99	9.21	10.6
3	0.0717	0.115	0.216	0.352	0.584	1.00	1.42	1.87	2.37	2.95	3.67	4.64	6.25	7.81	11.3	12.8
4	0.207	0.297	0.484	0.711	1.06	1.65	2.19	2.75	3.36	4.04	4.88	5.99	7.78	9.49	13.3	14.9
5	0.412	0.554	0.831	1.15	1.61	2.34	3.00	3.66	4.35	5.13	6.06	7.29	9.24	11.1	15.1	16.7
6	0.676	0.872	1.24	1.64	2.20	3.07	3.83	4.57	5.35	6.21	7.23	8.56	10.6	12.6	16.8	18.8
7	0.989	1.24	1.69	2.17	2.83	3.82	4.67	5.49	6.35	7.28	8.38	9.80	12.0	14.1	18.5	20.3
8	1.34	1.65	2.18	2.73	3.49	4.59	5.53	6.42	7.34	8.35	9.52	11.0	13.4	15.5	20.1	22.0
9	1.73	2.09	2.70	3.33	4.17	5.38	6.39	7.36	8.34	9.41	10.7	12.0	14.7	16.9	21.7	23.6
10	2.16	2.56	3.25	3.94	4.87	6.18	7.27	8.30	9.34	10.5	11.8	13.4	16.0	18.3	23.2	25.2
11	2.60	3.05	3.82	4.57	5.58	6.99	8.15	9.24	10.3	11.5	12.9	14.6	17.3	19.7	24.7	26.8
12	3.07	3.57	4.40	5.23	6.30	7.18	9.03	10.2	11.3	12.6	14.0	15.8	18.5	21.0	26.2	28.3
13	3.57	4.11	5.01	5.89	7.04	8.63	9.93	11.1	12.3	13.6	15.1	17.0	19.8	22.4	27.7	29.8
14	4.07	4.66	5.36	6.57	7.79	9.47	10.8	12.1	13.3	14.7	16.2	18.2	21.1	23.7	29.1	31.3
15	4.60	5.23	6.26	7.26	8.55	10.3	11.7	13.0	14.3	15.7	17.3	19.3	22.3	25.0	30.6	32.8
16	5.14	5.81	6.91	7.96	9.31	11.2	12.6	14.0	15.3	16.8	18.4	20.5	23.5	26.3	32.0	34.3
17	5.70	6.41	7.56	8.67	10.1	12.0	13.5	14.9	16.3	17.8	19.5	21.6	24.8	27.6	33.4	35.7
18	6.26	7.01	8.23	9.39	10.9	12.9	14.4	15.9	17.3	18.9	20.6	22.8	26.0	28.9	34.8	37.2
19	6.84	7.63	8.91	10.1	11.7	13.7	15.4	16.9	18.3	19.9	21.7	23.9	27.2	30.1	36.2	38.6
20	7.43	8.26	9.59	10.9	12.4	14.6	16.3	17.8	19.3	21.0	22.8	25.0	28.4	31.4	37.6	40.0
21	8.03	8.90	10.3	11.6	13.2	15.4	17.2	18.8	20.3	22.0	23.9	26.2	29.6	32.7	38.9	41.1
22	8.64	9.54	11.0	12.3	14.0	16.3	18.1	19.7	21.3	23.0	24.9	27.3	30.8	33.9	40.3	42.8

Contd.

Degree of freedom	0.005	0.010	0.025	0.05	0.10	0.20	0.30	0.40	0.50	0.60	0.70	0.80	0.90	0.95	0.99	0.995
23	9.26	10.2	11.7	13.1	14.8	17.2	19.0	20.7	22.3	24.1	26.0	28.4	32.0	35.2	41.6	44.2
24	9.89	10.9	12.4	13.8	15.7	18.1	19.9	21.7	23.3	25.1	27.1	29.6	33.2	36.4	43.0	45.6
25	10.5	11.5	13.4	14.6	16.5	18.9	20.9	22.6	24.3	26.1	28.2	30.7	34.4	37.7	44.3	46.9
26	11.2	12.2	13.8	15.4	17.3	19.8	21.8	23.6	25.3	27.2	29.2	31.8	35.5	38.9	45.6	48.3
27	11.8	12.9	14.6	16.2	18.1	20.7	22.7	24.5	26.3	28.2	30.3	32.9	36.7	40.1	47.0	49.6
28	12.5	13.6	15.3	16.9	18.9	21.6	23.6	25.5	27.3	29.2	31.4	34.0	37.9	41.3	48.3	51.0
29	13.1	14.3	16.1	17.7	19.8	22.5	24.6	26.5	28.3	30.3	32.5	35.0	39.1	42.6	49.6	52.3
30	13.8	15.0	16.8	18.5	20.6	23.4	25.5	27.4	29.3	31.3	33.5	36.3	40.3	43.8	50.9	53.7
35	17.2	18.5	20.6	22.5	24.8	27.8	30.2	32.3	34.3	36.5	38.9	41.8	46.1	49.8	57.3	60.3
40	20.7	22.2	24.4	26.5	29.1	32.3	34.9	37.1	39.3	41.6	44.2	47.3	51.8	55.8	63.7	66.8
45	24.3	25.9	28.4	30.6	33.4	36.9	39.6	42.0	44.3	46.8	49.5	52.7	57.5	61.7	70.0	73.2
50	28.0	29.7	32.4	34.8	37.7	41.4	44.3	46.9	49.3	51.9	54.7	58.2	63.2	67.5	76.2	79.5
100	67.3	70.1	74.9	77.9	82.4	87.9	92.1	95.8	99.3	102.9	106.9	111.7	118.5	124.3	135.6	140.2

Solution of sample calculation on 8.12 in tabular form.

Sl.No	Rainfall class, cm	Observed frequency, O_i	Standard normal variate, t_i of normal distribution	Cumulative probability of variate, t_i $[F(x_i)]$	Incremental probability, $p(x_i)$	Predicted frequency, P_{ei}	χ^2
(1)	(2)	(3)	(4)	(5)	(6)	(7)	(8)
1.	30-40	5	-1.351	0.0881	0.0881	3.964	0.270
2.	40-50	6	-0.950	0.1711	0.0830	3.735	1.373
3.	50-60	3	-0.549	0.2920	0.1209	5.440	1.095
4.	60-70	7	-0.147	0.4417	0.1497	6.736	0.010
5.	70-80	5	0.254	0.6004	0.1587	7.141	0.642
6.	80-90	5	0.655	0.7438	0.1434	6.453	0.327
7.	90-100	5	1.057	0.8542	0.1104	4.968	0.000
8.	100-110	6	1.458	0.9270	0.0728	3.276	2.265
9.	110-120	3	1.860	0.9686	0.0416	1.872	0.680
		45					6.663

$$t = \frac{(x - \mu)}{\sigma}$$

we calculate the values of t_i for various corresponding values of x and enter them in col. (4). The values of x in calculation of t is taken as the value of upper limit of each class interval. For example in row 4, the value of x for the class interval 60-70 is taken as 70 in computation of t for that row which is $t = (70 - 73.67)/24.91 = -0.147$ which is entered in row 4 and in col. (4).

Similarly values of t for other rows are calculated and entered in col. (4). For getting the values of $F(x_i)$ in col. (7), we use table for the area under standard normal curve (Table 8.2). For example for $t_i = -0.147$, we have $F(x_i) = 0.50 - 0.0583 = 0.4417$ which is entered in col. (5) and in row 4. Similarly for row 5, the value of $t = (80 - 73.67)/24.91 = 0.254$ and corresponding value of $F(x_i) = 0.50 + 0.1004 = 0.6004$ which is entered in col. (5). Similarly other rows are calculated and the values of t_i and $F(x)$ are entered in col. (5).

Predicted frequency of each row were computed by multiplying the incremental frequency with the total number of data i.e. N which is 45 in this case. Finally using the observed and predicted frequencies and using Eqn. (8.62), chi-square values were computed and entered in last column. The total chi-square value is estimated as 6.663.

Here we have $h = 2$ i.e. number of parameters used in fitting the distribution is 2 and they are mean and standard deviation. So degrees of freedom is, $v = (N - h - 1) = 9 - 2 - 1 = 6$

From Table 8.7 at 6 degree of freedom and 95% level of confidence i.e. level of confidence $= 0.95$, we have chi-square value 12.6. Since the computed chi-square is less than the tabulated chi-square i.e. $6.663 < 12.6$, the hypothesis cannot be rejected at 95% level of confidence. Hence, fitting of normal distribution is accepted.

8.6.1.2 Confidence Limits

The values of the statistical parameters like mean, standard deviation etc. are derived from the sample data. But at times the sample data may have some errors and consequently the statistical parameters may be wrong. Further the sampling period may be short which gives unreliable and erroneous result of statistical parameters. Under such condition, the predicted value of the variate for a given return period, x_T determined by the probability distribution can have some errors. Extrapolation of data to obtain the probability of occurrence of an event of larger return period may lead to errors due to the limitation of the sample record. In such cases, confidence limits can be laid on both upper and lower sides of the predicted distribution of the events. We can raw curves joining such confidence limit points at equidistant on both the sides of the data of various return periods. This gives rise to confidence bands. It is to be remembered that all data lying within the confidence bands are reliable to the extent of probabilities on which the confidence limits are based. Following Gumbel, for the confidence probability c, the confidence interval of an event x_T is bounded by values x_1 and x_2 which can be calculated as:

$$x_{1/2} = x_T \pm f(c) \, x S_e \qquad (8.63)$$

where, $f(c)$ is function of the confidence probability c determined by using the table of normal variate as given below and S_e is probable error given by Eqn. (8.64). In Eqn. (8.63), we take + sign for upper confidence interval i.e. x_1 and $-$ve sign for lower confidence interval i.e. x_2.

Confidence probability, c in %	50	68	80	90	95	99
$f(c)$	0.674	1.00	1.282	1.645	1.96	2.58

$$S_e = b \frac{\sigma_{N-1}}{\sqrt{N}} \qquad (8.64)$$

$$b = \sqrt{1 + 1.3\,K + 1.1\,K^2} \qquad (8.65)$$

where, K is frequency factor, N is total sample size and σ_{N-1} is standard deviation. It is observed that for a given sample and T, 80% confidence limits are twice as large as the 50% limits and 95% limit are thrice as large as 50% limits.

Procedure to draw confidence bands for the extreme value distribution is outlined as follows:

- Compute the mean and standard deviation of the sample data
- Compute the frequency factor, K or read the value of K from K-T table for the given probability distribution function of the sample
- Compute the values of x_T for different return periods T
- Compute the value of b (Eqn. 8.65) and S_e (Eqn. 8.64)
- Take a probability paper and plot the fitted line for x_T
- Calculate the values of x_1 and x_2 i.e. $x_{1/2}$ using Eqn. (8.63)
- In the same paper, plot x_1 and x_2 against the values of T.

Confidence bands are drawn on both sides of fitted line for difference confidence limits like 80% or 90% limits. It is to be remembered that 99% confidence bands will be the outermost band whereas the 50% band will be very close to the fitted line. Further at the value of T increases, the limit of confidence increases and thereby more spreading of confidence bands occurs forming a conical shape. Again we should note that 95% confidence level means out of 100 samples, 95 samples will be inside the interval.

Sample Calculation 8.13

The river Mahanadi has a gauging station which yields 92 years of flood peak data. The mean and standard deviation of these 92 years of data are obtained as 6437 and 2951 m³/sec, respectively. Using Gumbel distribution, compute the flood discharge at a return period of 500 years. What are the 80% and 95% confidence limits for the estimate?

Solution

Given, $N = 92$

Using Table 8.5 at $N = 92$, we get $y_n = 0.5589$ and using Table 8.6, $S_n = 1.2020$.

Now $y_{500} = -[ln.\ ln\ (500 - 499)] = 6.21361$.

$\quad K_{500} = (6.21361 - 0.5589) / 1.2020 = 4.7044$

$\quad X_{500} = 6437 + 4.7044 \times 2951 = 20320$ m³/sec

Using Eqn. (8.65) we get

$$b = \sqrt{1 + 1.3\ K + 1.1\ K^2}$$

$$b = \sqrt{1 + 1.3 \times 4.7044 + 1.1 \times (4.7044)^2}$$

$$= 5.61$$

Using Eqn. (8.64) we get

$$S_e = b\ \frac{\sigma_{N-1}}{\sqrt{N}}$$

$$S_e = 5.61 \times \frac{2951}{\sqrt{92}}$$

$$= 1726$$

(i)   For 80% confidence probability, $f(c) = 1.282$.

Using Eqn. (8.63),

$x_{1/2} = 20320 \pm 1.282 \times 1726$

$x_1 = 22533$ m³/sec and $x_2 = 18107$ m³/sec

This shows the computed discharge of 20320 m³/sec at 500 years return period has a 80% probability to lie between 22530 and 18107 m³/ses.

(ii)   For 95% confidence probability, $f(c) = 1.96$.

Using Eqn. (8.63),

$x_{1/2} = 20320 \pm 1.96 \times 1726$

$x_1 = 23703$ m³/sec and $x_2 = 16937$ m³/sec

This shows the computed discharge of 20320 m³/sec at 500 years return period has a 95% probability to lie between 23703 and 16937 m³/ses.

Likewise the flood data were computed at other return periods like 50, 100, 200 etc. and the confidence bands for each return period at 80 and 95% confidence limits were computed. In a probability graph paper the estimated discharges at various return periods were plotted and the fitted line is drawn to these plotted points (Fig 8.6). Similarly, the confidence bands for various return periods are calculated at 80 and 95% confidence limits and the points are plotted on the same probability paper. Lines are drawn through these plotted points both above and below the fitted line which shows the confidence bands (dotted line in Fig 8.6). It is seen from the figure that as the confidence probability increases the confidence interval also increases and the confidence bands are seen to spread more.

8.7 Regression and Correlation

8.7.1 Regression

In hydrologic design, it is required to evaluate the relationship between two or more variables simultaneously. This can be done by regressing one set of variable on the other set. For example we have rainfall and runoff data of a river basin. If we represent runoff by symbol y_i and rainfall by symbol x_i, then y_i is regressed upon x_i to get a relation between rainfall and runoff. In the above regression, we say x_i as independent variable whereas y_i is called as dependent variable. There are two methods available to fit a curve between the given set of data.

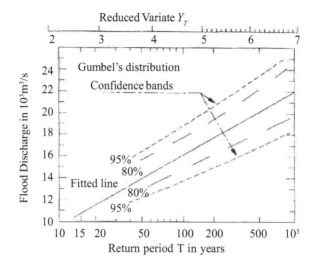

Fig. 8.6: Confidence bands for Gumbel's distribution

They are

- Graphical method

- Analytical method

Graphical Method: In graphical method, the independent data taken as abscissa of a graph paper (in X-axis) whereas the dependent data are taken as ordinate in Y-axis. Points are marked in the graph paper and the resulting plot is called as scatter diagram (Fig. 8.7). A best fit curve is drawn among these scattered points which passes through the mean of the spread of all the points. Knowing any value of x, the corresponding value of y can be read from the graph.

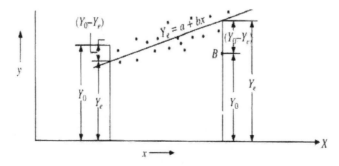

Fig. 8.7: A scatter diagram between variables x and y and the fitting of the scatter points with best fit line

Analytical Method: In analytical method, the best fit curve is drawn through the plotted points in such a way that sum of the squares of departures of the observed points from the fitted function is the minimum. In Fg.8.7, error between the observed and the estimated points (estimated point is obtained after fitting the line) is shown as $(y_o - y_e)$. While we fit a straight line between N sets of x_i and y_i data, the equation of straight line may be in the form

$$y_e = a + b\,x \tag{8.66}$$

There may have some points above this straight line whereas some points are likely to lie below as shown in Fig. 8.7. Hence the error value $(y_o - y_e)$ may be positive or negative. The procedure to obtain the best values of a and b *(a and b* in Eqn. 8.66) are called as intercept and slope of regression line) in Eqn. (8.66) should be such that the sum of the squares of the error values called as error functions also should be minimum. This is called as *method of least squares*. The error function given by method of least square is expressed as:

$$S_e = \sum_{i=1}^{N} \left(y_{oi} - y_{ei}\right)^2 \tag{8.67}$$

where, S_e is error function, y_{oi} is the observed data of i-th point and y_{ei} is the estimated data of i-th point and N is total number of data points.

$$S_e = \sum_{i=1}^{N} \left(y_i - a - b\,x_i\right)^2 \tag{8.68}$$

Since error function should be minimum, we have to differentiate Eqn. (8.68) with respect to a and equate it to zero which yields

$$\sum y_i - N\,a - b \sum x_i = 0 \tag{8.69}$$

Similarly differentiating Eqn. (8.68) with respect to b and equate it to zero yields

$$\sum x_i\,y_i - a \sum x_i - b \sum x_i^2 = 0 \tag{8.70}$$

Solution of Eqns. (8.69) and (8.70) gives

$$a = \frac{\sum y_i \sum x_i^2 - \sum x_i \sum (x_i\,y_i)}{N \sum x_i^2 - \left(\sum x_i\right)^2} = \frac{\sum y_i - b \sum x_i}{N} \tag{8.71}$$

$$\text{and } b = \frac{N \sum (x_i \, y_i) - \sum x_i \sum y_i}{N \sum x_i^2 - \left(\sum x_i \right)^2} \tag{8.72}$$

8.7.2 Correlation

Correlation refers to the degree of association between the variables. The statistical parameter determining their relation is known as coefficient of correlation (r). When two variables say x and y are involved we say it simple correlation. When more than two variables are involved, we say it multiple correlation. The value of coefficient of correlation ranges from -1 to 1. When the value is 1, it refers that there is a perfect relation between the two variables. Such case may occur when all the points lie on the straight line $y = a + b x$ where $a = \pm 1$ and $b = 1$. In Fig. 8.8 (a) the correlation coefficient of two sets of variables is shown as $r = 1$ and the equation of the straight line on which all the plotted points lie is $y = x - 1$. Value of $r = 1$ refers that there is a perfect linear dependence between x and y. In Fig. 8.8(b) the points are either on or slightly off of the line $y = x - 1$ and value of r is less than 1 and for the shown figure, it is 0.986. Some of the points deviate from the straight line and so there is no perfect linear dependence between the two variables. In Fig. 8.8 (c), the value of r is -0.671. The points in this case are scattered about the line. There is weak association between the two variables and the relation of association is negative. Though the value of r is less and negative yet there is some existence of dependence between the variables.

In Fig. 8.8 (d)) the scatter of the points is very great with a corresponding lack of a strong dependence. The value of r is this case is obtained as 0.211 which is very less indicating that there is little dependence of the two associating variables. Fig. 8.8 (e) and 8.8 (f) are examples of functionally dependence. Fig. 8.8 (e) refers parabolic relation between the two variables x and y and the relation is $y = x^2/4$. Fig. 8.8 (f) refers circular relation between the two variables x and y and the relation is $x^2 + y^2 = 9$. The value of r in Fig. 8.8 (e) is 0.963 indicating high degree of relation. In circular relation as in Fig. 8.8 (f), the value of r is zero. Fig. 8.8 (f) illustrates a situation where x and y are perfectly functionally related even though value of $r = 0$. However, the functional relation is not linear. This figure demonstrates that one cannot conclude that x and y are not related based on the fact that their coefficient of correlation is zero or small. The fact is often overlooked that high correlation does not necessarily mean there is a cause and effect relationship between the correlated variables.

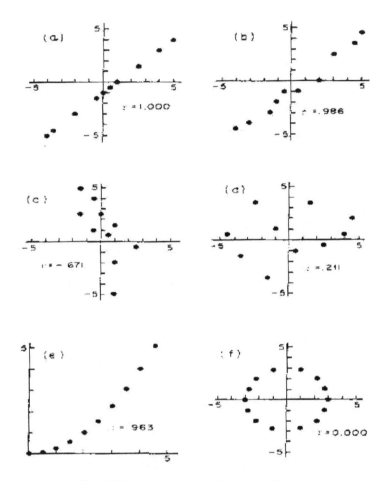

Fig. 8.8: Examples of correlation coefficient

8.8 Standard Error of Estimate

Standard error of estimate (S_{ee}) and coefficient of correlation are the two important statistical parameters that decide the goodness of fit of the curve. Standard error of estimate measures the spread of the points around the fitted curve. Lesser is the value of standard error of estimate, greater is the accuracy of the predicted values by the regression line. Value of standard error of estimate of zero indicates that there is a perfect linear dependence between x and y, all the plotted points lie on the straight line and in that case value of $r = \pm 1$. For N pairs of (x_i, y_i) data, the standard error of estimate is given as:

$$S_{ee} = \sqrt{\frac{\sum (y_{oi} - y_{ei})^2}{N}} = \sqrt{\frac{S_e}{N}} \qquad (8.73)$$

8.8.1 Covariance

The covariance of two variables x_i and y_i written as $Cov\ (x_i, y_i)$ is the second moment of the cross product of x_i and y_i. The population covariance is expressed as (Haan, 1977):

$$Pop.\ Cov\ (x_i, y_i) = \frac{1}{N}\sum_{i=1}^{N}\{(x_i - \overline{x})(y_i - \overline{y})\} \tag{8.74}$$

The sample covariance is expressed as (Haan, 1977):

$$Cov\ (x_i, y_i) = \frac{1}{N-1}\sum_{i=1}^{N}\{(x_i - \overline{x})(y_i - \overline{y})\} \tag{8.75}$$

For two independent variable i.e. if x_i and y_i are independent then their covariance is zero i.e. $Cov\ (x_i, y_i)$. Covariance has the units equal to the units of x times that of y.

8.8.2 Correlation Coefficient (r)

A normalized covariance called as correlation coefficient is obtained by dividing the covariance by the product of standard deviation of x and y. It is given as:

$$r = \frac{Cov(x_i, y_i)}{\sigma_x \cdot \sigma_y} \tag{8.76}$$

The correlation coefficient can also be written as:

$$r = \frac{N\left(\sum(x_i\ y_i)\right) - \left(\sum x_i\right)\left(\sum y_i\right)}{\sqrt{\left[N\left(\sum x_i^2\right) - \left(\sum x_i\right)^2\right] \cdot \left[N\left(\sum y_i^2\right) - \left(\sum y_i\right)^2\right]}} \tag{8.77}$$

It is to be noted that square of the coefficient of correlation is called as coefficient of determination (r^2). If the fitted line passes through all the plotted points, then $r = \pm 1$ and in that case coefficient of determination is 1 and $S_e = 0$. In statistical analysis, the hydrologists generally prefer to use coefficient of determination than coefficient of correlation. If $r = 0.80$, then $r^2 = 0.64$. This means 64% is explained due to their association and the rest are due to unexplained factors.

8.8.3 Significance of Parameter

The length of sample used for analysis is generally small in comparison to the population and so the value of coefficient of correlation (r) obtained in the analysis need to be tested for confidence limits. This can be achieved by computing standard deviation of the correlation coefficient as:

$$S_r = \frac{1-r^2}{\sqrt{N}}$$ (8.78)

Now compute $(r + 3 S_r)$ and $(r - 3 S_r)$ or $(r + 4 S_r)$ and $(r - 4 S_r)$. If both limits have the same sign as r, we consider r as significantly different from zero with respect to either of these empirical confidence limits. If these limits have different signs, then r is not considered as significantly different from zero (Patra, 2001).

In the regression equation of $y_e = a + b x$, the values of a and b computed by Eqns. (8.71) and (8.72), respectively should also be tested for their level of significance. For testing b, the t distribution is given as:

$$t = \left\{ r \frac{(b - \beta)}{b} \right\} \left\{ \frac{(N - 2)}{(1 - r^2)} \right\}^{1/2}$$ (8.79)

where, β is the true value of b and t following t –distribution with $(N - 2)$ degrees of freedom. The value of t can be obtained from standard t-distribution table. To estimate the test statistics take $\beta = 0$, and compute t for the value of r and degree of freedom as $(N - 2)$ from Eqn. (8.79). If the t value computed by Eqn. (8.79) is greater than the corresponding tabulated value at the tested level of significance *i.e.* at 90 or 95% level, then b is considered as significantly different from zero (Patra, 2001).

The intercept a of the regression equation can be tested as follows. The standardized variate of a is given as:

$$t = \left\{ r \frac{(a - \alpha)}{b} \right\} \left\{ \frac{(N - 2)}{(1 - r^2)(\sigma_n^2 + x_{av}^2)} \right\}^{1/2}$$ (8.80)

When the line is assumed to pass through the origin, value of is zero. At 95% level of significance, the value of t is obtained from the table of student's t distribution. At this level, both the values of a are calculated from the relation (Eqn. 8.80). If the value of a calculated from sample data and the value of a from Eqn.(8.80) are much different from zero, then the hypothesis of intercept being significantly different from zero is accepted. In other words, the hypothesis is tested by comparing the value of t obtained from Eqn. (8.80) and from student's t- distribution table for the probability level.

8.9 Standard Forms of Bivariate Equations

Followings are some standard forms of equations used in bivariate regression analysis in hydrology:

(i) Linear $y = a + b x$

(ii) Exponential $y = b\, e^{ax}$

(iii) Parabolic $y = a\, x^b$

(iv) Higher order equation $y = a_1 + a_2 + a_3x^2 + \ldots\ldots + a_{n+1}x^n$

(v) Other forms of equations are:

$y = a + (b/x)$

$y = a / (b + x)$

$y = x / + b\, (a + b\, x)$

$y = c + b\, e^{ax}$

$y = (c + b)/ ((x - a)$

$y = d\, e^{cx} + b\, e^{ax}$

$y = e^{ax}\, (d \cos bx + c \sin bx)$

Exponential and parabolic equations can be solved by taking logarithmic transformation in both sides so that the equations take the form of linear form. A sample calculation (given below) helps in transforming the power form or parabolic form into linear form and solving the equation.

Sample Calculation 8.14

Following observed data of rainfall and runoff of the month of September in a gauging site in river Brahmin in Odisha are collected. Develop a linear and parabolic relationship between the rainfall and runoff of the month of September. Calculate the coefficient of correlation and test its significance.

Year	2000	2001	2002	2003	2004	2005	2006	2007	2008	2009	2010
Rainfall, mm	250.5	140.5	180.5	143.6	100.3	120.4	100.5	135.8	140.4	125.8	150.5
Runoff, mm	137.5	73.6	90.9	78.3	55.8	64.8	55.0	70.7	75.5	66.9	82.1

Solution

Liner relation between rainfall and runoff is $y = a + b x$, where y is runoff and x is rainfall. To find out the value of a and b and coefficient of correlation, r the calculations are made and shown as in table below.

Year	Rainfall, x, mm	Runoff, y, mm	xy	x^2	y^2
(1)	(2)	(3)	(4)	(5)	(6)
2000	250.5	137.5	34443.75	62750.25	18906.25
2001	140.5	73.6	10340.8	19740.25	5416.96
2002	180.5	90.9	16407.45	32580.25	8262.81
2003	143.6	78.3	11243.88	20620.96	6130.89
2004	100.3	55.8	5596.74	10060.09	3113.64
2005	120.4	64.8	7801.92	14496.16	4199.04
2006	100.5	55.0	5527.5	10100.25	3025
2007	135.8	70.7	9601.06	18441.64	4998.49
2008	140.4	75.5	10600.2	19712.16	5700.25
2009	125.8	66.9	8416.02	15825.64	4475.61
2010	150.5	82.1	12356.05	22650.25	6740.41
Total	1588.8	851.1	132335.4	246977.9	70969.35

Eqn. (8.71) gives value of a which is:

$$a = \frac{\sum y_i \sum x_i^2 - \sum x_i \sum (x_i\, y_i)}{N \sum x_i^2 - \left(\sum x_i\right)^2}$$

$$a = \frac{851.1 \times 246977.9 - 1588.8 \times 132335.4}{11 \times 246977.9 - 1588.8 \times 1588.8} = -0.268$$

Thus, Eqn. (8.72) gives value of b which is:

$$b = \frac{N \sum (x_i\, y_i) - \sum x_i \sum y_i}{N \sum x_i^2 - \left(\sum x_i\right)^2}$$

$$b = \frac{11 \times 132335.4 - 1588.8 \times 851.1}{11 \times 246977.9 - 1588.8 \times 1588.8} = 0.537$$

The regression equation for prediction of runoff is $y = -0.268 + 0.537\, x$

Coefficient of correlation, r is given as (Eqn. 8.77):

$$r = \frac{N\left(\sum(x_i\,y_i)\right) - \left(\sum x_i\right)\left(\sum y_i\right)}{\sqrt{\left[N\left(\sum x_i^2\right) - \left(\sum x_i\right)^2\right] \cdot \left[N\left(\sum y_i^2\right) - \left(\sum y_i\right)^2\right]}}$$

$$r = \frac{11 \times 132335.4 - 1588.8 \times 851.1}{\sqrt{[11 \times 246977.9 - 1588.8 \times 1588.8] \cdot x\, [11 \times 70969.35 - 851.1 \times 851.1]}} = 0.993$$

Test for coefficient of correlation, r

Standard deviation of coefficient of correlation, r is:

$$S_r = \frac{1 - r^2}{\sqrt{N}}$$

$$= (1 - 0.993 \times 0.993) / (11)^{0.5} = 0.0042$$

$r + 3\,S_r = 0.993 + 3 \times 0.0042 = 1.0056$

$r - 3\,S_r = 0.993 - 3 \times 0.0042 = 0.9804$

$r + 4\,S_r = 0.993 + 4 \times 0.0042 = 1.0098$

$r - 4\,S_r = 0.993 - 4 \times 0.0042 = 0.9762$

Since all are positive, the value of r is significantly different from zero.

The parabolic relation between rainfall and runoff is expressed as:

$y = a\,x^b$

where y is runoff and x is rainfall. Taking logarithm of both sides we get:

$log\ y = log\ a + b\ log\ x$

which is linear in form $Y = A + b\,X$ with $log\ y = Y$, $log\ a = A$ and $log\ x = X$.

Calculations are made and shown in table below.

Now $A = log\ a$ is given as:

$$A = \frac{\sum Y_i \sum X_i^2 - \sum X_i \sum (X_i\,Y_i)}{N \sum X_i^2 - \left(\sum X_i\right)^2}$$

$$A = \frac{20.623 \times 50.765 - 23.601 \times 44.371}{11 \times 50.765 - 23.601 \times 23.601} = -0.194$$

Solution of sample calculation 8.14 in tabular form when relation between rainfall and runoff is parabolic

Year	Rainfall, x, mm	Runoff, y, mm	$\log x = X$	$\log y = Y$	XY	X^2	Y^2
(1)	(2)	(3)	(4)	(5)	(6)	(7)	(8)
2000	250.5	137.5	2.3988077	2.1383027	5.12937704	5.75427853	4.57233843
2001	140.5	73.6	2.1476763	1.86687781	4.00944928	4.61251359	3.48523277
2002	180.5	90.9	2.2564772	1.95856388	4.41945476	5.09168938	3.83597248
2003	143.6	78.3	2.1571544	1.89376176	4.08513659	4.65331528	3.58633361
2004	100.3	55.8	2.0013009	1.7466342	3.49554065	4.00520542	3.05073102
2005	120.4	64.8	2.0806265	1.81157501	3.76921094	4.32900658	3.281804
2006	100.5	55.0	2.0021661	1.74036269	3.48449511	4.00866894	3.02886229
2007	135.8	70.7	2.1328998	1.84941941	3.94462624	4.54926143	3.42035217
2008	140.4	75.5	2.1473671	1.87794695	4.03264151	4.6111855	3.52668475
2009	125.8	66.9	2.0996806	1.82542612	3.83281188	4.40865879	3.33218051
2010	150.5	82.1	2.1775365	1.91434316	4.1685521	4.74166521	3.66470972
Total	1588.8	851.1	23.601693	20.6232137	44.3712961	50.7654486	38.7852018

Hence, $a = anti\ log\ A = 0.639$

$$b = \frac{N \sum (X_i\ Y_i) - \sum X_i \sum Y_i}{N \sum X_i^2 - \left(\sum X_i\right)^2} = 0.964$$

Hence the parabolic equation is $y = 0.639\ x^{0.964}$

We can proceed with Eqn. (8.77) and can prove that $r = 0.993$. Proceeding in the same way as for the linear equation, we can also prove that the value of r is significantly different from zero. Since value of r is high and r is significantly different from zero, there is a good correlation between rainfall and runoff.

8.10 Multivariate Linear Regression and Correlation

In hydrology, sometimes, we come across two or more variables that affect the dependent variable. For example, a main river may receive discharge from two or more of its tributaries. If two tributaries say x_1 and x_2 contribute to the main river x, then the equation for discharge can be formulated as:

$$x = b_0 + b_1\ x_1 + b_2\ x_2 \tag{8.81}$$

where, b_0 is called as intercept, b_1 is the multiple regression coefficient between x and x_1 when x_2 is constant and b_2 is the multiple regression coefficient between x and x_2 when x_1 is constant. The values of b_0, b_1 and b_2 can be evaluated by least square as:

$$\sum x = N\ b_0 + b_1 \sum x_1 + b_2 \sum x_2 \tag{8.82}$$

$$\sum x\ x_1 = b_0 \sum x_1 + b_1 \sum x_1^2 + b_2 \sum x_1\ x_2 \tag{8.83}$$

$$\sum x\ x_2 = b_0 \sum x_2 + b_1 \sum x_1\ x_2 + b_2 \sum x_2^2 \tag{8.84}$$

Where, N is total number of samples. Solving Eqns. (8.82) to (8.84) simultaneously for the three unknowns, b_0, b_1 and b_2, respectively, the curve for the multivariate regression equation is formulated.

Sometimes the multivariate regression equation of the type

$$x = a\ x_1^{b_1} \cdot x_2^{b_2} \cdot x_3^{b_3} \ldots\ldots\ldots x_n^{b_n} \tag{8.85}$$

can be solved by transforming it into linear form by logarithmic transformation. The readers are requested to try themselves.

8.10.1 Multiple Correlation Coefficients

Multiple correlation coefficient, r refers to the association between the independent variables $(x_1, x_2$ etc., $)$ taken together with the variable x and is given as:

$$r = \frac{\sigma_{est}}{\sigma_x} \qquad (8.86)$$

where, σ_{est} is the standard deviation of the estimated values. It is obtained by substituting the observed x_1 and x_2 values in Eqn. (8.81) and estimating the series of x_i as x_{iest}. From the series of x_{iest}, standard deviation σ_{est} is computed. The symbol σ_x in Eqn. (8.86) represents the standard deviation of the x_i series.

8.10.2 Partial Correlation Coefficient of Multivariate Regression Equation

Sometimes, hydrologists are interested to know the correlation between the dependent variable x_i and a given independent variable say x_1 or x_2. This is achieved as:

$$r_{1-i}^2 = \frac{\left(\rho_1^2 - \rho_{1-i}^2\right)}{\left(1 - \rho_{1-i}^2\right)} \qquad (8.87)$$

where, σ_1 is the multiple correlation coefficient between the dependent variable x and all independent variables x_1 and x_2 etc and σ_{1-i} is the multiple correlation coefficient between the dependent variable x and all independent variables except x_1 with which the relation is required. The left hand side expression of the Eqn. (8.87) is called *partial correlation* and the term in left hand side of the equation is called as *partial correlation coefficient*. Eqn. (8.86) gives multiple correlation between the variables x and $(x_1, x_2 ... x_n)$. In the same way multiple correlation coefficient between the variables x and $(x_2, x_3 ... x_n$ except $x_1)$ can also be estimated. It is found that r_1 is always greater than r_{1-i}.

Sample Calculation 8.15

Following data are recorded in a gauging station. Assuming the rainfall of both August and September contribute to the runoff recorded in September, develop a multiple linear regression between the rainfall of both the months and runoff. Compute the multiple correlation coefficient of the regression equation.

Year	2000	2001	2002	2003	2004	2005	2006	2007	2008	2009	2010
Rainfall, August, mm	250.5	140.5	180.5	143.6	100.3	120.4	100	135.8	140.4	125.8	150.5
Rainfall, September, mm	240.1	123.6	167.8	140.5	110.5	116.3	96.4	130.2	135.7	130.5	155.8
Runoff, September, mm	137.5	73.6	90.9	78.3	55.8	64.8	55.0	70.7	75.5	66.9	82.1

Solution

Since rainfall of both August and September influences the runoff of September, the relation is of multiple regression type. The multiple regression equation is given as (Eqn. 8.81):

$$x = b_o + b_1 x_1 + b_2 x_2$$

where, x_1 and x_2 are the rainfall of August and September and x is the runoff of September. Using Eqns. (8.82 to 8.84) we have

$$\sum x = N b_o + b_1 \sum x_1 + b_2 \sum x_2$$

$$\sum x \, x_1 = b_o \sum x_1 + b_1 \sum x_1^2 + b_2 \sum x_1 x_2$$

$$\sum x \, x_2 = b_o \sum x_2 + b_1 \sum x_1 x_2 + b_2 \sum x_2^2$$

Calculations one made in tabular form as given in table below

Substituting the data from the above table in Eqns. (8.82 to 8.84) we have:

$11 b_o + 1588.8 \, b_1 + 1547.4 \, b_2 = 851.1$ or $b_o + 144.4 \, b_1 + 140.7 \, b_2 = 77.4$ (A)

$1588.8 \, b_o + 246977.9 \, b_1 + 239346.7 \, b_2 = 132335.37$ or $b_o + 155.4 \, b_1 + 150.6 \, b_2 = 83.3$ (B)

$1547.4 \, b_o + 239346.7 \, b_1 + 232521.4 \, b_2 = 128341.1$ or $b_o + 154.7 \, b_1 + 150.3 \, b_2 = 82.9$ (C)

Solution of Eqns. (A to C) gives $b_o = 0.742$, $b_1 = 0.603$ and $b_2 = -0.074$

Hence the multivariate regression equation is $x = 0.742 + 0.603 \, x_1 - 0.074 \, x_2$

Solution of sample calculation 8.15 in tabular form

Rainfall, Aug., x_1, mm	Rainfall, Sept., x_2, mm	Runoff, Sept., x, mm	$(x_2) \cdot (x)$	x_2^2	x^2	$(x_1) \cdot (x)$	x_1^2	$(x_1) \cdot (x_2)$
(1)	(2)	(3)	(4)	(5)	(6)	(7)	(8)	(9)
250.5	240.1	137.5	33013.75	57648.01	18906.25	34443.75	62750.25	60145.05
140.5	123.6	73.6	9096.96	15276.96	5416.96	10340.8	19740.25	17365.8
180.5	167.8	90.9	15253.02	28156.84	8262.81	16407.45	32580.25	30287.9
143.6	140.5	78.3	11001.15	19740.25	6130.89	11243.88	20620.96	20175.8
100.3	110.5	55.8	6165.9	12210.25	3113.64	5596.74	10060.09	11083.15
120.4	116.3	64.8	7536.24	13525.69	4199.04	7801.92	14496.16	14002.52
100.5	96.4	55.0	5302	9292.96	3025	5527.5	10100.25	9688.2
135.8	130.2	70.7	9205.14	16952.04	4998.49	9601.06	18441.64	17681.16
140.4	135.7	75.5	10245.35	18414.49	5700.25	10600.2	19712.16	19052.28
125.8	130.5	66.9	8730.45	17030.25	4475.61	8416.02	15825.64	16416.9
150.5	155.8	82.1	12791.18	24273.64	6740.41	12356.05	22650.25	23447.9
(1588.8)	(1547.4)	(851.1)	(128341.1)	(232521.4)	(70969.35)	(132335.37)	(246977.9)	(239346.7)

Using the developed multivariate regression equation and considering the values of rainfall in August and September, we estimate the runoff of the month of September which is given below.

Rainfall, Aug., x_1, mm	Rainfall, Sept., x_2, mm	Observed runoff, Sept., x, mm	Estimated runoff, Sep., x_{est}, mm
(1)	(2)	(3)	(4)
250.5	240.1	137.5	134.0261
140.5	123.6	73.6	76.3171
180.5	167.8	90.9	97.1663
143.6	140.5	78.3	76.9358
100.3	110.5	55.8	53.0459
120.4	116.3	64.8	64.737
100.5	96.4	55.0	54.2099
135.8	130.2	70.7	72.9946
140.4	135.7	75.5	75.3614
125.8	130.5	66.9	66.9424
150.5	155.8	82.1	79.9643
(1588.8)	(1547.4)	(851.1)	(851.7)

N.B. Values in parenthesis of the table (last row) gives sum total of the values.

Using the estimated runoff (values of last column of the above table), we estimate the standard deviation σ_{est} which is 22.43. Again using the observed values of runoff (column 3 of above table), we estimate the standard deviation σ_x which is 22.62. Now using Eqn. (8.86), we get multiple correlation coefficient, $r = 0.99$.

8.11 Analysis of Time series

Sometimes, we arrange a series of observations with respect to time of their occurrences in a systematic order. Such a series is called as *time series*. The statistical homogeneity of the series is evaluated with respect to time and space. A time series is said to be time homogeneous, if the similar or identical events which are under consideration in the series are likely to occur at all times. A hydrologic time series is hardly *time homogeneous* because of the effects of several variations in nature. Sometimes, the time series segments drawn from the same population have the same expected value of statistical parameters for each section. Such a phenomenon is called as *stationary*. In hydrology, the annual series may be stationary but the daily or monthly series are not stationary. Time homogeneity of the hydrologic data can be analysed by properties like trend, periodicity and persistence. They should be identified, quantifies and removed as they are deterministic in nature. A random stochastic component is

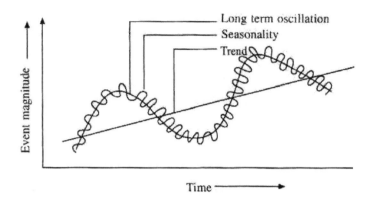

Fig. 8.9: Presence of trend, oscillation and seasonality in time series

said to be present when the serial correlation coefficients of different lags are zero. A time series x_t is modeled as:

$$x_t = x_{pt} + x_{Tt} + x_{ot} + x_{jt} + E_t \tag{8.88}$$

where, x_{Tt} is component due to *trend,* x_{pt} is the component due to *periodicity,* x_{ot} is the component due to *oscillations,* x_{jt} is the component due to *jump* and E_t is the stochastic component. Fig. 8.9 represents the presence of trend, oscillations and periodicity or seasonality for a time series.

A hydrologic time series can be classified to have either deterministic or stochastic component. In hydrologic time series, the components like trend, periodicity, oscillations and jump are deterministic in nature which can be quantified and removed. The stochastic components may be stationary or non-stationary. The ARIMA model is used to study the non-stationary component of the time series whereas the stationary component is studied by auto-regressive models or random distributed models like nomal, gamma etc.

8.11.1 Trend Analysis

A trend refers to the increasing or decreasing rate of change of mean value of the hydrologic variable

in one direction. The factors that are responsible for presence of trend in a time series are:

- Deforestation
- Industrialization
- Urbanisation
- Large scale land slide and

- Large changes in watershed characteristics

The following tests are used to detect the presence of trend in a time series:

- Turning Point Test
- Kendal's Rank-Correlation Test and
- Regression Test for Linear Trend

8.11.1.1 Turning Point Test

In turning point test, we test the data series to find out the number of turning points that exist in the series. The turning point exists if the data x_i is either greater or smaller than both of its preceding and succeeding data. If the conditions $x_{i-1} < x_i > x_{i+1}$ or $x_{i-1} > x_i < x_{i+1}$ then there exists the turning point in the data series. The procedure is outlined as under:

- Arrange the data in order of their occurrence.
- Apply either of the conditions $x_{i-1} < x_i > x_{i+1}$ or $x_{i-1} > x_i < x_{i+1}$ and find out how many turning point exists in the data series.
- Determine the total number of turning points in the series and let it be p.
- Compute the expected number of turning point in the series. Compute

$$E(p) = \frac{2(n-2)}{3} \tag{8.89}$$

where n is total number of data.

- Calculate the variance of p as:

$$Var(p) = \frac{(16n - 29)}{90} \tag{8.90}$$

Express p in standard normal form as:

$$Z = \frac{\{p - E(p)\}}{\{Var(p)\}^{1/2}} \tag{8.91}$$

- Test it at 5% level of significance i.e. consider the value of Z as ± 1.96 at 5% level of significance.
- If Z is within ± 1.96, we say that at 5% level of significance there is no reason to doubt that the sample events are other than a random sequence i.e. there is no trend.

8.11.1.2 Kendal's Rank-Correlation Test

The steps to conduct this test are as follows:

- Pick up the first value of the series x_i and compare it with respect to the series containing data except x_i. Find out how many times it is greater than the rest i.e. compare x_1 and x_2 and if $x_1 > x_2$ then mark it as 1 or say take $p_{1.1}$ as 1. Then compare it with x_3 and if again it is greater than x_3 then take $p_{1.2}$ as 2. Complete the process for x_1 with respect to the series and let p_{1ex} is all the expected value for x_1. Repeat it for x_2, x_3 ..... x_n and let $p_{2ex}, p_{3ex}....p_{nex}$ are the corresponding values.

- Find $p = p_{1ex} + p_{2ex} + p_{3ex}.........+p_{nex}$ (8.92)

- Maximum value of p can be $p_{max} = n\,(n\text{-}1)\,/2$ (8.93)

- Compute $E(p) = n\,(n\text{-}1)\,/4$ (8.94)

- Kendal's value is computed as:

$$\tau = \left[\left\{\frac{4\,p}{n\,(n-1)}\right\} - 1\right], \qquad E(\tau)\ should\ be\ 0.$$ (8.95)

- Variance of Kendal's value is:

$$Var\,(\tau) = \left[\frac{\{2\,(2n+5)\}}{9n\,(n-1)}\right]$$ (8.96)

- Standard test for statistics of Z is given as:

$$Z = \frac{\tau}{\{Var\,(\tau)\}^{1/2}}$$ (8.97)

- Test for the hypothesis at 5% level of significance of Z, i.e. $Z = \pm1.96$, if value of Z so computed as above (Eqn. 8.97) lies between $\pm$ 1.96, then there is no trend in the data series.

8.11.1.3 Regression Test for Linear Trend

We take the sample of annual series data. Fit a regression equation to these data say in the linear form of $y = a + bx$. If the value of b is found to be significantly different from zero, then trend is assumed to be present in the data. Then a trend line is fitted to the data and the residuals are estimated. These residuals are tested for their dependence.

8.11.1.4 Estimation and Removal of Trend

Following methods are normally used to quantify and remove any trend component present in a hydrologic data series.

(i) Moving average method and

(ii) Least square method

Moving Average Method

In this method, we first smoothen the irregularities of the time series and then subtract the smoothened ordinates from the corresponding sample events. These subtracted values form a trend free data. The procedure outlined is as follows:

(i) Write the years and the corresponding events against the years in a row.

(ii) A moving average of 3 or 5 is normally used to smoothen the series. For moving average of 3, the equation is:

$$x_2' = \frac{a_1\, x_1 + a_2\, x_2 + a_3\, x_3}{3}$$

$$x_3' = \frac{a_1\, x_2 + a_2\, x_3 + a_3\, x_4}{3}$$

$$x_{n-1}' = \frac{a_1\, x_{n-2} + a_2\, x_{n-1} + a_3\, x_n}{3} \qquad (8.98)$$

In this process the first and the last data of the series are lost. The coefficients a_1, a_2 and a_3 are computed by fitting a polynomial to the series. For this, data can be arranged in 3 columns as $(x_1\, x_2\, x_3)$, $(x_2\, x_3\, x_4)$, $(x_3\, x_4\, x_5)$ and so on. Coefficients a_1, a_2 and a_3 are computed by least square method.

(iii) By using the fitted equation, the smoothened values are written. For m = 3, (n-2) values of x_i' are obtained. The moving average values range from x_2 to x_{n-1}.

(iv) In the next row, events with trend removed are obtained by subtracting events of step (i) from (iii).

While applying moving average method to a time series, an oscillatory movement into the random component is introduced. Care must be taken to discuss the oscillatory property of the time series while ascertaining trend by moving average method.

Least Square Method: We take the sample of data. Then a trend line in the form of polynomial is fitted to the data and the residuals are estimated. The residual is the difference between the observed and estimated data by polynomial equation. These residuals are the trend free data. The polynomials are fitted to the data by least square method. If we try to fit a regression equation to these data say in the linear form of $y = a + b\,x$, then the value of b should not be significantly different from zero such that trend is assumed to be absent in the data. On the other hand a high value of b indicates the presence of trend in the data.

Sample Calculation 8.16

Following are the observed rainfall of a site in Ibb river basin of Odisha. Check the trend in the data series. If there is any trend in the data series, then remove it.

Year	2001	2002	2003	2004	2005	2006	2007	2008	2009	2010
Annual rainfall, mm	650	150	730	500	360	501	670	525	480	515

Solution

Turning point test

We compare each data with its preceding and succeeding values and find out the turning point of each data. We find there are five turning points in the data series i.e. $p = 5$.

$$E(p) = \frac{2(n-2)}{3} = \frac{2(10-2)}{3} = 5.33$$

$$Var(p) = \frac{(16\,n - 29)}{90} = \frac{(16 \times 10 - 29)}{90} = 1.455$$

$$Z = \frac{\{p - E(p)\}}{\{Var(p)\}^{1/2}} = \frac{(5 - 5.33)}{(1.455)^{1/2}} = -0.276$$

Since value of $Z = -0.276$ which lies between ± 1.96, we assume there is no trend in the data series at 5% level of significance.

Kendal's rank correlation test

We compare the first data of the series with rest of the data in the series and find it is greater than its own value by 7 times i.e. 650 is greater than the other data in the series (except 650) by 7 times. Hence $p_{1ex} = 7$. Thus, $p_{2ex} = 0$, $p_{3ex} = 7$, $p_{4ex} = 2$, $p_{5ex} = 0$, $p_{6ex} = 1$, $p_{7ex} = 3$, $p_{8ex} = 2$ and $p_{9ex} = 0$.

Total number of $p = 22$

Maximum value of $p = n (n - 1)/2 = (10 \times 9)/2 = 45$

$E (p) = n (n - 1)/4 = (10 \times 9)/ 4 = 22.5$

Kendal's value = $\tau = \left[\left\{ \dfrac{4 p}{n(n-1)} \right\} - 1 \right]$

$\left[\left\{ \dfrac{4 \times 22}{10 \times 9} \right\} - 1 \right]$

$= -0.022$

$Var\left(\tau \right) = \left[\dfrac{\left\{ 2\left(2n + 5 \right) \right\}}{9n\left(n - 1 \right)} \right]$

$\left[\dfrac{\left\{ 2 \times \left(2 \times 10 + 5 \right) \right\}}{9 \times 10 \times 9} \right]$

$= 0.061$

$Z = \dfrac{\tau}{\left\{ Var\left(\tau \right) \right\}^{1/2}}$

$= \dfrac{-0.022}{0.061^{1/2}} = -0.089$

Since value of $Z = -0.089$ which lies between ± 1.96, we assume there is no trend in the data series at 5% level of significance.

Hence by both the methods we find that there is no trend in the data series.

8.11.2 Oscillation

There may have some amount of small oscillation around the trend line. These oscillations sometimes may be cyclic. In cyclic type, maximum and minimum values occur at constant amplitude and at fixed time intervals. However, all oscillations are well distributed about their mean value. Monthly or seasonal hydrologic events exhibit a regular oscillatory form of variations due to the rotation of the earth around the sun. Oscillations may be long or short term types. This component is hardly found in hydrologic time series and so we are not discussing about it in this book.

8.11.3 Jump

If there is some sudden change in the watershed characteristics, then it may produce a jump in the statistical parameters of the hydrologic time series. This jump is deterministic in nature which can be quantified and removed from the time series. From inspection and trial, a jump in the mean can be located and quantified. For this, a long term data covering either side of the time of occurrence of jump is required. Fig 8.10 shows jump in a time series.

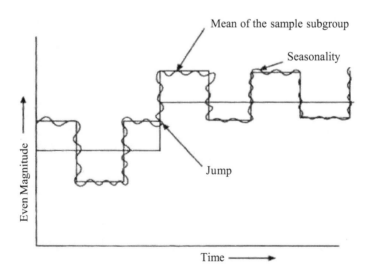

Fig. 8.10: Jump in time series

8.11.4 Periodicity or Seasonality

Sometimes some of the hydrologic events occur at regular time interval maintaining constant periodicity. For example the dry and wet seasons maintain a nearly constant length of time in their occurrence. A sinusoidal curve can be superimposed on the time versus magnitude curve. This helps in determining the periodic component or the component due to seasonality/periodicity by harmonic analysis in which Fourier series is used for representing the time series.

Let x_1, x_2, x_3,.... Are the observations of the time series with T as the total length of record. Then according to Fourier series we have

$$x_t = \frac{1}{2}A_o + \sum_{J=1}^{T/2}\left(A_J \cos\frac{360\,J\,t}{T} + B_J \sin\frac{360\,J\,t}{T}\right) \tag{8.99}$$

where, A_o is the constant, t is the time, T is total length of period, and A_J and B_J are the amplitudes of the time series given by

$$A_J = \frac{2}{N}\sum_{t=1}^{N}Y_t \cos\frac{360\,J\,t}{T} \tag{8.100}$$

$$B_J = \frac{2}{N}\sum_{t=1}^{N}Y_t \sin\frac{360\,J\,t}{T} \tag{8.101}$$

where, Y_t is the deviation of x_t from the trend line for the period taken into consideration. J is the number of years of record i.e. $J = 1, 2, 3 \ldots\ldots N$.

Sum of the squared amplitudes (R_J^2) is given as:

$$R_J^2 = A_J^2 + B_J^2 \tag{8.102}$$

If series is pure random of N term with normal distribution (i.e. without periodic fluctuation), the mean squared amplitude of time series can be expressed as:

$$R_m^2 = \frac{4\sigma^2}{N} \tag{8.103}$$

where, R_m is the mean squared amplitude, σ^2 is the variance of series Y_t.

If the series contains periodic fluctuation, then following tests are performed for determining the periodicity.

(i) **Schuster Test:** Schuster (1898) has given the following formula for computing the sum of squared amplitudes.

$$R_J^2 = K\,R_m^2 \tag{8.104}$$

where, $K = -\log P_s$.

The value of P_s is generally taken as 10% and the value of K is taken as 2.303. Using Eqn. (8.103) and Eqn. (8.104) and taking $K = 2.303$, we get

$$R_J^2 = \frac{9.21\,\sigma^2}{N} \tag{8.105}$$

After determining R_J and with known value of J (J = number of years of record), we can determine the periodicity as T/J.

(ii) **Walker Test:** Walker (1925) suggested a formula to determine the value of K which is:

$$P_w = 1 - \left(1 - e^K\right)^{N/2} \tag{8.106}$$

By considering any value of P_w (probability), value of K can be determined for given value of N. Using Eqn. (8.103) and Eqn. (8.104), we get

$$R_J^2 = \frac{4\,K\,\sigma^2}{N} \tag{8.107}$$

8.12 Serial Correlation

A serial correlation is the association between the successive terms in the same series x_i by lagging them suitably as per the requirements. The method of calculation of correlation coefficient between two variables $(x_i,\ y_i)$ can be extended to the individual series of either x_i or y_i lagged by k distance apart. Let us consider a runoff series of N periods. One may be interested to find the correlation between the same series with lag or spacing of say k time apart. This correlation is referred as auto-correlation and is defined as:

$$\rho_k = \frac{Cov\left(x_i,\ x_{i+k}\right)}{\rho_{xi}\ \rho_{x+k}} \tag{8.108}$$

The following simplified equation can be used to obtain the serial correlation coefficient of lag or spacing of k time apart.

$$r_k = \frac{\dfrac{1}{(n-k)}\sum x_i \cdot x_{i+k} - \dfrac{1}{(n-k)^2}\sum x_i \cdot \sum x_{i+1}}{\left\{\dfrac{1}{n-k}\sum x_i^2 - \dfrac{1}{(n-k)^2}\left(\sum x_i\right)^2\right\}^{1/2}\left\{\dfrac{1}{(n-k)}\sum x_{i+k}^2 - \dfrac{1}{(n-k)^2}\left(\sum x_{i+k}\right)^2\right\}^{1/2}} \tag{8.109}$$

where, n is the length of data, k is the lag distance in the serial correlation and x_i is the variate. Summation is carried out from $i = 1$ to n-k. If $k = 0$, then correlation coefficient, $r_0 = 1$. For other values of k, value of r lies between $+1$ to -1. For a random series, $r_k = 0$ for all lags of k.

A graphical representation of lag k (taken as abscissa) and serial correlation r_k (taken as ordinate), is called as *correlagram*.

Sample Calculation 8.17

The following data of annual peak discharge (m³/sec) of a river gauging station are recorded for 30 years. Compute the serial correlation coefficient of lag 1.

10.5, 20.0, 23.2, 18.4, 20.5, 33.8, 28.0, 11.8, 35.6, 35.0, 37.8, 26.9, 22.0, 24.6, 15.8, 17.3, 45.0, 40.2, 38.0, 36.5, 37.3, 33.9, 30.5, 29.5, 27.8, 26.3, 20.5, 50.2, 48.2, 46.3.

Solution

Detailed calculations are shown in table as below.

Sl. No.	x_i (given)	x_{i+1}	$x_i\, x_{i+1}$	x_i^2	x_{i+1}^2
1	2	3	4	5	6
1.	10.5	20.0	210	110.25	400
2.	20.0	23.2	464	400	538.24
3.	23.2	18.4	426.88	538.24	338.56
4.	18.4	20.5	377.2	338.56	420.25
5.	20.5	33.8	692.9	420.25	1142.44
6.	33.8	28.0	946.4	1142.44	784
7.	28.0	11.8	330.4	784	139.24
8.	11.8	35.6	420.08	139.24	1267.36
9.	35.6	35.0	1246	1267.36	1225
10.	35.0	37.8	1323	1225	1428.84
11.	37.8	26.9	1016.82	1428.84	723.61
12.	26.9	22.0	591.8	723.61	484
13.	22.0	24.6	541.2	484	605.16
14.	24.6	15.8	388.68	605.16	249.64
15.	15.8	17.3	273.34	249.64	299.29
16.	17.3	45.0	778.5	299.29	2025
17.	45.0	40.2	1809	2025	1616.04
18.	40.2	38.0	1527.6	1616.04	1444
19.	38.0	36.5	1387	1444	1332.25
20.	36.5	37.3	1361.45	1332.25	1391.29

Sl. No.	x_i (given)	x_{i+1}	$x_i \cdot x_{i+1}$	x_i^2	x_{i+1}^2
21.	37.3	33.9	1264.47	1391.29	1149.21
22.	33.9	30.5	1033.95	1149.21	930.25
23.	30.5	29.5	899.75	930.25	870.25
24.	29.5	27.8	820.1	870.25	772.84
25.	27.8	26.3	731.14	772.84	691.69
26.	26.3	20.5	539.15	691.69	420.25
27.	20.5	50.2	1029.1	420.25	2520.04
28.	50.2	48.2	2419.64	2520.04	2323.24
29.	48.2	46.3	2231.66	2323.24	2143.69
30.	46.3			2143.69	
Total	891.4	880.9	27081.21	29785.92	29675.67

We have to find out serial correlation coefficient of lag 1 i.e. $k = 1$.

Hence Eqn. (8.109) can be written as (for $k = 1$):

$$r_k = \frac{\dfrac{1}{(n-k)}\sum x_i \cdot x_{i+k} - \dfrac{1}{(n-k)^2}\sum x_i \cdot \sum x_{i+1}}{\left\{\dfrac{1}{n-k}\sum x_i^2 - \dfrac{1}{(n-k)^2}\left(\sum x_i\right)^2\right\}^{1/2}\left\{\dfrac{1}{(n-k)}\sum x_{i+k}^2 - \dfrac{1}{(n-k)^2}\left(\sum x_{i+k}\right)^2\right\}^{1/2}}$$

$$r_1 = \frac{\dfrac{1}{29} \times 27081.21 - \dfrac{1}{29 \times 29} \times 891.4 \times 880.9}{\left\{\dfrac{1}{29} \times 29785.92 - \dfrac{1}{29 \times 29} \times 891.4 \times 891.4\right\}^{1/2}\left\{\dfrac{1}{(29)} \times 29675.67 - \dfrac{1}{29 \times 29} \times 880.9 \times 880.9\right\}^{1/2}}$$

$= 0.0016$ which shows poor serial correlation at lag 1.

8.13 Dependence Models

The following dependence models are used in hydrology:

(i)   Auto-Regressive Integrated Moving Average Model (ARIMA Model)

(ii) Auto-Regressive Moving Average Model (ARMA Model)

The ARMA model consists of auto-regressive (AR) model + a moving average (MA) model. The auto-regressive model can be expressed as:

$$Z_i = \sum Q_i \, Z_{i-1} + a_t \qquad\qquad (8.110)$$

The moving average model in its general form is given as:

$$Z_t = f\left(Z_t, Z_{t-1}, Z_{t-2}\ldots\ldots\right) + a_t \qquad\qquad (8.111)$$

The two models when combined together, they give rise to ARMA model which is:

$$Z_t = \sum_{i=1}^{p} Q_i \, Z_{t-1} + a_t + \sum_{j=1}^{q} Q_j \, a_{t-j} \qquad\qquad (8.112)$$

In the above equation, the first two terms in the right hand side represents are auto-regressive *pth* model (AR Model) and the second and third terms represent moving average (MA) *qth* model. The AR model of *pth* order suggests that the value of Z_t at any time t can be computed from the weighted sum of p values at times *(t -1), (t -2), ... (t – p)* and a random component a_t. A value of $p =1$ represents that the value depends on just on its preceding value and so on.

8.13.1 Markov's Model

In addition to ARMA and ARIMA, Markov's model is also used as dependence model in hydrology. Markov introduced the concept of chain process for forecasting the flow. The Markov's model is a linear auto-regressive model in which the probability of being a particular state in a given time period is dependent on the actual state in the preceding time period. The first order Markov's model is given as:

$$x_t - \mu = \alpha_1 \left(x_{t-1} - \mu\right) + v_t \qquad\qquad (8.113)$$

where, x_t is the dependent stationary stochastic series, μ is the mean, α_1 is the auto-regressive coefficient and v_t is the independent stationary stochastic component. Julian (1961) used the concept of Markov process in generating annual flow series in the following ways:

$$x_t = r. x_{t-1} + E\left(y\right) e^t \qquad\qquad (8.114)$$

where, r is the first order serial correlation coefficient for the hydrologic event, say runoff, and $E(y)$ is the random uncorrelated component due to annual rainfall.

Thomas and Fiering (1962) used the Markov chain model for generating the monthly flows by taking into consideration the serial correlation of monthly flows. This model allows month to month correlation structure. The model is widely used to generate flow series for monthly or seasonal periods where persistence is present. For example, the river discharge at a particular gauging site in the month of August is dependent on July values. There is a definite dependence on sequences of discharges on monthly or weekly basis. For instance in forecasting hydrologic events like runoff on monthly or weekly basis, the model can be used since there is dependence of data and not suitable for annual values since in the annual values, the data of events are not dependent on each other.

Question Banks

Q1. How is the knowledge of statistics and probability helpful in field of hydrology?

Q2. Why is it essential to predict the hydrologic variables by using the concept of probability?

Q3. What are the purposes of using different probability distribution functions?

Q4. In hydrology what are the objectives of statistics?

Q5. Write the various forms of abstract models that are used in hydrology. Mention the characteristics of each of them.

Q6. Mention differences between deterministic and stochastic models? With example explain them.

Q7. What is a time series? What are the various forms of time series? Give examples of each of hem and explain.

Q8. Give the relationship between annual and partial duration series. Under what condition both the series become insignificant.

Q9. What is a stationary time series? With an example illustrate it.

Q10. Differentiate between continuous and discrete time series.

Q11. How can you make discrete variable continuous?

Q12. What does the central tendency represent? What are the various forms of central tendency?

Q13. Write notes on the followings:

   (i) Arithmetic mean (ii) Geometric mean (iii) Harmonic mean and (iv) Weighted mean

Q14. A catchment has seven stations named A, B, C, D, E, F and G. The area enclosed by each station and the rainfall recorded by each station is given as below. Compute the mean rainfall of the catchment by (i)

arithmetic mean, (ii) geometric mean, (iii) harmonic mean and (iv) weighted mean method.

Station	A	B	C	D	E	F	G
Rainfall, cm	30.0	50.5	67.2	45.8	36.9	40.0	35
Area, km^2	41.8	35.7	45.0	55.2	48.0	50.1	40.0

Q15. What is median? How can the median of a data series be calculated? Write the formula to compute the median of a group of data?

Q16. Find the median value of the following rainfall data recorded at 9 stations.

Station	A	B	C	D	E	F	G	H	I
Rainfall, cm	23.7	30.0	33.6	20.5	25.0	33.8	24.8	30.2	24.5

Q17. Calculate the median of the following data.

Rainfall, cm	8.5	3.5	7.5	10.0	9.5	7.0	8.6
Frequency	3	4	2	1	1	1	2

Q18. What is mode? How can the mode of a continuous data series be calculated? Write the formula to compute the mode of series containing group data?

Q19. Write the relationship between mean, mode and median.

Q20. The annual rainfall of a station in M.P. for 45 years are recorded as follows. Group the data into interval of 10 and estimate the mean, mode and median of the sample.

Rainfall data (cm): 50, 63, 77, 49, 64, 88, 64, 68, 45, 69, 39, 40, 53, 33, 80, 74, 47, 42, 33, 44, 66, 78, 87, 96, 91, 82, 84, 100, 39, 110, 112, 105, 69, 96, 119, 102, 100, 95, 78, 30, 73, 59, 111, 106, 109.

Q21. How can the variability of hydrologic data is measured? Mention the different parameters used to measure variability or dispersion of data series.

Q22. Write down the formula of the following parameters and explain their uses in hydrology data analysis.

(i) Standard deviation

(ii) Variance

Q23. With sketches, write short notes on the followings?

   (i)   Measures of skewness

   (ii) Measures of peakedness

Q24. Mention the formula to compute the followings:

   (i)   Pearson's first coefficient of skewness

   (ii) Pearson's second coefficient of skewness

Q25. What are the benefits of representing data in graphical forms?

Q26. What is a Ogive curve? Draw an Ogive curve considering some hypothetical data series.

Q27. What do you mean by class interval? How is the appropriate width of a class interval calculated?

Q28. Considering the data of 45 years as mentioned in Q. 20 above, plot a curve of frequency histogram and cumulative frequency. Compute the standard deviation, variance, coefficient of variation, coefficient of skewness and coefficient of kurtosis of the data.

Q29. Enumerate the importance of frequency analysis in study of hydrologic variables. How can the fitting of the frequency distribution be carried out?

Q30. Write down the probability density functions of the following discrete functions and explain how they can be used to compute the probabilityof a data series.

   (i)   Binomial distribution

   (ii) Poisson distribution

Q31. A hydraulic structure needs to be cast which needs no rainfall during 5 days of its casting. The month is August in which there is chance of 30% rainfall on any day. Compute the probability of having no rainfall during these casting periods of 5 days and one rainy day during the casting period.

Q32. Rainfall data of 100 years of a station were scanned. It was observed that the probability of occurrence of rainfall of 220 mm in a day is 8% *i.e.* 0.08. Find out the probability of three rainfall of one day magnitude exceeding 220 mm in the next 10 days.

Q33. What is a normal distribution? Explain how this is useful in study of frequency analysis. What are the properties of frequency distribution curve for this distribution?

Q34. Mention the probability distribution function of log-normal distribution? Explain how it is used in frequency analysis.

Q35. Following measurements of flood peak data for 24 years of a basin are recorded.

Year	1990	1991	1992	1993	1994	1995	1996	1997	1998	1999	2000	2001
Flood data, m^3/sec	185	189	230	215	162	172	148	154	150	159	140	144

Year	2002	2003	2004	2005	2006	2007	2008	2009	2010	2011	2012	2013
Flood data, m^3/sec	116	122	128	119	147	146	130	134	106	105	104	119

Estimate the flood data at 200 and 1000 years return periods by normal and log normal distribution.

Q36. Write down the probability distribution functions of the following disributiond and describe in details how they are used in frequency analysis study in hydrology.

(i) Pearson Type III distribution (ii) Log Pearson Type III distribution (iii) Exponential distribution (iv) Gamma distribution (v) Gumbel's distribution (vi) Weibull distribution.

Q37. Using the flood data of Q35, estimate the flood data at 100 and 1000 years return periods by Gumbel's, Pearson Type III and Log Pearson Type III distribution.

Q38. The flood data of a particular site in a basin has the following particulars.

Length of record for which data are recorded is 32.The observed flood at 50 and 100 years return periods are 40 and 47 m^3/sec, respectively. Assuming Gumbel's distribution holds god, compute the flood data at 200 years return period.

Q39. What is the use of a probability graph paper? How can it be constructed?

Q40. Mention the points which should be considered to select the appropriate distribution that fits to a particular type of hydrologic data series. Also enumerate the procedure to select the best fit distribution by Chi-square test.

Q41. The annual rainfall data of a station in Cuttack grouped in class interval of 10 along with the frequency are given as below. Use chi-square test to check whether the normal distribution can be fitted to the data sets or not.

Rainfall class, cm	30-40	40-50	50-60	60-70	70-80	80-90	90-100	100-110	110-120
Frequency, f	7	8	5	7	5	5	5	6	2

Q42. What are the importance of confidence limits and confidence bands in use of hydrology?

Q43. Enumerate the procedures to draw confidence bands for the extreme value distribution.

Q44. The river Brahmani has a gauging station which yields 90 years of flood peak data. The mean and standard deviation of these 90 years of data are obtained as 6435 and 2900 m³/sec, respectively. Using Gumbel distribution, compute the flood discharge at a return period of 50, 100, 200 and 500 years. What are the 80% and 95% confidence limits for the estimates? Draw the confidence bands for the data.

Q45. What is the difference between regression and correlation? How the regression constants in regression equation are estimated by method of least square?

Q46. Describe how the standard error of estimate and coefficient of correlation are helpful in to decide the goodness of fit of a curve.

Q47. Discuss the various test used in significance tests of parameters of regression equation.

Q48. Rainfall and runoff data of September in a gauging site in river Mahanadi are recorded and mentioned as below. Develop a linear and parabolic relationship between the rainfall and runoff of the month of September. Calculate the coefficient of correlation and test its significance.

Year	2001	2002	2003	2004	2005	2006	2007	2008	2009	2010	2011
Rainfall, mm	260.5	145.5	180.5	143.6	100.3	120.4	100.5	135.8	140.4	125.8	154.5
Runoff, mm	147.5	78.6	92.9	78.3	55.8	64.8	55.0	70.7	75.5	66.9	84.1

Q49. What is a multivariate linear regression equation? How can you compute the parameters of a multivariate linear regression equation? Also write the formula to estimate the correlation coefficient of a multivariate linear regression equation.

Q50. Following data are recorded in a gauging station. Assuming the rainfall of both July and August contribute to the runoff recorded in August, develop a multiple linear regression between the rainfall of both the months and runoff. Compute the multiple correlation coefficient of the regression equation.

Year	2001	2002	2003	2004	2005	2006	2007	2008	2009	2010	2011
Rainfall, July, mm	255.5	145.5	180.5	143.6	100.3	120.4	100	135.8	140.4	127.8	160.5
Rainfall, August, mm	245.1	129.6	167.8	140.5	110.5	116.3	96.4	130.2	135.7	130.5	165.8
Runoff, August, mm	157.5	80.6	90.9	78.3	55.8	64.8	55.0	70.7	75.5	66.6	88.1

Q51. Mention the various factors responsible for causing trend in a time series.

Q52. Describe the various tests required to detect the presence of trend in a time series.

Q53. How can you quantify and remove any trend component present in a hydrologic data series?

Q54. Following are the observed rainfall of a site in Ibb river basin. Check the trend in the data series by (i) Turning point test (ii) Kendal's rank correlation test. If there is any trend in the data series, then remove it.

Year	2001	2002	2003	2004	2005	2006	2007	2008	2009	2010	2011
Rainfall, mm	630	150	730	500	360	501	670	525	480	510	500

Q55. Write short notes on the followings:

(i) Jump

(ii) Oscillation

(iii) Serial correlation

(iv) Periodicity

Q56. The following data of annual peak discharge (m^3/sec) a river gauging station are recorded for 32 years. Compute the serial correlation coefficient of lag 1.

12.5, 24.0, 26.2, 19.4, 22.5, 33.8, 28.0, 11.8, 35.6, 35.0, 37.8, 26.9, 22.0, 24.6, 15.8, 17.3, 45.0, 40.2, 38.0, 36.5, 37.3, 33.9, 30.5, 29.5, 27.8, 26.3, 20.5, 50.2, 48.2, 48.3, 50.0, 34.9.

Q57. Discuss the following dependence models used in hydrology.

(i)   ARIMA Model

(ii) ARMA Model,

(iii) Markov Model

References

Haan, C.T. 1977. Statistical Methods in Hydrology. The Iowa State University Press, Ames, USA.

Mutreja, K.N. 1986. Applied Hydrology. Tata McGraw-Hill, New Delhi.

Patra, K.C. 2001. Hydrology and Water Resources Engineering. Narosa Publishing House. New Delhi, pp. 561.

Spiegel, M.R. 1961. Theory and Problems of Statistics. Schaum Publishing Company, New York.

Steel, R.G.D. and Torrie, J.H. 1960. Principles and Procedures of Statistics. McGraw-Hill Book Co., New York.

Sturges, H.A. 1926. The Choice of a Class Interval. Journal of American Statistical Association, 21: 65-61.

Subramanya, K. 2003. Enginering Hydrology, Tata McGraw-Hill Publishing Company Limited, New Delhi, pp. 392.

Suresh, R. 2008. Watershed Hydrology. Standard Publishers Distributors, Nai Sarak, Delhi, pp. 692.

Varshney, R.S. 1979. Engineering Hydrology. Nem Chand & Bros., Roorkee, U.P. India, pp. 917.

CHAPTER 9

Flood Routing

9.1 Introduction

Flood water which is in the form of a hydrograph when passes down a river section undergoes several changes in its form. There may be some addition of water to the river section when the flood wave passes down the river section. There may be some diversion of water from the section. Thus, depending on the addition or abstraction of water to the incoming flood water, shape of the flood hydrograph in the downstream section may change. Sometimes channel storage and resistance to flow of water in the channel section also affects the shape of the flood hydrograph. When the flood passes through a reservoir, its peak gets attenuated whereas the time base of the hydrograph gets enlarged. This is due to the effect of reservoir storage. Similarly when the flood is passing down a river section and there are no lateral inflows, the peak of the flood hydrograph gets attenuated and the time base gets enlarged. The reason of changing the shape of the flood hydrograph in this case is due to channel resistance in the form of friction. However, if there are some lateral inflows, the magnitude of attenuation may be decreased or there may be some amplification of flood peaks. It is essential to know the shape of a flood hydrograph at a location downstream of a reservoir or channel section. This can be known by a process called as *flood routing*. Prediction of outflow hydrograph at a section downstream of a reservoir or channel section from the knowledge of flood data at one or more upstream sections is called as *flood routing*.

9.2 Importance of Flood Routing

Followings are the importance of flood routing.

- It acts as a tool for flood forecasting.

- It helps to fix the capacity of spillway of reservoirs, design of water control structures including reservoirs etc.

- It provides a guideline for installation of the flood protection structures.

9.3 Types of Flood Routing

Flood routings are of two types. They are

- Reservoir routing

- Channel routing

9.3.1 Reservoir Routing

In reservoir routing, the interest is to know the shape of the flood outflow hydrograph, predict variations of reservoir elevations and outflow discharge with time. Following parameters are required for reservoir routing.

i.   Storage-elevation characteristics of the reservoir

ii.   Outflow-elevation characteristics of the spillway and other outlet structures of the reservoir

iii. Inflow hydrograph of the upstream section and

iv.   Initial values of storage, inflow and outflow rates.

Reservoir routing is essential in the design of capacity of spillways and other outlet structures in reservoirs and fixing the location and sizing the capacity of reservoir to meet specific requirements.

9.3.2 Channel Routing

In channel routing, the outflow flood hydrograph at any section downstream of a channel section is determined based on the knowledge of inflow flood hydrograph at the upstream section. The interest here is to determine the change in shape of a flood hydrograph as it passes down a channel section. Information on flood-peak attenuation and duration of high water levels obtained by channel routing are helpful in flood protection and flood forecasting operations.

9.4 Shape of Flood Hydrograph Before and After Flood Routing

When flood water passes through a reservoir or channel it gets moderated. When the head over the spillway increases either due to accumulation of flood or due to high rainfall in the reservoir and its upstream catchments, discharges passing over the spillway out of the reservoir increases. During the initial phase, rate of outflow is always less than the initial values. There is some accumulation of storage in the reservoir at the beginning stage. As storage volume increases, outflow gradually increases but at a reduced rate than the increase in inflow. After some time, the two curves i.e. inflow and outflow hydrograph intersects each other (point C in Fig. 9.1). At point C, maximum outflow occurs and this point C represents the peak of outflow hydrograph. The difference of two peaks of the inflow and outflow hydrograph is called as *attenuation*. The time difference between the two peaks of the inflow and outflow hydrographs is called as *lag time or translation of peak* (Patra, 2001*)*. After point C, accumulated storage gets depleted due to less inflow but high outflow from the storage system. In other words, there is loss from the storage system.

The shape of outflow flood hydrograph in channel section depends on (i) bed slope of the channel, (ii) geometry of the channel section, (iii) shape of inflow flood hydrograph (iv) length of reach and (v) initial storage of the channel section before the inflow hydrograph reaches the channel section. When the

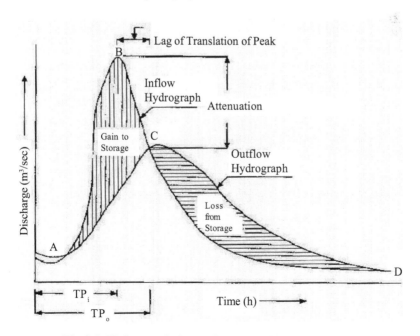

Fig. 9.1: Hydrograph shape after and before flood routing

inflow hydrograph that reaches a channel section has high peak, than the magnitude of attenuation is higher than the moderate flood hydrograph. Similarly when the length of channel reach is more or the channel section is wider at the outlet, the reduction in the peak is higher. For a steep channel or channel with high storage capacity, the reduction of peak will be less.

9.5 Routing Methods

The methods available for routing in channel and reservoirs are mainly classified into the following three types:

- Hydrologic routing
- Hydraulic routing and
- Routing machines

Each of the above mentioned method is again divided into several categories. Fig. 9.2 illustrates the flow chart describing various types of routing methods (Patra, 2001). Hydrologic routing method involves equation of continuity and is simpler in use than the hydraulic routing. Hydraulic routing involves equation of continuity together with equation of unsteady flow solved by St. Venant equations. Hydraulic routing uses two conservation equations. The model using conservation of mass and momentum equations to the flow problems always give better results. Since the flow in river section, reservoir or channel section is unsteady and gradually varied, the flow routing in this section is better solved by hydraulic routing technique. But this method requires high quality of input data, good computing technologies and use of complex mathematics and so is of little interest to the hydrologists. Routing machines take care of the problems by correlating the channel or reservoir system with mechanical gears or fluctuations of voltage in an instrument.

Basic Equations involved in Routing

Equation of continuity is the basic equation used in hydrologic routing. It states that difference of inflow and outflow rate to a system is equal to rate of change of storage and is listed as:

$$I - Q = \frac{dS}{dt} \tag{9.1}$$

where, I = inflow rate, Q = outflow rate and $\dfrac{dS}{dt}$ = rate of change of storage (S being storage with t referred as time).

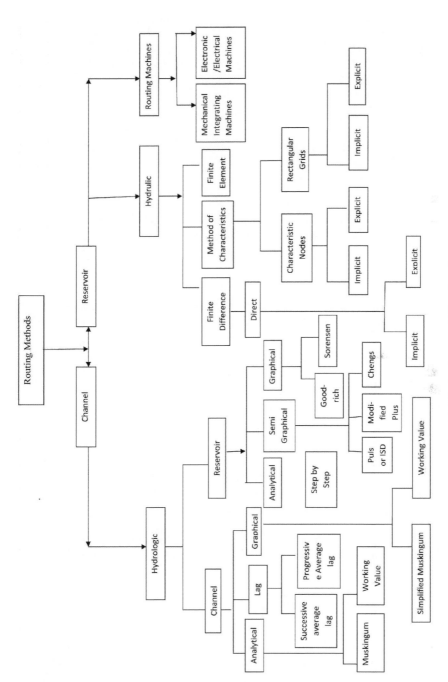

Fig. 9.2: Various types of flood acting

In a small time interval, Δt, the average volume of inflow and outflow would be equal to $\overline{I} \cdot \Delta t$ and $\overline{Q} \, \Delta t$, respectively and the change in storage volume would be designated as ΔS. Hence Eqn. (9.1) can be written as:

$$\overline{I} . \Delta t - \overline{Q} . \Delta t = \Delta S \tag{9.2}$$

where $\overline{I}$ is average inflow rate and $\overline{Q}$ is the average outflow rate. Considering $\overline{I} = (I_1 + I_2)/2$, $\overline{Q} = (Q_1 + Q_2)/2$ and $\Delta S = S_2 - S_1$, we can write Eqn. (9.2) as:

$$\left(\frac{I_1 + I_2}{2}\right) \Delta t - \left(\frac{Q_1 + Q_2}{2}\right) \Delta t = S_2 - S_1 \tag{9.3}$$

The time interval Δt should be very small such that the inflow and outflow hydrographs are assumed as straight lines in that time interval. Moreover, Δt should be smaller than the time of transit of the flood wave through the reach.

For unsteady flow, the continuity equation in the differential form is written as:

$$\frac{\partial Q}{\partial x} + T \frac{\partial y}{\partial t} = 0 \tag{9.4}$$

where, T = top width of the section and y = depth of flow.

Application of momentum equation gives equation of motion of flood wave which is:

$$\frac{\partial y}{\partial x} + \frac{V}{g} \frac{\partial V}{\partial x} + \frac{1}{g} \frac{\partial V}{\partial t} = S_o - S_f \tag{9.5}$$

Where, V = velocity of flow at any section, S_0 = bed slope of the channel, S_f = slope of energy grade line and y = depth of flow.

Eqns. (9.4) and (9.5) were proposed by Saint Venant (1871). Numerical solutions of the above two Saint Venant's equations by computers helps in solving the hydraulic flood routing technique.

9.5.1 Hydrologic Storage Routing

Hydrologic storage routing is also called as level pool routing. When a flood wave $I(t)$ enters a storage reservoir, the water level in it rises. Excess water from the reservoir passes out though the spillway and other outlet structures of the reservoir. In a reservoir, there is a definite relationship between the head over the spillway called as elevation, stage or gauge and discharge of the

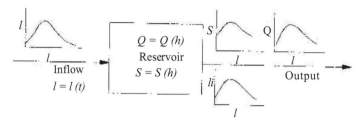

Fig. 9.3: Schematic diagram of storage routing

reservoir and also relationship exists between the stage and storage in the reservoir. These relationships can be represented in the forms of graphs of stage-storage and stage-discharge of the reservoir. The stage-discharge relationship is expressed as $Q = Q (h)$ and stage-storage relationship expressed as $S = S(h)$. Moreover due to passage of flood wave through the reservoir, water level in the reservoir changes with time i.e. $h = h(t)$. This causes the storage and discharge in the reservoir change with time. These relationships are explained in Fig. (9.3). In hydrologic storage routing, it is essential to find out the variations of S, h and Q with time i.e. find $S = S(t)$, $Q = Q(t)$, and $h = h(t)$ for a given $I = I(t)$.

Following data are essential for hydrologic storage routing.

- Storage volume *vs* elevation characteristics of the reservoir

- Outflow *vs* water surface elevation characteristics of the spillway and other outlet structures of the reservoir and hence storage *vs* outflow discharge

- Inflow hydrograph of the upstream section i.e. $I = I(t)$ and

- Initial values of storage, inflow and outflow rates at time $t = 0$.

A variety of methods are available for flood routing through reservoir. But all these methods involve continuity equation (Eq. 9.3) but the continuity equation in each method is used in various rearranged forms. Hydrologic storage routing is also called as *level pool routing* since in the reservoir storage, horizontal water surface is assumed which is level. Some important methods of hydrologic storage flood routing are described in this chapter.

9.5.1.1 Modified Pul's Method

The continuity equation (Eq. 9.3) can be rearranged as:

$$\left(\frac{I_1 + I_2}{2}\right)\Delta t + \left(S_1 - \frac{Q_1 \Delta t}{2}\right) = \left(S_2 + \frac{Q_2 \Delta t}{2}\right) \tag{9.6}$$

At the beginning of flood routing, the values of initial storage, S_1, discharge Q_1, inflows I_1 and I_2 and time interval Δt are known.

In Eq. (9.6) now all the terms in left hand side are known and so the right hand term $\left(S_2 + \dfrac{Q_2\, \Delta t}{2} \right)$ can be calculated at the end of time step Δt. Since the relationship $Q = Q\,(h)$ and $S = S(h)$ are known, $\left(S + \dfrac{Q\, \Delta t}{2} \right)_2$ will enable one to calculate the reservoir elevation and so the discharge at the end of time step Δt. The procedure is repeated till it covers the full inflow hydrograph.

Following semi graphical method is useful in hand computation.

- Decide the time step Δt which should be so chosen that peak of inflow hydrograph is not missed. Generally value of Δt is taken as 20 to 40% of the time of rise of inflow hydrograph (Subramanya, 2003).

- From the known storage-elevation and discharge-elevation data, prepare a plot by taking elevation as ordinate and in abscissa.

- On the same plot, prepare a graph of outflow vs elevation by taking elevation as ordinate and discharge $\left(S + \dfrac{Q\, \Delta t}{2} \right)$ in abscissa.

- Plot the inflow hydrograph and read the ordinates of it at time interval Δt. Let the ordinates be represented as I_i.

- Now use Eq. (9.6) and since all the terms in left hand side of this equation are known, one can determine. $\left(S_2 + \dfrac{Q_2\, \Delta t}{2} \right)$

- From the plot of step 2, reservoir elevation corresponding to value of $\left(S_2 + \dfrac{Q_2\, \Delta t}{2} \right)$ is determined.

- From the plot of same graph of step 2 between reservoir elevation and discharge, the value of outflow discharge corresponding to elevation of step 6 is determined. This outflow is the value of Q_2 at the end of time step Δt.

- Deducting $Q_2\, \Delta t$ from $\left(S_2 + \dfrac{Q_2\, \Delta t}{2} \right)$ gives $\left(S - \dfrac{Q\, \Delta t}{2} \right)_1$ for the beginning of the next time step.

- The procedure is repeated till the entire inflow hydrograph is routed.

The following sample calculation illustrates the use of the above mentioned steps using modified Pul's method.

Sample Calculation 9.1

A reservoir has the following elevation (stage), discharge and storage relationships.

Elevation, m	Storage (x 10^6 m³)	Outflow discharge (m³/sec)
100.00	3.380	0
100.40	3.450	12
100.9	3.880	25
101.5	4.320	40
101.8	4.880	50
102.5	5.350	60
103.0	5.550	65
103.5	5.850	70

When the reservoir elevation was 100.40 m, the following inflow hydrograph enters the reservoir. Route the flood hydrograph and calculate the (i) outflow hydrograph (ii) reservoir elevation vs time curve during the passage of the flood wave.

Time, hr	0	6	12	18	24	30	36	42	48	54	60
Discharge, m³/sec	10	22	50	82	70	55	40	30	15	10	5

Solution

A time interval for flood routing, Δt is assumed as 6 hrs which is 20 to 40% of time of rise of inflow hydrograph. Time of rise of given inflow hydrograph is 18 hr.

From the available data, elevation –discharge $-\left(S + \dfrac{Q\,\Delta t}{2} \right)$ table is prepared.

Value of $\Delta t = 6$ hr $= 0.0216$ M sec.

Elevation, m	Discharge Q, m³/sec	$\left(S + \dfrac{Q\,\Delta t}{2}\right)\left(x\,10^5\,m^3\right)$
100.00	0	33.8
100.40	12	35.8
100.9	25	41.5
101.5	40	48.1
101.8	50	56.4
102.5	60	64.3
103.0	65	67.9
103.5	70	72.3

Now graphs of discharge vs elevation and $\left(S + \dfrac{Q\,\Delta t}{2}\right)$ vs elevation are drawn on the same plot using the above data and shown in Fig. 9.4.

At the start of routing when elevation of the reservoir is 100.40 m, discharge, Q = 12 m³/sec and value of $(S - 0.5\,\Delta t\,Q)$ is 33.20 x 10⁵ m³. Starting from this value of $(S - 0.5\,\Delta t\,Q) = 33.20$ x 10⁵ m³, Eq. (9.6) is used to get $(S + 0.5\,\Delta t\,Q)$ at the end of first time step of 6 hour as:

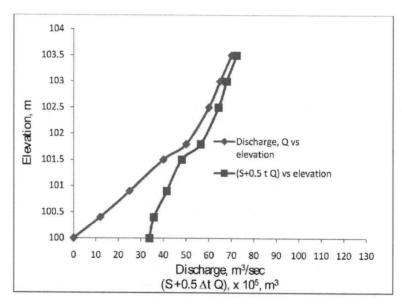

Fig. 9.4: Plot between elevation vs discharge and elevation vs $(S + 0.5\,\Delta t\,Q)$

$$\left(S + \frac{Q\,\Delta t}{2}\right)_2 = \left(I_1 + I_2\right)\frac{\Delta t}{2} + \left(S - \frac{Q\,\Delta t}{2}\right)_1$$

$= (10 + 22) \times (6 \times 3600)/2 + 3320000 = 3665600\ \text{m}^3 = 36.656 \times 10^5\ \text{m}^3.$

Using value of $(S + 0.5\ \Delta t\ Q) = 36.656 \times 10^5\,\text{m}^3$, we get elevation from Fig. 9.4 as 100.55 m and corresponding outflow discharge as 14.8 m^3/sec. For the next step, initial value of $(S - 0.5\ \Delta t\ Q)$ is computed as:

$(S - 0.5\ \Delta t\ Q) = (S + 0.5\ \Delta t\ Q)$ of the previous step $- Q.\ \Delta t$

$= 36.656 \times 10^5 - 14.8 \times 6 \times 3600 = 3345920\ \text{m3} = 33.459 \times 10^5\ \text{m}^3.$

The process is repeated for entire duration of inflow hydrograph in a tabular form as shown below (Table 9.1).

Using the data of columns 1, 2 and 8, inflow and outflow hydrographs are drawn (Fig. 9.5). Similarly, using the data of columns 1, and 7, a plot of reservoir elevations vs time is drawn (Fig. 9.6).

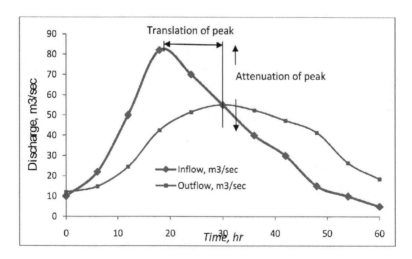

Fig. 9.5: Variations of inflow and outflow hydrograph

Table 9.1: Flood routing by modified Pul's method (Sample Calculation 9.1)

Time, hr	Inflow, I (m³/sec)	(m³/sec)	$\cdot \Delta t$ (× 10⁵ m³)	$(S - 0.5 \, \Delta t \, Q)$ (× 10⁵ m³)	$(S + 0.5 \, \Delta t \, Q)$ (× 10⁵ m³)	Elevation, m	Outflow, Q (m³/sec)
1	2	3	4	5	6	7	8
0	10	16	3.456	33.200	36.656	100.40	12.0
6	22	36	7.776	33.459	41.235	100.55	14.8
12	50	66	14.256	35.943	50.199	100.86	24.5
18	82	76	16.416	41.019	57.435	101.58	42.5
24	70	62.5	13.500	46.333	59.833	101.91	51.4
30	55	47.5	10.260	47.931	58.191	102.15	55.1
36	40	35	7.560	46.873	54.433	102.05	52.4
42	30	22.5	4.860	44.216	49.076	101.72	47.3
48	15	12.5	2.700	40.134	42.834	101.53	41.4
54	10	7.5	1.620	37.110	38.730	101.05	26.5
60	5			34.756		100.63	18.4

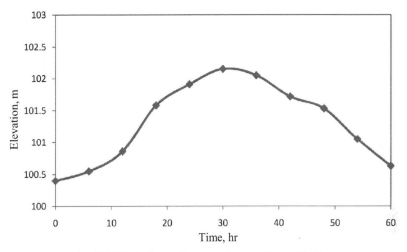

Fig. 9.6: Variations of reservoir elevation with time

9.5.1.2 Goodrich Method

This is another popular method used for hydrologic reservoir routing. Eq. (9.3) when rearranged we get

$$\left(I_1 + I_2\right) + \left(\frac{2\,S_1}{\Delta t} - Q_1\right) = \left(\frac{2\,S_2}{\Delta t} + Q_2\right) \tag{9.7}$$

Suffixes 1 and 2 are meant for denoting values t the beginning and end of a time step, respectively.

Following steps are employed in Goodrich method (Suresh, 2008).

- At the beginning of routing, left hand side terms of Eq. (9.7) are known from which the right hand side term of Eq. (9.7) is evaluated at the end of time step Δt.

- From the known storage-elevation-discharge relation, two graphs are drawn between elevation-discharge and elevation – $\{(2\,S/\Delta t) - Q\}$ on the same paper or plot an outflow - $\{(2\,S/\Delta t) + Q\}$ curve.

- From $\{(2\,S/\Delta t) + Q\}$-elevation and elevation-discharge relation or outflow - $\{(2\,S/\Delta t) + Q\}$ curve compute Q_2, corresponding to the value of $\left(\frac{2S_2}{\Delta t} + Q_2\right)$ of first step.

- From $\left(\dfrac{2S_2}{\Delta t} + Q_2 \right)$ subtract $2Q_2$ of third step to obtain $\left(\dfrac{2S}{\Delta t} - Q \right)$

- The new $\left(\dfrac{2S}{\Delta t} - Q \right)$ is used as the second right hand side term of Eq. (9.7) to proceed for the next step of routing for the second time step Δt.

- The process continues till entire inflow hydrograph is routed.

In this method if smaller values of Δt are chosen, then accuracy will be more.

Following sample calculation helps to understand the use of Goodrich method using above mentioned steps.

Sample Calculation 9.2

A reservoir has the following elevation (stage), discharge and storage relationships.

Elevation, m	Storage ($\times 10^6$ m³)	Outflow discharge (m³/sec)
100.00	3.380	0
100.40	3.450	12
100.9	3.880	25
101.5	4.320	40
101.8	4.880	50
102.5	5.350	60
103.0	5.550	65
103.5	5.850	70

When the reservoir elevation was 100.40 m, the following inflow hydrograph enters the reservoir. Route the flood hydrograph by Goodrich method.

Time, hr	0	6	12	18	24	30	36	42	48	54	60
Discharge, m³/sec	10	20	50	80	70	50	40	30	15	10	5

Solution

Using the data given in sample calculation, we prepare a table as below. Use this table to draw a curve (Fig. 9.7) between discharge, Q vs. $\left(\dfrac{2S}{\Delta t} + Q \right)$

Assume a time interval, Δt = 6 hr = 0.0216 M sec.

Elevation, m	Discharge Q, m³/sec	Storage, S (x 10⁶ m³)	$\left(\dfrac{2S}{\Delta t}+Q\right)$, m^3 / sec
100.00	0	3.380	312.96
100.40	12	3.450	331.44
100.9	25	3.880	384.26
101.5	40	4.320	440.00
101.8	50	4.880	501.85
102.5	60	5.350	555.37
103.0	65	5.550	578.89
103.5	70	5.850	611.67

Routing of flow through the reservoir is carried out as shown in Table 9.2.

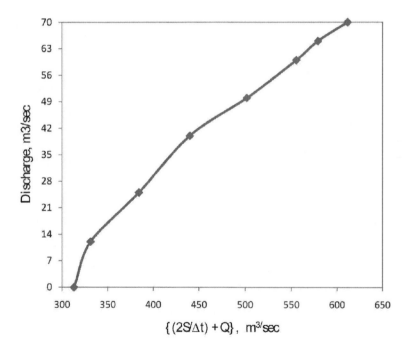

Fig. 9.7: Plot between discharge vs {(2 S/"t) + Q} of sample calculation 9.2

Table 9.2 Flood routing by Goodrich method (Sample calculation 9.2)

Time, hr	Inflow, I (m³/sec)	$I_1 + I_2$, (m³/sec)	4	5	Outflow, Q (m³/sec)
1	2	3	4	5	6
0	10			331.44	12.0
		30	307.44	337.44	
6	20			380.44	13.5
		70	310.44		
12	50			461.44	24.5
		130	331.44		
18	80			526.24	42.6
		150	376.24		
24	70			537.44	54.4
		120	417.44		
30	50			514.04	56.7
		90	424.04		
36	40			479.04	52.5
		70	409.04		
42	30			431.04	46.5
		45	386.04		
48	15			380.84	37.6
		25	355.84		
54	10			348.04	23.9
		15	333.04		
60	5				16.2

At initial time, t = 0, reservoir elevation is given as 100.4 m. Outflow, Q corresponding to reservoir elevation of 100.4 m is given as 12.0 m³/sec. Now using the given storage and discharge data, at this initial time of t = 0, we calculate value of $\{(2\ S/\Delta t) + Q\}$ which is 331.44 m³/sec. These values are entered in columns 5 and 6 in Table 9.2 as above.

Now $\left(\dfrac{2S}{\Delta t} - Q\right)_1$ = 331.44 – 2 x 12 = 307.44 m³/sec

For the first time interval of 6 hr, $I_1 = 10$, $I_2 = 20$ and $Q_1 = 12$ and

$\left(\dfrac{2S}{\Delta t} + Q\right)_2$ = (10 + 20) + 307.44 = 337.44 m³/sec

From Fig. 9.7, at $\left(\dfrac{2S}{\Delta t} + Q\right)$ = 337.44, $Q = 13.5$ m³/sec

For the next time increment, $\left(\dfrac{2S}{\Delta t} - Q\right)_1$ = 337.44 – 2 x 13.5 = 310.44 m³/sec

The process is repeated till the entire duration of routing is completed. All the calculations are shown in Table 9.2.

Plot of inflow and outflow hydrograph of sample calculation 9.2 is shown in Fig. 9.8.

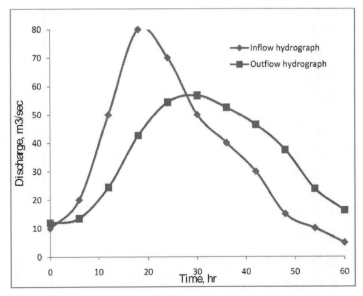

Fig. 9.8: Variations of inflow and outflow hydrograph

9.5.1.3 Pul's Method

This is a semi graphical method of flood routing. It was developed by I.G. Puls. It is also known as inflow-storage-discharge (ISD) method. It employees the basic equation:

$$\left(\frac{I_1 + I_2}{2}\right) \Delta t + \left(S_1 - \frac{Q_1 \, \Delta t}{2}\right) = \left(S_2 + \frac{Q_2 \, \Delta t}{2}\right) \tag{9.8}$$

Following steps are involved in this method (Patra, 2001).

- Plot a graph between storage (taken in abscissa) vs reservoir elevation (taken in ordinate) using the data given in problem.

- Compute the discharge passing over the spillway using the discharge formula and reservoir elevation from which compute the value of $(0.5 \, \Delta t. \, Q)$. Then add the value of $(0.5 \, \Delta t. \, Q)$ so calculated with the given value of storage, S. Thus, step 2 gives values of $(S + 0.5 \, \Delta t \, Q)$.

- On the same graph paper where storage vs reservoir elevation plot is drawn, plot another graph between $(S + 0.5 \, \Delta t \, Q)$ (taken in abscissa) vs. elevation (taken in ordinate).

- As in step 2, compute the value of $(S - 0.5 \, \Delta t \, Q)$ and plot another graph between $(S - 0.5 \, \Delta t \, Q)$ vs elevation on the same graph paper where other two graphs are drawn.

- At the beginning of flood routing, the values of initial storage, S_1, discharge Q_1, inflows I_1 and I_2 and time interval Δt are known.

- Compute the left hand side terms of Eq. (9.8) since all the terms are known at the beginning of flood routing. This also gives the value of $(S_2 + 0.5 \, \Delta t \, Q_2)$ which is the right hand side term of Eq. (9.8).

- From the plots of steps (1), (3) and (4), knowing $(S_2 + 0.5 \, \Delta t \, Q_2)$, the reservoir elevation and therefore $(S_2 - 0.5 \, \Delta t \, Q_2)$ and Q are read. This Q is the outflow Q_2 here.

- Adding the value of $(S_2 - 0.5 \, \Delta t \, Q_2)$ of step 7 with $(I_2 + I_3)$. 0.5. Δt, right hand side term of Eq. (9.8) for second step of routing is evaluated after second time interval Δt. This gives $(S_3 + 0.5 \, \Delta t \, Q_3)$.

- The process from steps (6) to (8) are continued till the entire inflow hydrograph is routed.

Sample Calculation 9.3

Following data were recorded at a station Naraj in the river Mahanadi.

Elevation, m	360	361	362	363	364	365	366	367	368	369
Storage, Mm³	50.5	100.6	190.6	250.5	350.5	451.0	600	750	915.8	990
Outflow, m³/s	2.4	6.8	15.2	22.0	40.2	65.9	90.0	150	162	201

Route the following inflow hydrograph for the reservoir given below.

Time, hr	0	6	12	18	24	30	36	42	48	54	60
Inflow, m³/s	12.5	35	50	100	120	102	96	75	35	21	15

Solution

Chose a time interval, $\Delta t = 6$ hr $= 0.0216$ M sec.

Calculate the values of $(S - 0.5 \Delta t \, Q)$ and $(S + 0.5 \Delta t \, Q)$ using the given data as shown in tabular form below and use these data to draw plots among elevation, storage, $(S - 0.5 \Delta t \, Q)$ and $(S + 0.5 \Delta t \, Q)$ (Fig. 9.9).

Elevation, m	360	361	362	363	364	365	366	367	368	369
Storage, Mm³	0.5	0.705	0.905	1.505	2.560	3.605	4.55	5.78	6.58	7.45
Outflow, m³/s	2.4	6.8	15.2	22.0	40.2	65.9	90.0	150	162	201
$(S- 0.5 \Delta t \, Q)$ Mm³	0.474	0.631	0.741	1.267	2.125	2.893	3.578	4.16	4.83	5.279
$(S +0.5 \Delta t \, Q)$ Mm³	0.526	0.778	1.069	1.743	2.994	4.316	5.522	7.40	8.33	9.621

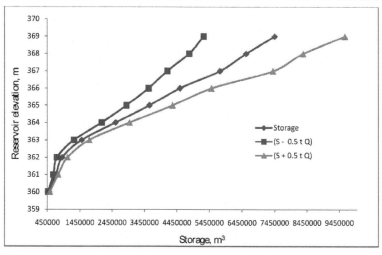

Fig. 9.9 Plot between elevation vs storage, elevation between $(S - 0.5 \Delta t \, Q)$ and $(S + 0.5 \Delta t \, Q)$ for sample calculation 9.3

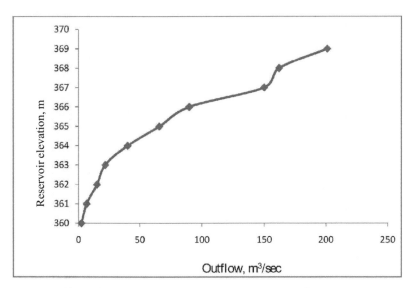

Fig. 9.10: Plot between reservoir elevation vs discharge

Another curve between elevation vs discharge is drawn using the given data (Fig. 9.10). Calculations of routings are shown in tabular form (Table 9.3). At the beginning of routing, reservoir elevation is 360 m and outflow from reservoir is 2.4 m³/sec. Values of $(S - 0.5 \, \Delta t \, Q)$ corresponding to 360 m elevation is read from Fig. 9.9 as 474080 m³ = 0.4741 Mm³.

Now using Eq. (9.8), we get right hand side term $(S + 0.5 \, \Delta t \, Q)$ = 0.513 + 0.4741 = 0.9871 Mm³.

For $(S + 0.5 \, \Delta t \, Q)$ = 0.9871 Mm³ = 987100 m³ and using Figs. 9.9 and 9.10, we get the new elevation, outflow and $(S - 0.5 \, \Delta t \, Q)$ as 361.6 m, 11.1 m³/sec and 0.6942 Mm³, respectively. These values are now entered in Table 9.3 as shown in the table. The process continues till entire inflow hydrograph is routed.

Now using the tabular data (Table 9.3), we draw inflow as well as outflow hydrograph as shown in Fig. 9.11.

Table 9.3: Flood routing by ISD (Pul's) method

Time, hr	Inflow, I (m³/sec)	$(I_1+I_2).0.5.\Delta t$, Mm³	$(S - 0.5 \, \Delta t \, Q)$, Mm³	$(S + 0.5 \, \Delta t \, Q)$, Mm³	Elevation, m (Fig. 9.9)	Outflow, (m³/sec) (Fig. 9.10)
0	12.5			0.9871	360	2.4
6	35	0.513	0.4741	1.6122	361.6	11.1
12	50	0.918	0.6942	2.6301	362.2	16.6
18	100	1.62	1.0101	3.7760	363.0	22.0
24	120	2.376	1.4000	4.3480	364.1	41.8
30	102	2.398	1.9500	4.3380	364.6	53.5
36	96	2.138	2.2000	3.9970	364.5	52.8
42	75	1.847	2.1500	3.2380	364.3	48.7
48	35	1.188	2.0500	2.3050	363.6	32.6
54	21	0.605	1.7000	1.6890	362.8	20.8
60	15	0.389	1.3000		362.2	16.6

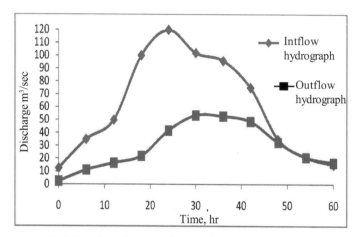

Fig. 9.11: Variations of inflow and outflow hydrograph

9.5.1.4 Standard Fourth –Order Runge-Kutta Method (SRK)

Pul's method, Modified Pul's method and Goodrich method of hydrologic reservoir routing uses continuity equation which is a differential equation of first order in the form $I - Q = dS/dt$ in which I is inflow which is a function of time, t i.e. $I = I(t)$ and Q is outflow which is a function of head, H over the spillway crest i.e. $Q = Q(H)$. Reservoir storage is the product of reservoir area, A and head of water in the reservoir, H. Again reservoir area is a function of head, H.

Now change in storage, dS with time dt can be written as:

$$dS/dt = A\ (H).\ dH/dt \qquad (9.9)$$

By using continuity equation and Eq. (9.9),

$$\frac{dS}{dt} = I(t) - Q(H) = A(H).\frac{dH}{dt}$$

$$\frac{dH}{dt} = \frac{I(t) - Q(H)}{A(H)} = \text{Function of } (t,\ H) = F\ (t,\ H) \qquad (9.10)$$

Eq. (9.10) can be solved numerically by employing standard *Fourth –Order Runge-Kutta Method* which is a more efficient computational method of flood routing which is described as follows (Subramanya, 2003):

At the initial condition: $t = t_0,\ I = I_0,\ Q = Q_0,\ H = H_0,\ \text{and } S = S_0.$

If the routing is conducted from the initial condition and in time step Δt, the water surface elevation H at $(i+1)^{th}$ step is given in SRK method as:

$$H_{i+1} = H_i + 1/6 \left(K_1 + 2K_2 + 2K_3 + K_4 \right) \Delta t \tag{9.11}$$

Where, $K_1 = F \left(t_i, H_i \right)$

$\quad\quad K_2 = F \left(t_i + \Delta t/2 \ . \ H_i + 0.5 \ K_1 \ . \ \Delta t \right)$

$\quad\quad K_3 = F \left(t_i + \Delta t/2 \ . \ H_i + 0.5 \ K_2 \ . \ \Delta t \right)$

$\quad\quad K_4 = F \left(t_i + \Delta t \ . \ H_i + K_3 \ . \ \Delta t \right)$

In Eq. (9.11), the suffix "i" denotes values at i^{th} step and the suffix $(i+1)$ denotes that at $(i+1)^{th}$ step. At $i = 1$, the initial condition I_0, Q_0, H_0, and S_0 prevail. Starting from these known values at the initial condition, and knowing $Q = F (H)$, and $A = F (H)$, a given hydrograph $I = I(t)$ is routed by selecting a time step Δt. At any time, $t = (t_0 + i \ . \ \Delta t)$, value of H_i is known and the coefficients K_1, K_2, K_3 and K_4 are determined by repeated appropriate evaluation of the function $F (t, H)$. It is seen that SRK method directly determines H_{i+1} by four evaluations of the function $F(t, H)$.

Knowing the values of H at various time intervals, other variables like $Q(H)$ and $S(H)$ can be evaluated to complete the routing operation.

In addition to the above mentioned four methods i.e. ISD (Pul's) method, modified Pul's method, Goodrich method and SRK method, there are several other methods available for reservoir flood routing. Some are graphical whereas others are analog methods (Varshney, 1979; Mahmood and Yevjevich, 1975).

9.5.2 Hydrologic Channel Routing

In the hydrologic reservoir routing, storage is a function of outflow discharge only. But in channel routing, storage depends on two factors i.e. outflow discharge and inflow discharge. The flow in a river is an example of such type where the storage always changes depending on the inflow and outflow to it. In fact the velocity of flow in a river section changes with time as well as space and the flow is said to be unsteady and non-uniform type. The addition of lateral inflow to the storage makes the situation more complex. For easy in computation of hydrologic channel routing, it is assumed that the flow is gradually varies unsteady and there is no lateral inflow to the system. In hydrologic channel routing, water surface in the channel reach remains never parallel to the channel bed. The flood wave propagates in the channel and during the rising stage, the inflow is always higher than the outflow. This gives rise to additional storage between sections $3 - 3'$ and $4 - 4'$. The total storage comprises (i) prism storage and (ii) wedge storage (Fig. 9.12, a).

Prism Storage

Prism storage is the storage between any two sections of a channel by an imaginary plane passing parallel to the channel bottom. It is represented by the section AB34 4'4' (Fig. 9.12, b).

Wedge Storage

Wedge storage is the storage between top of the prism storage and the actual water surface during the passage of flood wave in a channel reach. It is represented by the section AB(Fig. 9.12, b).

At a fixed depth at a downstream section of a river, the prism storage (S_p) is a function of outflow discharge and is expressed as:

$$S_p = f(Q) \tag{9.12}$$

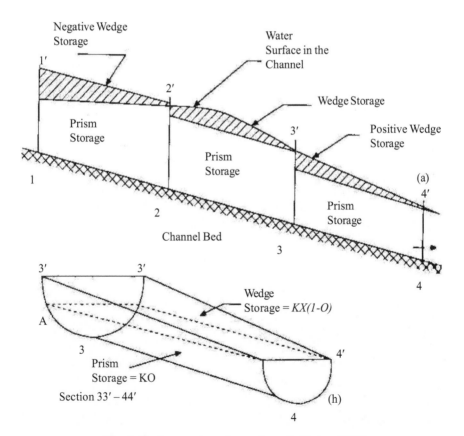

Fig. 9.12: Storage in a channel reach (Patra, 2001)

The wedge storage depends on difference of inflow and outflow rates of the stream flow. It changes from a positive value at an advancing flood to a negative value during a receding flood. The wedge storage is expressed as:

$$S_w = f(I) \tag{9.13}$$

Total storage, $S = S_p + S_w$ and is expressed as:

$$S = K\left[x\,I^m + (1-x)\,Q^m\right] \tag{9.14}$$

Where, K and x are coefficients and m is exponent whose value depends on type of the channel. Value of m varies from 0.6 for rectangular channel to about 1.0 for natural channel.

In eq. (9.14), x is the *"weighing factor"* which is taken as a value varying from 0 to 0.5 and K is the *"storage time constant"* which is very close to travel time of flood wave in channel reach. *Storage time constant* has the dimension of time.

Muskingum Equation

Considering a natural channel for which $m = 1.0$, Eq. (9.14) is simplified as:

$$S = K\left[x\,I + (1-x)\,Q\right] \tag{9.15}$$

Eq. (9.15) is called as Muskingum equation which shows a linear relationship for S in terms of I and Q. If we consider the value of the *weighing factor,* $x = 0$, then Eq. (9.15) reduces to

$$S = K\,Q \text{ i.e. } S = f(Q) \tag{9.16}$$

Eq. (9.16) gives relationship between storage and outflow discharge and represents the equation of a linear storage reservoir.

Eq. (9.15) can be rearranged and written as

$$S = \left[K\,x\,(I - Q) + K\,Q\right]$$

Where, the factor $[K\,x\,(I - Q)]$ is the *wedge storage* and $K\,Q$ is the *prism storage* and S = total channel storage.

Determination of K and x

Following steps are used to evaluate the values of K and x in Muskingum equation.

- Select any two sections of a channel reach. Let I be the inflow hydrograph of section 1-1 and Q be the corresponding outflow hydrograph of section 2-2.

- Chose a trial value of weighing factor, x

- Calculate the value of S for different time periods by Eq. (9.15) i.e.

$$S = K\left[x I + (1-x)Q\right]$$

- In an ordinary graph paper, plot a graph between the storage, S calculated for various time periods as mentioned in step 2 (taken as abscissa) vs

$$\left[x I + (1-x)Q\right]$$

- If the plot forms a loop then change the value of x and again compute the value of S for different time periods using Eq. (9.15). Repeat sep 4 as mentioned above and check whether the graph is linear or not. If the graph is not linear then again change the value of x and repeat the procedure till the loop becomes linear or almost linear. Select this value of

  x for the linear relation of S vs $\left[x I + (1-x)Q\right]$.

- Storage time constant, K is the inverse of slope of the line formed for the selected value of x. Thus, if S is plotted in abscissa and $[x I + (1-x) Q]$ as ordinate then reciprocal of the slope of the line i.e. $S/[x I + (1-x) Q]$ is the value of K.

Following sample calculation helps in determination of values of K and x.

Sample Calculation 9.4

Following inflow and outflow data were recorded at a station Mundali, Odisha of the river Mahanadi. Estimate the value of K and x applicable for this station for use in Muskingum equation.

Time, hr	0	6	12	18	24	30	36	42	48	54	60
Inflow, I, m³/s	6	10	24	45	35	20	12	10	8	6	6
Outflow, Q, m³/s	6	8	12	25	40	25	15	12	10	7	7

Solution

Chose a time interval, $\Delta t = 6$ hr. Calculations of various parameters are shown in tabular form (Table 9.4).

Table 9.4 Determination of values of _K_ and _x_ of sample calculation 9.4.

Time, hr	I, m³/s	Q, m³/s	$(I-Q)$, m³/s	Average $(I-Q)$ m³/s	$\Delta S = $ Col.5 x Δt, m³/s. hr	$S = \Sigma S$, m³/s.hr	$[xI + (1-x)Q]$, m³/s		
							$x = 0.35$	$x = 0.30$	$x = 0.20$
1	2	3	4	5	6	7	8	9	10
0	6	6	0	6	6	0	6.00	6.00	6.00
6	20	8	12	25	150	6	8.70	8.60	10.4
12	50	12	38	29	174	156	16.20	15.60	19.6
18	50	30	20	7	42	330	32.00	31.00	34.00
24	32	38	-6	-9.5	-57	372	38.25	38.50	36.80
30	22	35	-13	-13.5	-81	315	23.25	23.50	32.40
36	14	28	-14	-13	-78	234	13.95	14.10	25.20
42	10	22	-12	-11	-66	156	11.30	11.40	19.60
48	8	18	-10	-8	-48	90	9.30	9.40	16.00
54	6	12	-6	-3.5	-21	42	6.65	6.70	10.8
60	6	7	-1			21	6.65	6.70	6.80

Now in a graph paper, three plots are drawn taking storage in abscissa and *[x I + (1 – x) Q]* as ordinate. In first pot, values of *[x I + (1 – x) Q]* are computed with x = 0.35 whereas in other two plots the values are at x = 0.30 and x = 0.20, respectively. The graph drawn is shown in Fig. 9.13. From this figure, it is seen that at x = 0.20, the loop is almost a straight line and so x is selected as 0.20 for this sample calculation.

From Fig. 9.13, the plot of *S* vs *[x I + (1 – x) Q]* corresponding to x = 0.20 is chosen. The inverse of the slope of this curve is found to be 12.65.

Hence the value of *K* for this sample calculation is 12.65 hr.

9.5.2.1 Muskingum Method of Routing

Consider a channel reach of two sections 1-1 and 2-2, respectively. Let the inflows, outflows and storage at section 1-1 at any instant of time be designated as I_1, Q_1 and S_1, respectively. Similarly let the inflows, outflows and storage at section 2-2 at routing time interval of Δt be designated as I_2, Q_2 and S_2, respectively. Applying Muskingum equation (Eq. 9.15) to each of the section we have:

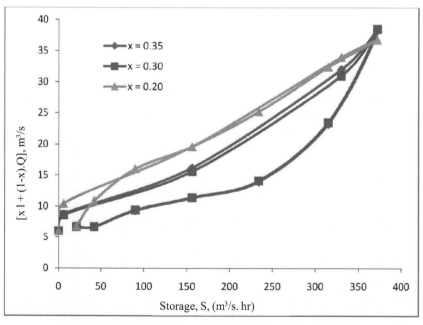

Fig. 9.13: Determination of *K* and *x* for a channel reach

$$S_1 = K\left[x\,I_1 + (1-x)Q_1\right]$$

and $S_2 = K\left[x\,I_2 + (1-x)Q_2\right]$

Now subtracting the above two equations we have change in storage in the sections 1-1 and 2-2,

$$S_2 - S_1 = K\left[x(I_2 - I_1) + (1-x)(Q_2 - Q_1)\right] \tag{9.17}$$

In the similar way we can use the continuity equation at the above two sections in the same reach and can write change in storage in time interval of Δt as

$$S_2 - S_1 = \left(\frac{I_1 + I_2}{2}\right)\Delta t - \left(\frac{Q_1 + Q_2}{2}\right)\Delta t \tag{9.18}$$

Combining Eqns. (9.17) and (9.18) and solving for Q_2 we have

$$Q_2 = C_0 I_2 + C_1 I_1 + C_2 Q_1 \tag{9.19}$$

Where, $C_0 = \dfrac{-K\,x + 0.5\,\Delta t}{K - K\,x + 0.5\,\Delta t}$ $\tag{9.20}$

$$C_1 = \frac{K\,x + 0.5\,\Delta t}{K - K\,x + 0.5\,\Delta t} \tag{9.21}$$

$$C_2 = \frac{K - K\,x - 0.5\,\Delta t}{K - K\,x + 0.5\,\Delta t} \tag{9.22}$$

If we add Eqns. (9.20), (9.21) and (9.22) we get $C = C_0 + C_1 + C_2 = 1.0$

Eq. (9.20) is popularly called as *Muskingum routing equation*. Accuracy of routing depends on time interval Δt. Smaller time interval gives more accurate result. It has been found that for best results, Δt should be so chosen that $K > \Delta t > 2\,K\,x$. The weighing factor x should always be less than 0.5 since for values greater than 0.5, Q_2 becomes negative.

Routing Procedure by Muskingum Method of Channel Routing

Following steps are used in Muskingum method of channel routing:

- Select any two sites of a straight channel section say A and B and let there be no addition or subtraction of lateral flow to this section bounded by A and B. The site A is in the upstream side where data of the inflow hydrograph is available. The site B is in the downstream end where outflow hydrograph corresponding to inflow hydrograph of section A is to be determined.

- Using the measured data of inflow and outflow hydrographs of sections A and B, respectively, compute the parameters K and x as discussed earlier.

- Select a time interval Δt of flood routing such that $K > \Delta t > 2 K x$.

- Using Eqns. (9.20), (9.21) and (9.22), compute the value of C_0, C_1 and C_2, respectively and check the sum of the values of C_0, C_1 and C_2 becomes 1.0.

- For the initial time i.e. $t = 0$, I_1, I_2 and Q_1 are known. Now use Eq. (9.19) to compute Q_2.

- For the next step after time interval of Δt, Q_2 so computed in step 5, becomes Q_1. Then routing is continued till entire inflow hydrograph is routed.

Sample calculation 9.5

A reservoir reach has the value of storage time constant, $K = 12$ hr and weighing factor, $x = 0.25$. Route the following inflow hydrograph which passes through it. Assume at the initial time of inflow hydrograph, the outflow discharge is 10 m³/s.

Time, hr	0	6	12	18	24	30	36	42	48	54
Inflow, m³/s	10	22	48	58	55	44	35	25	20	12

Solution

It is given that $K = 12$ hr and $x = 0.25$. Hence $2 K. x = 6$ hr

Select a time interval Δt of flood routing such that $K > \Delta t > 2 K x$. Hence we can assume Δt a value between 6 and 12 hr and let it be 6 hr since it suits the data of the ordinate interval of the given inflow hydrograph.

$$C_0 = \frac{-12 \ x \ 0.25 + 0.5 \ x \ 6}{12 - 12 \ x \ 0.25 + 0.5 \ x \ 6} = 0$$

$$C_1 = \frac{12 \ x \ 0.25 + 0.5 \ x \ 6}{12 - 12 \ x \ 0.25 + 0.5 \ x \ 6} = 0.5$$

$$C_2 = \frac{12 - 12 \ x \ 0.25 - 0.5 \ x \ 6}{12 - 12 \ x \ 0.25 + 0.5 \ x \ 6} = 0.5$$

Check $C_0 + C_1 + C_2 = 0 + 0.5 + 0.5 = 1.0$

For the first time interval of 0 to 6 hr, $I_1 = 10$, $I_2 = 22$ and $Q_1 = 10$

Using Eq. (9.19) we have:

$Q_2 = C_0 I_2 + C_1 I_1 + C_2 Q_1 = 0$ x $22 + 0.5$ x $10 + 0.5$ x $10 = 10$ m³/s

For the second time step i.e. 6 to12 hr, Q_2 becomes Q_1 and is equal to 10 m³/s

The procedure continues till the entire duration of inflow hydrograph is routed. The computations are shown in Table 9.5. The inflow and outflow hydrograph are drawn (Fig.9.14) and the attenuation and peak lag are found to be 8.0 m³/s and 12 hr, respectively.

Table 9.5: Muskingum method of flood routing (Sample calculation 9.5)

Time, hr	I, m³/s	$C_0 I_2 = 0$ x I_2	$C_1 I_1 = 0.5$ x I_1	$C_2 Q_1 =$ 0.5 x Q_1	$Q = (0$ x $I_2 + 0.5$ x $I_1 + 0.5$ x $Q_1)$, m³/s
1	2	3	4	5	6
0	10				10.0
		0	5	5	
6	22				10.0
		0	11	5	
12	48				16.0
		0	24	8	
18	58				32.0
		0	29	16	
24	55				45.0
		0	27.5	22.5	
30	44				50.0
		0	22	25	
36	35				47.0
		0	17.5	23.5	
42	25				41.0
		0	12.5	20.5	
48	20				33.0
		0	10	16.5	
54	12				26.5

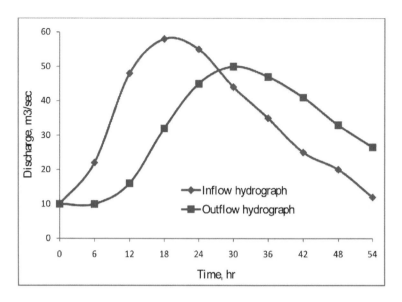

Fig. 9.14: Flood routing by Muskingum routing method (sample calculation 9.5)

9.5.2.2 Muskingum Crest Segment Routing Method

In this method of flood routing, portion of inflow hydrograph which is very close to the crest or peak segment of the hydrograph is routed. Computations are performed by using Eq.(9.19) as:

$$Q_n = K_1 I_n + K_2 I_{n-1} + K_3 I_{n-2} + \ldots\ldots + K_n I_1 \qquad (9.23)$$

Where, $K_1 = C_0$, $K_2 = C_0 C_2 + C_1$, $K_3 = K_2 C_2$ and $K_n = K_{n-1} C_2$

Eq. (9.23) is valid for $n > 2$ (Suresh, 2008).

Eq. (9.23) can be used to route the inflow hydrograph at any time without routing the entire hydrograph. Sample calculation 9.6 illustrates the use of the method for flood routing.

Sample Calculation 9.6

Using the inflow hydrograph and other data of sample calculation 9.5, compute the outflow at 30 hr duration by using Muskingum crest segment method.

Solution

Looking the duration of the inflow hydrograph it is seen that for duration of 30 hr, $n = 6$

Outflow at 30 hr duration i.e. at $n = 6$ is given from Eq. (9.23) as;

$$Q_6 = K_1 I_6 + K_2 I_5 + K_3 I_4 + K_4 I_3 + K_5 I_2 + K_6 I_1$$

We have already computed the values of C_0, C_1 and C_2 in solution of sample calculation of 9.5 as 0, 0.5 and 0.5, respectively.

Hence, $K_1 = C_0 = 0$

$$K_2 = C_0 C_2 + C_1 = 0 + 0.5 = 0.5$$

$$K_3 = K_2 C_2 = 0.5 \ x \ 0.5 = 0.25$$

$$K_4 = K_3 C_2 = 0.25 \ x \ 0.5 = 0.125$$

$$K_5 = K_4 C_2 = 0.125 \ x \ 0.5 = 0.625$$

and $K_6 = K_5 C_2 = 0.625 \ x \ 0.5 = 0.3125$

Now putting these values of K_1 to K_6 and using the inflow data of I_1 to I_6 from the sample calculation 9.5, we have, Q_6 as:

$$Q_6 = K_1 I_6 + K_2 I_5 + K_3 I_4 + K_4 I_3 + K_5 I_2 + K_6 I_1$$

$= 0 \ x \ 44 + 0.5 \ x \ 55 + 0.25 \ x \ 58 + 0.125 \ x \ 48 + 0.625 \ x \ 22 + 0.3125 \ x \ 10 = 64.875 \ \text{m}^3/\text{s}$

9.5.3 Hydraulic Channel Routing

Hydraulic channel routing is embodies on solution of basic St Venant equations [(9.4) and (9.5)] which are simultaneous, quasi-linear, first order partial differentiation equations of hyperbolic type. Analytical solution of these equations is not easy except in special simplified case the solution is possible. However, modern high speed digital computer can provide sophisticated technologies which may make it possible to provide numerical solution of the equations. There are two broad categories of numerical methods available to solve the St Venant equations. They are as follows:

(i)   Approximate method and

(ii) Complete numerical method

9.5.3.1 Approximate Method

Approximate method is based mostly on equation of continuity only. Hydrological method of storage routing and Muskingum channel routing which are presented earlier belong to this category.

9.5.3.2 Complete Numerical Method

There are three numerical methods used for hydraulic method of channel routing. They are

(i)   Finite difference method

(ii)   Finite element method and

(iii) Method of characteristics

Method of characteristics again comprise two methods i.e. (i) characteristics nodes and (ii) rectangular grid. Finally finite difference, finite element, characteristics nodes and rectangular grid each comprise two techniques as (i) implicit method and (ii) explicit method. All these methods are mentioned in the flow chart of Fig. 9.2.

Finite difference method is also called as direct method. In direct method, all the partial derivatives are replaced by finite differences and the resulting algebraic equations are then solved. In finite difference, three approximations are employed. They are (i) backward difference approximations, (ii) forward difference approximations and (iii) central difference approximations. At each point in a curve, all these three approximations are employed. Further, all these approximations can be used in both z-x and z-t plane of flow in three dimensions where x represents the spatial component, t refers to temporal component and z is the direction which is perpendicular to both x and t cartisan co-ordinate direction.

In method of characteristics, St Venant equations are converted into a pair of ordinary differential equations and then solved by finite difference method. In finite element method, the whole system is divided into a number of elements and partial differential equations are integrated at the nodal points of the elements.

In explicit method, the algebraic equations are linear and the dependent variables are extracted explicitly at the end of each time step. In implicit method, the dependent variables occur implicitly and the equations are non-linear. Details of hydraulic flood routing are available at references, Mahmood and Yevjevich (1975), Streeter and Wylie (1967) and Subramanya (1991).

9.5.4 Flood Routing Machines

9.5.4.1 Mechanical Flood Router

Machines capable of solving the storage equation can save considerable time and efforts in routing operations. There are two types of routing machines. They are mechanical and electronic type. A large number of mechanical machines have been developed for flood routing in reservoirs and/or channels. Amongst them, integrating flood router is popular. Integrating flood router consists of five drum mounted graph charts driven continuously by motor arrangement. Out of five drums, in three drums, input data like inflow hydrograph, elevation vs storage and elevation vs discharge curves are fed as input graph chart. Other two drums plot time vs elevation and time vs outflow graphs continuously during routing operation. Pointers kept in position touching the graphs or charts help in reading the input data from the first three drums. To plot the output graphs, pen pointers are attached to the last two drums. Movement of the pointers, drum and pen are synchronized such that as the drums rotate all the five pointers move and the two output graphs are automatically plotted. However, here the machine requires calibration to fit to the given system.

The rolling type mechanical flood router developed by F.B. Harkness is another important mechanical device which is used for channel routing where Muskingum method is used with time step of $\Delta t = 2\,Kx$. In this instrument, an undercarriage with wheel, pen and pointer arrangement moves over the plotted inflow hydrograph. The undercarriage is attached to a T-shaped frame at such distance that as the undercarriage moves over the plotted inflow hydrograph, the outflow hydrograph is automatically plotted by a pen arrangement attached to it at fixed distance to the inflow hydrograph pointer reader. A sketch of rolling type mechanical flood router developed by F.B. Harkness in 1951 is shown in Fig. 9.15.

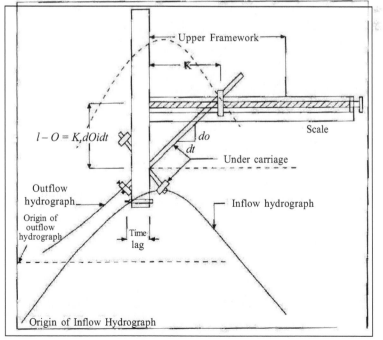

Fig. 9.15: Rolling type mechanical flood router developed by F.B. Harknes (Patra, 2001)

9.5.4.2 Electric Analog Routing Machine

In electric analog routing machine, the basic electrical circuit theory is compared with the routing equation in channel. The inflow, outflow and storage of water in a reservoir or channel is made analogous with that of electric current. A condenser having some capacitance is capable of storing current is made analogous to storage, S. The inflow and outflow from the reservoir or channel is made similar to the inflow and outflow of current from the condenser. The figure of an electric analog routing machine developed by U.S. Weather Bureau (1955) is shown in Fig. 9.16. This machine contains two condensers C_1 and C_2, having some particular capacitances. The circuit containing condenser C_1 has three resistances R_1, R_2 and R_3. The circuit containing condenser C_2, has two resistances R_1 and R_4. In between the two condensers C_1 and C_2, the resistance is R_5 (Fig. 9.16). In addition, the machine contains two self balancing potentiometers, two phototube bulbs and a variac transformer. During operation, inflow hydrograph is plotted on the chart of potentiometer P_1 whereas the outflow hydrograph is traced automatically on the chart fixed over the potentiometer P_0. The total charge across the two condensers C_1 and C_2 are compared with the Muskingum equation which is analogous to it. The machine is used for flood routing at several flood forecasting sites.

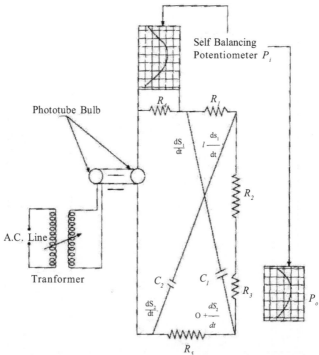

Fig. 9.16: Electric analogue routing machine (Patra, 2001)

9.5.4.3 Digital Computers

The computation of flood routing has become more simplified with the invention of digital computers. A unique computer program can be written for the system of channel or reservoir for routing various inflow hydrographs. The program can be validated with the previous recorded inflow and outflow data. Once the program is validated, it takes a few seconds to compute the outflow hydrograph from the given information of inflow hydrograph. A large number of watershed simulation models have been developed which takes into account the rainfall, meteorological data, and watershed characteristics as input data and computes runoff and unit hydrograph. The unit hydrograph is convoluted with effective rainfall hyetograph to obtain flood hydrograph which is routed through the channel or reservoir giving directly the outflow for the system. Use of a validated computer program along-with telemetry gauges and digital computers at a desired site of channel or river makes the process more quick and accurate.

9.6 Flood Routing in Conceptual Hydrograph Development

Flood routing can be used in study of development of conceptual hydrograph. The routing through a reservoir which gives attenuation and channel routing which gives translation to an input hydrograph are treated as two basic modifying operators. Following two fictitious items are used in the studies of development of instantaneous unit hydrographs through conceptual models.

- *Linear reservoir*: It is a fictitious reservoir in which storage is directly proportional to discharge i.e. $S = K Q$. This relation between storage and discharge is conceptual. In actual case, the relation between them is non-linear in the form $S = K Q^a$ where K is coefficient and a is exponent. Linear reservoir is used to provide attenuation to a flood wave.

- *Linear channel*: It is a fictitious channel in which the time required to translate a given discharge Q through a given reach is constant. A linear channel delays the arrival of discharge t the outlet without affecting peak. An inflow hydrograph passes through such a channel with only translation and no attenuation.

Conceptual modeling for instantaneous unit hydrograph (IUH) development has undergone rapid advances since Zoch (1937) proposed his work simulating a linear channel in series with a linear reservoir. The catchment action is considered as analogous to the response of linear reservoir or linear channels. Detail discussions on all these models are beyond the scope of this book. However, in this chapter, a simple method i.e Clark's method (1945) which uses Muskingum method of linear reservoir routing is discussed for development of conceptual model.

9.6.1 Clark's Method for IUH

Clark's method is also called as time-area histogram method is used to develop an IUH due to an instantaneous rainfall excess over a basin by considering a linear channel in series with a linear reservoir. It is assumed that the rainfall excess first undergoes pure translation and then attenuation. The translation is achieved by a travel time-area histogram (TAH) and attenuation by routing the results of the above through a linear reservoir at the watershed outlet. Attenuation is achieved due to storage affect of the catchment. Routing of rainfall excess is done by using Muskingum method routing through a linear reservoir

The basis of development of IUH is the distribution of arrival time of the volume of rainfall excess or surface runoff at watershed outlet. This time duration is assumed as maximum translation time of the surface runoff because it is the time required to move the unit volume of rainfall excess from remotest point to the watershed outlet. Thus, the time here refers to the time of concentration. The unit hydrograph follows the assumption that there is uniform distribution of unit rainfall excess over the entire watershed. The time distribution of rainfall excess results the time-area histogram which is considered as the inflow hydrograph to a hypothetical storage reservoir at the outlet of the reservoir.

9.6.1.1 Development of Time-Area Histogram (TAH)

Before development of TAH, we should understand two terms i e. channel travel time and time of concentration.. Channel travel time refers to the time taken by the water to move along a channel from one point to another along it. This time period is less than the time of concentration. In ungauged watershed, time of concentration can be determined by Kirpich (1940) formula or US Practice formula as discussed in earlier chapter. For gauged watershed, the time interval between the end of rainfall excess and the point of inflection of the surface runoff (Fig. 9.17, a) provides a way to estimate the time of concentration from known rainfall runoff records.

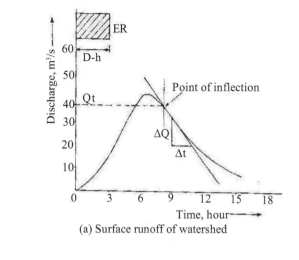

(a) Surface runoff of watershed

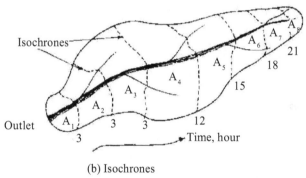

(b) Isochrones

Fig. 9.17: Development of time area histogram for Clark's model

Time-area histogram is prepared using the following steps.

- At first locate several points on the toposheet of watershed. Compute their travel time and see that this computed travel time is less than the time of concentration of the watershed.

- Now draw isochrones by joining the points of equal travel time. These isochrones divide the whole watershed into different segments having equal travel time. For drawing isochrones, surface profile of the longest water course is plotted in terms of elevation vs distance curve (Fig. 9.17, b) which is started from the outlet of the watershed. The total length of the longest watercourse is divided into N parts and their respective elevations and distance between each consecutive part are determined. Then these distance and elevation readings are transformed on the contour map of the watershed. Fig. 9.17 (b) represents a watershed being divided into 8 ($N = 8$) parts by isochrones having an equal time of travel.

- In the third step, inter-isochrone areas are determined by a planimeter. Let them be A_1, A_2, A_3....A_N.

- Finally, construct the TAH by plotting inter-isochrone areas against time interval in the form of bar chart or histogram (Fig. 9.18).

If a rainfall excess of 1 cm occurs instantaneously and uniformly over the watershed, this time area histogram represents the sequence in which the volume of rainfall will be moved out of the watershed and arrive at the outlet. In Fig. (9.18), a sub area A_r km² represent a volume of A_r km² . cm which is equal to A_r x 10^4 m³ moving out in a time of $\Delta t_c = tc/N$ hr where t_c is time of concentration, hr. The hydrograph of outflow obtained by this figure do not provide for the storage properties of the watershed. To overcome this deficiency, Clark assumed hypothetical linear reservoir which is made available at the outlet to provide the required attenuation.

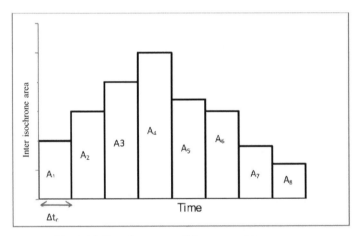

Fig. 9.18: Development of time-area-histogram

9.6.1.2 Routing TAH

In Clark's model, we assume a linear reservoir for which the weighting factor, x = 0. Under such context The Muskingum equation $S = K\left[xI + (1-x)Q \right]$ reduces to $S = KQ$. The value K can be estimated by considering the point of inflection of a surface runoff hydrograph ((Fig. 9.17, a). It is assumed that at the point of inflection, the inflow into the channel has ceased ($I = 0$) and the outflow beyond this point is due to the withdrawal from the storage. The continuity equation is then written as:

$$I - Q = \frac{dS}{dt} = -Q = \frac{dS}{dt} = K \frac{dQ}{dt}$$

Which gives $K = -Q / \dfrac{dQ}{dt}$ (9.24)

Value of K can be determined from a known surface runoff hydrograph of the catchment. Knowing the value of K, the inflows at various times are routed by Muskingum method. The inflow rate between an inter-isochrone area A_r km^2 with a time interval Δt_c (hr) can be calculated as follows:

$I = (A_r \times 10^4)/(3600 \, \Delta t_c) = 2.78 \, A_r/\Delta t_c \ (m^3/s)$

The Muskingum equation would now be

$$Q_2 = C_0 I_2 + C_1 I_1 + C_2 Q_1 \tag{9.25}$$

Where, $C_0 = \dfrac{0.5 \, \Delta t_c}{K + 0.5 \, \Delta t_c}$

$C_1 = \dfrac{0.5 \, \Delta t_c}{K + 0.5 \, \Delta t_c}$

$C_2 = \dfrac{K - 0.5 \, \Delta t_c}{K + 0.5 \, \Delta t_c}$

The above equations represent $C_0 = C_1$. Since inflows are derived from the histogram, $I_1 = I_2$. Now Eq. (9.25) reduces to

$$Q_2 = 2 \, C_1 I_1 + C_2 Q_1 \tag{9.26}$$

Eq. (9.26) helps in computation of ordinates of IUH. Following sample calculation helps in determination of ordinates of IUH.

Sample Calculation 9.7

A watershed has the following characteristics.

(i)   Area = 100 km^2

(ii)   Time of concentration =20 hr

(iii) Storage time constant = 15 hr and

(iv) Inter-isochrone area distribution are as follows:

Travel time, hr	0 - 3	3 - 6	6 - 9	9 - 12	12 - 15	1 5- 18	18 - 21
Inter-isochrone area, km²	2.0	5.0	15.0	10.0	8.0	5.0	3.0

Derive IUH using Clark model.

Solution

Given $K = 15$ hr, $t_c = 20$ hr and $\Delta t_c = 3$ hr

Value of the coefficient $C_1 = \dfrac{0.5\,\Delta t_c}{K + 0.5\,\Delta t_c}$

$$= \frac{0.5 \times 3}{15 + 0.5 \times 3} = 0.091$$

Similarly value of $C_2 = \dfrac{K - 0.5\,\Delta t_c}{K + 0.5\,\Delta t_c}$

$$\frac{15 - 0.5 \times 3}{15 + 0.5 \times 3} = 0.818$$

Putting the values of C_1 and C_2 in Eqn. (9.26)

$Q_2 = 2\,C_1\,I_1 + C_2\,Q_1$

$Q_2 = (2 \times 0.091)\,I_1 + 0.818\,Q_1$

$Q_2 = 0.182\,I_1 + 0.818\,Q_1$

The inflow, I is computed as $I = 2.78\,A_r/\Delta t_c$ (m^3/s)

At $t = 0$, $Q_1 = 0$

Computations of IUH ordinates are shown in table below.

Time, hr	Inter isochrones area, A_r, km²	Inflow, I^*, m³/s	0.182 I_1	0.818 Q_1	Ordinates of IUH, m³/s
1	2	3	4	5	6
0	0	0	0	0	0
3	2	1.85	0.34	0	0.34
6	5	2.32	0.42	0.278	0.698
9	15	4.63	0.84	0.571	1.411
12	10	2.32	0.42	1.154	1.574
15	8	1.48	0.27	1.287	1.557
18	5	0.77	0.14	1.273	1.413
21	3	0.40	0.07	1.156	1.226
24	0	0	0	1.003	1.003
27	0	0	0	0.820	0.820
30				.	.
33				.	.
36				.	.
39				so on	so on

9.6.2 Nash's Conceptual Model

Like Clark's model, Nash (1957) proposed a conceptual model for development of IUH for a watershed. In his model, the watershed is assumed to be made up of a series of n number of identical linear reservoirs. The storage constant, K of all these reservoirs is assumed to be same. In this case, the outflow of any reservoir is the input to the next reservoir. The first reservoir is assumed to receive a unit volume equal to 1 cm of effective rainfall from a given watershed having some particular area instantaneously. This inflow is routed through the first reservoir to get the outflow hydrograph. The outflow from the first reservoir is the inflow to the second reservoir. This way, the routing continues for all the n reservoirs which are in series. The outflow hydrograph from the last reservoir is considered as the IUH of the watershed. The conceptual cascade of reservoirs as discussed above and the shape of the outflow hydrograph are shown in Fig. 9.19.

From the continuity equation, $I - Q = \dfrac{dS}{dt}$

For a linear reservoir, $S = K\,Q$

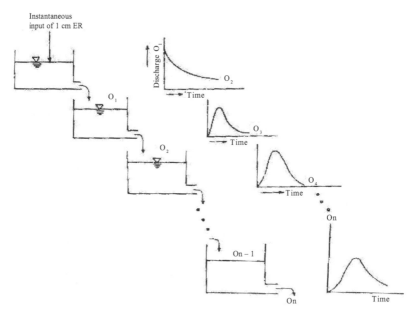

Fig. 9.19: Nash Model: Cascade of linear reservoirs

Hence, $\dfrac{dS}{dt} = K\dfrac{dQ}{dt}$ (9.27)

Substituting value of dS/dt from Eq. (9.27) in the above mentioned continuity equation and simplifying, we get

$$I = Q + K.\dfrac{dQ}{dt}$$ (9.28)

Solving the differential equation (Eq. 9.28) where both inflow, I and outflow, Q are functions of time, t, we get

$$Q = \dfrac{1}{K} e^{-\frac{t}{K}} \int e^{\frac{t}{K}} I\, dt$$ (9.29)

For the first reservoir, the input is supplied instantaneously i.e. for time, $t > 0$, inflow $I = 0$. Moreover at time, $t = 0$, = instantaneous volume inflow = 1 cm of effective rainfall.

Therefore, for the first reservoir, Eq. (9.29) becomes

$$Q_1 = \dfrac{1}{K} e^{-\frac{t}{K}}$$ (9.30)

The outflow of the first reservoir given by Eq. (9.30) is the inflow of the second reservoir. The outflow of the second reservoir is given as

$$Q_2 = \frac{1}{K}e^{-\frac{t}{K}} . \int e^{\frac{t}{K}} . \frac{1}{K}e^{-\frac{t}{K}} \, dt$$

$$Q_2 = \frac{1}{K^2} . t . e^{-\frac{t}{K}} \tag{9.31}$$

Q_2 of Eq. (9.31) replaces I of Eq. (9.29) and the outflow of the third reservoir is obtained as:

$$Q_3 = \frac{1}{2}\frac{1}{K^3}t^2 \, e^{-\frac{t}{K}} \tag{9.32}$$

Continuing this way, we get the outflow from the nth reservoir i.e. Q_n as

$$Q_n = \frac{1}{(n-1)! \, K^n} t^{n-1} \, e^{-\frac{t}{K}} \tag{9.33}$$

This outflow from the nth reservoir is the result of 1 cm of excess rainfall falling instantaneously over the whole watershed and therefore, it represents the IUH of the watershed.

Replacing Q_n by the notation of the ordinate of IUH as $u(t)$ we have

$$u(t) = \frac{1}{(n-1)!} \frac{t^{n-1}}{K^n} e^{\frac{-t}{K}} \tag{9.34}$$

In Eq. (9.34), $u(t)$ is in cm/h, t is in hr and K and n are constants. In Eq. (9.34), value of n is an integer and not a fraction. To make $(n-1)!$ to be valid both for n as integers and fractional numbers, the term $(n-1)!$ is replaced by gamma function . Accordingly Eq. (9.34) is

$$u(t) = \frac{1}{K\,\Gamma(n)}\left(\frac{t}{K}\right)^{n-1} e^{\frac{-t}{K}} \tag{9.35}$$

When n is an integer, $\Gamma(n) = (n-1)!$ which is easy to be evaluated. But when it is a fraction, its value is determined from Table 9.6 (Panigrahi, 2013).

9.6.2.1 Determination of n and K of Nash's Model

The parameters n and K are determined by taking moments about the origin i.e. $t = 0$, are given as follows.

Table 9.6: Value of gamma function, K

K	$\dot{\Gamma}K$	K	$\dot{\Gamma}K$	K	$\dot{\Gamma}K$	K	$\dot{\Gamma}K$	K	$\dot{\Gamma}K$
1.00	1.000000	1.205	0.916857	1.405	0.8870028	1.605	0.894088	1.805	0.932720
1.01	0.994325	1.21	0.915576	1.41	0.886764	1.61	0.894680	1.81	0.934076
1.02	0.988844	1.22	0.913106	1.42	0.886356	1.62	0.895924	1.82	0.936845
1.03	0.983549	1.23	0.910754	1.43	0.886036	1.63	0.897244	1.83	0.939690
1.04	0.978438	1.24	0.908521	1.44	0.885805	1.64	0.898642	1.84	0.942612
1.05	0.973504	1.25	0.906402	1.45	0.885661	1.65	0.900116	1.85	0.945611
1.06	0.968744	1.26	0.904397	1.46	0.885694	1.66	0.901668	1.86	0.948687
1.07	0.964152	1.27	0.902503	1.47	0.885633	1.67	0.903296	1.87	0.951840
1.08	0.959725	1.28	0.900718	1.48	0.885747	1.68	0.905001	1.88	0.955071
1.09	0.955459	1.29	0.899041	1.49	0.885945	1.69	0.906781	1.89	0.958379
1.10	0.951351	1.30	0.897471	1.50	0.886227	1.70	0.908639	1.90	0.961765
1.11	0.947395	1.31	0.896004	1.51	0.886591	1.71	0.910571	1.00	0.961766
1.12	0.943590	1.32	0.894640	1.52	0.887039	1.72	0.912581	1.92	0.968774
1.13	0.939931	1.33	0.893378	1.53	0.887567	1.73	0.914665	1.93	0.972376
1.14	0.936416	1.34	0.892216	1.54	0.888178	1.74	0.916826	1.94	0.976099
1.15	0.933040	1.35	0.891151	1.55	0.888868	1.75	0.919062	1.95	0.979880
1.16	0.929803	1.36	0.890185	1.56	0.889639	1.76	0.912375	1.96	0.983743
1.17	0.926699	1.37	0.889313	1.57	0.890489	1.77	0.923763	1.97	0.987684
1.18	0.923728	1.38	0.888537	1.58	0.891420	1.78	0.926227	1.98	0.991708
1.19	0.920885	1.39	0.887854	1.59	0.892428	1.79	0.928767	1.99	0.995813
1.20	0.918169	1.40	0.887264	1.60	0.893515	1.80	0.931384	2.00	1.000000

- First moment about the origin (M_1) is

$$M_1 = n. \; K \tag{9.36}$$

- Second moment about the origin (M_2) is

$$M_2 = n \; (n + 1) \; K^2 \tag{9.37}$$

Based on above two equations, values of n and K can be determined provided effective rainfall hyetograph (ERH) and direct runoff hydrograph (DRH) are available.

If M_{Q1} = first moment of the DRH about the origin divided by total direct runoff

M_{Q2} = second moment of the DRH about the origin divided by total direct runoff

M_{I1} = first moment of the ERH about the origin divided by total effective rainfall and

M_{I2} = second moment of the ERH about the origin divided by total effective rainfall

then, $M_{Q1} - M_{I1} = n \; K$ \hfill (9.38)

and $M_{Q2} - M_{I2} = n \; (n + 1) \; K^2 + 2 \; n \; K \; M_{I1}$ \hfill (9.39)

For a watershed with known values of M_{Q1}, M_{Q2}, M_{I1} and M_{I2}, Eqns. (9.38) to (9.39) can be employed to determine the values of n and K.

9.6.2.2 Computations of Ordinates of IUH

Once the Nash parameters n and K are determined, Eq. (9.35) i.e.

$$u(t) = \frac{1}{K \, \Gamma(n)} \left(\frac{t}{K} \right)^{n-1} e^{\frac{-t}{K}}$$ can be used to determine the values of ordinates of

IUH i.e. $u(t)$. In the Eq. (9.35), values of n, K and t are known. Here time, t is time interval of ordinates of IUH. If values of t and K are in hr, then value of $u(t)$ is obtained in cm/h which may be converted to m^3/s by the relation

$$u(t)_{m^3/s} = u(t)_{cm/h} \; x \; 2.78 \; x \; A \tag{9.40}$$

Where, A is area of the watershed in km^2.

Once the ordinates of IUH are determined, we can compute the ordinates of a D-hr unit hydrograph by the superposition or S-curve method as discussed in earlier Chapter "Hydrograph".

Sample Calculation 9.8

A watershed of area 120 km² has the following effective rainfall hyetograph (ERH) and the corresponding direct runoff hydrograph which are given below. Determine the values of n and K by using Nash conceptual model. Also determine the ordinates of IUH.

Time, hr				0 - 3			3 - 6
Intensity of effective rainfall, cm/h				4.0			3.0

Time, hr	0	3	6	9	12	15	18
Ordinates of DRH, m³/s	0.0	14.0	40.0	55.0	22.0	5.0	0.0

Solution

Solution is given in tabular form in next page containing data of evaluation of M_{11} and M_{12}.

In the table, incremental rainfall excess is computed by multiplying col.2 with col. 3.

Col. 5 i.e. moment arm is computed as the time interval between the centroid of each excess rainfall hyetograph with respect to the time $t = 0$. For the first time interval of 0 – 3 hr, it is 1.5 and for the next time interval of 3 -6 hr, it is $3 + 1.5 = 4.5$.

Col. 6 is obtained by multiplying col. 4 with col. 5 i.e. first moment = incremental rainfall excess x moment arm

Col. 7 is the second moment of incremental rainfall excess about origin which is equal to col. 4 x (col. 5)²

Col. 8 is the second moment of incremental rainfall excess about its own centroid = $(1/21)$ x (col. 3)³ x col. 4

Col. 9 is the summation of col. (7) and col. (8).

Now M_{11} = first moment of the ERH about the origin divided by total effective rainfall

$$= \frac{\sum col.6}{\sum col.4} = \frac{58.5}{21} = 2.78$$

Similarly, M_{12} = second moment of the ERH about the origin divided by total effective rainfall

$$= \frac{\sum col.9}{\sum col.4} = \frac{236.25}{21} = 11.25$$

Evaluation of M_{I1} and M_{I2} are given in the following table.

Time, hr	Incremental rainfall excess rate (cm/h)	Time interval, Δt, hr	Incremental rainfall excess, cm	Moment arm, hr	First moment, (col.4) x (col. 5), cm hr	Second moment		
						Incremental rainfall excess (moment arm)^2 (col.4) × (col.5)^2	Second moment of the incremental rainfall excess about its own centroid	Total (col. 7) + (col. 8)
1	2	3	4	5	6	7	8	9
0	0.0	0	0	0	0	0	0	0
3	4.0	3	12.0	1.5	18.0	27.0	15.43	42.43
6	3.0	3	9.0	4.5	40.5	182.25	11.57	193.82
Total			21.0		58.5	209.25	27.00	236.25

Evaluation of M_{Q1} and M_{Q2} are given in the following table.

Time, hr	Ordinates of DRH, m³/s	Average rate of DR in time Δt , m³/s	Interval (Δt) , hr	Incremental vol. of runoff (i.e. area of histogram), col. 3 x col. 4	Moment arm, hr	First moment, col. 5 x col. 6	Incremental vol. of runoff x (moment arm)² (col. 5) x (col.6)²	Second moment of the incremental runoff about its own centroid	Total, (col. 8 + col. 9)
1	2	3	4	5	6	7	8	9	10
0	0	0	0	0	0	0	0	0	0
3	14	7.0	3	21.0	1.5	31.5	47.25	27	74.25
6	40	27.0	3	81.0	4.5	364.5	1640.25	104.14	1744.39
9	55	47.5	3	142.5	7.5	1068.7	8015.625	183.21	8198.84
12	22	38.5	3	115.5	10.5	1212.7	12733.88	148.50	12882.4
15	5	13.5	3	40.5	13.5	546.75	7381.125	52.07	7433.2
18	0	2.5	3	7.5	16.5	123.75	2041.875	9.64	2051.52
				408.0		3348.0			32384.6

Evaluation of MQ1 and MQ2 given in table in previous page. In this table. Col.8 is the second moment of incremental rainfall excess about origin which is equal to col. 5 x (col. 6)^2

Col.9 is the second moment of incremental rainfall excess about its own centroid = (1/21) x (col. 4)^3 x col. 5

Col. 10 is the summation of col. (8) and col. (9).

Now M_{Q1} = first moment of the DRH about the origin divided by total direct runoff

$$= \frac{\sum col.7}{\sum col.5} = \frac{3348}{408} = 8.21$$

Similarly, M_{Q2} = second moment of the DRH about the origin divided by total direct runoff

$$= \frac{\sum col.10}{\sum col.5} = \frac{32384.6}{408} = 79.37$$

From Eq. (9.38), $M_{Q1} - M_{I1} = n\,K$

$8.21 - 2.78 = n\,K$

$n\,K = 5.43$ (a)

From Eq. (9.39), $M_{Q2} - M_{I2} = n\,(n + 1)\,K^2 + 2\,n\,K\,M_{I1}$

$79.37 - 11.25 = (n\,K)^2 + K.\,nK + 2\,n\,K\,M_{I1}$

$68.12 = (5.43)^2 + 5.43\,K + 2\;x\;5.43\;x\;2.78$ (b)

Solution of Eq. (b) gives $K = 1.55$ hr

Substituting the value of $K = 1.55$ in Eq. (a), we get $n = 3.5$

Computation of ordinates of IUH is carried out using Eq. (9.35) which is

$$u(t) = \frac{1}{K\,\Gamma(n)} \left(\frac{t}{K}\right)^{n-1} e^{\frac{-t}{K}}$$

Where the values of n and K are 3.5 and 1.55, respectively.

Value of $\Gamma(3.5) = 2.5\,\Gamma(2.5) = 2.5\;x\;1.5\,\Gamma(1.5)$

From Table 9.6, $\Gamma(1.5) = 0.886227$

= 2.5 x 1.5 x 0.886227 = 3.323

Substituting the values of n, K and in Eq. (9.35),

$$u(t) = \frac{1}{1.55 \times 3.323} \left(\frac{t}{1.55}\right)^{2.5} e^{\frac{-t}{1..55}}$$

$$u(t) = 0.194 \; x \left(\frac{t}{1.55}\right)^{2.5} e^{\frac{-t}{1..55}} \qquad\qquad (c)$$

Ordinates of 3 hr IUH for various values of Δt" are calculated using above Eq. (c) and are mentioned in table below.

Time, hr	$(t/1.55)^{2.5}$	$e^{-t/1.55}$	$u(t) = 0.194 \times$ col. 2 $\times$ col. 3, cm/hr	3-hr unit hydrograph		
				$u(t) = 2.78 \times 120 \times$ col. 4 m^3/s	$u(t)$ lagged by 3 hr, m^3/s	Unit hydrograph, {col. 5 + col. 6}/2, m^3/s
1	2	3	4	5	6	7
0	0	0	0	0	0	0
3	5.21	0.144	0.1455	48.554	0	24.277
6	29.48	0.021	0.1201	40.066	48.554	44.31
9	81.24	0.003	0.0473	15.773	40.066	27.92
12	166.77	0.0004	0.0129	4.3172	15.773	10.045
15	291.33	6.27E-05	0.0035	1.1822	4.3172	2.7497
18	459.57	9.05E-06	0.0008	0.2692	1.1822	0.7257
21	675.64	1.31E-06	0.0002	0.0573	0.2692	0.1632
24	943.40	1.89E-07	3E-05	0.0115	0.0573	0.0344
27	1266.43	2.72E-08	7E-06	0.0022	0.0115	0.0069
					0.0022	0.0011

Question Banks

Q1.   What are the various factors that affect the shape of a flood hydrograph?

Q2.   What do you mean by flood routing? What is its importance?

Q3.   Mention the different types of flood routing with a flow chart.

Q4.   What is the difference between a reservoir and a channel flood routing?

Q5.   Mention the parameters required for reservoir routing.

Q6.   Write down the different uses of reservoir routing.

Q7.   What do you mean by channel routing? How is it helpful in the field of hydrology?

Q8.   With a sketch, explain how does the flood routing changes the shape of a flood hydrograph?

Q9.   Define the following terms.

(i)   Attenuation of flood peak

(ii) Translation of flood peak

Q10.   What are the parameters on which the shape of outflow flood hydrograph of a channel section depends?

Q11.   What are the various routing methods? Differentiate between hydrologic and hydraulic flood routing.

Q12.   Which one of the following gives more accurate results and why?

(i)   Hydrologic flood routing (ii) Hydraulic flood routing

Q13.   Mention the different difficulties in using hydraulic flood routing.

Q14.   For hydrologic reservoir routing, what are the data required? With a sketch, explain the principle of working of hydrologic reservoir routing.

Q15.   Hydrologic reservoir routing is also called as level pool routing. Mention why?

Q16.   Describe the various steps used in Modified Pul's method of hydrologic reservoir routing.

Q17.   A reservoir has the following elevation (stage), discharge and storage relationships.

Elevation, m	Storage (x 10^6 m³)	Outflow discharge (m³/sec)
100.00	3.400	0
100.50	3.460	10
100.8	3.880	24
101.6	4.420	43
101.8	4.980	55
102.6	5.390	63
103.2	5.650	69
103.7	5.950	75

When the reservoir elevation was 100.50 m, the following inflow hydrograph enters the reservoir. Route the flood hydrograph and calculate the (i) outflow hydrograph (ii) reservoir elevation vs time curve during the passage of the flood wave.

Time, hr	0	6	12	18	24	30	36	42	48	54	60	66
Discharge, m³/sec	10	20	52	85	70	57	40	30	15	10	7	5

Q18. Explain the Goodrich method of hydrologic reservoir flood routing. Mention the various steps used in this method of flood routing.

Q19. A reservoir has the following elevation (stage), discharge and storage relationships.

Elevation, m	Storage (x 10^6 m³)	Outflow discharge (m³/sec)
100.00	3.480	0
100.60	3.550	11
100.9	3.890	25
101.6	4.520	38
101.8	4.780	48
102.4	5.350	58
103.1	5.550	68
103.8	5.850	73

When the reservoir elevation was 100.60 m, the following inflow hydrograph enters the reservoir. Route the flood hydrograph by Goodrich method.

Time, hr	0	6	12	18	24	30	36	42	48	54	60
Discharge, m³/sec	10	24	54	85	68	47	40	30	15	10	6

Q 20. Pul's method of flood routing is also called as inflow-storage-discharge method. Why? Explain the use of this method in hydrologic reservoir routing.

Q21. Following data were recorded at a station in the river Brahmani.

Elevation, m	361	362	363	364	365	366	367	368	369	370
Storage, Mm³	55.5	105.6	180.6	245.5	350.5	450.0	604	750	915.8	993
Outflow, m³/s	2.5	6.9	15.5	22.4	40.5	65.9	90.0	150	162	204

Route the following inflow hydrograph for the reservoir given below by Pul's method.

Time, hr	0	6	12	18	24	30	36	42	48	54	60
Inflow, m³/s	13.0	36	54	104	120	100	96	75	35	21	12

Q22. Write short notes on use of Fourth Order Runge-Kutta method of flood routing.

Q23. What are the factors on which the storage of a channel routing depends?

Q24. The flow in a river section belongs to which of the following category of routing and mention why?

(i) Hydrologic reservoir routing (ii) Hydraulic channel routing.

Q25. With sketch differentiate between prism and wedge storage.

Q26. Write the Muskingum equation and explain how is it used in reservoir flood routing.

Q27. How are the weighing factor and storage time constant of a hydrologic channel routing by Muskingum method determined?

Q28. Explain in details the Muskingum method of hydraulic channel routing. Also mention the various steps used by this channel routing.

Q29. Following inflow and outflow data were recorded at a station in Mahanadi, Odisha. Estimate the value of K and x applicable for this station for use in Muskingum equation.

2001 Time, hr	0	6	12	18	24	30	36	42	48	54	60	66
Inflow, I, m³/s	5	10	24	45	35	20	12	10	8	7	6	5
Outflow, Q, m³/s	5	8	12	25	40	25	15	12	10	7	7	5

Q30. A reservoir reach has the value of storage time constant, $K = 13$ hr and the weighing factor, $x = 0.20$. Route the following inflow hydrograph which passes through it. Assume at the initial time of inflow hydrograph, the outflow discharge is 12 m³/s.

Time, hr	0	6	12	18	24	30	36	42	48	54
Inflow, m³/s	12	20	50	60	55	44	35	25	20	10

Q31. Using the inflow hydrograph and other data of Q. 30, compute the outflow at 30 hr duration by using Muskingum crest segment method.

Q32. Write short notes on the followings.

    (i) Weighing factor

    (ii) Storage time constant

    (iii) Muskingum routing equation

    (iv) Muskingum crest segment routing method

    (v) Finite difference method of channel routing

    (vi) Finite element method of channel routing

    (vii) Method of characteristics of channel routing

Q 33. With diagram explain how the following instruments are used in flood routing.

    (i) Mechanical flood router

    (ii) Electric analog routing machine

Q34. How does the digital computer help in flood routing?

Q35. Explain how the conceptual hydrograph development is done by flood routing.

Q36. In which way a linear reservoir varies from a linear channel.

Q37. What do you mean by travel-time-area histogram? With figure, explain how is it useful in development of instantaneous unit hydrograph? Also describe how is it prepared?

Q38. Enumerate the procedures of development of instantaneous unit hydrograph by Clark's method.

Q39. A watershed has the following characteristics:

(i)   Area = 120 km²

(ii) Time of concentration =24 hr

(iii) Storage time constant = 15 hr and

(iv) Inter-isochrone area distribution are as follows:

Travel time, hr	0 - 3	3 - 6	6 - 9	9 - 12	12 - 15	1 5- 18	18 - 21
Inter-isochrone area, km²	2.5	5.6	15.3	10.0	8.0	5.0	3.2

Derive IUH using Clark model.

Q 40. Describe in details how does the Nash's model help in derivation of instantaneous unit hydrograph?

Q41. A watershed of area 150 km² has the following effective rainfall hyetograph (ERH) and the corresponding direct runoff hydrograph which are given below. Determine the values of n and K by using Nash conceptual model. Also determine the ordinates of IUH.

Time, hr				0 - 4			4-8
Intensity of effective rainfall, cm/h				4.0			3.0

Time, hr	0	4	8	12	16	20	24
Ordinates of DRH, m³/s	0.0	15.0	40.0	50.0	20.0	5.0	0.0

References

Clark, C.O. 1945. Storage and the unit hydrograph. Trans. Am. Soc., Civil Engrs., Vol. 110: 1419-1446.

Kirpich, Z.P. 1940. Time of concentration of small agricultural watersheds. Civil Engineering, 10: 362.

Mahmood, K. and Yevjevich, V. (Ed.). 1975. Flow in Open Channels, Vol. 1 & 2,Water Resources Publication, Fort Collins,Colo, USA.

Nash, J.E. 1957. The form of instantaneous unit hydrograph. IASH, Publication No. 45, Vol. 3 & 4: 114-121.

Patra, K.C. 2001. Hydrology and Water Resources Engineering. Narosa Publishing House. New Delhi, pp. 561.

Panigrahi, B. 2013. A Handbook of Irrigation and Drainage. New India Publishing Agncy, New Delhi, pp. 602.

Streeter, V.L. and Wylie, E.B. 1967. Hydraulic Transients. McGraw-Hill, New York, NY.

Subramanya, K. 1991. Flow in Open Channels. Tata McGraw-Hill, New Delhi.

Subramanya, K. 2003. Enginering Hydrology. Tata McGraw-Hill Publishing Company Limited, New Delhi, pp. 392.

Suresh, R. 2008. Watershed Hydrology. Standard Publishers Distributors, Nai Sarak, Delhi, pp. 692.

Varshney, R.S. 1979. Engineering Hydrology. Nem Chand & Bros., Roorkee, U.P. India, pp. 917.

Zoch, R.T. 1937. On the relation between rainfall and stream flow. Monthly Weather Rev, Vol. 65: 135-147.

CHAPTER 10

Groundwater, Wells and Tubewells

10.1 Introduction

During rain or heavy irrigation, a part of the rainfall or irrigated water moves below the effective root zone of the crop which moves further below and adds to the water table lying above the impermeable bed or hard rock which is called as ground water reservoir. At times the accumulated surface water is also recharged to the soil reservoir both naturally as well as artificially that add to the underlying ground water reservoir. This groundwater is an important source of water for irrigation as well as domestic uses. It is the most abundant fresh water resource of the earth. Underground water reservoirs contain large stock of fresh water resources. Total quantity of groundwater within 800 m distance below the ground surface is more than 30 times the total water in all fresh water lakes, more than 60 times the total water in soil and other unsaturated rock materials, more than 300 times the water vapour in the atmosphere and 3000 times the average volume of all rivers and rivulets in the world (Lenka, 2001). But the distribution of groundwater is not uniform. At some places, it lies below a few metres whereas at other places, it is a thousand metres or more below the ground surface. In some places, there is abundant groundwater potential whereas at other places there is meager potential.

The above mentioned groundwater is the water reaching the groundwater reservoir from the infiltrated water of precipitation which constitutes an important component of the hydrologic cycle. It is only the part of the subterranean water that occurs in the saturated pores of the containing rock materials underlain by water held in a zone of aeration. It should be remembered that all the water present in soil below the ground surface is not groundwater.

10.2 Sub-surface Water Distribution:

The soil profile lying below the soil surface up to the impervious strata/bed rock (called as sub-surface) can be divided in to (i) zone of aeration (also called as unsaturated zone) and (ii) zone of saturation. Zone of aeration is also called as vadose zone where soil water is held in suspension. Zone of aeration contains three sub zones i.e. soil water zone, intermediate zone and capillary zone. Soil water zone contains soil water, intermediate zone holds pellicular and gravitational water and the capillary zone holds capillary water. Water stored in the zone of saturation is called as groundwater. Saturated zone has the upper boundary as groundwater table and lower boundary as the impervious/bed rock. The distribution of soil profile containing different distributions of sub-surface water is depicted in Fig. 10.1.

Water table, in general, lies approximately parallel to the topographical surface above it. Near the source of recharge, the groundwater level is raised upward forming groundwater mound whereas the level gets decreased gradually as one goes away from the recharge source. This form a cone of elevation which is just opposite to the cone of depression in case of a pumping/discharge well. The action of gravity tends to make the watertable a level surface after some period. However, frequent addition and withdrawal of ground water do not permit the water table to assume a level surface. It may coincide approximately

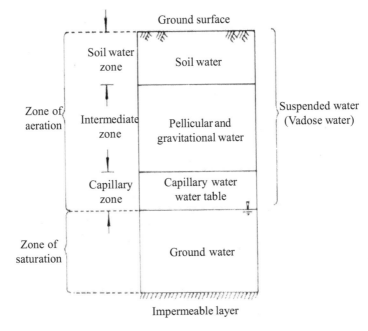

Fig. 10.1: Distribution of sub-surface water

with the top of zone of saturation in coarse gravel but a few centimeters to several meters below it in fine grained soils.

10.3 Geologic Formations for Groundwater Supply

Rock is the most important geologic formation that holds and supply groundwater for various uses. These rocks are of two types i.e. (i) unconsolidated type and (ii) consolidated type. In consolidated type, the particles are held tightly by compaction and cementation whereas in unconsolidated type, materials are loose. Examples of consolidated rocks are granite, sandstone, limestone *etc.* Examples of unconsolidated rocks are gravel, sand and clay. The geologic formation, depending on the quantity of water they hold and yield to a well when discharged, may be divided into four types. They are (i) aquifer, (ii) aquifuse, (iii) aquiclude and (iv) aquitard. Of these four formations, aquifer is very important.

10.3.1 Aquifer

The term aquifer is originated from the Latin word "Aquiferre" in which "aqui" stands for water and "ferre" stands for to bear. Thus, the term aquifer stands for "water bearing formation". It is defined as the saturated geologic formation that contains sufficient amount of groundwater and also transmits and yields appreciable amount to a well constructed in it. Among all kinds of geologic formations, gravel is the best and next to gravel, the best aquifer materials are sand, sandstone and limestone. Clay and fine silts are the most unproductive materials. Porous and permeable materials form good aquifers. Various types of aquifers are discussed subsequently in this chapter.

10.3.2 Aquiclude

This is the formation which contains sufficient amount of water due to porosity, but has very small capacity to transmit it being impermeable to flow of water. Clay is an example of aquiclude.

10.3.3 Aquifuge

This is the geologic formation which is neither porous nor permeable. There are no interconnected openings and hence it cannot transmit water. Examples of aquifuge are the massive compact rock, granites etc. without any fractures in it.

10.3.4 Aquitard

It is a semi-pervious type formation through which seepage is possible. It is the formation through which water moves at very small rate and thus yields insignificantly compared to an aquifer. Examples of aquitard are sandy clay soil, loamy clay soil etc.

It is to be noted that the definitions of aquifer, aquitard and aquiclude as mentioned above are relative. A formation that is considered as aquifer at a particular place say in arid zone may be considered as aquitard or aquiclude in another place where water is available in plenty.

10.4 Types of Aquifers

Groundwater aquifers are classified into the following four types.

- Unconfined aquifer

- Confined aquifer

- Semi confined aquifer and

- Perched aquifer

10.4.1 Unconfined Aquifer

The top most water bearing stratum, having no confined impermeable over burden lying over it, is called as unconfined aquifer. The unconfined aquifer is also called as non-artesian aquifer or water table aquifer. Water in these aquifers is at atmospheric pressure. When wells are constructed in these aquifers, the level of the water in these wells represents the watertable in the aquifer. Wells in these aquifers are called as gravity well or watertable wells or simply wells. Only saturated zone of this aquifer is of importance in groundwater studies. Recharge of this aquifer takes place through infiltration of precipitation from the ground surface.

10.4.2 Confined Aquifer

When an aquifer is confined at both its upper and lower surface by impermeable stratum such as aquifuges or aquicludes, it is called as confined aquifer. It is inclined so as to expose the aquifer somewhere to the catchment area at a higher level for the creation of sufficient hydraulic head. Confined aquifer is also called as artesian aquifer. Water in these aquifers is at pressure than the atmospheric pressure. Therefore, when a well is dug in these aquifers, water will rise to a level above the bottom of the upper confining stratum because of pressure under which water is held. The imaginary level to which water will rise in a well located in an artesian aquifer is known as piezometric level. Water is fed to the confined aquifer through recharge area which is exposed at the ground surface. The well constructed in these aquifers are called as artesian wells. Flowing well is a special type of artesian well and is found in the areas where the piezometric surface of the confined aquifer is at higher elevation than the natural ground surface. In these well, water automatically flows out and even gush out of surface for a reasonable height. In fact the term "artesian" is derived from the fact that a large number of such free flow wells were found in Artois, a former province in north France

10.4.3 Semi Confined Aquifer

A semi confined aquifer, also called as leaky aquifer, is a completely saturated aquifer that is bounded at above or at both above and bottom by a semi pervious layer or aquitards. Lowering of the piezometric head in leaky aquifer will generate a vertical flow of water from the semi pervious layer into the pumped aquifer.

10.4.4 Perched Aquifer

A special type of unconfined aquifer is perched aquifer. Perched aquifer occurs within the unconfined aquifer. It occurs when the groundwater body is separated from the main groundwater body by a relatively impermeable stratum of small areal extent. It contains small amount of groundwater. It generally lies above the main groundwater body. The top surface of the water held in the perched aquifer is known as the perched watertable. In groundwater exploration, a perched water table is often confused with a general groundwater body.

Different types of aquifers are illustrated in Fig. 10.2.

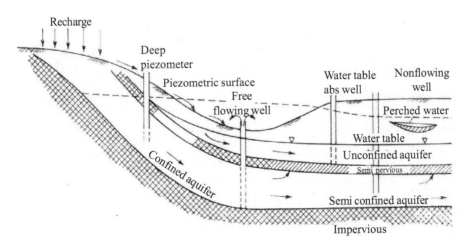

Fig. 10.2: Different types of aquifers and wells

10.5 Groundwater Yield

During the process of rainfall or irrigation, some amount of water infiltrates into the soil. This water moves downward and fills up the pore space present in the sub-soil in the formation. When all the pore spaces are filled up with water, then it is called as saturated soil. The water contained in these pores is drained by digging wells under the action of gravity drainage. However, it is found out that not all the water hold by the pores are drained out or yielded by the wells. Some amount of water is held by them due to their molecular attraction and this water is called as pellicular water.

10.6 Specific Yield

Volume of groundwater extracted by the gravity drainage from a saturated water bearing material is known as yield. Specific yield is defined as the ratio of yield to total volume of the material drained or dewatered and is given by the formula as:

$$Specific\ yield = \frac{Volume\ of\ water\ obtained\ by\ gravity\ drainage}{Total\ volume\ of\ material\ drained\ or\ dewatered} \times 100$$

10.7 Specific Retention

As discussed above, not all the water contained by the saturated material is drained out by the action of gravity drainage and some amount is retained by the material which is called as pellicular water. This quantity of water retained by the material against the action of gravity is termed as specific retention or field capacity. Like specific yield, this is expressed as the percentage of total volume of material drained. Specific retention is given by the formula:

$$Specific\ retention = \frac{Volume\ of\ water\ held\ against\ gravity\ drainage}{Total\ volume\ of\ material\ drained\ or\ dewatered} \times 100$$

Thus, the sum of specific yield and specific retention is equal to porosity.

10.8 Specific Capacity

Specific capacity of a well is the rate of flow of water from a well per unit of drawdown. It is discussed vividly in Chapter 11 of this book.

10.9 Storage Coefficient

The storage coefficient of an artesian aquifer is equal to the volume of water released from the aquifer of unit cross-sectional area and of the full height of the aquifer when the piezometric surface declines by unity. In a more common way it can be defined as the volume of water that an aquifer releases or stores per unit surface area of the aquifer per unit change in the component of head normal to that surface.

10.10 Well Irrigation

In India, ground water withdrawal from wells for irrigation has been practiced since long and it is still now an important source of irrigation. Most of the wells draw supplies from the first water bearing structure and discharge about 5 litres/sec (lps) depending on the static water table position and geological formations.

10.10.1 Advantages of Well Irrigation

Followings are the advantages of well irrigation (Panigrahi, 2011).

- It is owned by the farmers and as such has full control on sources of irrigation. It is also a handy source of irrigation.

- It is located nearer to the place where the irrigation water has to be served to crops and so conveyance loss is minimum.

- Particularly suitable for intensive irrigation to raise cash crops and is suitable to grow 2 to 3 crops from the same piece of land in a year.

- Low cost of construction.

- Reliable source of irrigation.

- It can be used as a conjunctive source with canal irrigation

- Supply is more uniform in quality and chemical composition and as such more helpful to crop growth and yield.

- Supply is free from weeds, turbidity and bacterial pollution which is associated with canal water

- Economical water use having higher delta and low duty.

- Benefit of rainfall can be availed. If there is chance of rainfall then well irrigation can be withheld.

10.10.2 Disadvantages of Well Irrigation:

The disadvantages of well irrigation are (Panigrahi, 2011):

- Associated higher repair with maintenance and hence higher operating cost.

- It has low discharge and consequently small area is irrigated by it.
- Water is debris / silt free. So more fertilizer is required to be added to crop field to raise crop yields.
- If the velocity of water from well exceeds critical velocity of water in the soil that convey water to well, then there is chance of well subsidence.

10.10.3 Location of Well

- Site of open well should be in the central position of the farm so that water can be served to all parts of farms easily without more conveyance loss.
- It should be located at a higher elevation so that water can flow by gravity to all parts of farms.
- Site should have good water bearing aquifer to supply sufficient yield to well.
- If there exists any well, then the near well should be at least 75 -175 m away from it to avoid any interference of wells and avoid low yield.
- It should be very much away from any source of pollution.
- There should not have any adjoining boundary lines within the circle of reference of the well.

10.10.4 Open Wells

Open wells are bigger in size than that of tubewell having less depth compared to tubewell and hence is associated with fewer yields. The yield of open well varies from 1 to 5 litres/sec. The diameter of the open well generally varies from 2 to 9 m. It draws water from any one pervious stratum. However, it is recommended that the well be constructed to a depth below ground water table such that the well does not dry during summer. From financial point of view, the depth is limited to 30 m. Contribution of water to well is from both bottom and sides of wells in unlined case. If well is lined, contribution is only from bottom.

10.10.5 Types of Open Wells

Open wells may be dug, bored, driven, jetted or drilled type.

Dug Wells

Dug wells are constructed manually or by semi mechanical means. They are dug down to the water bearing strata of the aquifer. They derive water from the formation close to the ground surface. Water percolates into the wells from all sides and bottom. The earthen sides are protected by circular cement rings. The depth of wells may vary from a few metres (shallow dug well) to several meters (deep dug well) and the size varies from 0.75 to 1.2 m in diameter. In large dug wells, the size may be 6 to 10 m (Lenka, 2001). Fig. 10.3 presents the view of a shallow dug well.

Fig. 10.3: Picture of a shallow open well

Bored Wells

Bored wells are constructed by hand or by power augers. At times, it means same as the drilled wells. In order to increase the yield of an open well, a bored well is set up at the centre of dug well. Such an arrangement is called as dug-cum-bored well. For such wells to be successful, a confined aquifer should be available below the dug well.

Driven Wells

Driven wells are wells that are constructed by driving a pointed screened cylindrical section also called as strainer and the pipes are connected to it directly into the water bearing formation. Water enters the well through the strainer. Diameter of the driven well varies from 3 to 10 cm. Depth is limited to 30 m. These wells are suited to domestic use since the yield is less (80 to 200 lit/min).

Jetted Wells

Jetted wells are constructed by the cutting action of a high velocity water jet. Small size wells of diameter 3 to 8 cm is constructed with this method. Construction of wells can be done either by hand operated or power operated auger.

Drilled Wells

Drilled wells are made constructed either by percussion or rotary drills and the excavated materials are brought to surface by bailer, sand pump, suction bucket or by hydrostatic pressure. They are constructed by means of drilling a deep bore of small diameter passing through many strata. Drilled wells are also called as tubewells.

Open well can be further classified as unlined or lined well. It may also be shallow or deep well.

Unlined Wells

It is also called as kacha well having no lining either at bottom or at sides. These wells have depth 3 - 3.5 m. They are constructed in stable pervious soil so that side subsidence will not occur and groundwater can be easily tapped. These wells are cheap and easy to construct. But their longevity is less since there is chance of side collapse. The flow enters the well both from sides and bottom. Fig. 10.4 represents the view of an unlined well.

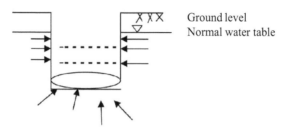

Fig. 10.4: Unlined well

Lined Wells

Depending on the types of lining materials used, lined wells may be pervious lined type or impervious lined type.

Pervious Lined Wells

Such wells have dry bricks or stone lining with open joints through which water enters the well. In sandy formation in order to avoid flow of sand surrounding the lining into the well, the space behind the lining is packed with brick ballast of size 20 -25 mm. These wells are comparatively small depth and are suitable for shallow water table and for gravel and sand formation. The flow of water into the well is both from bottom and sides. Flow is of both radial and spherical pattern. Wherever pervious wells are plugged at bottom, flow pattern is radial type only. Fig. 10.5 gives the view of pervious lined wells.

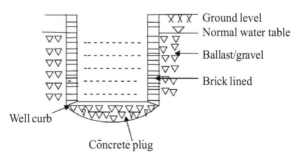

Fig .10.5: Pervious lined well

Impervious Lined Wells

In non alluvial soil, if the soil is sandy or soft and pervious, then there is chance of side collapse. So lining of sides of wells is necessary. Such wells need an impervious lining with brick masonry or stone masonry in cement or lime mortar or with reinforced concrete. Steining of 30-60 cm thick is supported on well curb which is also constructed of reinforced concrete. Impervious lined well have depth ranging from 5 to 30 m depending on water bearing strata. The flow towards well is through bottom and hence spherical. A better type of impervious well draws water from deeper sandy formation through a pipe sunk down from centre of the well (Fig. 10.6).

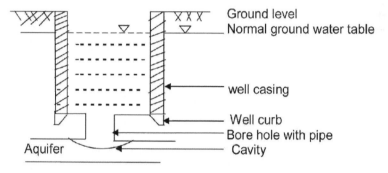

Fig .10.6: Impervious lined well

Shallow Well

A shallow well penetrates into the uppermost pervious stratum only from which it draws water. It is of less depth; yield is less and dries up quickly when discharged for a long period. It covers less area and is suitable for non rice crops.

Deep Well

It draws water from deeper layers and has high yield. It covers more area. It supplies water sufficient for paddy and other crops for irrigation. Like impervious lined well, a hole may be made at the centre of the well which draws up water from the deeper aquifer. Table 10.1 gives difference between the shallow and deep wells.

Table 10.1: Difference between shallow wells and deep wells

Sl.No.	Shallow well	Deep well
(i)	Initial cost low	Initial cost high
(ii)	Cross sectional area at bottom of well less which limits high supply to well	Cross sectional area high due to cavity formation and so supply more water to well
(iii)	Suitable where ground water table is very near to the ground surface	Suitable even if ground water table is deep below ground surface.
(iv)	Yield is less	Yield is high
(v)	Quality of water is not good	Better quality water

10.10.6 Yield of an Open Well

Safe yield of a well is the yield taken out as discharge from a well without causing any mining. It is expressed as m^3/ hr or lit / min. Under stable condition, water level in the well coincides with that in the ground water table in the adjoining aquifer. But as the pumping in the well continues, water level in the pumping well falls immediately where- as the levels in adjoining aquifer remains at higher elevation. This forms a cone of depression around the well. Difference between the water level in the well and that in the aquifer is known as draw down or depression head (Slichter, 1989). The more the rate of withdrawal, the more is the depression head. The depression head is maximum at the centre of the well and gradually decreases as the radial distance from well increases. At radius of influence, it is zero. There are two tests for determining the yield of an open well. They are pumping test and recuperation test.

10.10.6.1 Pumping Test

It is an accurate method to determine the safe yield of an open well. In this method, water level in the well is decreased through heavy withdrawal till maximum safe depression head is reached. At this stage, the pump speed is regulated so that the rates of withdrawal equal the recharge. Thus constant discharge water level is maintained in the well which creates equilibrium condition. The ratio of pump discharge per hour under this equilibrium condition gives the yield of the wells at a particular draw- down.

Applying Darcy's law

$$V = K\,i = K.\,h\,/r \tag{10.1}$$

where V = velocity of flow of water into the well at equilibrium condition, i = hydraulic gradient equal to ratio of drawdown (h) to radial distance (r) from wells and K = constant called as hydraulic conductivity of the aquifer.

Discharge (Q) is given as:

$$Q = A.V = A.\ K.\ h/r \qquad (10.2)$$

Hence, considering unit width of aquifer, we get $A = r$

$$Q = r.\ K.\ h\ /\ r$$

$$Q = K.\ h \qquad (10.3)$$

The yield of a well can also be expressed in another way.

Equation (10.3) can be written as:

$$Q = (K/A)\ A.\ h \qquad (10.4)$$

where K/A is expressed as specific yield of the well (m³ /hr) per m² of the aquifer through which water enters into the well from aquifer through unit depression head. The values of K/A for different soils are given in Table 10.2 (Sharma, 1987).

Table 10.2: Values of K/A for different types of soils

Soil type	Value of K/A, (m³ / hr per m² / unit depression)
Clay	0.25
Fine sand	0.5
Coarse sand	1.0

Sample Calculation 10.1

Design an open well in fine sandy soil with safe yield 0.5 m³/hr /m²/ unit depression head so that an yield of 0.004 m³/sec is discharged from well with depression head h = 4 m

Solution

Using Eq (10.4) we have $Q = (K/A).\ A.\ h$

$Q = 0.5$ (m³ /hr /m² / m depression) $x\ A\ x\ 4$ m

  $= 2$ A m³ /hr/m²

Given $Q = 0.004$ m³ /sec $= 14.4$ m³ /hr

So 14.4 m³ /hr $= 2$A m³ /hr /m²

Hence, A $= 7.2$ m²

Area $A = (\pi / 4).\ d^2$ where $d =$ diameter of well

$(\pi/4)\ d^2 = 7.2$ m²

$$d = \frac{\sqrt{4 X 7.2}}{\sqrt{3.142}}$$

Hence, diameter, $d = 3$ m

10.10.6.2 Recuperation Test

It is used where it is difficult to maintain constant water level in the well. In this method the water level is decreased to any level below the normal level by pumping and then pumping is stopped. When pumping is stopped then water level in the well starts rising (Fig. 10.7). Time taken by water in well to attain any draw down below ground water table normal position is recorded. Then the well discharge formula (Eq 10.5) as given below is used.

$$Q = \frac{2.3}{t} \log_{10} \frac{h1}{h2} A. h \qquad (10.5)$$

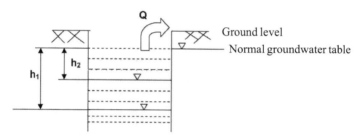

Fig. 10.7: Yield of an open well by recuperation test

where t = time taken by water to attain drawdown below normal ground water table position, h_1 and h_2 are initial and final ground water table position, h = average depression head and A = area of cavity of well bottom through which water enters the well and is given as:

$$A = \frac{4}{3} \left(\frac{\pi}{4}\right) d^2 \qquad (10.6)$$

where d = diameter of well

Sample Calculation 10.2

Calculate the average yield of a well of 3 m diameter by recuperation test. The water level in well is decreased to an extent of 2.5 m. The rate of rise of water is 1.0 m/hr. Assume average depression head, $h = 3$ m.

Solution

Given $h_1 = 2.5$ m

After one hour $t = 1$ hr, $h_2 = 1.5$ m

$A = (4/3) \cdot (\Pi/4) \cdot 3^2$ and $h = 3$ m

$$Q = \frac{2.3}{t} \times \log_{10}\left(\frac{2.5}{1.5}\right) \times \frac{4}{3} \times \frac{\pi}{4} \times 9 \times 3$$

$$= 9 \times 3.142 \times 2.3 \log_{10}(1.67)$$

$$= 26 \text{ m}^3/\text{hr} \approx 7 \text{ lit/sec}$$

10.11 Tubewell Irrigation

10.11.1 Advantage of Tubewell Irrigation

Advantages of tubewell irrigation are:

(i)   It is an independent source of irrigation which a farmer can at his convenience dig and maintain.

(ii)   Some patches of lands which cannot be easily irrigated by other sources can be irrigated by it.

(iii)   Volumetric rate of irrigation water is possible resulting in economical and optimum utilization of water.

(iv)   Water loss in conveyance system is less since conveyance channels are generally lined.

(v)   Since seepage and conveyance losses are minimum there is less water logging problems.

(vi)   Irrigation capital investment is less compared to canal irrigation.

(vii)   There is quick return from well irrigation. Canal irrigation takes 5-6 years to supply irrigation water where as tubewell irrigation is possible in the same year where tubewell is setup.

(viii) Optimum utilization of rain water is possible.

(ix)   Water is suitable for domestic use as quality of water is good.

10.11.2 Disadvantages of Tubewell Irrigation

Following are the disadvantages of tubewell irrigation.

(i)   Life of tubewell is limited to 10-15 years.

(ii)   Operating cost is more and so costlier than canal water.

(iii) Cover less area compared to canal irrigation.

(iv)   Frequent repair and maintenance is required.

(v)   Not well suited in areas where elevation is not available since water moves by gravity.

(vi)   Tube well water carries less silts and hence crop yield is less. In order to increase crop yields, crops needs more fertilizers.

(vii) Chances of theft of well components are more.

(viii) Not suitable for high water requiring crops.

10.11.3 Suitability of an Area for Tubewell Irrigation

Tube well irrigation is suitable where the sub soil formation is good for storing water and the aquifer is suitable to supply ample ground water to the well. Water bearing stratum may be confined or unconfined. If aquifer has impervious layer at the bottom and top is open to atmosphere, then it is unconfined. A confined aquifer is confined between two impervious layers. The best place for a tube well digging is sandy or sandy aquifer. If the subsoil strata has a 30-50 m thick coarse sand stratum, then it is ideal to dig well. The tube well has a low yield capacity generally $0.042 \, m^3$ / sec., i.e 42 lps which can generally command irrigation to 160 ha of non paddy crop at 100% irrigation intensity after taking care of rainfall. A deep tube well can supply irrigation water 300 ha of non paddy crop. While selecting a site for tubewell, care should be taken so that it is dug in a high elevation to facilitate in gravity supply of water. It should also be located in the central position of the farm so that water can easily be conveyed to all parts of the farm with minimum conveyance loss. In addition to above mentioned points, lithological test, aquifer test and location of aquifer boundary should also be taken into account before selection of site of tube well construction (Ahrens, 1958; Benison, 1947; Dylla and Muckel, 1967).

10.11.4 Types of Tubewells

For irrigation point of view tubewells are of two types (i) Direct irrigation tube well and (ii) Augmentation tube well.

10.11.4.1 Direct Irrigation Tubewell

This is the tubewell which provides irrigation directly to crops in the farm. It may be classified as shallow or deep tube well.

10.11.4.2 Augmentation Tubewell

In rainy season, canals run almost full and there is no shortage of supply to crops. But in winter and summer the canal supply is reduced. At this time the ground water can be pumped in conjunction with canal water. Tube wells are used to feed to canal supply which can be used for irrigation. The augmented tube well can serve for augmentation of canal water supply or as a direct source for irrigation. It can also be used as a viable source for ground water recharge.

Tube wells can also be classified as shallow and deep tubewells.

10.11.4.3 Shallow Tubewells

As the name suggests, these are shallow in depth, usually they tap water from a few layered aquifers. They have diameter usually same from top to bottom. They cover less irrigation areas as their discharge capacity is less.

10.11.4.4 Deep Tubewells

These well are deeper in depth penetrating several multilayered aquifers. They draw water from deep confined aquifers. As augmentation tube well, they can be installed in adjacent to rivers. They are spaced at about 500 m from one another. For a given drawdown the deep tubewell yields more discharge than a shallow tubewell (Fig. 10.8).

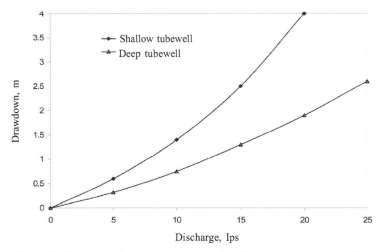

Fig. 10.8: Discharge drawdown curves of shallow and deep tubewells

Tubewells can further be classified as (i) cavity type wells, (ii) strainer type wells and (iii) gravel packed wells or shrouded well.

10.11.4.5 Cavity Type Tubewells

A cavity type tubewell draws water from the bottom of the well. Water does not enter into the well through the sides as is done by a screen or strainer well. The flow, thus, in a cavity well is spherical not radial. Cavity well draws water from only one water bearing stratum. Such a well is economical as it requires only plain well pipe not screened pipe which is very costly. The plain pipe is introduced into the bore hole up to a depth of water bearing stratum from which water is to be drawn.

A cavity type tubewell essentially consists of a pipe bored through the soil and resting on the bottom of a strong clay layer that acts as aquiclude. A cavity is formed in water bearing sand stratum lying below the clay layer from where water enters into the pipe (Fig. 10.9). The cavity is developed by drawing out apart of the sand from this layer so that such a cavity works as the storage reservoir for pumping. The cavity of such a tubewell needs to be developed carefully and slowly by using a centrifugal pump. Cavity type tubewell has low discharge and is suited for domestic purposes.

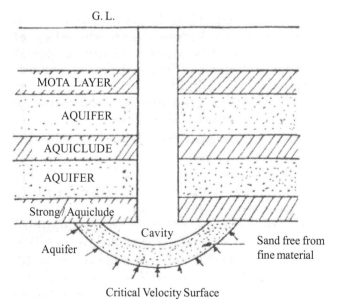

Fig. 10.9: Cavity type tubewell

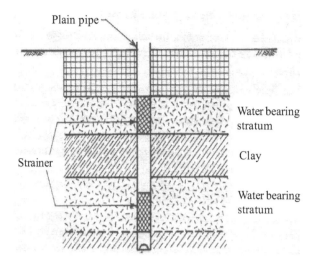

Fig. 10.10: Strainer type tubewell

10.11.4.6 Strainer Type Tubewell

In strainer type tubewells, drilling continues through several types of layers. Drilling continues till a suitable water bearing strata is obtained. Plain pipes are introduced into the bore hole and placed in the strata having less permeability like clay and strainers are located in good water bearing strata having high permeability like sand and gravel as shown in Fig. 10.10. A bail is provided at the bottom. Water enters into the well through the strainers from the sides and the flow is radial not spherical as is in case of cavity wells. For the strainer type tubewells, drilling is to be started with pipes of larger diameter (called as casing of pipes) than the strainer and these are taken out after the strainers are installed.

A strainer essentially consists of a perforated or slotted pipe with a wire mesh wrapped around it with a small annular space between the two. The wire screen prevents the sand and other coarser particles larger than its mesh size from entering the well. This facilitates in obtaining higher flow velocities and getting clear water. A strainer type tubewell is generally unsuitable for fine sandy strata.

10.11.4.7 Types of Strainers

There are various types of strainers available for use in tubewells. Each type has its own advantages depending on its cost and efficiency to allow water to enter into the well. Some of the common types used in field are:

- Coir rope strainers

- Agricultural strainers or Ashford strainers

- Cook-Tej brass slotted strainers
- Phoenix strainers
- *P.V.C.* strainers
- Bamboo strainer
- Porous concrete pipe strainer
- Pre-packed filter strainer
- Coconut coir cord strainer
- Brownlie strainer
- Layne and Bowel strainer
- Legget strainer
- Polythene strainer

Coir rope strainer: Coir rope strainers are low cost strainers, generally used for shallow tubewells. Such a strainer consists of a cylindrical frame made of iron rods or bamboo strips, wound round by coir rope of 3 to 5 mm diameter. However, this strainer has low durability (3 to 5 years) and needs to be changed repeatedly. Coating the iron rods with bitumen and replacing coir rope by nylon rope can increase the life of the strainer.

Brownlie strainer: Brownlie strainer is considered to be the best type of strainer. It consists of a polygonal convoluted perforated steel pipe surrounded by a wire mesh consisting of heavy parallel copper wires woven with copper ribbons with a little space in between the two.

Layne and Bowel strainer: Layne and Bowel strainer is a strong but costly type strainer. It consists of a perforated steel or wrought iron pipe over which wedge-shaped steel wire is wound. The strainer pipe joints are made by screwed collars.

Legget strainer: In legget strainer a cleaning device, in the shape of cutters which can be turned in the slits, is provided. These cutters operate from the ground surface and clean the perforations of the strainer clogged by any solid matters.

Phoenix strainer: Phoenix strainer is made of mild steel tubes with slots of desired width. Chromium plating is done to prevent the chemical action of salts.

Agricultural strainers or Ashford strainer: It is a very delicate type of strainer and has long life. It essentially consists of a perforated pipe with a wire wound around it, to provide annular space between the perforated pipe and the wire mesh which is shouldered to the wire. The wire mesh is further protected and strengthened by a wire net over the mesh.

Cook-Tej brass slotted strainers: This type of strainer is the costliest strainer since brass is used for their construction. They are not easily corroded by water and hence have longer lives. The slots on these strainers are about 2.5 cm in length and are about 0.5 cm apart. Width depends on the type of sand strata. The patterns of slots may be horizontal or vertical and may be in line or staggered. (Fig. 10.11).

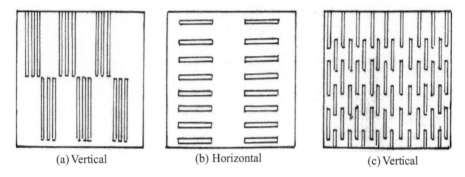

| (a) Vertical | (b) Horizontal | (c) Vertical |

Fig. 10.11: Different patterns of slots in strainer

P.V.C. **strainer:** Strainers made with *P.V.C.* material are useful in case of aquifers with poor quality water. Slot size in such strainers should be carefully selected.

Porous concrete pipe strainer: These types of strainers are prepared from 1:6 cement gravel mixture. It is reinforced with six equally spaced longitudinal M.S. bars to the ends of which flanges are welded for making joints of the pipes. These types of strainers are not hydraulically efficient.

Pre-packed filter strainer: Its cost is very high. However, it gives satisfactory performance. It consists of a slotted pipe to the outer surface of which pea gravel is bonded by means of a special adhesive.

Coconut coir cord strainer: It is a comparatively cheaper type of strainer. Generally it is 10 cm in internal diameter. It consists of steel frame fabricated by welding the ends of eight numbers of 6 mm thick M.S. bars to the periphery of 13 cm long pieces of steel pipe, around which a coconut coir cord of 6 mm diameter is wound.

Bamboo strainer: Bamboo strainer is a cheaper version of steel casing and brass strainer. In bamboo strainer case the brass strainer is replaced by bamboo strips made as a cylindrical frame by fixing strips of bamboo with nails around iron rings. Instead of nails, thin wires can be used to tie the bamboo strips with the rings. The rings are placed inside and the bamboo strips are tied over it.

10.11.4.8 Gravel Packed Tubewells

These wells are improved form of strainer type tubewells. These wells are also called as shrouded wells. Shrouding refers to the filling of the surrounded strata of the strainer with a material coarser than that of the natural stratum which is usually gravel of suitable size (Fig. 10. 12). The purpose of gravel packing is to increase the effective diameter of the well as a result of which higher discharge can be obtained. In natural gravel packed well, gravels are available in the strata communicating water to the well whereas in artificial gravel packed well, gravels of suitable size are put in the space between the casing pipe and the strainer.

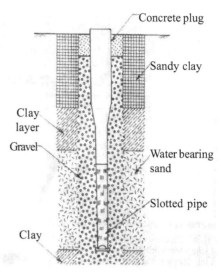

Fig. 10.12: Gravel packed well

10.11.5 Design of Tubewells

Good design of a tubewell aims to ensure an optimum consideration of performances, long service life and reasonable installation and operation cost. The tubewells should give highest yield from the aquifer and should give optimum efficiency in terms of specific capacity. Various design parameters of a tubewell/water well are (i) Well structure (ii) Well diameter (iii) Well depth (iv) Length of well screen (iv) Well screen slot opening (v) Well screen diameter (vi) Gravel pack design and (vii) Pumping arrangement. Gravel pack design is the most important part of design of tubewell. Gravel pack design ensures stability of well so that side collapse as tubewell boring is avoided. At the same time, it filters water from aquifer so that good quality water is obtained from the well.

10.11.5.1 Well Structure

It consists of two main elements (i) the upper portion that houses the pumping equipments and the vertical section through which water flows upward from the aquifer to the level where it enters the pump and (ii) lower intake portion of the well. In lower intake portion, a well screen is used which allows water to enter into the well at low velocity, prevents sand and other fine materials and prevents caving-in of the loose formation materials. The intake portion consists of a bore hole drilled into the aquifer to an adequate depth.

10.11.5.2 Well Diameter

Well diameter is generally governed by the size of the pump. Diameter of the well should satisfy the following condition

(i)   Opening should be large enough to accommodate pump and pipes so that it will be easy for installation and efficient operation. The opening of the well should be such that after inserting the casing and housing pipe, 2-3 cm annular space is left for gravel packing.

(ii)   Diameter of the intake section should be large enough as to ensure good hydraulic efficiency of the well. However, the diameter of the welland well pipe is decided from the considerations of permissible flow velocities through the tube. Since, strainers are installed at different levels, the velocity of water in the well pipe of fixed size will not be constant but will be increasing towards the top. Hence, it is theoretically possible to reduce the size of the tube from the top towards the bottom such that there is a uniform velocity throughout the entire length of pipe. It is thus not necessary to have a uniform size of well diameter. In deep wells, well diameter below lowest anticipated pump setting, may be reduced to affect some economy. The lowest anticipated pump setting should be carefully estimated keeping in view future decision of ground water table. In actual practice, it is difficult to install varying size pipes, and hence, a single sized or at the most a two sized pipe may be used. Table 10.3 (Das Gupta, 1987) gives the recommended size of well casing /pipe for different well yields

Table 10.3 Recommended well casing for different wells yields

Anticipated well yield, lpm	Nominal size of pump bowl, cm	Size of well casing	
		Minimum (cm)	Optimum(cm)
Up to 400	10	12.5	15
400-600	12.5	15	20
600-1400	15	20	25
1400-2200	20	25	30
2200-3000	25	30	35
3000-4500	30	35	40
4500-6000	35	40	50
6000-10000	40	50	60

The diameter should be so chosen that it gives desired percentage of open area in the screen (15 – 18%) so that entrance velocity near the screen is within 3 to 6 cm /sec. This will reduce the well losses and hence draw down. Table 10.4 gives representative open area of well screen for various screen sizes (Das Gupta, 1987).

Table 10.4: Representative open area of well screen for various screens sizes

Nominal screen size, inches	Intake area /ft length of screen, sq inches				
	Slot size				
	10	20	40	60	100
3	10	19	32	42	55
10	36	65	110	143	166
20	47	88	156	209	252
30	57	108	192	268	379

Screen size is designated by the size of steel pipe through which screen may be set. Slot number denotes width of openings in thousandths of an inch.

10.11.5.3 Well Depth

Expected depth of a well can be determined from well log of the place where well will be dug or from well log of a nearby well. It is desirable to have full penetration of aquifer so that specific capacity would be higher. Under such a case, the well is called as a fully penetrating well

However if the available screen length is not large enough to fully penetrate the aquifer, it should be centered between top and bottom of the aquifer. Also when the lower portion of the aquifer contains poor quality of water, well screen should be deeper to avoid pumping poor quality water.

10.11.5.4 Length of Well Screen

Optimum length of well screen depends on thickness of aquifer available, draw down and nature of aquifer materials as described below.

(a) Homogeneous Artesian/Confined Aquifer

In this case maximum draw down is the distance from initial piezometric level to top of confined aquifer. Screen length of the aquifer depends on the thickness of the aquifer as provided in Table 10.5 (Das Gupta, 1987).

Table 10.5. Length of screen as varies with thickness of aquifer

Thickness of aquifer	Length of screen
<7.5 m	70% of aquifer thickness
<7.5 – 15 m	75% of aquifer thickness
>15 m	80% of aquifer thickness

The best result is obtained by centering the screen section in the aquifer. However to obtain high yield, screen length or strainer may be provided in the entire thickness of confined aquifer.

(b) Non Homogeneous Artesian Aquifer

Most permeable section of the aquifer should be screened.

(c) Homogeneous Unconfined / Water Table Aquifer

To obtain higher specific capacity, a screen as long as possible may be used. It is preferred to use screen in the bottom one third length. In some cases, bottom half may be screened.

(d) Non Homogeneous Unconfined Aquifer

Principles of design are same as case (ii). i.e. non homogeneous artesian aquifer except that screen should be located in the lower most portion of permeable section.

10.11.5.5 Well Screen Slot Opening

For homogeneous aquifer slot sizes that contain 40-50% sand are preferable. For non-corrosive water, select higher size opening and for corrosive water, select lower opening. For homogeneous aquifer consisting of gravels and sand, slot openings that retain 30-50% aquifer materials should be selected.

10.11.5.6 Well Screen Diameter

Diameter of the intake portion of the well can be varied without affecting specific capacity of well yield. Doubling the diameter of intake section can increase the well yield by about 10%. Screen length depends on thickness of water bearing strata. Screen opening depends on gradation of sand, where as screen diameter depends on factor that total area of screen opening must be such that the entrance velocity of water is 3 – 6 cm/sec. If velocity is 3-6 cm/sec, then friction loss in screen opening will be less, rate of incrustation will be minimum and rate of corrosion will be minimum.

10.11.5.7 Gravel Pack Design

Design of gravel pack of well consists of the following steps (Anonymous, 1966; Das Gupta, 1987; Ellithorpe, 1970):

(i)   Prepare sieve analysis curves for all strata of aquifer. Determine the stratum composed of the finest sand and select the grading of gravel pack on the basis of sieve analysis.

(ii)   Multiply 70% size of sand by a factor in between 4-6: 4 if formation is fine and uniform and 6 if it is coarse and non uniform. If formation sand has highly non-uniform gradation and include silt, then multiplier may be 6-9. If 70% sand is retained by a mesh of 0.013 cm, then multiply 0.013 with 5. Hence $D_{70} = 0.065$ cm.

(iii)   Construct grading curve for gravel pack by trial and error. Through initial point on gravel pack curve and D_{70}, draw a smooth curve representing a material with a uniformity coefficient of 2.5 or less.

Uniformity coefficient (C_u) is expressed as:

$$C_u = \frac{D_{60} \ (40 \ percent \ retained)}{D_{10} \ (90 \ percent \ retained)} \qquad (10.7)$$

(iv)   Prepare specification of gravel pack material by selecting 4 or 5 sieve sizes that cover the spread of curve.

(v)   As a final step, select a size of well screen size openings that retains 90% of gravel pack. Width of slot ranges from 1.5-4.0 mm and length from 5-12.5 m. Thickness of gravel pack $\approx$ 10-20 cm.

10.11.6 Pumping Arrangements

To lift water from the tubewell, three types of pumps are generally used. They are (i) Centrifugal pump, (ii) Bore hole type pump and (iii) Jet pump.

A centrifugal pump lifts water from the lower level to a higher level by creating required pressure with the help of centrifugal action. The maximum suction head under which the pump practically works effectively is about 6 to 8 m. So, this pump is used in places where the fluctuations of water table plus the depression head is limited to a maximum value of 8 m. as shown in Fig. 10. 13.

Generally monoblock centrifugal pumps are used to lift water from the well. The pump may be placed below the ground level and just above the highest water level in the well. It should not be submerged in water. Highest water level in the well is the maximum rise level of water in the well which is very near to the ground surface and it occurs in the rainy season. Centrifugal pump arrangement can be done for shallow wells not for deep wells where water is at greater depth. Such type of pumping arrangement is not now used. Rather bore hole type pump are now used to lift water from the well.

Bore hole type pump are mounted on a vertical shaft and are driven by a motor. They are small in size and can be easily lowered in the casing pipe itself. The top 20 to 30 m of the bore hole and the casing pipe is generally kept wider than the remaining normal bore so as to accommodate the pump bowl in the casing pipe.

Bore hole pumps are of two type i.e. (i) submersible pump and (ii) turbine type pump. In submersible pump, the motor and the pump are both attached together and lowered inside the bore, whereas in turbine type, the pump is driven by a direct coupled electric motor and is placed at the top of the line shaft at the ground level. A submersible type bore hole pump is very popular and relatively cheaper than the monoblock centrifugal pump.

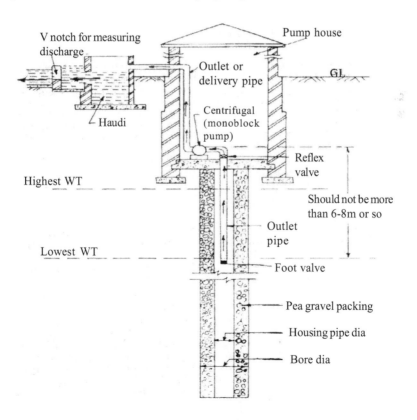

Fig. 10.13: Section of a tubewell using monoblock centrifugal pump

Jet pumps are also used for lifting water from the smaller tubewells installed for individual domestic supplies where water is not available within 8 m suction lift. Jet pumps can be used for suction lift of 6 to 30 m or so. A jet pump consists of centrifugal pump and a jet mechanism or ejector. In this assembly, motor and pump constitute a small unit like a centrifugal monoblock pump, and is placed at the ground level. Discharge of jet pump is about 20 lit/min against a head of 15 to 30 m above the ground level. The efficiency of jet pumps are not high and so are not generally used in irrigation tubewells. The efficiency is about 35 percent as compared to 65 to 85 percent for other types of pumps (Garg, 2012).

Horse power of the pump motor is computed as:

$$H.P_{motor.} = \frac{\omega \cdot Q \cdot H}{75 \times Pump\ efficiency \times Drive\ efficiency \times Motor\ efficiency} \quad (10.8)$$

where, ω = unit weight of water, 1000 kg/m³; Q = discharge to be delivered, cumecs and H = total head against which the pump has to work, m.

Sample Calculation 10.3

Design a tubewell to deliver 0.042 cumecs of water at a depression head of 5 m. The average water level is 10 m below the ground level in post monsoon season and 15 m in pre-monsoon season. The well log data obtained during investigation is presented below.

Depth, below ground level, m	0 - 13	13 - 18	18 - 30	30 - 50	50 - 65	65 -75	Below 75
Types of strata	Clay	Sand	Clay + kankar	Coarse sand	Clay	Medium sand	Clay + sans stone

There is a horizontal length of pipe 0f 10 m connected to the pump at ground level as deliver pipe to discharge water to the crop field.

Solution:

Design of well diameter and well pipe: Assume the velocity of flow in the well pipe to be 2.5 m/sec.

Given discharge = Q = 0.042 m³/sec

Hence, area of cross section of the pipe, A = Discharge/Velocity = 0.042/2.5 = 0.0168 m²

Diameter of the pipe = $d_w = \sqrt{\dfrac{4\,A}{\pi}} = \sqrt{\dfrac{4 \times 0.0168}{\pi}}$ = 0.1462 m = 14.62 cm

Use 15 cm diameter well pipe.

For easy in operation of entering the casing, housing pipe etc and for gravel packing add another 5 cm to the well pipe diameter so that the well size becomes 20 cm. Thus a bore whole of 20 cm diameter need to be dug.

Length of the strainer: From the well log data is found that the aquifers 30 – 50 m and again 65 – 75 m are good aquifers and they are confined between the clay strata. Thus, the data reveals that it is confined aquifer.

The total thickness of the aquifer = (50 – 30) + (75 – 65) = 30 m

Assuming the homogeneous confined aquifer and since the total thickness of the aquifers is 30 m (> 15 m), 80 percent of the total aquifer thickness may be provided with well screens (Table 10.5). Thus, the length of the strainer/well screen = 80 percent of 30 m = 24 m.

Check for entrance velocity: Assume 15 percent of the opening area in the screen as slot openings which comes to π x 0.15 x 15% = 0.0707 m².

So entrance velocity = $\dfrac{Discharge}{Length\ of\ strainer \times slot\ opening\ area} = \dfrac{0.042}{24 \times 0.0707} = 0.024\ m/sec$

Entrance velocity is computed as 0.024 m/sec = 2.4 cm/sec.

Since this value lies between 2 to 3 cm/sec, the entrance velocity checked is okay.

Type of pump required: The fluctuation of water table during pre-monsson and post monsoon season is 15 – 10 = 5 m and depression head given is 5 m. Thus total head comes to 5 + 5 = 10 m which is more than 8 m. Therefore, centrifugal pump will not work. Use a bore hole type pump to lift water from the tubewell.

Horse power requirement of the pump motor: Eq. (10.8) is used to estimate the horse power requirement of the pump motor.

Total head against which motor has to work = H = Maximum depth of water table + Depression head + Velocity head + Losses

Maximum depth of water table = 15 m

Depression head = 5 m

With discharge of 0.042 m³/sec, and using 15 cm diameter well pipe, the actual velocity of flow in pipe is $= V = \dfrac{0.042 \times 4}{\pi \times (0.15)^2} = 2.38 \; m/sec$

Velocity head $= \dfrac{V^2}{2g} = \dfrac{(2.38)^2}{2 \times 9.81} = 0.289 \; m$

Losses of head in pipe = Loss of head due to friction in pipe (h_f) + Losses at bends in strainers etc. which may be assumed as 25 percent of h_f where h_f is given as:

$$h_f = \dfrac{f'.l.V^2}{2.g.d}$$

where $f = 0.024$ and l = length of the pipe = total vertical length of pipe including blind pipe and strainer + horizontal length of delivery pipe = 75 + 10 = 85 m

Hence, $h_f = \dfrac{0.024 \times 85 \times (2.38)^2}{2 . \, x \, 9.81 \, x \, 0.15} = 3.93 \; m$

Losses in strainer and bends = 0.25 x 3.93 = 0.98 m

Total head losses = 3.93 + 0.98 = 4.91 m

Hence, total head, H = 15 + 5 + 0.289 + 4.91 = 25.199 = 25.2 m

Assume pump, drive and motor efficiencies as 0.65, 0.80 and 0.80, respectively.

So horse power requirement of the motor s calculated ad:

$$H.P_{motor} = \dfrac{\omega . Q . H}{75 \times Pump \; efficiency \times Drive \; efficiency \times Motor \; efficiency}$$

$$= \dfrac{1000 \times 0.042 \times 25.2}{75 \, x \, 0.65 \, x \, 0.80 \, x \, 0.80} = 33.9 \equiv 35 \; H.P.$$

Hence, use a 35 H.P. motor to lift water from the deep tubewell.

10.12 Investigations for Groundwater Development

For groundwater development of an area, it is essential to have knowledge on occurrence and movement of groundwater. Moreover, the hydrologic parameters of the aquifers and geological conditions of the area are required to decide the location and types of wells. Groundwater exploration studies will help in determining the prospective areas for setting up tubewells. This is often referred as groundwater prospecting. Methods of exploration of groundwater are given in the chart (Fig. 10.14) as below:

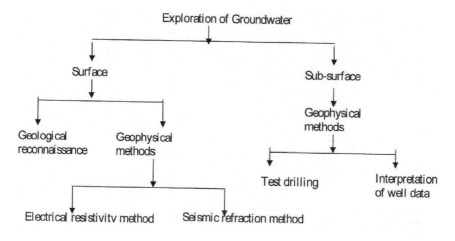

Fig. 10.14: Chart showing methods of exploration of groundwater

10.12.1 Geological Reconnaissance

It helps in understanding the geology of the area in relation to possible occurrence of groundwater. A study of structural geology in conjunction with stratigraphy to locate possible water bearing formations is required. Stratigraphy helps in locating the position and thickness of water bearing formations and confining beds. Sand, gravel etc. are the best aquifers whereas clay, shale and crystalline rocks do not yield sufficient water for explorations.

The geological studies by which the groundwater may be investigated would comprise of making a detailed geological mapping of the area , which may give accurate knowledge of the existing geological structures, including the lithological character and petrological features of various underground rocks and strata.

10.12.2 Surface Geophysical Methods

Surface geophysical methods also help to estimate where groundwater occurs. It also helps in assessing the possible areas where good quality groundwater exists. Though these methods are not always successful but they are less costly than the sub-surface investigations. These methods were first used in exploration of possible sites for petroleum and mineral industries. There are mainly two types of surface geophysical methods used for groundwater exploration. They are (i) electrical resistively method and (ii) seismic refraction method. Principles of these methods are outlined here. For detailed procedures for the tests, any standard Groundwater Hydrology Book may be referred.

10.12.2.1 Electrical Resistivity Method

Electrical resistivity makes use of the fact that water increases the conductivity of the rocks and thereby decreasing their resistivity. Hence, if it can be established geologically, that the same rock formation is existing for a certain depth and by electrical testing it is found that the resistivity is decreasing below this depth, then it can be easily concluded that the water is present below this later depth as decided by electrical testing. The resistivity of rocks at various depths can be calculated based on the following principles. .

In this method, four electrodes are inserted into the ground. Out of these four electrodes, the outer two electrodes are called current electrodes (C_1 and C_2) and the inner two are called as potential electrodes (E_1 and E_2). Current electrodes consist of metal rods driven into the ground. Potential electrodes are porous cups filled with a saturated solution of copper sulphate. Because of potential difference between the inner two potential electrodes, current will flow forming current lines. Similarly, equipotential lines are formed which are the lines joining points of equal potentials. The current lines are orthogonal to the equipotential lines (Fig. 10.15).

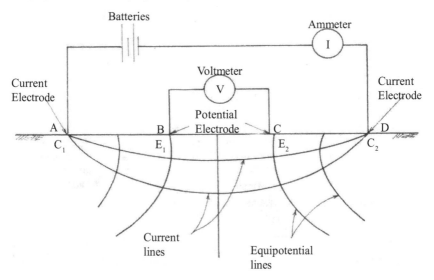

Fig. 10.15: Potential difference between two electrodes

Certain amount of current (as measured by ammeter) is allowed to flow through the outer two electrodes (C_1 and C_2) and the resulting potential difference between the inner two electrodes (E_1 and E_2) is measured by a voltmeter. When the distance AB = BC = CD = a (known as Wenner arrangement), the resistivity is given as:

$$\rho = 2\,\pi\,a\,\frac{V}{I} \qquad\qquad (10.9)$$

where, V = potential or voltage, I = current, ρ = resistivity and a is as defined above.

When the distance BC = b and AD = L, (known as Schlumberger arrangement), the resistivity is given as:

$$\rho = \pi \frac{(L/2)^2 - (b/2)^2}{b} \frac{V}{I} \qquad\qquad (10.10)$$

D.C. current is applied between the electrodes and for different spacing of the electrodes, values of V and I are measured. When the spacing between the electrodes is increased, a deeper penetration of the electrical field occurs and a different resistivity value id obtained. The values of the resitivity are plotted against electrode spacing for various spacing at one location and a smooth curve is plotted. Comparing the nature of this curve with the curves obtained under similar conditions, the nature of the strata and also the depth of the water table is obtained.

10.12.2.2 Seismic Refraction Method

In this method, a small wave called as seismic wave or shock wave is created at the earth's surface either by using a dynamite charge or by the impact of a heavy instrument. The time required for the sound or the shock wave to travel an unknown distance is measured. Knowing the velocity of the sound and time required for the wave to move down, strike the geologic formation and then to come back to the surface is measured. From these two known observations, one can interpret at what distance the geologic formation is present and hence gives an indication of water bearing strata. The principle underlying these seismic methods is that the velocity of the elastic waves which is caused by the shock is different in moist deposits than in the dry deposits of the same composition. Since the velocities of the shock waves depend on the type of formation and the presence of water, it may be possible to detect and estimate the presence of water table depth below ground surface from the differences in the indicated velocities recorded at several geophones.

10.12.3 Sub-surface Geophysical Methods

Geophysical well logging is an accurate and convenient way of obtaining many sub-surface data. This provides information about the nature of formations in their nearly undisturbed conditions. This information is useful in designing tubewells in an area.

Sub-surface geophysical methods consists of test drilling and obtaining well logs or bore hole logs. Logs may be defined as the records of the sub-surface investigation and provide useful interpretation regarding nature and properties of the materials occurring at various depths below the ground surface. These records may be presented in the form of tables or graphic plots with symbolic descriptions. The data making the basis of these records may be obtained by different methods and accordingly there are different types of bore hole logs. Some of them are:

- Geological logs or well logs

- Resistivity logs

- Sonic logs

- Thermal logs and

- Radiometric logs

Geological logs are those which represent the type of the strata existing at different depths and encountered during direct digging or boring of the wells (Fig. 10.16 a).

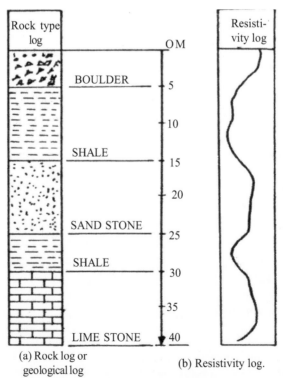

(a) Rock log or geological log

(b) Resistivity log.

Fig. 10.16: Bore hole logs

Resistivity logs are those which indicate the values of electrical resistivity of the rocks at different depths from the surface downwards in bore holes (Fig. 10.16 b).

Sonic logs indicate the values of velocities of compressional waves at different depths. Thermal logs indicate the variation in temperature with depths whereas radiometric logs represent the variations of radioactivity with depths. Details of geophysical aspects of groundwater are given in Sharma and Chawla (1977) and Roscoe Moss Company (1990).

10.13 Construction of Tubewell

The construction of tubewell consists of the following operations:

- Exploratory drilling
- Drilling operations
- Shrouding of tubewell and
- Development of tubewell

10.13.1 Exploratory Drilling

It is desirable to drill a test hole first to the desired depth and collect preliminary information regarding (i) static water level, (ii) types and nature of strata met with, (iii) quality of groundwater and (iv) probable yield.

When such test drilling is done, the following procedures must be followed.

(i) The test hole drilled should be at least 10 cm in size.

(ii) A complete and accurate bore hole log of the test hole should be maintained.

(iii) Static water level in the well after some time when the rise of water level is stabilized is recorded.

(iv) The total depth of drilling shall be sufficient to determine the depth and other characteristics of the lowest aquifer as anticipated from other local considerations.

(v) During drilling, water samples from different strata are collected for chemical analysis to ascertain the quality.

(vi) In case of percussion drilling, static water level in the bore hole is measured daily in the morning before commencement of drilling operations and in the evening at close of drilling operations, to give an indication of behaviour of the aquifers met with during drilling operations.

10.13.2 Drilling Operations

The term drilling also called a boring refers to the process used in making opening in the ground for the purpose of well construction. The important methods used for drilling are as given below:

(A)   Percussion drilling (also called as cable toll method) which is again of two types i.e. hand boring by rope or rod and mechanical boring with power rigs.

(B)   Hydraulic rotary drilling or simply rotary drill which is again of three types i.e. direct circulation hydraulic oratory, (ii) reverse circulation hydraulic rotary and (iii) rotary air percussion drilling.

10.13.2.1 Percussion Drilling

This method of drilling is also called as cable tool method or standard method of drilling. This is called as percussion since the well hole is made by percussion i.e. by hammering and cutting. This method is best suited to well drilling in any formation from soft clay to the consolidated hard rock materials and is generally unsuitable for loose formations. The drilling can be accomplished either by hand boring by rope or by mechanical means using power rigs.

Hand boring by rope or rod: Hand boring by manual process is the simplest, cheapest and most widely used process. In this method, a pit is dug at the centre of the site which measure about 2 to 2.5 m in diameter and 2 to 2.5 m depth. The boring tube also called a casing pipe is lowered into the centre of the pit. A cutter shoe is fixed at the bottom of the casing pipe to facilitate in cutting the bed materials. After lowering the first casing pipe in the pit, it is clamped in position with wooden blocks. Casing pipe is then filled with water.

A wire rope passing over the pulley which is fixed to a tripod stand is tied to a bailer which does the actual cutting of the bed materials. For raising and lowering the bailer, the rope is tied to a winch so that the wire rope can be wound or unwound as required (Fig. 10. 17). The purpose of the tripod stand is to see that the bailer moves up and down in the casing centrally. The bailer used in drilling is a steel pipe of about 6 mm thickness and is about 5 cm less in diameter than the casing pipe. The length varies from 2 m depending on the size of the bore. The bailer has a cutting edge at the bottom and a hook at the top. A flap is fitted inside the bailer just above the cutting edge. The bailer moves up and down in the casing pipe and a slightly circular motion is automatically imparted to it by the torsion in the wire rope. During the downward stroke, the flap valve gets open and the cutting bed materials get pushed inside the bailer. During the upward stroke, the flap valve in the bailer is closed due to the weight of the

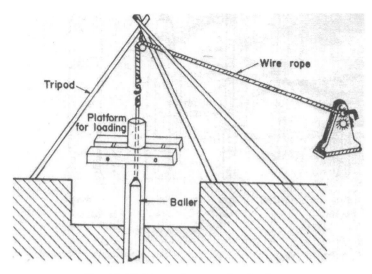

Fig. 10.17: Manual percussion boring

material inside and thus the materials already entered into the bailer cannot come out. After about 30 to 40 strokes the bailer is lifted out of the casing pipe by winding the rope on the winch and the loose material retained inside the bailer is emptied out.

The casing pipe is clamped with wooden sleepers at a convenient point and is then loaded with sufficient weight so that the weight helps the casing pipe to move down automatically in the bore hole as the bailer cuts the bed materials at the bottom. The boring continues and the casing pipes are screwed one after the other till the pipes reach the bottom end of the required depth. Water is poured into the casing pipe at top so that it cleans the bore hole and comes out at the end of the pipe into the bore hole. Hand boring by rope is suitable for soft soil strata or alluvial strata. In hard rock strata, rods are used instead of wire rope to connect the drilling tools. The rod helps in rotary drilling instead of percussion drilling in hard rock strata.

Mechanical boring with power rigs: Mechanical equipments for percussion drilling consist of a portable rig mounted on a truck chasis or trailer (Fig. 10.18). The important components of this equipment are:

- A folding mast of at least 10 m high
- A two line hoist
- A line for operating the drilling tools
- A line for extracting the bailer
- A spudder for raising and dropping the tools

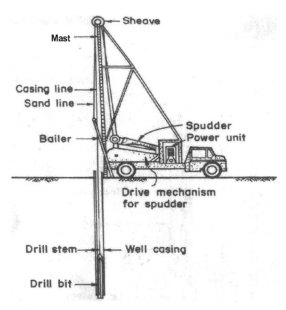

Fig. 10.18: Truck mounted drilling equipment

A diesel engine to provide power to continue the different operations during drilling.

The drilling tools consist of (i) drilling bit, (ii) drilling stem, (iii) drilling jars and (iv) a rope socket with mandrel. The drilling bit is the actual cutting machine which breaks the hard strata and has chisel shaped straight edge. The mandrel holds the cable rigidly in the rope socket. The rope socket is joined with drilling jar. The drilling jar helps the wire rope not to get stuck up in sticky formation. The drilling stem is used to connect the drilling jars on one end and the drilling bit on the other. It also increases the weight of the drilling bit.

The process of boring is same as that in case of hand boring method. But in this case the speed of drilling is high since power rigs are used in boring. Moreover, large sized and deep boring is possible by using power rigs. The progress of percussion drilling depends on the nature of strata which is encountered. In sandy or clayey strata, progress is as much as 25 to 30 m whereas in hard rock, it may be just 1 to 2 m.

10.13.2.2 Rotary Drilling

Rotary drilling is a powerful and quick method of drilling which is suitable to almost all types of formations. In case of alluvial soil, the arte of drilling is fast and in hard rock strata, the arte is slow. The operation consists of rapidly rotating a column of pipes at the lower end of which a cutting bit or drill bit is attached.

The bit cuts a hole slightly larger than the size of the drill pipe so that the cut materials can be removed through the annular space of the bore and drill pipes. As the drill bits rotate, simultaneously a downward thrust is also applied on the bit through the drill pipe so that the cutting points penetrate through the strata. The drilling bits have hollow shanks and have one or more centrally located openings for the flow of the drilling mud.

The rotary drill is usually stronger as it has to withstand the turning effect of the rotating drill.

The important parts of the machinery are:

- Revolving table or turn table
- Hydraulic pumps
- Line shaft and
- Engine or electric motor

The drill pipe also called as drill rod carries the drill bit at the bottom and is screwed at the top to a pipe of square in section known as Kelley. The Kelley slides down along with the drilling rod as the bore deepens. The turn table is revolved by a bevel gear arrangements connected with the line shaft and is controlled by a clutch.

In the rotary drilling method no casing pipes are used. Instead the materials are cut, made into small pieces and are removed by maintaining a constant circulation of drilling fluid under hydraulic pressure. For removing the cut materials, two methods are used. They are (i) direct circulation hydraulic rotary method and (ii) reverse circulation hydraulic rotary method.

Direct circulation hydraulic rotary method: In this case, mud laden water is forced down under pressure through the drilling pipe and the bore hole and while it comes up, it brings along with it drill cuttings. In mud mixing, local mud or special chemical like "Aquagel" or bentonite are used. The mud used has several purposes as follows:

- It brings up the drill cuttings.
- It plasters the inside walls of the bore hole so that the walls are not collapsed
- It keeps the drilling cool at the time of drilling operation.
- It also prevents cuttings to get settle down to the bottom of the bore if circulation is interrupted.
- It speeds up the rate of drilling by softening the formation.

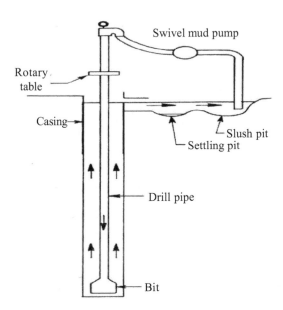

Fig. 10.19: Direct circulation hydraulic rotary method

A slush pit is used to prepare mud and a swivel mud pump is installed for pumping the mud (Fig. 10.19). A stand by pump is used in case of rotary drilling if first pump fails to work. This is done to ensure that there will be continuous drilling till the completion of bore. During drilling if a clay stratum is encountered, than no mud is required to be added to water. Rather water used for drilling may be clear. If the formation has loose soils, than immediately casing pipes are to be entered into the bore hole so that the sides of the bore hole do not collapse. Further the pumps used in forcing the mud into the bore hole should work continuously without any break. If the pump is stopped during operation than the cut materials may settle at the bottom of the bore.

Reverse circulation hydraulic rotary method: Direct circulation hydraulic rotary method is suitable for soft aquifer materials and for small size bores. Instead, a reverse circulation hydraulic rotary method is employed for drilling in heavy gravel or for drilling large sized bores. In this method, the water is pumped down through the space between the drill pipe and the bore hole. The water is drawn out through the drill pipe by means of a centrifugal pump (Fig. 10.20). The water brings along with it the drill cuttings. In case of reverse rotary method, more water is required than that required by direct rotary method.

Rotary air percussion drilling: In conventional rotary drilling, energy is applied in two ways. They are by rotation of the bit and by static force exerted on the bit by means of weight on the bit applied by drill string. A third source of supply of energy to the bit in the form of percussion impact is added by rotary air percussion drilling. Rate of penetration of the drill bit is increased several times due to this third source of energy supply. The energy is applied by an air actuated single piston hammer. Percussion is accomplished by forcing compressed air through the hammer.

In addition to the above methods of drilling tubewells, there are some other methods like auger method, core drilling and water jet method. These methods have limited applications and are to be used in certain specific conditions. Hence, they are not discussed in this chapter.

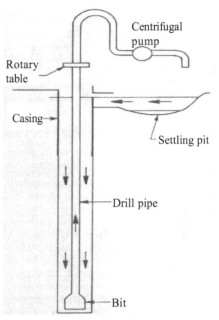

Fig. 10.20: Reverse circulation hydraulic rotary method

Choice of boring method depends on the many factors including the simplicity in construction and operation, cost of construction, convenience in handling and transporting the equipment, rate of drilling and type of formation materials. The advantages and disadvantages of each method are discussed as above. The specific advantages of each method are outlined as below.

Advantages of Cable Tool Method: The advantages are (Garg, 2012):

- A more accurate sample of formation can be obtained.

- Lesser amount of water is required during operation.

- Cable tool rig is lighter and hence easy to transport.

- Very much helpful for consolidated rocks.

- Cheaper compared to rotary method.

Advantages of Hydraulic Rotary Method: The advantages are (Garg, 2012):

- Can be used for larger size bores up to 1.5 m diameter.

- Can be best used for drilling test holes, because the hole can be abandoned with minimum cost.

- Rotary drilled wells can be gravel packed which increases its specific capacity and causes less sand trouble in getting water from the tubewell.

- Casing is to be driven only after the hole completion, and hence, can be set at any depth.

- It is a faster method of drilling and is essential suitable for unconsolidated formations.

- It can handle alternate hard and soft formations with ease at low danger of accident.

10.13.3 Shrouding of Tubewells

Shrouding of tubewells refer to gravel packing of the constructed tubewells. In this process, the annular space around the strainer is filled with gravels of size coarser than that of the natural aquifer so that the area of contact between the aquifer and the tubewell is increased which increases the specific capacity of the tubewell. The main advantages of the shrouding are (Sharma, 1987):

- It permits utilization of aquifer with poor yields characteristics through the increased effective diameter.

- It decreases the head loss of water that enters into the well.

- It prevents collapse of well due to creation of cavity around the casing.

- Drawdown is decreased due to reduced entrance friction.

- It acts as a storage basin which is of importance where most of the supply is coming from thick sandy aquifer.

Gravel packed well may be natural packed well or artificial packed well. The artificially gravel packed well differs from naturally developed well in the sense that the zone immediately surrounding the well screen is made more permeable by removing the formation materials and then packing with artificially graded coarse material. By doing so, effective diameter of well increases.

In naturally developed well, slot opening retain 40-50% of aquifer materials and allow 60-50% to enter into well during development. In gravel packed well, graded gravel is so chosen that it will retain all formation materials; a well screen opening is then chosen so that it will retain all gravels. Gravel packed well costs more than a naturally developed well. Artificial gravel pack is necessary for formation like fine uniform sand, thick artesian aquifer, loosely cemented sandstone etc. In general artificial gravel pack is needed when aquifer material is homogeneous with uniformity coefficient <3 and effective grain size <0.25 mm.

Detail procedures of design of gravel packed wells have been reported earlier in this chapter.

10.13.4 Development of Tubewells

Development of tubewell refers to operations of flushing, testing and equipping the tubewell before it is put in actual service so that good quality groundwater of sufficient quantity can be obtained from tubewell. During drilling operation, muds clog the water bearing formation which decreases the entry velocity of water into the well. Moreover, clogging increases the head loss and decreases the well efficiency. Development of tubewell clears the clogging and eradicates the above mentioned problems. The specific capacity of the well increases i.e. more discharge is obtained with less drawdown. The development causes the gravel pack and the formation around the well to settle and compact itself against the screen, thereby leading to greater stability.

The various methods of development of tubewell are based on the principle of agitating water in the formation with a view to remove the finer material near the casing and to prevent bridging of the sand particles across the screen openings and in the formation around the well.

Depending on the formation characteristics of the aquifers, a well may be developed by the following ways:

- Over-pumping
- Back-washing
- Surging
- Compressed air
- Jetting

10.13.4.1 Over-Pumping

It is the simplest and the most commonly used method of well development. In this method, the well is pumped at a rate higher than the designed rate by large capacity and variable speed pump. Starting with slow speed, the rate of pumping is gradually increased. The rate of speed goes on increasing till sand free good quality groundwater is obtained. At this stage, sand and gravels surrounding the screen get stabilized. The discharge of water during development should be free from sand during the operating test run, with a maximum tolerance of 20 parts of sand in one million of water by volume after 20 minutes of pumping, but should at no stage, exceed 50 parts of sand per million parts of water by volume for irrigation wells.

Step draw-down test is conducted by installing a test pump in the tubewell temporarily and pumping out water at various speeds. At each rate of discharge, pumping is carried out at least for 30 minutes. If the water level and discharge are found to be fluctuating, development is carried out for some more time till discharge becomes steady and water contain tolerable limit of sands.

10.13.4.2 Back-Washing Method

Commonly used methods of back-washing development are:

(i)   **Back-washing with bailer:** It is the simplest and most effective method of well development. In this method water is fed into the well as fast as possible and is removed at a faster rate with sand pump or bailer. The movement of water agitates the formation around the well. The faster the reverse movement of water through the screen, the more is the agitation in the formation. Similarly, the faster the water is bailed out, the more vigoursly the fine particles are sucked into the well from the formation.

(ii) **Back-washing under pressure:** In this method, water is fed into the well to cause outward pressure in the formation around the well. This is done either by connecting a pump or hose line from a source of supply directly to the casing pipe, or by providing top connection with a side valve outlet and suspending a pipe line to the bottom of the well for feeding water to the aquifer through the screen. Large quantity of water is fed from the hose line into the well under high pressure. After about 2 to 5 minutes, the connection is removed and the well is pumped vigorously.

(iii) **Back-washing by raw-hiding**: In this method, air lift pump is used. The pump discharges water for sometime and then it is stopped. Again it is operated and stopped. Thus pump operates intermittently to produce rapid changes in the pressure heads in the well. It is desirable to start raw-hiding at low capacity.

Procedures for Back-washing: Following procedures may be adopted for back washing.

(i)   In the first procedure, the well is pumped to maximum capacity and then is suddenly stopped so that water level in the well falls down and causes reversal of flow. This agitates the formation materials around the screen. When water level in the well comes up to its original level, it is pumped again and the process continues till maximum discharge and sand free discharge is obtained.

(ii) In the second procedure, the pump is discharged at maximum rate and then is suddenly stopped. Water level in the well goes down. Before the water level in the well regains its original position, it is again pumped at maximum speed. The process causes rapid reversal of flow under a high pressure which agitates the formation materials around the screen. The alternate starting and stopping of the pump causes vigorous shaking of the aquifer materials adjacent to the screen and thereby resulting more effective well development.

(iii) In the third procedure, the pump is started and as soon as water reaches the ground level, the pump is stopped. This causes some reversal of flow in the well and agitates the formation materials around the screen. Where static water level is very deep so that water level in it falls quickly, the method works well.

10.13.4.3 Surging

Well development by surging is done by raising and lowering a plunger. The movement of the plunger is restricted in the casing pipe. Casing pipe is provided above the screened position of the well. The up and down movement of the plunger forces water alternatively into the surrounding formation and then allows it to move back into the well. The outflow portion of the surge cycle on the down stroke forces water outward through the screen to break the bridging of finer particles. The inflow portion of the surge cycle moves the fines towards the screen and into the well from where they were removed. Two types of surge plungers are used. They are (i) solid surge plunger and (ii) valve surge plunger.

Solid surge plunger is made from circular belting and plyboard blocks tightened between heavy steel washers and two faced couplings. The belting discs are cut to fit freely in the casing pipe. The valve surge plunger is convertible to solid type plunger by replacing the washers of small diameter by solid plates of diameter equal to that of wooden block. Solid type plungers do not work well in the low yielding aquifers whereas a valve type plunger works well there. The plunger should not be stopped at the bottom to avoid its becoming sand locked in the screen.

10.13.4.4 Compressed Air Method

Development of tubewell with an air compressor is commonly used method and is effective for small diameter wells. There are two general techniques. They are (i) surging and (ii) back-washing. Sometimes a combination of these two methods is used.

Back-washing techniques: It involves forcing of the well water back into the aquifer by means of compressed air introduced into the well through the top of the casing after it has been closed. When the pressure is released, the water flows back into the well through the screen to bring the fine particles from the area surrounding the well, thus ensuring their removal. The process is continued till no sand is brought in.

Surging techniques: It involves the principles of both the pumping method and the surging method as described earlier. In this method, a large quantity of air is released in the water to produce a strong surge which consequently dislodges the aquifer materials. Air compressors mounted on the trolleys and run by diesel engines are normally used for this technique. A 5 cm diameter air pipe is lowered into a 15 to 20 cm diameter discharge pipe (also called as education pipe or drop pipe). The lower portion of the air line has perforations up to a length of 2m and the end is sealed. In the beginning the air line extends below the discharge line and is used to release a large volume of air. This causes a strong surge in the strata and dislodges the aquifer materials surrounding the well screen. The air line is raised into the discharge pipe and consequently an air jet pumping action is created causing the water to flow out. When sand free water comes out, the end of the discharge pipe is raised and the above process is repeated till the entire length of the strainer is subjected to surging and pumping action.

10.13.4.5 Jetting

Well development by jetting with water at high velocity is an effective method. The method involves operating a horizontal water jet inside the well in such a way that the high velocity water stream shoots out through the screen opening. This washes out the fine materials of the aquifer. The turbulence created by the jet brings these fine materials back into the well through the screen openings above and below the point of operation. By slowly rotating the jetting tool, and by gradually raising and lowering it, the entire surface of the screen can be covered.

10.14 Life of Tubewells and Causes of its Failure

A normal tubewell lasts for about 15 to 20 years. Useful life of a tubewell depends on its (i) design, (ii) construction, (iii) types of aquifers tapped, (iv) quality of water, (v) discharge and drawdown at which it is operated and (vi) maintenance.

After a tubewell is constructed and used for some years, there may occur some problems including obtaining low yield. If the problems continue than after some years the well may completely fail to give any yield. A well which suffers from reduction in yield is termed as sick well. A sick well is inefficient and costly. The common well problems occur due to (Edward, 1966; Michael, 1978):

(i)   Inflow of fine sand, silt and /or muddy water into the tube well

(ii)   Low discharge due to improper well development

(iii) Low discharge and muddy water due to improper gravel pack design and screen

(iv)   High rate of pumping which result in the depletion of water in the aquifer

(v)   Corrosion or incrustation problem

(vi)   Mechanical failure while lowering and setting tube well assembly

(vii) Well screen not provided in good aquifer as per litholog study

(viii) Interference of well

(ix)   Wells constructed nearer to barriers like mountain

The two main causes for failure of a tubewell are due to incrustation or corrosion which are discussed as below.

10.14.1 Incrustation

Incrustation of the well pipe occurs due to the deposition of alkali salts on the inside walls of the pipes. The salts that cause incrustation are calcium carbonate, calcium and magnesium sulphates and silicates. Because of incrustation, the effective diameter of the well pipe gets reduced which results in obtaining lower well yield.

Incrustation can be controlled by:

- Reducing the drawdown and hence the pumping rate.

- By using screens having larger areas of openings (or larger size pipes) so as to allow some allowances for the future incrustation.

- Some chemicals like acids may be used to remove the incrustation, but care should be taken such that it does not affect the strainer materials.

- By properly maintaining and periodically cleaning the well screens.

10.14.2 Corrosion

The well pipe is gradually destroyed by corrosion due to the action of acidic water on the pipe material. When chlorides and sulphates or carbon dioxide are present in water, the well pipe will get corroded. The aquifer sand, surrounding the well pipe, finds a way out into the corroded pipe through the worn out pipe walls, thus bringing sand along with water. Thus, corrosion results in excessive withdrawal of sand along with well water.

Stainless steel strainers, though costly, are good to be used as well pipes to avoid corrosion. However, cast iron or ordinary steel screens may be used but they need galvanizing to avoid corrosion. Other measures that can be used to control corrosion are:

- By reducing the drawdown and hence the pumping rate

- By reducing the flow velocity by increasing the percentage of open area or size of the well pipe

- By using thicker pipes

- By using corrosion resistant materials for the pipes and

- By using corrosion resistant coatings on the pipes such as galvanization of pipes.

10.15 Verticality of Tubewells

If vertical turbine pumps are purposed to be installed in tubewells, it should be ensured that the tubewell is vertical by means of a verticality test. This is essential to avoid the misalignment of the line shaft assembly which otherwise will result in excessive wear and tear of the shaft bearings. Installation of tubewell in vertical position will reduce the power consumption. The equipment to check the verticality of tubewell consists of (i) a tripod stand having at the top a metallic disc with a small rotating pulley at the centre, (ii) a plumb of about 40-45 cm long with diameter 0.5 cm smaller than the inside diameter and (iii) a reel of flexible steel wire. For details of verticality test of tubewells, any book on Groundwater Hydrology may be referred.

10.16 Groundwater Resources of India

Groundwater is one of the most important water sources in India. It accounts for over 400 km^3 of the annual utilizable resource in the country. It has become a popular alternative for irrigation and domestic water use across the country. In addition to being accessible, groundwater quality is generally excellent in most areas and presents a relatively safe source of drinking water.

The presence and availability of groundwater varies greatly with changes in topography, subsurface geology and the prevailing climate in the region. In some areas, groundwater exists in deep aquifers at a depth even more than 800 m while in other areas, the water is stored near the surface. The location of the aquifer also affects its recharge rate and its susceptibility to pollution and overuse.

India has a vast groundwater resources but it is widely distributed. It can be found in every tract of the country. However, its distribution is not uniform in all parts of the country. The distribution of the resources depend on the hydro-geological environment, climatic conditions, physiographic features including geologic conditions of the aquifer. From the viewpoint of yield of groundwater, the soil and rock formations of the country can be considered under three classes. They are (i) consolidated hard rock formation (120 M ha), (ii) semi-consolidated formations (56 M ha) and (iii) unconsolidated formations (153 M ha). Consolidated hard rock formation gives low yields of groundwater. Semi-consolidated formations consist of sedimentary rocks like sandstones, limestones and conglomerates and yield moderate amount of groundwater. Unconsolidated formations constitute large areas and consist of sand, gravel, boulders, laterites and alluvial tracts and are found in Indus basin, Ganga and Brahamputra basin and coastal belts in the country. This formation yields substantial amount of groundwater. The alluvial tract of the Gangetic plain, which extends over 2000 km across central and northern India has the best potential for groundwater extraction in the country.

In general, the mountainous and hilly regions in the north and west do not allow adequate infiltration and as a consequence, groundwater is mostly limited to valleys and other lower lying areas. In the peninsular part of the country, the underlying geology limits the formation of large continuous aquifers. Groundwater is therefore scattered where fissures permit adequate storage or is found in shallow depressions near the surface. As a result, the overall yield potential in this region is low although some areas may see medium to high potential depending on the local hydrogeology. Coastal regions are usually rich in groundwater owing to the largely alluvial terrain. However, the aquifers in these areas have chances of being easily contaminated by saltwater ingress due to over pumping.

The total annual replenishable ground water resources of the Country have been reassessed as 433 Billion Cubic Meters (BCM) and the net annual ground water availability is estimated as 399 BCM. Existing gross ground water draft as on March 2004 for all uses is 231 BCM per year. The stage of ground water development is 58%. The development of ground water in different areas of the country has not been uniform. Highly intensive development of ground water in certain areas in the country has resulted in over exploitation leading to decline in the levels of ground water and sea water intrusion in coastal areas. Table 10.6 represents groundwater potential in different river basin of India.

Table 10.6 Ground water potential in river basins of India (pro rata basis) (unit: km³/year)

S. N.	Name of the Basin	Total replenishable ground water resources	Ground water potential available for use
1.	Brahmani with Baitarni	4.05	3.16
2.	Brahmaputra	26.55	21.80
3.	Chambal Composite	7.19	3.66
4.	Cauvery	12.30	4.67
5.	Ganga	170.99	96.37
6.	Godavari	40.65	24.94
7.	Indus	26.49	5.22
8.	Krishna	26.41	14.50
9.	Kutch & Saurashtra Composite	11.23	4.64
10.	Madras and South Tamil Nadu	18.22	6.55
11.	Mahanadi	16.46	13.02
12.	Meghna	8.52	6.95
13.	Narmada	10.83	7.18
14.	Northeast Composite	18.84	13.26
15.	Pennar	4.93	2.66
16.	Subarnarekha	1.82	1.40
17.	Tapi	8.27	3.97
18.	Western Ghat	17.69	11.18
	Total	431.43	245.13

Source: www.india.gov.in

10.17 Groundwater Recharge

Groundwater recharge is also known as deep drainage or deep percolation. It is a hydrologic process where water moves downward from surface water to groundwater. This process usually occurs in the vadose zone below effective root zone of plants. Groundwater recharge may occur both naturally and through anthropogenic processes like artificial groundwater recharge where rainwater and or reclaimed water is routed to the subsurface. Groundwater is recharged naturally by rain and snow melt and to a smaller extent by surface water. Recharge may be impeded somewhat by human activities including paving, development, or clogging. These activities can result in loss of topsoil resulting in reduced water infiltration, enhanced surface runoff and reduction in recharge. Use of groundwaters, especially for irrigation, may also lower the water tables. Groundwater recharge is an important process for sustainable groundwater management, since the volume-rate abstracted from an aquifer in the long term should be less than or equal to the volume-rate that is recharged. Groundwater recharge is the enhancement of natural ground water supplies using man-made conveyances such as infiltration basins, trenches, dams, or injection wells. Aquifer storage and recovery (ASR) is a specific type of groundwater recharge practiced with the purpose of both augmenting ground water resources and recovering the water in the future for various uses.

Rain water harvesting plays a major role in augmenting the groundwater aquifers, which would cater to the needs of the future generations. Contribution of rainfall is the main source to recharge to groundwater. However, this recharging capacity of the soil depends on the type of soil formation. For consolidated rocks, only 5 to 10 percent of the rainfall is recharged to groundwater. For hard rocks, the contribution is about 10 percent. For the river alluvium, the recharge contribution is 10 to 15 percent and for the Indo-Gangetic alluvium, the recharge contribution is the highest (20%). A part of rainfall, canal water in the form of seepage also adds to groundwater. Return flow from irrigation, seepage from waste water from industries etc. also contribute some amount of recharge water to the groundwater reservoir. Table 10.7 (Suresh, 2008) represents state-wise annual groundwater recharge from various sources in India. The data of Table 10.7 indicates that about 37.111 M ha m of groundwater resources are generated from recharge of rainwater in the country out of which undivided Madhya Pradesh and undivided Uttar Pradesh contributes maximum amount with figures of 5.318 and 4.257 M ha m, respectively. Seepage flow from canal irrigation contribute 5.463 M ha m. Thus, total recharge from both rainwater and canal water constitute 42.574 M ha m. Out of this recharge groundwater, some amount is lost as sub surface flow and the net recharge available is 26.996 H ≈ 27 M ha m.

Table 10.7 State wise annual groundwater recharge from various sources (M ha m)

State	Rainfall	Canal	Total recharge	Net recharge
Andhra Pradesh	2.468	0.567	3.035	2.122
Bihar	3.319	0.543	3.862	2.702
Delhi	0.061	-	0.061	0.037
Gujarat	1.739	0.049	1.788	1.258
Haryana	0.370	-	0.567	0.432
Himachal Pradesh	0.283	-	0.283	0.111
J & K	1.184	0.049	1.233	0.493
Karnataka	1.826	0.222	2.048	1.234
Kerala	1.172	0.148	1.320	0.666
Madhya Pradesh	5.318	0.173	5.491	3.295
Maharastra	2.443	0.148	2.591	1.555
North eastern states	5.022	0.144	5.170	2.061

Contd.

State	Rainfall	Canal	Total recharge	Net recharge
Odisha	2.394	0.419	2.813	1.974
Punjab	0.629	0.693	1.122	0.851
Rajasthan	0.493	0.222	0.714	0.419
Tamilnadu	1.715	0.432	2.147	1.419
Uttar Pradesh	4.257	1.234	5.491	4.382
West Bengal	2.418	0.419	2.837	1.987
Total	37.111	5.463	42.574	26.996

10.17.1 Artificial Groundwater Recharge

Artificial recharge is the planned, man-made increase of groundwater levels. By improving its natural replenishment capacities and *percolation* from *surface waters* into *aquifers,* the amount of *groundwater* available for abstraction is increased. *Artificial surface groundwater recharge* refers to different *groundwater recharge* techniques that release *effluent* from above the ground into the *groundwater aquifer* via soil *percolation.* Artificial groundwater recharge is becoming increasingly important in India, where over-pumping of groundwater by farmers has led to underground resources becoming depleted.

Artificial recharge methods can be further classified into two broad groups: direct methods and indirect methods. Direct methods can again be classified into surface spreading techniques and subsurface techniques. The most widely practiced methods of artificial recharge of groundwater employ different techniques of increasing the contact are and resident time of surface water with the soil so that maximum quantity of water can infiltrate and augment the groundwater storage. Under the surface spreading techniques various methods available are flooding, ditch and furrows, surface irrigation, stream modifications and finally the most accepted one and suitable for small community water supplies are runoff conservation structures. Under the subsurface techniques injection wells and gravity head recharge wells are common ones. Indirect methods of artificial recharge adopts the technique of induced recharge by means of pumping wells, collector wells and infiltration galleries, aquifer modifications and groundwater conservation structures, which require highly skilled manpower and other resources.

Question Banks

Q1. Differentiate between surface water and groundwater.

Q2. With a neat label sketch, describe the distribution of sub-surafce groundwater.

Q3. Differentiate between the followings:

(i) Zone of aeration and zone of saturation

(ii) Confined aquifer and unconfined aquifer

(iii) Perched aquifer and confined aquifer

(iv) Artesian well and flowing well

(v) Observation well and piezometer.

Q4. How does a flowing well originate?

Q5. Define aquifer. Give some examples of good and bad aquifer.

Q6. Why coarse sand is a good aquifer whereas clay is a bad aquifer?

Q7. With the help of hydrologic cycle, describe the occurrence of groundwater?

Q8. Define specific yield and specific retention. What is the difference between them?

Q9. Porosity of an aquifer is 50 percent. If volume of water obtained by gravity drainage is 0.3 time the total volume of aquifer drained, find out the specific yield and specific retention of the aquifer. If the total volume of the aquifer drained is 500 m³, then calculate the volume of water held against gravity drainage.

Q10. What is storage coefficient and how is it estimated?

Q11. What is a well irrigation? What is its advantages and disadvantages?

Q12. What points need to be taken into consideration for selection of ideal site of design of well.

Q13. What is an open well? Write notes on various types of open well?

Q14. Differentiate between impervious and previous lined well.

Q15. What are the differences between shallow and deep well.

Q16. Describe the pumping test and recuperation test for determining the yield of an open well.

Q17. Design an open well in fine sandy soil with safe yield 0.45 m³/hr /m²/ unit depression head so that an yield of 0.0045 m³/sec is discharged from well with depression head h = 3 m

Q18. Calculate the average yield of a well of 2,5 m diameter by recuperation test. The water level in the well is decreased to an extent of 3.0 m. The rate of rise of water is 1.0 m/hr. Assume average depression head, h =3 m.

Q19. What is tubewell irrigation? Describe the various advantages and disadvantages of tubewell irrigation.

Q20. Write short notes on various types tubewells.

Q21. Draw a typical discharge drawdown curve of shallow and deep tubewells and explain how the discharge varies with types of tubewells for a given discharge.

Q22. Describe in details about the various design parameters of tubewells.

Q23. With a label diagram describe about the structure of a tubewell.

Q24. What are the factors on which optimum length of well screen depend on?

Q25. What is difference between naturally developed well and artificial gravel packed well? Narrate the different design procedure of gravel packed well.

Q26. What are the differences between cavity type tubewell and strainer type tubewells?

Q27. How does a cavity well works? With a diagram, narrate the working and construction of cavity type tubewells.

Q28. Describe in details the construction and working mechanism of a strainer type tubewells.

Q29. What are the uses of strainer in tubewell. Describe various types of strainers used in tubewell irrigation.

Q30. Design a tubewell to deliver 0.045 cumecs of water at a depression head of 7 m. The average water level is 10 m below the ground level in post monsoon season and 17 m in pre-monsoon season. The well log data obtained during investigation is presented below.

Depth, below ground level, m	0 - 15	15 - 20	20 - 30	30 - 50	50 - 65	65 - 75	Below 75
Types of strata	Clay	Sand	Clay + kankar	Coarse sand	Clay	Medium sand	Clay + sand stone

There is a horizontal length of pipe of 8 m connected to the pump at ground level as deliver pipe to discharge water to the crop field.

Q31. What do you mean by groundwater prospecting? Enumerate the various methods used for investigation of groundwater development.

Q32. Write notes on the following methods of investigation of groundwater development.

(i) Electrical resistivity method

(ii) Seismic refraction method

(iii) Well logs

(iv) Resistivity logs

Q33. Describe the various operations involved in the construction of tubewell.

Q34. How is the well construction done by percussion drilling?

Q35. What is the difference between direct circulation hydraulic rotary method and reverse circulation hydraulic rotary method?

Q36. Differentiate between the cable tool method and hydraulic rotary method of drilling of tubewell.

Q37. What do you mean by shrouding of tubewell? Enumerate the main advantages of shrouding.

Q38. What does well development refer to? What are the advantages of well development?

Q39. Describe the different methods of development of tubewell?

Q40. How is the well development by surging done? Describe the surging technique of well development.

Q41. What is the average life of a tubewell? How can the life span be increased?

Q42. What is a sick well? How can the failure of a tubewell be controlled?

Q43. What for the verticality test of tubewell is done? How is it accomplished?

Q44. Write short notes on the followings:

(i) Aquifer

(ii) Aquitard

(iii) Aquifuge

(iv) Aquiclude

Q45. Write descriptive notes on groundwater resources of India.

Q46. What do you mean by artificial groundwater recharge? What are its advantages?

Q47. What are the different ways by which groundwater resources of a region can be enhanced?

Q48. What is the difference between a natural and artificial groundwater recharge?

Q49. Enumerate the importance of groundwater for irrigation and other purposes.

Q50. How does the rain water help in groundwater recharge? Describe the various means by which it can be achieved.

References

Ahrens, T.P. 1958. Water Well Engineering. Water Well Jl. 12, December.

Anonymous. 1966. Procedure used in gravel packing wells. Johnsons Drillers Jl. 38:2.

Benison, E.W. 1947. Ground Water. E. E. Johnson St. Paul, Minn. 500p.

Das Gupta, A. 1987. Handouts on Ground Water Hydrology, Asian Institute of Technology, Bangkok, pp. 282.

Dylla, A.S. and Muckel, D. C. 1967. Experimental development of shallow ground water wells. Pub. University Nevada and U. S. Dept. of Agr. 23 p.

Edward, E. 1966. Ground Water and Wells. Pub. Edward E. Johnson Inc., Saint Paul, Minnesota, U.S.A., 440 p.

Ellithorpe, G. L. 1970. Current practices relative to the design and placement of artificial gravel pack for tubewells. Proc. Summer Institute on Tube Well Technology I.A.R.I., New Delhi.

Garg, S.K. 2012. Irrigation Engineering and Hydraulic Structures. Khanna Publishers, Delhi.

Lenka, D. 2001. Irrigation and Drainage. Kalyani Publishers, New Delhi.

Michael, A.M. 1978. Irrigation Theory and Practices. Vikash Publishing House Pvt. Ltd., New Delhi.

Panigrahi, B. 2011. Irrigation Systems Engineering. New India Publishing Agency, New Delhi.

Roscoe Moss Company. 1990. Handbook of Groundwater Development. John Wiley & Sons. Singapore.

Sharma, H.D. and Chawla, A.S. 1977. Mannual on Ground Water and Tubewells. Central Board of Irrigation and Power, New Delhi.

Sharma, R.K. 1987. A Text Book of Hydrology and Water Resources. Dhanpat Rai and Sons, Delhi, pp. 818.

Slichter, C. S. 1989. Theoretical Investigation of the Motion of Ground-waters. 19th Annual Report, U.S.G.S., 295-384.

Suresh, R. 2008. Watershed Hydrology. Standard Publishers Distributors, Delhi, pp. 692.

CHAPTER 11

Well Hydraulics

11.1 Introduction

The laws governing the flow of water in wells are referred to as the hydraulics of wells. Just like surface water flows from a higher elevation to a lower elevation by gravity, groundwater also flows from a region of higher water table to a region of lower water table. This difference of water table measured by any two piezometers divided by the longitudinal distance between them is called as hydraulic gradient. Hydraulic gradient plays a dominant role in deciding the flow of groundwater. This hydraulic gradient together with the hydraulic conductivity of the aquifer material forms the main law of well hydraulics called as Darcy's law. An understanding of Darcy's law together with some other laws as discussed subsequently in this chapter is essential in order to know the discharge of a well in a particular formation. Darcy's law is valid for studying groundwater flow in alluvial formations. However, it is not valid for aquifers having hard rock areas.

11.2 Properties of Aquifer Materials

A soil mass is a three phase system consisting of soil particles called as soil grains, water and air. The void space between the soil grains is field partly with water and partly with air. In case of a dry soil mass, void spaces fully occupy air and water. In case of a fully saturated soil, voids are completely filled with water. In general the soil mass has three constituent parts which are blended together forming a complex material [Fig. 11.1 (a)]; the properties of which depend upon the relative proportion of these constituent, their arrangement and a variety of other factors. For calculation purpose, it is more convenient to show these constituents occupying separate space [Fig 11.1(b) (i) and (ii)].

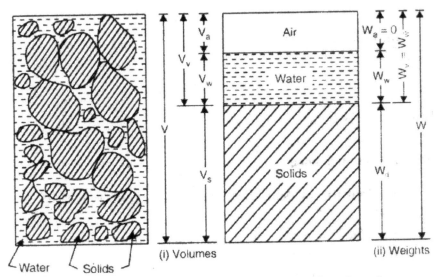

(i) Volumes   (ii) Weights

(a) Elements of natural soil   (b) Elements separated into three phases

Fig. 11.1: Soil as a three phase system

As shown in Fig.11.1 (b) (i), the total volume (V) of the soil mass consists of (i) volume of air (V_a) (ii) volume of water (V_w) and (iii) the volume of solids (Vs). The volume of voids (V_v), is, therefore, equal to volume of air plus the volume of water. Fig. (Fig. 11.1 (b) (ii) also represents the constituents occupying separate spaces in terms of weights. The weight of air is considered to be negligible. So the weight of total voids is equal to the weight of water (W_w).The weight of solids is represented by W_d or W_s, which is evidently equal to the dry weight of soil sample. The total weight (W) of the moist sample is, therefore, equal to ($W_w + W_d$)

The important aquifer properties that influence the water holding and water transmitting characteristics of aquifers are discussed below.

11.2.1 Porosity

This property is related to the pore space of the formation. Porosity is expressed as the ratio of the volume of interstices/voids (V_v) to the total volume of aquifer materials or soil mass (V) and is given as:

$$\eta = \frac{V_v}{V} \tag{11.1}$$

Porosity is expressed in percent when the above mentioned equation is multiplied by 100. Values of porosity depend on the arrangement of the particles, shape of the grains and degree of assortment.

Eq. (11.1) can also be written as:

$$\eta = \frac{V_v}{V} = \frac{V - V_s}{V} = 1 - \frac{V_s}{V}$$

(11.2)

where V_s = volume of soil mass or solid.

Porosity can also be expressed in terms of density as:

$$\eta = 1 - \frac{\gamma_d}{\gamma m}$$

(11.3)

Eq. (11.3) when multiplied by 100, it gives porosity in percent. In Eq. (11.3). and are called as bulk density and particle density, respectively.

Porosity is of two types i.e. (i) primary porosity and (ii) secondary porosity. Primary porosity refers to the porosity developed during formation of rocks. Secondary porosity refers to that porosity which is developed during weathering and development of joints caused due to expansion, contraction and deformation of formation because of change in temperature, pressure etc in the consolidated sedimentary rocks, igneous rocks etc. Porosity different aquifer materials are mentioned Table 11.1 below.

Table 11.1: Porosity of various aquifer materials

Aquifer materials	Porosity (percent)
Lime stone, shale	1 – 10
Sand stone	10 – 20
Sand, gravel	30 – 40
Clay	45 - 55

11.2.2 Void Ratio

Void ratio (e) of a given soil sample is the ratio of the volume of voids/pores/interstices (V_v) to the volume of soil solids (V_s) in a soil mass. Thus

$$e = \frac{V_v}{V_s}$$

(11.4)

Void ratio is generally expressed in fraction where as porosity is expressed in percent. Fig 11.2 (a) and (b) represent void ratio and porosity of a given soil sample respectively.

If the volume of voids is taken equal to e, then volume of solid (V) is equal to 1 and the total volume is equal to ($1+e$) (Fig 11.2 (a)). Similarly if the volume of

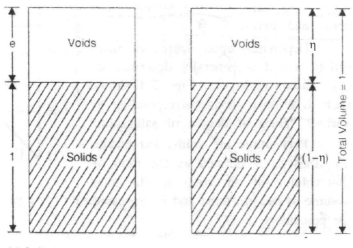

(a) Soil element in terms of e   **(b) Soil element in terms of η**
Fig. 11.2: Void ratio and porosity

voids is taken equal to ç (Fig. 11.2 (b)), the total volume of the element will be
1 (unity) and hence the volume of solid would be equal to (1- η).

From Fig. 11.2 (a), we have by definition of porosity,

$$\eta = \frac{V_v}{V} = \frac{e}{1+e} \tag{11.5}$$

Similarly from Fig. 11.2 (b) and using definition of void ratio, we have

$$e = \frac{V_v}{V_s} = \frac{\eta}{1-\eta} \tag{11.6}$$

Combining Eqs. (11.5) and (11.6), we get

$$1 - \eta = \frac{1}{1+e} \tag{11.7}$$

11.2.3 Degree of Saturation

In a given volume of soil, some space is occupied by water and rest is occupied
by air. The degree of saturation (S) is defined as the ratio between volume of
water (V_w) in a given soil sample to the total volume of voids (V_v) present in it.

$$S = \frac{V_w}{V_v} \tag{11.8}$$

Degree of saturation when expressed in percent, the term in Eq. (11.8) is multiplied by 100. In fully saturated soil (some times fully saturated soil is also called saturated soil where as unsaturated soil is the soil which partly occupied by water and rest are by air), all the voids are filled with water and there is no air i.e. $V_v = V_w$ and hence from Eq (11.8), degree of saturation (S) is 1.0 (100%). Thus, for a perfectly dry soil sample $V_w = 0$ and hence $S = 0$ i.e. zero degree of saturation.

Sample Calculation 11.1

A soil sample has 50% porosity. Find its void ratio?

Solution

Given porosity $(\eta) = 50\% = 0.50$.

So void ratio e (Eq 11.6) is

$$e = \frac{\eta}{1-\eta} = \frac{0.50}{1-0.50} = 1.0$$

Sample Calculation 11.2

What will be the porosity of a soil sample having void ratio 0.4?

Solution

From Eq (11.5), we have porosity (η)

$$\eta = \frac{e}{1+e} = \frac{0.4}{1+0.4} = 0.286 = 28.6\%$$

Sample Calculation 11.3

What will be the degree of saturation of a given mass of soil sample that contain volume of water half to volume of voids?

Solution

Given $V_w = \dfrac{V_v}{2}$

$$S = \frac{V_w}{V_v} = \frac{V_v}{2 \times V_v} = 0.5 = 50\%.$$

11.2.4 Specific Yield

Specific yield is defined as the ratio of volume of water drained by gravity to total volume of the material drained and is given by the formula as:

$$Specific\ yield = \frac{Volume\ of\ water\ obtained\ by\ gravity\ drainage}{Total\ volume\ of\ material\ drained\ or\ dewatered} \times 100 \qquad (11.9)$$

The term effective porosity and specific yield are synonymous. Specific yield can be computed by the following methods:

- Pumping test
- Recharge method
- Field saturation and drainage method and
- Sampling after lowering the water table method

In pumping method, draw down data is recorded by installation of observation wells at various radial distances away from the main pumping well and this data are used to compute the specific yield. Details of the methods mentioned as above for determination of specific yields are available in (Suresh 2008).

11.2.5 Specific Retention

It is expressed as ratio of volume of water held against gravity drainage to the total volume of material drained and is given as:

$$Specific\ retention = \frac{Volume\ of\ water\ held\ against\ gravity\ drainage}{Total\ volume\ of\ material\ drained\ or\ dewatered} \times 100 \qquad (11.10)$$

Thus, the sum of specific yield and specific retention is equal to porosity.

Values of porosity and specific yield of some of the aquifer materials are presented in Table 11.2.

Table 11.2 Values of porosity and specific yield of some selected aquifer materials

Aquifer material	Porosity, percent	Specific yield, percent
Clay	45-55	1-10
Sand	30-40	10-30
Gravel	30-40	15-30
Sandstone	10-20	5 - 15
Shale	1-10	0.5 - 5
Limestone	1-10	0.5 - 5

Specific retention is determined by centrifuge technique. In this technique, a given soil sample is saturated with water and centrifugal force of about 1000 times of gravity force is applied to the saturated soil sample for 1 hour. When the force is applied, some amount of water from the saturated soil will come out or drain out and some will be retained by it. The retained amount is held in the voids and is called as centrifuge moisture equivalent.

Fig. 11.3: illustrates the values of porosity, specific yield and specific retention of various types of geologic formations.

11.2.6 Permeability

Permeability of a material is a measure of its ability to transmit fluid like water under a given hydraulic gradient. A medium has coefficient of permeability (K) of unit length per unit time if it transmits in a unit time a unit volume of groundwater, at the prevailing viscosity through a unit cross sectional area measured at right angle to the direction of flow under a unit hydraulic gradient. It is generally expressed as m/day, m/sec and cm/sec. It is to be remembered that porosity is not a measure of permeability since most of the soil formations are even more porous but they are unable to yield water such as dense clay. Coefficient of permeability is also termed as hydraulic conductivity. Approximate values of K for different aquifer materials are mentioned in Table 11.3 (Garg, 2012).

Table 11.3: Values of coefficient of permeability of selected aquifer materials

Sl.No.	Type of aquifer materials	Approximate values of K, cm/day
1.	Granite, Quartzite	0.6×10^{-5}
2.	Slate, Shale	4×10^{-5}
3.	Limestone	4×10^{-5}
4.	Sandstone	0.004
5.	Sand and gravel	0.40
6.	Only gravel	4.0
7.	Only sand	0.04
8.	Clay	0.04×10^{-5}

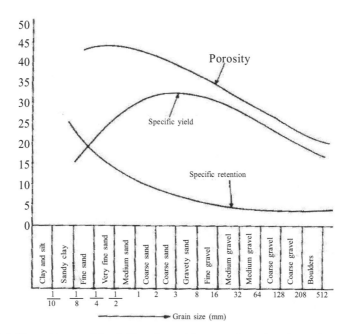

Fig. 11.3: Porosity, specific yield and specific retention of geologic formations

11.2.7 Hydraulic Conductivity

Coefficient of permeability is also termed as hydraulic conductivity. It is the ability of the soil to transmit water. In Darcy's law, hydraulic conductivity, K is the proportionality factor which essentially expresses an interaction between the porous media (soil) and flowing fluid (water). Value of hydraulic conductivity depends on the soil matrix and properties of water flowing through the soil. The soil matrix includes properties like porosity, pore size distribution and tortuosity of the soil pores. Fluid properties that affect the hydraulic conductivity are density and viscosity of fluid. Tortuosity is defined as the ratio of capillary length (L_c) in a given soil column to the length of soil column (L). The capillary length is the actual path travelled by the average parcel of water through the soil column.

11.2.8 Measurement of Hydraulic Conductivity

Measurement of hydraulic conductivity can be done by the following methods:

(i) Laboratory method

- Constant head method
- Falling head method

(ii) Field method

(a)   Below water table

- Piezometer method

- Auger hole method

(b)   Above water table

- Shallow well pumping method

- Cylinder permeameter method

- Double tube method

(iii) Tracer method

11.2.8.1 Constant Head Method

This method is suitable for that sol which has medium to high permeability range. Using a core sampler, a given soil sample from a given layer whose permeability is to be determined is collected in such a manner that the soil sample is least disturbed. Then it is allowed for saturation. The saturated soil sample is taken in a permeameter or cylindrical container which is provided with porous plates at both top and bottom. The inflow water is maintained at constant level in a manometer connected to a tank which is connected to the permeameter or cylindrical container. The collected water from the cylindrical container is collected in pot as shown in Fig.11.4 below. Following observations are recorded during the experiment.

(i)   Amount of water discharged through the soil sample *(Q)*

(ii)   Length of soil sample *(l)*

(iii)   Cross sectional area of the cylindrical container *(A)*

(iv)   Duration of flow *(t)*

(v)   Head of flow (h) and

(vi)   Temperature of water

Following formula is used to compute hydraulic conductivity (K):

$$K = C\frac{Q\,l}{A\,t\,h}$$
(11.11)

Where C = correction factor which is the ratio of viscosity of water at observed temperature during the experiment to the viscosity of water at 60 °F temperature and other terms are defined as above.

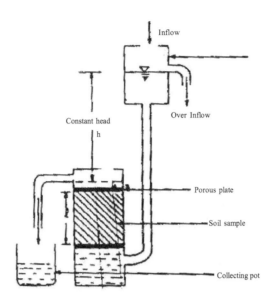

Fig. 11.4: Hydraulic conductivity by constant head method

Sample Calculation 11.4

A given soil sample was collected from a layer of 0-15 cm in a field. The hydraulic conductivity of the sample is to be determined. Following data were collected.

(i)   Diameter of soil sample = 8 cm

(ii)   Length of soil sample = 4 cm

(iii) Hydraulic head = head of flow = 35 cm

(iv)   Volume of water flow collected in a pot in 1 minute = 200 cc.

Assuming correction factor, $C = 0.85$, determine the value of hydraulic conductivity.

Solution

Given, d = diameter of soil sample = 8 cm.

Cross sectional area of soil sample = $A = \pi \, d^2/4 = 50.272 \text{ cm}^2$

Length of soil sample, $l = 4$ cm

Hydraulic head = head of flow = $h = 35$ cm

Volume of water flow collected in a pot in 1 minute = 200 cc i.e. $Q = 200$ cc and $t = 60$ sec.

Putting these values in Eq. (11.11) we get

$$K = C\frac{Q\,l}{A\,t\,h} = 0.85x\frac{200\ x\ 4}{50.272\ x\ 60\ x\ 35}$$

$$= 0.00644 \text{ cm/sec}$$

11.2.8.2 Variable Head Method

In this method, soil of low permeability is used to determine the permeability. As in constant head method, undisturbed or least disturbed soil sample is collected and is fully saturated. Then the saturated soil sample is taken in a cylindrical container which is provided with porous plates at both top and bottom (Fig. 11.5). Other experimental procedures maintained here are same as discussed in constant head method. In addition to all data which are collected in constant head method, the following extra data are collected in this method.

(i)   Head indicated by the manometer (h_o) at initial time i.e. when $t = 0$

(ii)   Head at any elapsed time, t, (h_e)

(iii)   Elapsed time, t_e

(iv)   Cross sectional area of the manometer tube (a)

The formula used to determine hydraulic conductivity, K is

$$K = 2.3\ C \log\left(\frac{h_0}{h_e}\right)\frac{a\,l}{A\,t_e} \tag{11.12}$$

where, all the terms are defined earlier.

Sample Calculation 11.5

A given soil sample was collected from a field. The hydraulic conductivity of the sample is to be determined. Following data were collected.

(i)   Diameter of soil sample = 8 cm

(ii)   Length of soil sample = 4 cm

(iii)   Initial hydraulic head in the manometer tube = 350 cm and after 5 hour, the hydraulic head in the manometer = 280 cm

(iv)   Diameter of the manometer tube =1.5 cm

Assuming correction factor, $C = 0.85$, determine the value of hydraulic conductivity.

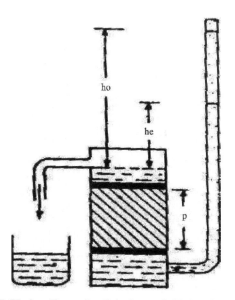

Fig. 11.5: Hydraulic conductivity by variable head method

Solution

Given d = diameter of soil sample = 8 cm.

Cross sectional area of soil sample = $A = \pi d^2/4 = 50.272$ cm²

Length of soil sample, $l = 4$ cm

Initial hydraulic head at $t = 0$, $h_0 = 350$ cm and final head, $h_e = 280$ cm

Elapsed time, $t_e = 5$ hour = 300 sec

Cross sectional area of manometer, $a = \pi d^2/4 = 3.142$ x $(1.5)^2/4 = 1.767$ cm²

$C = 0.85$

Putting these values in Eq. (11.12) we get

$$K = 2.3\ C \log\left(\frac{h_0}{h_e}\right) \frac{a\,l}{A\,t_e}$$

= 2.3 x 0.85 log (350/280) x (1.767 x 4) / (50.272 x 300)

= 8.87901 x 10⁻⁵ cm/sec.

11.2.8.3 Piezometer Method to Determine Hydraulic Conductivity

Piezometer method is suitable for homogeneous and isotropic soil layer. In this method, a piezometer is installed in a field to a depth below water table in the field. A cylindrical shaped unlined cavity is developed at the lower end of the piezometer: diameter of which is same as the diameter of the piezometer (usually taken as 4.9 cm) and length is taken as 10.2 cm. Bottom of the cavity should not touch the impervious layer. Before the experiment is started, all water from the cavity is drained out. Water then starts rising in the piezometer through the unlined cavity. Initial water level in the piezometer say h_1 is measured and then after some time say "t, final water level, h_2 is measured (Fig. 11.6). Like this several observations of rising water level and time are recorded. The formula used to determine, K is

$$K = \frac{\pi\, R^2}{A\, \Delta t} \ln\left(\frac{h_1}{h_2}\right)$$
(11.13)

where, K = hydraulic conductivity, cm/hr and R = radius of cavity, cm and A = constant factor taken as 43.2 cm and other terms are defined earlier. The factor A actually depends on the length of the cavity, diameter of cavity, position of lower end of the cavity from the impervious bed (S), and distance (d) measured from the water table position.

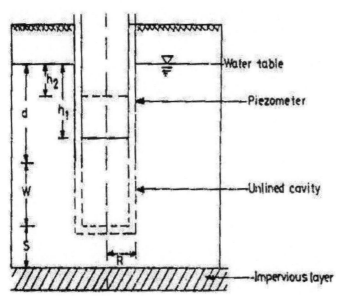

Fig. 11.6: Piezometer method to determine hydraulic conductivity

11.2.8.4 Auger Hole Method

In this method, an auger hole is bored upto a depth below the water table. Then water is taken out of the hole when the water table is in equilibrium. Then the rise of water table with respect to various time interval in the auger hole is recorded from which value of hydraulic conductivity is determined. Let at any time, t_1, the water table in the hole be h_1 from the bench mark level of water table as shown in Fig. 11.7. Similarly at time t_2, let the water table in the hole be h_2 from water table. Let d be the depth of auger hole and r is its radius. Several readings of depth of water table and time are taken which are used in computation of hydraulic conductivity.

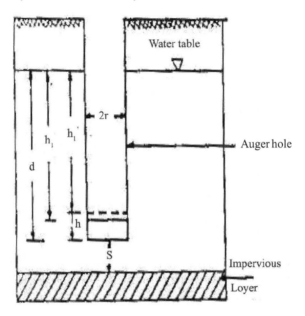

Fig. 11.7: Measurement of hydraulic conductivity by auger hole method

Hydraulic conductivity is determined by the formula:

$$K = \frac{\pi \, r^2 \cdot \Delta h}{A' \, \Delta t \left(\dfrac{h_1 + h_2}{2} \right)}$$ (11.14)

Where K is in cm/sec, r is in cm, h_1 and h_2 are in cm, Δt is in sec, $\Delta h = h_1 - h_2$ and is in cm and A' is constant factor which depends on radius of the hole, depth of hole below the water table (d), and position of lower end of the hole from the impervious bed (S). Value o can be read from the nomograph of Fig. 11.8.

Out of the different methods, auger hole method is the simplest method to determine hydraulic conductivity. However, for layered soils and for soils having roots and holes, and for gravelly soil, this method is not very suitable since it gives inaccurate readings.

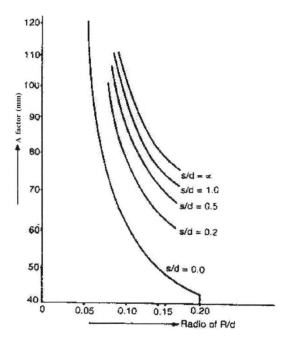

Fig. 11.8: Nomograph for determination of A'

11.2.8.5 Determination of Hydraulic Conductivity above Water Table

In the field, hydraulic conductivity of the soil can also be determined by drilling an auger hole which is not drilled below the water table but some distance above it. Water is filled in the hole and is allowed to flow out from the hole into the surrounding soil. The rate of declination of water level in the hole is taken as observation to evaluate the value of K. Different methods used to evaluate the value of K are (i) shallow well pumping method, (ii) cylinder permeameter method and (iii) double tube method. Details of all these three methods for determination of value of K are available in Suresh (2008).

11.2.8.6 Hydraulic Conductivity by Tracer Method

In tracer method, some tracer materials like salt, dye, or radioactive isotope is mixed with groundwater at a place. After some time say Δt, the distance covered by the tracer material in groundwater at another place is measured. Knowing the value of the distance covered by the tracer say, ΔL in time Δt, average value of velocity of groundwater is determined as $V_{av} = \Delta L/\Delta t$.

Then applying Darcy's law and knowing the value of V_{av}, one can estimate the value of K.

11.2.9 Intrinsic Permeability

Intrinsic permeability *(k)* of a rock or soil is a property of the medium only. It is independent of properties of the fluids. It is related to the hydraulic conductivity, K as:

$$k = \frac{K \mu}{\rho g} \qquad (11.15)$$

where, μ = dynamic viscosity, ρ = density of fluid, and g = acceleration due to gravity and other terms are as defined earlier.

Intrinsic permeability, k has units of m^2 or darcy, equal to 0.987 $(\mu m)^2$.

11.2.10 Transmissivity/Transmissibility

Transmissivity/transmissibility is the rate at which water is transmitted through a unit width of aquifer and with thickness, b under a unit hydraulic gradient. The term transmissibility introduced by Theiss is measured by the coefficient of transmissibility (T) and is defined as the rate of flow of water through a vertical strip of the water bearing material (aquifer) of unit width and thickness, b (depth of a fully saturated zone) under a unit hydraulic gradient.

The relationship between hydraulic conductivity (K) and transmissibility (T) is:

$$T = K \cdot b \qquad (11.16)$$

Unit of T is m^2/sec, lit/day/metre width (lpd/m) and has dimension of L^2/T. A well with value of $T = 1 \times 10^5$ lpd/m is considered satisfactory for irrigation purpose. Typical values of T lie between 1×10^4 to 1×10^6 lpd/m.

11.2.11 Storage Coefficient/Storativity

It is the volume of water that an aquifer releases from or takes into storage per unit surface area of the aquifer per unit change in head. It is dimensionless. In case of unconfined aquifer, storage coefficient is equivalent to specific yield. The storage coefficient of most of the confined aquifers ranges from 10^{-5} to 10^{-3}. On the other hand, the specific yield of unconfined aquifer ranges from 0.1 to 0.3. Sometimes, the term storativity is also used to represent the term storage coefficient.

11.2.12 Hydraulic Resistance

Hydraulic resistance is defined as the resistance offered by the semi-pervious layer on vertical leakage of water. It is reciprocal to leakage factor. It is expressed as:

$$C = \frac{b'}{K'} \tag{11.17}$$

where, C = hydraulic resistance, days; b' = thickness of semi-pervious layer, cm and K' = hydraulic conductivity of semi-pervious layer. Value C indicates whether the aquifer is good or bad. If value of C is zero or near zero, then it is a good aquifer having less resistance to yield of groundwater flow. For a semi-pervious layer, it ranges from 10^2 to 10^6 minutes.

11.2.13 Leakage Factor

Leakage factor is defined as:

$$\lambda = \sqrt{K \cdot b \cdot C} = \sqrt{T \cdot C} \tag{11.18}$$

where, λ = leakage factor, m; K = hydraulic conductivity of the aquifer, m/day; C = hydraulic resistance, day; b = thickness of the aquifer, m and T = transmissibility of aquifer, m^2/day.

If the aquifer is covered at both top and bottom by semi-pervious layers having hydraulic resistances C_1 and C_2, respectively, then the leakage factor is modified as:

$$\lambda = \sqrt{K\, b\, C_1\, C_2} \, / \, \sqrt{(C_1 + C_2)} \tag{11.19}$$

where, K and b are same as in Eq. (ZZZ).

11.2.14 Drainage Factor

Drainage factor, B is defined as:

$$B = \sqrt{\frac{(K \cdot b)}{(\alpha \cdot S_y)}} \tag{11.20}$$

where, S_y = specific yield defined earlier, α = reciprocal of Boulton's delay constant and other terms are defined earlier.

11.2.15 Hydraulic Diffusivity

Hydraulic diffusivity (D) is a parameter defined as:

$$D = \frac{T}{S} = \frac{K\, b}{S_s\, b} = \frac{K}{S_s} \tag{11.21}$$

where, T = transmisibility = $K. b$ (b is thickness of aquifer) and S = storage coefficient = $S_s . b$ (S_s being specific storage).

Sample Calculation 11.6

Compute the value of hydraulic resistance of a semi-pervious layer having thickness 5 m and hydraulic conductivity 10^{-3} m/day.

Solution

Given thickness of the aquifer, = 5 m and hydraulic conductivity, = 10^{-3} m/day.

Hydraulic resistance is computed as,

$$C = \frac{b'}{K'} = 5/10^{-3}$$

$$= 5.0 \times 10^3 \text{ days.}$$

Sample Calculation 11.7

An aquifer is covered at top and at bottom by semi-pervious layers having hydraulic resistances of 5000 and 4000 days, respectively. Hydraulic conductivity of the aquifer is 10^{-3} m/day. If the thickness of the aquifer is 5 m, then compute the leakage factor.

Solution

Given hydraulic resistance of upper semi-pervious layer, C_1 = 5000 days.

Hydraulic resistance of lower semi-pervious layer, C_2 = 4000 days.

Hydraulic conductivity of aquifer, K = 10^{-3} m/day

Thickness of aquifer, b = 5 m.

Leakage factor is computed as:

$$\lambda = \sqrt{K \, b \, C_1 \, C_2} / \sqrt{(C_1 + C_2)}$$

$$= \sqrt{10^{-3} \times 5 \times 5000 \times 4000} / \sqrt{(5000 + 4000)}$$

$$= 3.33 \text{ m}$$

11.3 Groundwater Movement

Darcy's Law: The movement of groundwater is based on the principle of flow through the porous media. The governing equation of flow is described by hydraulic principles reported in 1856 by Henri Darcy, who investigated the flow through porous sand beds. He observed that the flow rate through the porous beds is proportional to the head loss and inversely proportional to the length of the flow path. Darcy's law forms the basis for the governing groundwater flow equations.

The experimental set up of flow of water in a sand column which is inclined to the ground surface is shown in Fig. 11.9. Two piezometers are located in the column at a longitudinal distance of L and the elevation heads of these two piezometers with reference to a datum are z_1 and z_2, respectively. Let the pressure head in these two piezometers be $\dfrac{p_1}{\gamma}$ and, $\dfrac{p_2}{\gamma}$ respectively.

If the velocity of flow at the section 1 (inlet) is V_1 and that at the section 2 (outlet) is V_2, then the velocity head at sections 1 and 2 will be $\dfrac{V_1^2}{2g}$ and $\dfrac{V_2^2}{2g}$, respectively. If the head loss due to friction in the pipe is h_l, then total energy for this system can be expressed by Bernoulli equation as:

$$\frac{p_1}{\gamma}+\frac{V_1^2}{2g}+z_1 = \frac{p_2}{\gamma}+\frac{V_2^2}{2g}+z_2+h_l \tag{11.22}$$

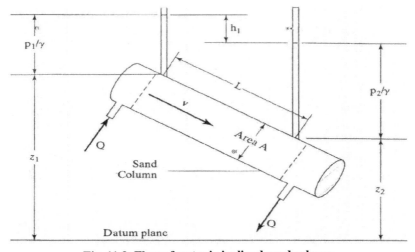

Fig. 11.9: Flow of water in inclined sand column

where, p = pressure, γ = specific weight of water, V = velocity of water, z = elevation head and h_l = head loss due to friction of flow in pipe.

Since the velocity of flow in porous beds are very small and the cylindrical sand column is of regular size (area of cross section of the pipe at both inlet and outlet sections are same), the velocity head at both the sections may be neglected. Under this assumption, the head loss is expressed as:

$$h_l = \left(\frac{p_1}{\gamma} + z_1 \right) - \left(\frac{p_2}{\gamma} + z_2 \right) \tag{11.23}$$

From Eqn. (11.23), it is revealed that the head loss is independent of the inclination of the pipe/sand column.

In Eq. (11.23), the term in right hand side within parenthesis can be written as:

$$H = \frac{p}{\gamma} + z \tag{11.24}$$

where, H = piezometric head and other terms are defined earlier.

Darcy related the flow rate to the head loss and length of column through a proportionality constant referred to as K, hydraulic conductivity which is a measure of the ability of the porous media to transmit water. Darcy's law can be stated as:

$$V = \frac{Q}{A} = -K \frac{dh}{dL} = -K \frac{(H_1 - H_2)}{dL} \tag{11.25}$$

The negative sign indicates that the flow of water is in the direction of decreasing head. For the above mentioned sand column, the head loss $dh = h_l = (H_1 - H_2)$ and is given by Eqn. (11.23) and the length of the sand column, $dL = L$.

Hence, Eqn. (11.25) can be expressed as:

$$V = \frac{Q}{A} = -K \frac{\left(\dfrac{p_1}{\gamma} + z_1 \right) - \left(\dfrac{p_2}{\gamma} + z_2 \right)}{L} \tag{11.26}$$

Eqn. (11.26) represents the average velocity of flow in the entire cross section of sand column of area, A. The actual velocity of flow is limited to the pore channel only, so that the seepage velocity V_s is equal to the Darcy velocity divided by the porosity (η).

Thus, the actual seepage velocity is:

$$V_s = \frac{Q}{\eta \cdot A} = -K \frac{\left(\frac{p_1}{\gamma} + z_1\right) - \left(\frac{p_2}{\gamma} + z_2\right)}{L \cdot \eta} \tag{11.27}$$

It is to be noted that the seepage velocity is much higher than the actual Darcy velocity.

Darcy's law is related to Reynold's number, Re which is defined as:

$$Re = \frac{\rho V d_m}{\mu} \tag{11.28}$$

where, ρ = density of water, μ = dynamic viscosity of water, V = velocity of groundwater and d_m = effective size of particles representing 10% of the finer materials. Reynold's number for groundwater flow varies from 1 to 10. Groundwater flow in nature is mostly in laminar flow condition and thus, it obeys Darcy's law.

Sample Calculation 11.8

An unconfined aquifer comprising sandy soil over lays a compact clay layer. Two piezometers A and B are put in the aquifer at a horizontal distance of 100 m. The piezomeric head readings of A and B are 50 and 40 m, respectively measured above some datum (datum is below the ground level). Value of permeability of the soil in the layer where two piezometers are put is measured to be 1.1 m/day. Find out the velocity of flow. Determine the quantity of flow if the section of flow is 1000 m wide and the thickness of the layer is 20 m.

Solution

Given permeability, $K = 1.1$ m/sec.

Piezometric head readings at A (H_1) and B (H_2) are 50 and 40 m, respectively. So, the piezomtric head difference is 50-40 = 10 m.

Since, distance between the two piezometers are 100 m, the hydraulic gradient, $i = 10/100 = 0.1$.

Using Darcy's law, velocity of flow is given as, $V = K i = 1.1 \times 0.1 = 0.11$ m/day.

Cross sectional area of flow = A = 1000 x 20 = 20000 m².

Hence, the quantity of flow is, $Q = A V = 0.11 \times 20000 = 2200$ m³/day = 91.7 m³/hr = 25.5 l/sec.

11.4 Flow of Water in Horizontal and Vertical Soil Column

11.4.1 Horizontal Soil Column

Flow of liquid occurs in soil column when there is some difference in piezometric /hydraulic head causing hydraulic gradient. As stated earlier, the piezometric head comprises the pressure and gravitational head. The elevation of the point relative to some reference point called as datum determines the gravitational head whereas the pressure head is determined by the height of water column in the piezometer put at that point. In case of horizontal soil column (Fig. 11.10), the reference point is same and so the gravitational head is ignored. The difference of the pressure head/hydraulic head determines the flow. If the pressure head is same at two points, then the hydraulic gradient will be zero and under such case, there will be no flow (static flow). Referring Fig (11.10), the hydraulic gradient is given as:

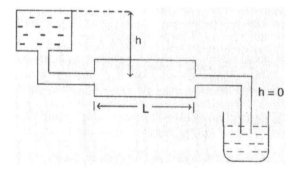

Fig. 11.10: Flow through horizontal soil column

$$i = \frac{h}{L} \tag{11.29}$$

Flow velocity is (by Darcy's law):

$$V = K\frac{h}{L} \tag{11.30}$$

11.4.2 Vertical Soil Column

In case of vertical column (Fig. 11.11), piezometric head at the outlet point is the sum of the hydraulic head and the length of the soil column. If the length of the soil column is L and the hydraulic head difference is h, then the hydraulic gradient in a vertical column of length L is given as:

$$i = \frac{(h + L)}{L} \text{ and} \qquad (11.31)$$

Flow velocity is (by Darcy's law):

$$V = K\,i = K\,\frac{(h + L)}{L}$$

$$V = K + K\,\frac{h}{L} \qquad (11.32)$$

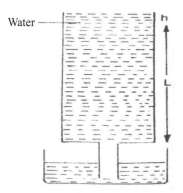

Water

Fig. 11.11: Flow through vertical soil column

It is observed from Eqn. (11.30 and 11.32) that the flow velocity in case of a vertical column is more than that of a horizontal soil column.

It is to be remembered that the soil permeability is not same in all the layers. So, in case of layered soil with varying values of soil permeability, the equivalent value should be considered to determine the flow velocity and flow quantity. Determination of equivalent soil permeability in layered soil with flow direction in both horizontal and vertical direction of soil layers is discussed below.

11.5 Hydraulic Conductivity of Layered Soil

11.5.1 Flow Parallel to Stratification

Let there be n layers of soil profiles of thicknesses $B_1,\ B_2,\ B_3\dots\ B_n$. Let the hydraulic conductivity of these layers be $K_1,\ K_2,\ K_3\dots\ K_n$. Let these layers be parallel to each other. Suppose the flow is parallel to these layers i.e. flow is in the parallel direction to the stratified layers and there is no flow across the boundary of individual layers (Fig. 11.12). In this case hydraulic gradient causing flow in each layer will be same and say it be i.

Let the flow in first, second, third ..... n^{th} layer per unit width be $q_1,\ q_2,\ q_3\dots q_n$.

If total flow per unit width of aquifer is q then

$$q = q_1 + q_2 + q_3 + \dots\dots\dots\dots + q_n \qquad (11.33)$$

Now applying Darcy's law to flow in each layer, one get (for unit width of aquifer)

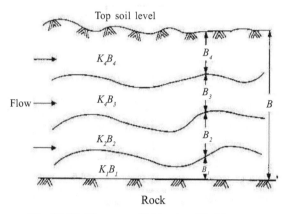

Top soil level

Flow

$K_4 B_4$

$K_3 B_3$

$K_2 B_2$

$K_1 B_1$

B_4

B_3

B_2

B_1

B

Rock

Fig. 11.12: Flow parallel to stratified layers

$q_1 = K_1.\ B_1.\ i$

$q_2 = K_2.\ B_2.\ i$

$q_3 = K3.\ B_3.i$

... .

....

$q_n = K_n.\ B_n.i$

Let K be the equivalent hydraulic conductivity of all the soil layers. Applying Darcy's law to the entire layers,

$$q = K.\ (B_1 + B_2 + B_3 + \ldots\ldots\ldots + B_n).\ i \qquad (11.34)$$

Again from Eq. (11.33), $q = K_1.\ B_1.\ i + K_2.\ B_2.\ i + K_3.\ B_3.i + \ldots\ldots\ldots\ldots\ldots + K_n.\ B_n.i$ $\qquad (11.35)$

Using Eqns. (11.34) and (11.35),

$$K.\ (B_1 + B_2 + B_3 + \ldots.. + B_n).\ i = K_1.\ B_1.\ i + K_2.\ B_2.\ i + K_3.\ B_3.i + \ldots\ldots + K_n.\ B_n.i$$

Simplifying the above equation,

$$K = \frac{K_1\ B_1 + K_2\ B_2 + K_3\ B_3 + \ldots\ldots\ldots + K_n\ B_n}{B_1 + B_2 + B_3 + \ldots\ldots\ldots + B_n}$$

$$= \frac{\sum KB}{\sum B} \qquad (11.36)$$

11.5.2 Flow Perpendicular to Stratification

Let us consider the same (n) number of stratified soil layers which ar parallel to each other each of thickness say $L_1, L_2, L_3 .... L_n$. Let the hydraulic conductivity of these layers be $K_1, K_2, K_3 ..... K_n$. Let the flow be perpendicular to these layers as shown in Fig. 11.13. If there is no lateral flow in the stratified layers then the same discharge will pass in all the layers i.e. in this case $q = q_1 = q_2 = q_3 .... = q_n$.

Total thickness of the aquifer, $L = L_1 + L_2 + L_3 + ... + L_n$

Let $i_1, i_2, i_3 ....i_n$ be the hydraulic gradients of layers one, two, three etc.

If $h_1, h_2, h_3, ...h_n$ are the hydraulic heads in layers one, two, three etc.. then hydraulic gradient

$i_1 = (h_1 - h_2) /L_1, \ i_2 = (h_2 - h_3) /L_2, \ i_3 = (h_3 - h_4) /L_3 \ ...... \ i_n = (h_n - h_{n+1}) /L_n$

Considering unit width of aquifer and applying Darcy's law,

$$q = q_1 = K_1. \ i_1. \ B = K_1. \ B. \ (h_1 - h_2) /L_1$$

where, B = length of flow path.

Thus, $q = q_2 = K_2. \ B. \ (h_2 - h_3) /L_2$

$q = q_3 = K_3. \ B. \ (h_3 - h_4) /L_3$

$q = q_n = K_n. \ B. \ (h_n - h_{n+1}) /L_n$

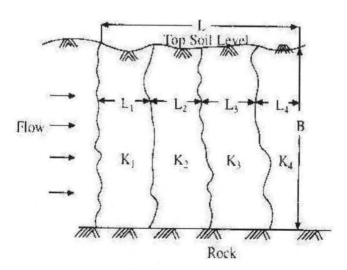

Fig. 11.13: Flow perpendicular to stratified layers

From above equations, we can get

$(h_1 - h_2) = q. L_1/K_1. B, (h_2 - h_3) = q. L_2/K_2. B$......$(h_n - h_{n+1}) = q. L_n/K_n. B$

And adding all the above terms and simplifying,

$(h_1 - h_n). B = q (L_1/K_1) + q (L_2/K_2) + q (L_3/K_3 + ... q (L_n/Kn)$ (11.37)

Considering the entire thickness of the aquifer, $q = K. (h_1 - h_n). B/L$

where, K = equivalent hydraulic conductivity of all layers and rearrangement of this equation gives

$(h_1 - h_n). B = q. L/K$ (11.38)

From Eqns (11.37) and (11.38),

$q. L/K = q (L_1/K_1) + q (L_2/K_2) + q (L_3/K_3 + ... q (L_n/Kn)$ (11.39)

Eq. (11.39) gives, $K = L/ \{(L_1/K_1) + (L_2/K_2) + (L_3/K_3) + ... + (L_n/Kn)\}$

Or, $K = \dfrac{\sum L_i}{\sum \dfrac{L_i}{K_i}}$ (11.40)

Where, $\sum L_i = L_1 + L_2 + L_3 + ...L_n$

Sample Calculation 11.9

Following data were recorded from a site when soil exploration was done. Find out the equivalent soil hydraulic conductivity of all the layers considering (i) flow parallel and (ii) perpendicular to the stratified layers.

Layers	Thickness, cm	Hydraulic conductivity, m/day
A	1.0	1.2
B	2.4	1.5
C	1.5	0.90
D	2.0	1.0
E	1.7	1.5

Solution

When flow is parallel

Using Eq. (11.36), equivalent hydraulic conductivity is given as

$$K = \frac{\sum KB}{\sum B} = \frac{1.2 \, x \, 1.0 + 1.5 \, x \, 2.4 + 0.90 \, x \, 1.5 + 1.0 \, x \, 2.0 + 1.5 \, x \, 1.7}{1.0 + 2.4 + 1.5 + 2.0 + 1.7}$$

$$= 1.244 \text{ m/day}$$

When flow is perpendicular

Using Eq. (11.40), equivalent hydraulic conductivity is given as

$$K = \frac{\sum L_i}{\sum \dfrac{L_i}{K_i}}$$

$$= \frac{1.0 + 2.4 + 1.5 + 2.0 + 1.7}{(\dfrac{1}{1.2} + \dfrac{2.4}{1.5} + \dfrac{1.5}{0.90} + \dfrac{2.0}{1.0} + \dfrac{1.7}{1.5})}$$

$$= 1.189 \text{ m/day}$$

11.6 Hydraulics of Tubewell

In order to have proper design and construction of a tubewell, knowledge of well hydraulics is essential. Well hydraulics is governed by Darcy's law. When water is pumped by a well, water level in the pumping well is lowered down. Thus lowering of water level at any point in the aquifer is termed as drawdown or depression head. The drawdown varies with the distance from the well. It is maximum at centre of tube well and decreases as the distance from tube well increases forming a cone of depression. Draw down is zero at certain distance at which draw down of tube well is zero. This distance at which draw down of tube well is zero is called as radius of influence; R. Values of R depend on the well discharge, aquifer materials, time of pumping etc. In general value of R is maximum for a confined aquifer and minimum for an unconfined aquifer (Campbell and Lehr, 1973).

The lowering of water level in tube well causes head difference towards the well which induces water flow into the well from the surrounding water bearing formations of aquifers. The velocity of flow gradually increases towards the well. With increasing velocity, the hydraulic gradient increases as the flow converges towards the well, as a result the lowered water surface has a steeper slope towards the wells. The form of the surface resembles cone shaped depression. The main equation that governs the flow path in a tubewell is generally given by Darcy's law which states that velocity of flow through a porous media (V) is directly proportional to hydraulic gradient (i) i.e $V \acute{a}\, i$, or $V=K.\, i$, where K is hydraulic conductivity of aquifer. Since $Q = A\, V$, we can write $Q = A.\, K\, I$ (Panigrahi, 2011).

Velocity of flow through a soil depends on soil type, arrangement of grains and viscosity of water. Darcy's law is valid for laminar flow for which Reynolds number is less than 1 and perhaps as high as 10. It should be remembers that at the vicinity of well, the flow has high velocity and high hydraulic gradient that causes flow not to be laminar but turbulent. Under such case Darcy's law cannot be used.

11.7 Groundwater Flow Directions

Flow Nets

Just like surface water flows from higher elevation to the lower elevation, in a much similar way groundwater flows from higher head to the lower head i.e. in the direction of decreasing head. Groundwater flows along the flow lines also called as stream lines. For a given set of boundary conditions, flow lines/stream lines and equipotent lines can be mapped in two dimensions to form a flow net (Fig. 11.14). The two sets of lines form an orthogonal pattern of small squares for isotropic system. In a few simplified cases, the differential equation governing flow can be solved to obtain the flow net (Michael, 1978).

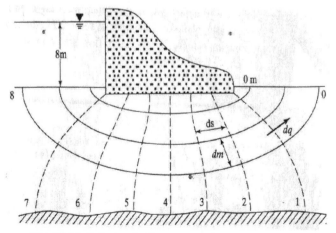

Fig. 11.14: Schematic diagram of a typical flow net

For an isotropic aquifer, equipotential lines are prepared based on the water levels in the observation wells penetrating the aquifer. Flow lines are then drawn at right angles (orthogonal) to them. For the flow net of Fig. 11.14, hydraulic gradient, i is given by

$$i = \frac{dh}{ds} \qquad (11.41)$$

where, ds = distance between two consecutive equipotential lines and dh = head loss between these two consecutive equipotential lines.

If dm is the distance between two consecutive equipotential lines, then per unit thickness of the aquifer, the constant flow rate, q between two adjacent flow lines is:

$$q = K.\frac{dh}{ds} dm \qquad (11.42)$$

If we assume, $ds = dm$ for a square net, then for n squares between two flow lines over which total head, H is divided then we get, $h = H/n$.

If there are m numbers of flow channels, then total discharge, Q per unit thickness of the aquifer is:

$$Q = m\,q = \frac{K\,m\,H}{n} \qquad (11.43)$$

The different notations of Eqn. (11.43) are already mentioned earlier.

It is to be noted that stream lines are parallel to the impermeable bed of the aquifer. Stream lines are also usually horizontal through high value of hydraulic conductivity and vertical through low values of hydraulic conductivity, because of refraction of lines across a boundary between different values of hydraulic conductivity media.

From field measurements of static water levels in wells within a basin, a water level contour map can be prepared. Flow lines sketched perpendicular to contours, show direction of movement. Contour maps of groundwater levels, together with flow lines, are useful data for locating new wells. Convex contours represent areas of groundwater recharge whereas concave contours are associated with areas of groundwater discharge. It is further to be noted that areas with wider contour spacing (flat gradients) possess higher hydraulic conductivities than those with narrow spacing (steep gradient). This means, the prospects for a productive well is better near the section having flat gradient i.e. wider contour spacing.

Sample Calculation 11.10

Estimate the total flow underneath a dam of thickness 20 m as shown in Fig. 11.14. Hydraulic conductivity of the aquifer beneath dam, K is 10^{-4} cm/sec.

Solution

Given $K = 10^{-4}$ cm/sec, m = number of flow channels = 4, n = number of squares over the direction of flow = 8, and H = total head = 8 m.

Using Eqn. (11.43), we get $Q = \dfrac{K \, m \, H}{n}$

$$= \frac{10^{-4} \, x \, 4 \, x \, 8}{8} = 4 \, x \, 10^{-4} = 0.0004 \, m^2 \, / \sec.$$

For 20 m thickness of the dam, total flow = 20 x 0.0004 = 0.008 m³/sec = 691.2 m³/day.

11.8 General Flow Equations

Partial differential equation for flow in porous media is obtained from Darcy law combined with equation of continuity (equation of mass balance). Both steady state and transient flow equations can be derived.

11.8.1 Transient Saturated Flow

Consider a unit volume of porous media (Fig. 11.15) which is called as an elemental control volume. Mass balance equation which is based on law of conservation of mass involves inflow, outflow and change in groundwater storage. As per law of conservation of mass,

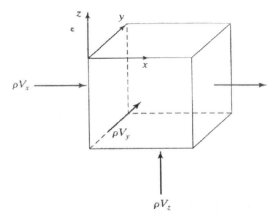

Fig. 11.15: Elemental control volume

Mass in – Mass out = Change in storage per time.

Let Darcy velocity in x-direction be V_x and that in y and z-directions are V_y and V_z, respectively. Density of water is noted as ρ and porosity of media is η.

Mass in – mass out is referred as net flow.

Net flow in x-direction per unit volume of porous media $= -\dfrac{\partial}{\partial x}(\rho V_x)$

Thus, the net flow in y-direction per unit volume of porous media $= -\dfrac{\partial}{\partial y}(\rho V_y)$

and

that in z-direction per unit volume of porous media $= -\dfrac{\partial}{\partial z}(\rho V_z)$

Change in fluid mass storage in the control volume in unit time $= \dfrac{\partial}{\partial t}(\rho \eta)$

As per mass balance equation,

$$\frac{\partial}{\partial x}(\rho V_x) - \frac{\partial}{\partial y}(\rho V_y) - \frac{\partial}{\partial z}(\rho V_z) = \frac{\partial}{\partial t}(\rho \eta) \tag{11.44}$$

Right hand side of Eqn.(11.44) can be expanded as:

$$\frac{\partial}{\partial t}(\rho \eta) = \rho\left(\frac{\partial \eta}{\partial t}\right) + \eta\left(\frac{\partial \rho}{\partial t}\right) \tag{11.45}$$

Using Eqns. (11.44) and (11.45), we get

$$\frac{\partial}{\partial x}(\rho V_x) - \frac{\partial}{\partial y}(\rho V_y) - \frac{\partial}{\partial z}(\rho V_z) = \rho\left(\frac{\partial \eta}{\partial t}\right) + \eta\left(\frac{\partial \rho}{\partial t}\right) \tag{11.46}$$

Eqn. (11.46) represents the transient saturated flow equation for a confined aquifer.

The first term on the right hand side of Eqn. (11.46) represents mass rate of water produced by compaction of the porous media (change in η) where as the second term on the right hand side of Eqn. (11.46) represents mass rate of water produced by an expansion of water under a change in ρ. The first term refers to the aquifer compressibility (α) and the second term relates to the compressibility of the fluid (β).

A change in ρ and η produces a change in hydraulic head, h and that the volume of water produced by the two mechanisms for a unit decline in head, h is S_s where S_s is the specific storage.

Specific storage is given by equation:

$$S_s = \rho g (\alpha + \eta \beta) \tag{11.47}$$

The mass rate of water produced (time rate of change of fluid mass storage) is given as:

$$\text{Mass rate of water produced} = (\rho S_s \, \partial h / \partial t) \tag{11.48}$$

From Eqns. (11.46) and (11.48) we get,

$$\frac{\partial}{\partial x}(\rho V_x) - \frac{\partial}{\partial y}(\rho V_y) - \frac{\partial}{\partial z}(\rho V_z) = \rho S_s \, \partial h / \partial t \tag{11.49}$$

Applying Darcy law and considering an incompressible fluid, we can write Eqn. (11.49) as:

$$\frac{\partial}{\partial x}\left(K_x \frac{\partial h}{\partial x}\right) + \frac{\partial}{\partial y}\left(K_y \frac{\partial h}{\partial y}\right) + \frac{\partial}{\partial z}\left(K_z \frac{\partial h}{\partial z}\right) = S_s \frac{\partial h}{\partial t} \tag{11.50}$$

For a homogeneous and isotropic media, $K_x = K_y = K_z = K$ and Eqn. (11.50) can be written as:

$$\frac{\partial^2 h}{\partial x^2} + \frac{\partial^2 h}{\partial y^2} + \frac{\partial^2 h}{\partial z^2} = \frac{S_s}{K} \cdot \frac{\partial h}{\partial t} \tag{11.51}$$

Eqn. (11.51) can also be written as:

$$\nabla^2 h = \frac{S_s}{K} \cdot \frac{\partial h}{\partial t} \tag{11.52}$$

For case of a horizontal confined aquifer of thickness, b, the storativity or storage coefficient, S is equal to $S_s b$ and transmissivity, T is equal to $K.b$.

Hence, for horizontal isotropic homogeneous confined aquifer, Eqn. (11.52) can be written as:

$$\nabla^2 h = \frac{S}{T} \cdot \frac{\partial h}{\partial t} \quad \text{in two dimensions.} \tag{11.53}$$

11.8.2 Steady State Saturated Flow

For a steady state condition, change in storage in the elemental control volume is zero and hence right hand side of Eqn. (11.53) is zero.

Eqn. (11.53) then can be written as

$$\nabla^2 h = 0 \; i.e. \; \frac{\partial^2 h}{\partial x^2} + \frac{\partial^2 h}{\partial y^2} + \frac{\partial^2 h}{\partial z^2} = 0 \tag{11.54}$$

Eqn. (11.54) is called as Laplace equation used for steady, incomprehensible and isotopic flow case and is one of the best understood partial differential equation. The solution of Eqn. (11.54) in two dimensions can be done by graphical flow nets.

Dupuit Assumptions

For the case of unconfined groundwater flow, Dupuit developed a theory that allows for a simple solution based on several important assumptions. These assumptions are:

(i)   The water table or free surface is only slightly inclined.

(ii)   Streamlines may be considered horizontal and equipotential lines are vertical.

(iii) Slopes of the free surface and hydraulic gradient are equal.

11.9 Steady State Well Hydraulics

In case of steady state flow, the variation in head occurs only in space and not in time. The general flow equations as discussed above can be solved for pumping wells in both unconfined and confined aquifers under steady as well as unsteady conditions. However, for easy in computation, the boundary conditions must be kept relatively simple and aquifers must be assumed as homogeneous and isotropic.

Steady One-Dimensional Flow

Steady state groundwater flow in one dimension (x-direction) for isotropic and homogeneous aquifer can be written from Laplace equation (Eqn. 11.54) as:

$$\frac{\partial^2 h}{\partial x^2} = 0 \tag{11.55}$$

Solution of Eqn. (11.55) leads to:

$$h = \frac{-v\,x}{K} \tag{11.56}$$

where, $h = 0$ and $x = 0$ and $dh/dx = -v/K$ according to Darcy's law. This states that head varies linearly with flow in x-direction. The steady one-dimensional flow in an unconfined aquifer is derived with Dupuit's assumptions. The resulting variation in head with x is called the Dupuit Parabola which represents the approximate shape of the water table for relatively flat slopes. If steep slopes near the well exits, Dupuit's assumptions do not hold good and may lead erroneous result if used as such. However, under such case complex mathematical models are available for solution (Hansen *et al.*, 1979).

11.10 Steady Radial Flow to a Well

Assumptions

Following assumptions are required before deriving the flow equations in well whether it is in confined or unconfined aquifer.

(i)   The well is pumped at constant rate (steady state).

(ii) The well is fully penetrating the aquifer.

(iii) Aquifer is homogeneous, isotropic, horizontal and of infinite horizontal extent.

(iv) Water is released from the storage in the aquifer or other underground materials in immediate response to drop in water level or piezometric surface.

(v)   Well storage is neglected.

(vi) Piezometric surface is also horizontal.

11.10.1 Confined Aquifer

Well discharge, Q at any radial distance, r from the centre of the well for steady state case with thickness of aquifer being b is given by Darcy's law as:

$$Q = 2\pi\,r\,b\,K\,\frac{dh}{dr} \tag{11.57}$$

Let the radius of the well be r_w.

At the periphery of the well when $r = r_w$, the head, $h = h_w$ (Fig. 11.16)

Integrating Eqn. (11.57) with boundary conditions of $r = r_w$, when head, $h = h_w$ and then simplifying we get:

$$Q = 2\pi \ b K \frac{h - h_w}{\ln\left(r / r_w\right)} \tag{11.58}$$

Since transmissivity, $T = K.b$

Eqn. (11.58) can be written as:

$$Q = 2\pi \ T \frac{h - h_w}{\ln\left(r / r_w\right)} \tag{11.59}$$

Eqn. (11.59) represents that as radial distance, r increases, the head, h_w also increases, yet the maximum head is h_o when the radial distance, $r = r_o$ (r_o is called as radius of influence where the draw down because of well pumping is zero).

Now Eqn. (11.59) can be written as:

$$Q = 2\pi \ T \frac{h_0 - h_w}{\ln\left(r_0 / r_w\right)} \tag{11.60}$$

Eqn. (11.60) is called as Dupuit's equation.

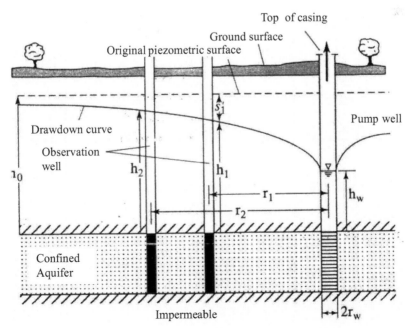

Fig. 11.16: Radial flow to a well in confined aquifer

It is to be remembered that as the head increases, drawdown decreases. The drawdown at any radial distance from the well is the difference of the original piezometric surface and the water level of an observation well at that radial distance in the aquifer.

If we consider two observation wells in the aquifer in the vicinity of the main pumping well at radial distances r_1 and r_2 where the heads are h_1 and h_2, respectively we can write Eqn. (11.59) as:

$$Q = 2\pi \, T \frac{h_2 - h_1}{\ln\left(r_2 / r_1\right)} \tag{11.61}$$

Let the drawdown at observation wells at radial distances r_1 and r_2 be s_1 and s_2, respectively.

Now we can write, $s_2 = h_o - h_2$ and $s_1 = h_o - h_1$ and using this hypothesis, we can write Eqn. (11.61) in terms of drawdown as:

$$Q = 2\pi \, T \frac{s_1 - s_2}{\ln\left(r_2 / r_1\right)} \tag{11.62}$$

Eqns. Eqn. (11.61) and (11.62) are also called as Thiem's equation. The difference of Thiem's and Dupuit'equations are that in Thiem's equation we consider two observation wells to get well discharge whereas in Dupuit's equation, no observation wells are taken and the well discharge is calculated in terms of radial distances and heads at the boundary limiting conditions of near the well and at the radial distance of radius of influence.

11.10.2 Unconfined Aquifer

Applying Darcy's law for radial flow to a fully penetrating well in unconfined, isotropic, horizontal and homogeneous aquifer and using Dupuit's assumptions, we get well discharge, Q at any radial distance, r where head is h (Fig. 11.17) as:

$$Q = 2\pi \, r \, h \, K \frac{dh}{dr} \tag{11.63}$$

Let us consider two observation wells at radial distances r_1 and r_2 where heads are h_1 and h_2, respectively (respective drawdown at r_1 and r_2 are s_1 and s_2, respectively).

Eqn. (11.63) can be integrated within the boundary limits of r_1 and r_2 where heads are h_1 and h_2, and then simplifying we get:

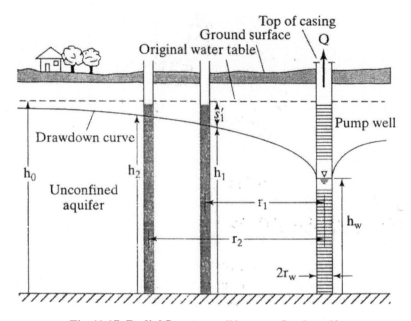

Fig. 11.17: Radial flow to a well in unconfined aquifer

$$Q = \pi K \frac{h_2^2 - h_1^2}{\ln\left(r_2 / r_1\right)} \qquad\qquad 11.64)$$

In terms of drawdown, Eqn. (11.64) can be written as follows:

$$h_2^2 - h_1^2 = \left(h_2 - h_1\right)\left(h_2 + h_1\right)$$
$$= \left(s_1 - s_2\right)\left(h_2 + h_1\right)$$

If the amount of drawdown is small compared to the saturated thickness of the aquifer, then h_2 and h_1 are nearly equal and each approximately equals to the saturated thickness of aquifer h_o.

Hence, $h_2^2 - h_1^2 = \left(s_1 - s_2\right)\left(2h_0\right)$

Putting this value in Eqn. (11.64) we get,

$$Q = 2\pi K h_0 \frac{s_1 - s_2}{\ln\left(r_2 / r_1\right)}$$

$$= 2\pi T \frac{s_1 - s_2}{\ln\left(r_2 / r_1\right)} \qquad\qquad (11.65)$$

Eqns. (11.64) and (11.65) are called as Thiem's formula for unconfined aquifer.

If we consider radial distance as radius of influence, r_o where the head is h_0 (this head is also equal to the saturated thickness of the aquifer) and a radial distance r_w where the head is h_w, Eqn. (11.64) becomes

$$Q = \pi \ K \ \frac{h_0^{\ 2} - h_w^{\ 2}}{\ln \left(r_0 / r_w \right)} \tag{11.66}$$

Eqn. (11.66) is called as Dupuit's formula for unconfined aquifer.

11.10.3 Unconfined Aquifer with Uniform Recharge

When there is some recharge due to rainfall or infiltration of excess irrigation water that causes deep percolation, the flow towards the well will increase.

Let us consider a cylinder of thickness dr and uniform recharge rate as W.

The increase of discharge dQ, through this cylinder situated at a radial distance of r from the centre of the main pumping well is given as:

$$dQ = -2 \pi r \ dr \ W \tag{11.67}$$

Integrating above equation we get,

$$Q = -\pi r^2 \ W + C \tag{11.68}$$

At the well, $r = 0$, $Q = Q_w$, so that Eqn. (11.68) becomes

$$Q = -\pi r^2 \ W + Q_w \tag{11.69}$$

Substituting this flow in the equation for flow to the well gives

$$2 \pi r \ K \ h \frac{dh}{dr} = -\pi r^2 \ W + Q_w \tag{11.70}$$

Integrating Eqn. (11.70) and considering the limit that at $r = r_0$, $h = h_0$ we get

$$h_0^{\ 2} - h^2 = \frac{W}{2K}\left(r^2 - r_0^{\ 2} \right) + \frac{Q_w}{\pi \ K} \ln \frac{r_0}{r} \tag{11.71}$$

It follows that as $r = r_0$, $Q = 0$ and then Eqn. (11.69) gives

$$Q_w = \pi \ r_0^{\ 2} \ W$$

Thus, the total flow of the well equals the recharge within the circle defined by the radius of influence, r_0.

11.11 Interference Among Wells

If two or more wells are constructed in such a way that they are near to each other and their cones of depressions interact, they are said to interfere. Such interference of wells decreases the discharges of such interfering wells. Principle of linear superposition can be applied in finding the total drawdown of a number of wells (multiple well systems) such that drawdown of each well is interfering with that of the other.

Thus,

$$s_t = s_1 + s_2 + s_3 + \ldots \ldots s_n$$

where, s_t is the total drawdown at a given point, $s_1, s_2 \ldots \ldots s_n$ are the drawdown at the point caused by each well individually.

Considering a well field in confined aquifer having n number of wells, the drawdown, s at a given point in the area of influence is given by the equation as (Murty, 2004):

$$s = \sum_{1}^{n} \frac{Q_i}{2 \pi K b} \ln \frac{R_i}{r_i} \qquad (11.72)$$

w here, R_i = distance from the ith well to a point at which the drawdown becomes negligible and r_i = distance from the ith well to the given point.

Muskat has proposed the following formulae for computation of discharge from such interfering wells. These formulae are found to give reliable results.

(a) For Confined Aquifer (Artesian Well)

For two artesian identical wells (same diameters) at a distance B apart,

$$Q_1 = Q_2 = 2\pi \ T \frac{h_0 - h_w}{\ln \left(\dfrac{r_0^2}{r_w B} \right)} \qquad (11.73)$$

For three artesian identical wells (same diameters) at a distance B apart in a pattern of equilateral triangle:

$$Q_1 = Q_2 = Q_3 = 2\pi \ T \frac{h_0 - h_w}{\ln \left(\dfrac{r_0^3}{r_w B^2} \right)} \qquad (11.74)$$

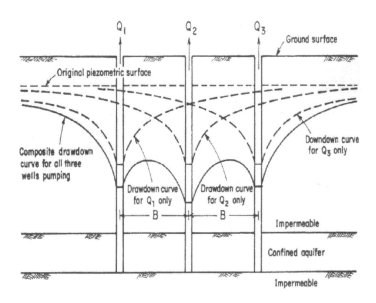

Fig. 11.18: Discharge of well when three wells are in a line

For three artesian identical wells (same diameters) at a distance B apart in a line (Fig. 11.18), the discharge in the first and the third well is (Murty, 2004):

$$Q_1 = Q_3 = \frac{2 \pi T \left(h_0 - h_w \right) \ln \left(\dfrac{B}{r_w} \right)}{2 \ln \left(\dfrac{R}{B} \right) \ln \left(\dfrac{B}{r_w} \right) + \ln \left(\dfrac{B}{2 r_w} \right) \ln \left(\dfrac{R}{r_w} \right)} \tag{11.75}$$

whereas the middle well gives discharge, Q_2 as (Murty, 2004):

$$Q_2 = \frac{2 \pi T \left(h_0 - h_w \right) \ln \left(\dfrac{B}{2 r_w} \right)}{2 \ln \left(\dfrac{R}{B} \right) \ln \left(\dfrac{B}{r_w} \right) + \ln \left(\dfrac{B}{2 r_w} \right) \ln \left(\dfrac{R}{r_w} \right)} \tag{11.76}$$

(b) For Unconfined Aquifer (Gravity Well)

For two identical gravity wells (same diameters) at a distance B apart, discharge is (Sharma and Sharma, 2002):

$$Q_1 = Q_2 = \frac{2\pi\, K \left(\dfrac{h_0^2}{2} - \dfrac{h_w^2}{2} \right)}{\ln \left(\dfrac{r_0^2}{r_w\, B} \right)}$$

$$\text{or } Q_1 = Q_2 = \frac{\pi\, K \left(h_0^2 - h_w^2 \right)}{\ln \left(\dfrac{r_0^2}{r_w\, B} \right)} \tag{11.77}$$

For three identical gravity wells (same diameters) at a distance B apart in a pattern of equilateral triangle, discharge is (Sharma and Sharma, 2002):

$$Q_1 = Q_2 = Q_3 = \frac{\pi\, K \left(h_0^2 - h_w^2 \right)}{\ln \left(\dfrac{r_0^3}{r_w\, B^2} \right)} \tag{11.78}$$

The various terms of the Eqns. (11.73) to (11.78) have already been defined earlier.

11.12 Well Loss and Specific Capacity

The drawdown in a pumped well represents the hydraulic head loss consisting of

(i) Aquifer head loss (also called as formation loss) and

(ii) Well loss

Well loss is caused due to flow through the well screen and flow inside the well to the pump intake. Well loss, because of turbulence in flow near the well can be considered proportional to. Q^n where n is a constant greater than 1. Value of n will be 1 for sufficiently low pumping rate so that the flow in the vicinity of the well is laminar and for discharge condition with turbulence case, n is greater than 1 (Das Gupta, 1987; Sharma, 1987).

The drawdown because of well loss can, hence, be written as:

$$s_I = C\, Q^n$$

where C is well loss coefficient.

For a confined aquifer, the drawdown in the aquifer (s_2) because of a pumping well of radius, r_w and radius of influence, r_o having a discharge of Q is given as:

$$s_2 = \frac{Q}{2 \pi T} \ln\left(\frac{r_0}{r_w}\right) = B\,Q \qquad (11.79)$$

where, $B = \dfrac{1}{2 \pi T} \ln\left(\dfrac{r_0}{r_w}\right)$

B is called as formation loss coefficient.

Total head loss (s_w) considering both aquifer loss and well loss is:

$$s_w = s_1 + s_2 = B\,Q + C\,Q^n \qquad (11.80)$$

The total drawdown, s_w as shown in Fig. 11.19 is summation of drawdown due to formation loss, $B\,Q$ and well loss $C\,Q^n$.

The size of the well has little significance on well discharge. However, it has significant influence on well loss. For simplicity, in turbulence case, sometimes the value of n is taken as 2. Considering $n = 2$ Eqn. (11.80) is written as:

$$s_w = B\,Q + C\,Q^2 \qquad (11.81)$$

Eqn. (11.81) can be written as:

$$\frac{s_w}{Q} = B + C\,Q \qquad (11.82)$$

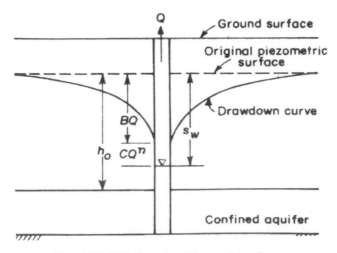

Fig. 11.19: Relation of well loss to drawdown

Plotting s_w/Q versus Q, a linear relationship is obtained. The y-intercept of the linear line gives value of B whereas the slope gives value of C.

Specific capacity of a well is a measure of productivity of well. Larger the specific capacity better is the performance of the well. It is given by ratio of discharge to drawdown i.e. Q/s_w. It can be written as:

$$Specific\ capacity = \frac{Q}{B\,Q + C\,Q^n}$$

Considering $n = 2$ we get,

$$Specific\ capacity = \frac{Q}{B\,Q + C\,Q^2}$$

$$or\ Specific\ capacity = \frac{1}{B + C\,Q} \tag{11.83}$$

Eq. (11.83) clearly indicates that the specific capacity of a well is not constant but decreases as the discharge increases. An increase in well loss due to clogging or deterioration of the well screen lowers the specific capacity which indicates low performance of the well.

11.13 Efficiency of Well

From the discussions as above, we may conclude that if the well loss is neglected, the discharge of a well will vary with drawdown. The discharge of the well per unit drawdown is called as specific capacity which is a measure of performance of the well. This specific capacity of a well will be different for different well designs. For determining the well efficiency, the well is pumped for varying drawdown conditions and then a graph is plotted with percent discharge (percent yield) as ordinate and percent drawdown as abscissa as shown in Fig. 11.20. The curve is called as yield-drawdown curve. The curve will be straight line up to certain value of drawdown and then it increases disproportionately to the yield. This places an efficient and optimum limit to the drawdown which may be allowed to be created in a well. This is found to be 70 percent of the maximum drawdown which can be created in a well.

The well efficiency (E_w) is a ratio of formation loss to total drawdown and is:

$$E_w = \frac{B\,Q}{B\,Q + C\,Q^n} \times 100 \tag{11.84}$$

For a small discharging with laminar flow case, n is taken as 1 and then Eqn. (11.84) is written as:

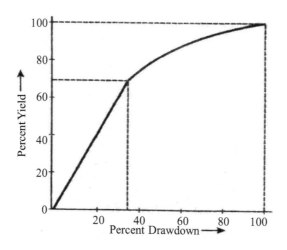

Fig. 11.20: Yield-drawdown curve

$$E_w = \dfrac{1}{1 + \dfrac{C}{B}} \, x \, 100 \tag{11.85}$$

For turbulence case, the well efficiency is:

$$E_w = \left(\dfrac{1}{1 + \dfrac{C \, Q^{n-1}}{B}} \right) x \, 100 \tag{11.86}$$

Eqn. (11.85) reveals that the well efficiency for a laminar flow case with artesian well is independent of pumping whereas it decreases with increasing discharge for turbulence case in artesian well (Eqn. 11.86).

Efficiency of a well is generally less than 100 percent. For an adequately developed well with its screen not affected by clogging or incrustation, the well efficiency is about 80 percent.

Sample Calculation 11.11

In a pumping test in medium sand and gravel, impermeable bed was found to lie at a distance of 14 m below the ground level. The normal groundwater level was at the surface. Two observation wells were located at a distance of 7.5 and 3m, respectively from the main pumping well where the drawdown were measured to be 0.3 and 1.5 m, respectively. If the discharge of the main pumping well was 5 lit/sec, calculate the value of coefficient of permeability of the soil.

Solution

The well is present in unconfined aquifer. The saturated thickness of the aquifer, h_0 is 14 m.

At radial distance, $r_1 = 3$ m, drawdown, $s_1 = 1.5$ m and hence, head, $h_1 = 14 - 1.5 = 12.5$ m.

Similarly at radial distance, $r_2 = 7.5$ m, the drawdown, $s_2 = 0.3$ m and hence, head, $h_2 = 14 - 0.3 = 13.7$ m.

The discharge is, $Q = 5$ lit/sec = 0.005 m³/sec.

Using Thiem's formla (Eq. 11.64), coefficient of permeability can be computed as

$$Q = \pi\, K\, \frac{h_2^{\,2} - h_1^{\,2}}{\ln\left(r_2 / r_1\right)}$$

$$0.005 = (\pi\, K)\left[\frac{(13.7)^2 - (12.5^2)}{\ln\left(7.5/3\right)}\right]$$

which gives $K = 0.0000463$ m/sec. = 0.00463 cm/sec.

Sample Calculation 11.12

A well penetrates fully to an unconfined aquifer of thickness 100 m. The drawdown in the main pumping well at a discharge of 200 lit/min is 15 m. Estimate the discharge of the same main pumping well at a drawdown of 20 m assuming steady state condition. Assume homogeneous and isotropic aquifer. The radius of influence is same for both the cases.

Solution

Given, discharge of the main pumping well, $Q = 200$ lit/min when the drawdown, $s_w = 15$ m.

Saturated thickness of the unconfined aquifer, $h_0 = 100$ m.

Drawdown in the pumping well, $s_w = 15$ m and hence head $h_w = 100 - 15 = 85$ m.

Drawdown in the second case in the same well, $s_1 = 20$ m and hence head $h_1 = 100 - 20 = 80$ m. We have to find out the new discharge at this drawdown i.e. at $s_1 = 20$ m.

Let radius of influence be r_0 and radius of well be r_w.

Now using Dupuit's formula (Eqn. 11.66) we have

$$Q = \pi K \frac{h_0^2 - h_w^2}{\ln (r_0 / r_w)}$$

In the first case: $h_w = 85$ m

Hence, $\dfrac{\pi K}{\ln (r_{0/r_w})} = \dfrac{Q}{\left(h_0^2 - h_w^2\right)}$

$$= \frac{200}{(100)^2 - (85)^2} = 0.0721 \tag{i}$$

In the second case: $h_1 = 80$ m and h_0 is same i.e. 100 m.

Now we can write the Dupuit's formula for the second case to find discharge as:

$$Q = \pi K \frac{h_0^2 - h_1^2}{\ln (r_0 / r_w)} \tag{ii}$$

Putting the value of $\dfrac{\pi K}{\ln (r_{0/r_w})}$ from Eqn. (i) in Eqn. (ii) (since for both the cases, K, r_0 and r_w are same)

$Q = 0.0721 \times [(100)^2 - (80)^2]$

$= 259.60 = 260$ lit/min.

Sample Calculation 11.13

A 30 cm size well penetrates 20 m below the ground surface. Static water level in the aquifer is 2 m below the ground surface. After 24 hours of pumping at a discharge of 5400 lit/min, the water level in the test well at 100 m is lowered by 0.50 m and in a second well at distance of 30 m away, water level is lowered by 1 m. Estimate the drawdown in the main pumping well and transmissivity of the aquifer.

Solution

The well penetrates 20 m below the ground surface and the piezometric level i.e. static water level is 2 m. Hence, the saturated thickness of the unconfined aquifer, $h_0 = 20 - 2 = 18$ m.

The Thiem's equilibrium formula for unconfined aquifer (Eq. 11.64) is:

$$Q = \pi K \frac{h_2^2 - h_1^2}{\ln\left(r_2 / r_1\right)} \tag{i}$$

Here, at distance $r_2 = 100$ m, head, $h_2 = 18 - 0.5 = 17.5$ m and

At $r_1 = 30$ m, head, $h_1 = 18 - 1 = 17$ m

Given discharge, $Q = 5400$ lit/min $= 0.09$ m³/sec.

Substituting the above values in Eqn. (i), we have:

$$0.09 = (\pi K)\left[\frac{(17.5)^2 - (17^2)}{\ln\left(100 / 30\right)}\right]$$

$$= 54.20 \, K$$

$K = 0.001661$ m/sec $= 0.1661$ cm/sec.

Transmissivity, $T = K. \, h_o = 0.001661 \times 18 = 0.03$ m²/sec.

To determine the drawdown in the main well, use the above Eqn. (i) as:

$$Q = \pi K \frac{h_1^2 - h_w^2}{\ln\left(r_1 / r_w\right)} \text{ [using } h_1 \text{ in place of } h_2 \text{ and } h_w \text{ in place of } h_1\text{]}$$

$$\text{or } 0.09 = \pi \times 0.001661 \times \left(\frac{17^2 - h_w^2}{\ln\left(\dfrac{30}{0.15}\right)}\right)$$

On solution, it gives $h_w = 14.07$ m.

Sample Calculation 11.14

A fully penetrating tubewell in confined aquifer has thickness 24 m. The aquifer is met at 25 m below the ground level. The static water table is 15 m below the ground level. The discharge of the main well is measured to be 3000 m³/day, when the depression head (drawdown) in it is 6 m. Permeability of the aquifer is 24.5 m/day. Assuming radius of influence of 300 m, find out the size of the tubewell.

Solution

Given the case of a fully penetrating well in confined aquifer.

Thickness of the aquifer, $b = 24$ m.

The aquifer is met at 25 m below the ground level and the static water table is 15 m below the ground level. Hence, head of water at radius of influence, $r_0 = 300$ m is:

$h_o = 25 - 15 = 10$ m.

The drawdown at the main well, $s_w = 6$ m and so head $h_w = h_0 - s_w = 10 - 6 = 4$ m.

Using Dupuit's formula (Eqn. 11.60) we have:

$$Q = 2\pi \ T \ \frac{h_0 - h_w}{\ln\left(r_0 / r_w\right)} \qquad\qquad (i)$$

Given $Q = 3000 \text{m}^3/\text{day}$ and $K = 24.5$ m/day.

Transmissivity, $T = K$. $b = 24.5 \times 24 = 588 \text{ m}^2/\text{day}$.

Radius of influence, $r_0 = 300$ m

Putting these values in Eqn. (i) above we have,

$$3000 = 2\pi \ x \ 588 \ \frac{(10 - 4)}{\ln\left(300 / r_w\right)}$$

Solving the above equation, we get, radius of the well, $r_w = 0.185$ m $= 18.5$ cm.

Size of the well = diameter of the well = 37 cm.

Sample Calculation 11.15

A well of radius 30 cm is pumped at a rate of 81.6 m³/hr. Two observation wells were dug at a distance of 6 and 15 m, respectively where the drawdown were measured to be 6 and 1.5 m, respectively. The bottom of the impermeable bed up to which the well is being pumped is 90 m below the ground surface. Piezometric surface lies on the ground surface too. (a) Compute the coefficient of permeability. (b) If all the observed points were on the Dupuit's curve, what was the drawdown in the well being pumped? (c) Find out the specific capacity of the well. (d) Find out also the maximum rate at which the well can be pumped?.

Solution

Given distance $r_1 = 6$ m where drawdown, $s_1 = 6$ m and at $r_2 = 15$ m, $s_2 = 1.5$ m.

Clearly it is the case of an unconfined aquifer and saturated thickness, h_o of the aquifer is 90 m.

Hence, head $h_1 = 90 - 6 = 84$ m and $h_2 = 90 - 1.5 = 88.5$ m

Discharge, $Q = 81.6$ m³/hr $= 1.36$ m³/sec

(a) Using Thiem's formula (Eqn. 11.64) we have:

$$Q = \pi K \frac{h_2^2 - h_1^2}{\ln(r_2/r_1)}$$

$$1.36 = \pi K \frac{(88.5)^2 - (84)^2}{\ln(15/6)}$$

Solution of the above gives K = 0.00051 m/min = 0.051 cm/min.

(b) Radius of the well, $r_w = 0.3$ m., $r_2 = 15$ m, $h_2 = 88.5$ m and we have to find out h_w.

Thiem's formula can now be written as:

$$Q = \pi K \frac{h_2^2 - h_w^2}{\ln(r_2/r_w)} \qquad (i)$$

Putting the above values in Eqn. (i), we have

$$1.36 = \pi \times 0.00051 \frac{(88.5)^2 - h_w^2}{\ln(15/0.3)}$$

Solution of the above equation gives $h_w = 67.4$ m.

Hence, drawdown in the well $= 90 - 67.4 = 22.6$ m.

(c) Specific capacity of the well is the ratio of discharge per unit drawdown in the pumped well. .

In this case we have to first find out the radius of influence (r_0) of the well.

Using Dupuit's equation (11.66) we have:

$$Q = \pi K \frac{h_0^2 - h_w^2}{\ln\left(r_0 / r_w\right)}$$

$$1.36 = \pi \times 0.00051 \frac{(90)^2 - (67.4)^2}{\ln\left(r_0 / 0.3\right)}$$

Solution of the above equation gives $r_0 = 20.01$ m $= 20$ m.

Now, specific capacity is discharge per unit drawdown. Hence, we have to find out the discharge when the drawdown in the well, $s_w = 1$ m i.e. head, $h_w = 90 - 1 = 89$ m.

Thiem's formula is used to find out the specific capacity as:

$$Specific\ capacity = \pi \times 0.00051 \frac{(90)^2 - (89)^2}{\ln\left(20 / 0.3\right)}$$

$$= 0.0683\ \text{m}^3/\text{min} = 68.3\ \text{lit/min}.$$

(d) Maximum discharge will occur when head in the well, h_w is zero

Now using Thiem's formula:

$$Q_{max.} = \pi \times 0.00051 \frac{(90)^2 - (0)^2}{\ln\left(20 / 0.3\right)} = 3.09\ \text{m}^3/\text{min} = 3090\ \text{lit/min}.$$

Sample Calculation 11.16

Two tubewells penetrating fully a 10 m thick confined aquifer are located 200 m apart. Tubewell has diameter equals to 30 cm, depression head = 5 m and radius of influence is 300 m. Coefficient of permeability is 0.1 cm/sec. (a) Find out if the wells are interfering? (b) Compute the discharge of the tubewell when one well is pumping. (c) Also find out the percentage decrease in discharge of the well if both the wells are working under a depression head of 5 m. (d) If a third well is also located 200 m away from both the wells so that these three wells form a pattern of equilateral triangle, what will happen to the discharge of each well.

Solution

(a) Given the radius of influence of the wells is 300 m but the two wells are located at 200 m apart. Hence, the wells are interfering.

(b) When one tubewell is working, the discharge is, Q

$$Q = 2\pi \ T \frac{h_0 - h_w}{\ln\left(r_0 / r_w\right)}$$

$$= 2\pi \ x \ 10 \ x \ 0.001 \frac{5}{\ln\left(300 / 0.15\right)} \text{ (given depression head, } h_o - h_w = 5 \text{ m)}$$

$$= 0.041 \text{ m}^3/\text{sec}$$

(c) When both the wells are working, the discharge of each well is (Eqn. 11.73):

$$Q_1 = Q_2 = 2\pi \ T \frac{h_0 - h_w}{\ln\left(\dfrac{r_0^2}{r_w \ B}\right)} \tag{i}$$

Given, depression head, $(h_o - h_w) = 5$ m. and distance between two tubewells, $B = 200$ m

Using these values and Eqn. (11.73) as above we get,

$$Q_1 = Q_2 = 2\pi \ x \ 0.001 \ x \ 10 \ x \frac{5}{\ln\left(\dfrac{300^2}{0.15 \ x \ 200}\right)} = 0.039 \text{ m}^3/\text{sec}$$

Percentage decrease in discharge is given as:

$$= \frac{Q_1 - Q_2}{Q_1} \ x \ 100$$

$$= \frac{0.041 - 0.039}{0.041} \ x \ 100 = 4.87 \text{ say 5 percent.}$$

(d) If a third well is also located 200 m away from both the wells so that these three wells form a pattern of equilateral triangle, then the discharge of each well is given by Eqn. (11.74) as:

$$Q_1 = Q_2 = Q_3 = 2\pi \ T \ \frac{h_0 - h_w}{\ln\left(\dfrac{r_0^3}{r_w \ B^2}\right)}$$

$$= 2\pi \ x \ 0.001 \ x \ 10 \ x \ \frac{5}{\ln\left(\dfrac{300^3}{0.15 \ x \ 200^2}\right)}$$

$$= 0.037 \ \text{m}^3/\text{sec}$$

Hence discharge of each well decreases.

Percentage decrease in discharge is given as:

$$= \frac{Q_1 - Q_2}{Q_1} \ x \ 100$$

$$= \frac{0.041 - 0.037}{0.041} \ x \ 100 = 9.8 \ \text{say 10 percent.}$$

Sample Calculation 11.17

In a large basin, one tubewell having 5 cm radius is working. The permeability of the soil is 300 cm/day. Radius of influence is 1.5 km. Watertable lies 50 m above the bottom of the impermeable bed. Maximum drawdown in the well is 15 m. After some time, two more wells were dig such that all the three wells form an equilateral triangle with 250 m distance apart. Assuming size of all the three wells is same and they discharge equally, calculate percentage loss of discharge in the first well due to interference of other two wells.

Solution

Given $r_w = 5$ cm $= 0.05$ m

Radius of influence, $r_0 = 1.5$ km $= 1500$ m.

Head, $h_o = 50$ m and $h_w = 35$ m.

Permeability of the soil $= 300$ cm/day $= 3$ m/day.

Considering unconfined aquifer, discharge of the well is given (Eqn. 11.66) as:

$$= Q = \pi K \frac{h_0^2 - h_w^2}{\ln (r_0 / r_w)}$$

$$= \pi x 3 x \frac{50^2 - 35^2}{\ln (1500 / 0.05)} = 1165.6 \text{ m}^3/\text{day}$$

When three wells are working in equilateral fashion, the discharge of each well is given by Eqn. (11.78) as:

$$Q_1 = Q_2 = Q_3 = \frac{\pi K \left(h_0^2 - h_w^2 \right)}{\ln \left(\dfrac{r_0^3}{r_w B^2} \right)}$$

$$= \frac{\pi x 3 x \left(50^2 - 35^2 \right)}{\ln \left(\dfrac{1500^3}{0.05 \, x \, 250^2} \right)}$$

$$= 852.8 \text{ m}^3/\text{day}$$

Percentage decrease in discharge is given as:

$$= \frac{Q_1 - Q_2}{Q_1} x 100$$

$$= \frac{1165.6 - 852.8}{1165.6} x 100 = 26.8 \text{ percent}$$

Question Banks

Q1.   What do you mean by hydraulics of wells? State its important in the field of groundwater hydrology?

Q2.   What is hydraulic gradient? How is it determined?

Q3.   Define the following terms:

(i) Specific yield (ii) Specific retention (iii) Coefficient of permeability (iv) Storage coefficient (v) Intrinsic permeability (vi) Transmissivity (vii) Specific capacity (viii) Well loss (ix) Well efficiency

Q4.   Though clay has more porosity, yet it is a bad aquifer. Justify why?

Q5. If a given soil sample has porosity 40 percent and specific yield is 10 percent, then what will be its specific retention?

Q6. Write down the units of coefficient of permeability and intrinsic permeability.

Q7. What is the range of the values of storage coefficient and specific yield?

Q8. What is Darcy's law? Derive the law from fundamental principle of flow through porous media.

Q9. What do you mean by depression head? How does it vary with distance from a well?

Q10. Define radius of influence of a well. What are the factors on which it depends?

Q11. What is the shape of the drawdown curve caused by pumping a well in an aquifer? Explain the nature of the curve.

Q12. Derive groundwater flow equations for both steady state and transient saturated state of a confined aquifer.

Q13. Differentiate between the followings.

   (i)   Isotropic and anisotropic aquifer

   (ii) Homogeneous and non-homogeneous aquifer

   (iii) Confined and unconfined aquifer

   (iv) Stream line and equipotential line

Q14. Derive Laplace equation for steady flow in incomprehensible, horizontal and isotropic confined aquifer.

Q15. Enumerate the procedure of development of flow net and describe its significance.

Q16. A dam has saturated thickness 30 m. Coefficient of permeability of the aquifer is 10^{-5} m/sec. The flow net under the dam reveals that there are 4 numbers of flow channels and 8 numbers of squares over the direction of flow. The total head causing the flow is 8 m. Compute the total seepage flow underneath the dam.

Q17. What is an equipotential line? How is it prepared?

Q18. How is the direction of groundwater flow decided?

Q19. Enumerate the different assumptions proposed by Dupuit.

Q20. How is Dupuit's parabola prepared? What does it signify?

Q21. Derive the flow to a fully penetrating well in isotropic, horizontal, homogeneous, confined aquifer. Write down the various assumptions also.

Q22. Derive the flow to a fully penetrating well in isotropic, horizontal, homogeneous, unconfined aquifer. Write down the various assumptions also.

Q23. What is the basic difference on which the discharge equations are used by Dupuit's and Thiem's formula.

Q24. Derive the equations for discharge of a well with uniform recharge in an unconfined aquifer.

Q25. What do you mean by interference of tubewells? How is it determined? How does it affect the yield of well?

Q26. State how the total drawdown in a well field having n number of wells pumping at a steady state with equal discharges is computed.

Q27. Write the formulae to find out the discharge of fully penetrating wells in the following cases:

(i) For two artesian identical wells at a distance B apart in confined aquifer,

(ii) For three artesian identical wells at a distance B apart in a pattern of equilateral triangle in confined aquifer

(iii) For three artesian identical wells at a distance B apart in a line in confined aquifer

(iv) For two identical gravity wells at a distance B apart in unconfined aquifer

(v) For three identical gravity wells at a distance B apart in a pattern of equilateral triangle in unconfined aquifer

Q28. Draw a diagram for three identical fully penetrating wells in confined aquifer spaced equal distances apart in a line and show the composite drawdown curve when all the tree wells are discharging at a time.

Q29. With a neat sketch, explain how the aquifer head losses and well losses are determined. Also describe how the total drawdown is computed.

Q 30. How the efficiency of a well is is determined? Narrate how the efficiency of the well can be enhanced.

Q31. In a pumping test in medium sand, impermeable bed was found to lie at a distance of 16 m below the the normal groundwater level. Two observation wells were located at a distance of 7 and 3 m, respectively from the main pumping well where the drawdown were measured to be 0.25 and 1.0 m, respectively. If the discharge of the main pumping well was 6 lit/sec, calculate the value of coefficient of permeability of the soil.

Q32. A 20 cm diameter well penetrates 19 m below the ground surface. Static water level in the aquifer is 1 m below the ground surface. After 24 hours of pumping at a discharge of 6000 lit/min, the water level in the test well at 90 m is lowered by 0.50 m and in a second well at distance of 30 m away, water level is lowered by 1 m. Estimate the drawdown in the main pumping well and transmissivity of the aquifer.

Q33. A well of radius 20 cm is pumped at a rate of 80 m³/hr. Two observation wells were dug at a distance of 7 and 14 m, respectively where the drawdown were measured to be 5 and 1.3 m, respectively. The bottom of the impermeable bed up to which the well is being pumped is 90 m below the ground surface. Piezometric surface lies on the ground surface too. (a) Compute the coefficient of permeability. (b) Find out the specific capacity of the well. (d) Find out also the maximum rate at which the well can be pumped?.

Q34. Two tubewells penetrating fully a 12 m thick confined aquifer are located 250 m apart. Tubewell has diameter equals to 20 cm, depression head is 6 m and radius of influence is 400 m. Coefficient of permeability is 0.15 cm/sec (a) Compute the discharge of the tubewell when one well is pumping. (b) Also find out the percentage decrease in discharge of the well if both the wells are working under a depression head of 5 m. (c) If a third well is also located 200 m away from both the wells so that these three wells form a pattern of equilateral triangle, what will happen to the discharge of each well.

Q35. Write short notes on

(i) Hydraulic resistance

(ii) Drainage factor

(iii) Leakage factor

(iv) Hydraulic diffusivity

Q36. Find out the equivalent hydraulic conductivity of a layered soil when the flow is (i) parallel to stratified layers and (ii) flow perpendicular to stratified layers.

Q37. Following data were recorded from a site when soil exploration was done. Find out the equivalent soil hydraulic conductivity of all the layers considering (i) flow parallel and (ii) perpendicular to the stratified layers.

Layers	Thickness, cm	Hydraulic conductivity, m/day
A	1.0	1.2
B	2.4	1.5
C	1.5	0.90
D	2.0	1.0
E	1.7	1.5
F	1.1	1.7

References

Campbell, M.D. and Lehr, J.H. 1973. Water Well Technology, McGraw-Hill Book Company, New York, pp. 681.

Das Gupta, A. 1987. Handouts on Ground Water Hydrology, Asian Institute of Technology, Bangkok, pp. 282.

Garg, S.K. 2012. Irrigation Engineering and Hydraulic Structures. Khanna Publishers, New Delhi.

Hansen, V.E., Israelsen, O.W. and Stringham, G.E. 1979. Irrigation Principles and Practices. John Wiley and Sons Inc., New York.

Michael, A.M. 1978. Irrigation Theory and Practices. Vikash Publishing House Pvt. Ltd., New Delhi.

Murty, V.V.N. 2004. Land and Water Management Engineering. Kalyani Publishers, Cuttack.

Panigrahi, B. 2011. Irrigation Systems Enginering. New India Publishinh Agency, New Delhi.

Sharma, R.K. 1987. A Text Book of Hydrology and Water Resources. Dhanpat Rai and Sons, Delhi, pp. 818.

Sharma, R.K. and Sharma, T.K. 2002. Irrigation Engineering including Hydrology, S. Chand and Company Ltd. New Delhi.

Suresh, R. 2008. Watershed Hydrology. Standard Publishers Distributors, Delhi, pp. 692.

Index

S